Monographs in Mathematics
Vol. 95

Franz Georg Timmesfeld

Abstract Root Subgroups and Simple Groups of Lie-Type

Birkhäuser Verlag
Basel · Boston · Berlin

Author:
Franz Georg Timmesfeld
Mathematisches Institut
Justus-Liebig-Universität Giessen
Arndtstr. 2
35392 Giessen
Germany
e-mail: Franz.Timmesfeld@math.uni-giessen.de

2000 Mathematics Subject Classification 20G15, 20E42, 20D05, 20D06, 20F32, 20H20, 51A50, 51E24, 51F25

A CIP catalogue record for this book is available from the Library of Congress, Washington D.C., USA

Deutsche Bibliothek Cataloging-in-Publication Data
Timmesfeld, Franz Georg:
Abstract root subgroups and simple groups of Lie type / Franz Georg Timmesfeld. - Basel ; Boston ; Berlin : Birkhäuser, 2001
    (Monographs in mathematics ; Vol. 95)
    ISBN 3-7643-6532-3

ISBN 3-7643-6532-3 Birkhäuser Verlag, Basel – Boston – Berlin

© 2001 Birkhäuser Verlag, P.O. Box 133, CH-4010 Basel, Switzerland
Member of the BertelsmannSpringer Publishing Group
Printed on acid-free paper produced from chlorine-free pulp. TCF ∞
Printed in Germany
ISBN 3-7643-6532-3

9 8 7 6 5 4 3 2 1   www.birkhauser.ch

to Helga

Contents

Introduction

It was already in 1964 [Fis66] when B. Fischer raised the question:

> Which finite groups can be generated by a conjugacy class D of involutions, the product of any two of which has order 1, 2 or 3?

Such a class D he called a class of *3-transpositions* of G. This question is quite natural, since the class of transpositions of a symmetric group possesses this property. Namely the order of the product $(ij)(kl)$ is 1, 2 or 3 according as $\{i,j\} \cap \{k,l\}$ consists of 2, 0 or 1 element. In fact, if $|\{i,j\} \cap \{k,l\}| = 1$ and $j = k$, then $(ij)(kl)$ is the 3-cycle (ijl).

After the preliminary papers [Fis66] and [Fis64] he succeeded in [Fis71], [Fis69] to classify all finite "nearly" simple groups generated by such a class of 3-transpositions, thereby discovering three new finite simple groups called $M(22)$, $M(23)$ and $M(24)$. But even more important than his classification theorem was the fact that he originated a new method in the study of finite groups, which is called "internal geometric analysis" by D. Gorenstein in his book: Finite Simple Groups, an Introduction to their Classification. In fact D. Gorenstein writes that this method can be regarded as second in importance for the classification of finite simple groups only to the local group-theoretic analysis created by J. Thompson.

Shortly after the appearance of the 3-transposition classification the notion of 3-transpositions was generalized to $\{3,4\}^+$-transpositions [Tim70], [Tim73] by the author, odd transpositions by M. Aschbacher [Asc72] and finally to root involutions [Tim75a] again by the author. The classification of groups generated by root involutions gives a uniform classification of all (finite) simple groups of Lie type in characteristic two (and some other groups), and played an important role for the classification of finite simple groups. Of course the original method of B. Fischer had to be refined and strengthened for these papers, but as in Fischer's paper the arguments have a very strong geometrical flavor. But of even greater importance to the classification than these papers, which are direct generalizations of B. Fischer's notion of 3-transpositions, are subsequent papers by M. Aschbacher and the author, which rely on the same geometric method, for example Aschbacher's theory of "tightly embedded sub-groups" [Asc75], [Asc76] culminating in his famous "classical involution" the-orem [Asc77a], for which he won the Cole prize in Algebra in 1980, and also the work of the author on TI-subgroups [Tim75b], [Tim78b]. Also the classi-fication of groups with "large extra special 2-subgroups" [Asc77b], [Tim78a], [Smi79], [Smi80a], which was considered as one of the main open problems for the classification, since most of the sporadic simple groups satisfy this prop-erty, relies on the $\{3,4\}^+$-transposition classification. Moreover J. Thompson's important classification of finite groups, generated by quadratically acting el-ements on some $\mathbb{Z}_p$-vector space, $p \geq 5$, [Tho], [Tho70] is through his so-called root structure theorem closely related to such a "geometric" hypothesis.

Now, in contrast to the local group theory, which depends on Sylow's the-orem and hence is restricted to finite groups, I always believed that the "geo-metric analysis" can be extended to arbitrary groups. For example Chevalley's commutator relations, see [Car72, (5.2)], which were obtained already in C. Chevalley's seminal paper [Che55] in 1955, are closely related to a hypothe-sis like "$\{3,4\}^+$-transpositions" or "root involutions" and are of fundamental importance to the study of Chevalley groups. Moreover, there is an extensive theory of "root groups" on spherical or more general Moufang buildings devel-oped by J. Tits, see [Tit74, Addenda] or [Ron89, chapter 6], which generalizes the notion of root groups in Chevalley groups and which is closely connected, as will be seen in § 5 of Chapter II, to the notion of "abstract root subgroups".

The classification theory of arbitrary groups generated by a class of sub-groups satisfying certain properties corresponding to cases in Chevalley's com-mutator relations started in 1990, 1991 with the papers on k-transvections [Tim90a] and k-root subgroups [Tim91]. Indeed the hypothesis of "k-trans-vections" can be considered as a direct generalization of 3-transpositions, since the 3-transposition condition can be expressed as: If d and e are non-commuting involutions, then the subgroup generated by d and e is isomorphic to $SL_2(2)$. Unfortunately I discovered later on, that the k-root group property is not general enough to obtain all groups of "Lie type", i.e. all simple normal subgroups of the automorphism group of spherical Moufang buildings. (Also I had some finite dimensionality condition in [Tim90a] and [Tim91].)

This led in 1999 to the notion of groups generated by "abstract root sub-groups" [Tim99]. It turned out, [Tim00b], that all groups of Lie type of rank ≥ 2 in some broad sense (different from $^2F_4$) possess such a class of abstract root subgroups. Hence the classification theorems of [Tim99] give a uniform classification of all simple groups of Lie type and all (possibly infinite dimen-sional) isotropic classical groups. Moreover, it turned out that this notion was tailormade to the demands of quadratic pairs, i.e. to generalize Thompson's theorem to not necessarily finite groups and arbitrary fields k with $k \neq GF(3)$ and char $\neq 2$. (In fact the case of "abstract transvection subgroups" of cardinality ≤ 3 is not contained in [Tim99]. This omission is overcome by the classification of infinite 3-transposition groups by Cuypers and Hall [CH92] and the classification of $GF(3)$-transvection subgroups by H. Cuypers.)

This book is divided into five chapters. Chapter I is devoted to an extensive study of so-called *rank one groups*, which are the principal building blocks of Lie type groups and which are of fundamental importance to the theory of abstract root subgroups. Here a *rank one group* X is a group generated by two different nilpotent subgroups A and B satisfying:

For each $1 \neq a \in A$ there exists a $b \in B$ satisfying $A^b = B^a$ (and vice versa). It turns out that this definition is equivalent to what is called a group with a *split BN-pair of rank one*. But for us the above conjugation equation will be the most important tool for the treatment of such groups. The importance of rank one groups for us comes from the following fact, see I(4.12): If r and $-r$ are opposite roots of some Moufang building with corresponding root subgroups A_r and A_{-r}, then $X_r = \langle A_r, A_{-r} \rangle$ is a rank one group.

Now arbitrary rank one groups might be too general. Hence of particular importance is the connection between arbitrary rank one groups and so called *special rank one groups*, see I§2, where X is special if A and B are abelian and

$$a^b = b^{-a}, \quad \text{where} \quad a \in A, \quad b \in B \quad \text{with} \quad A^b = B^a.$$

Examples of such special rank one groups are SL_2's or, more general, 2-dimensional classical groups of Witt index 1.

Special rank one groups are studied in detail in Chapter I. Here classification results are possible. One result in this direction is I(5.6), which says that if X is special, $1 \neq a \in A$ and $b \in B$ satisfying $A^b = B^a$ above, then there exist $a \in A_0 \leq A$, $b \in B_0 \leq B$ such that $X_0 = \langle A_0, B_0 \rangle$ is a special rank one group, which is a quotient of the covering group of $SL_2(k)$, k a prime field.

In Chapter II we introduce the notion of a group generated by abstract root subgroups, discuss the examples and develop the more elementary theory. For example in II §1 we show that the transvection groups corresponding to point-hyperplanes, the isotropic transvection groups or the Siegel transformation groups of orthogonal space are classes of abstract root subgroups of the corresponding classical groups. In §5 we show that if G is a group of Lie type (in some general sense) h a highest root of $\widetilde{\Phi}^+$, where Φ is the root system of G and $\widetilde{\Phi}$ an extension of Φ (see II(5.3).) and if $A = A_h$ is the root group corresponding to h, then $\Sigma = \{A^g \mid g \in G\}$ is a class of abstract root subgroups of G.

Among the other interesting results of Chapter II are II(2.14), which gives a simplicity criterion for groups generated by abstract root subgroups and II(4.14) in which the existence of a nilpotent radical $R(G)$ is shown, such that $G/R(G)$ is (essentially) simple.

Chapter III is the main part. Here the classification of groups generated by abstract root subgroups is given. This classification contained in [Tim90a], [Tim91] and [Tim99] is quite difficult to read in the original papers, since in [Tim99] I often quote the proof of [Tim90a] or [Tim91]. To make this classification accessible to a larger group of readers is the intention of Chapter III. Also the original proofs are simplified in several instances, for example: The classification of groups generated by abstract transvection subgroups, the classification of groups of type F_4, or the classification of groups of type E_6, E_7 and E_8.

The final §9 of Chapter III describes the overall subdivision in the classification of groups generated by abstract root subgroups and formulates main theorems, which together give a complete classification. So it might be advisable for the reader to start with this section before going into the details of the classification.

Chapter IV is devoted to a revision of the root involution classification. Here we show that if G is a finite group generated by a class D of root involutions (see IV(1.1) for definition) such that $O_2(G) = 1 = Z(G)$ and $D \cap D^2 \neq \emptyset$, where $D^2 = \{ab \mid a, b \in D \text{ and } ab = ba \neq 1\}$, then there is either a class Σ of abstract root subgroups of G naturally connected with D or G is one of a small number of exceptional cases. (See the introduction to Chapter IV for an exact statement). So under the condition $D \cap D^2 \neq \emptyset$, which is always satisfied in the so-called non-degenerate case, the root involution classifications can be obtained as a consequence of the classification of groups generated by abstract root subgroups. The proof of this theorem in Chapter IV is largely self contained, i.e. does not depend on deeper theorems about finite simple groups (the only results used are Bender's classification of groups with a strongly embedded subgroup [Ben71], a theorem of Aschbacher giving a condition for the existence of a strongly embedded subgroup [Asc73] and a theorem of Suzuki classifying $L_3(2^n)$) [Suz65]. But of course contrary to the other chapters a certain familiarity with finite group arguments is required.

Finally Chapter V is devoted to applications. Here, contrary to the previous chapters, I sometimes only sketch proofs. §1 deals with quadratic pairs, i.e. pairs (G, V) where V is an irreducible kG-module, Char $k \neq 2$ and $k \neq GF(3)$, and G is generated by elements acting quadratically on V. Here again the aim of the section is to describe how this hypothesis can be reduced to abstract root subgroups. In §2 we discuss the theory of subgroups of classical and Lie type groups generated by long root elements. This topic has a long history. For finite groups such subgroups have been determined by Kantor [Kan79] and Cooperstein [Coo79], [Coo81] and for algebraic groups (over algebraically closed fields) by Liebeck and Seitz [LS94]. (Of course there should also be mentioned the classical results of Mc Laughlin [McL67], [McL69].) We will concentrate mainly on results over arbitrary fields (or even division rings). First we will discuss conditions showing that such a subgroup is again generated by abstract root subgroups and then describe some of the results of A. Steinbach [Stb97], [Stb00] and H. Cuypers and A. Steinbach [CS99]. Finally in §3 we describe a new application of the theory of abstract root subgroups to the determination of certain chamber transitive subgroups of Lie type groups. Since such results are still in progress, we will just describe the ideas in a special case.

It is my belief that there will be more applications of the theory of abstract root subgroups still to come and also other classification theorems proved by "geometric methods". For example the fixed point subgroups of Lie type groups under endomorphisms can be understood with this theory, so long as the subgroup is generated by "root elements". Hence I think this book should be interesting to anybody interested either in classical groups, Lie type groups or abstract simple groups. Most of the sections are relatively elementary and can be understood with a reasonable knowledge of abstract group theory and classical groups. Only §4 in Chapter I and §5 in Chapter II need some background in the theory of (spherical) buildings.

Since I'm a finite group theorist by education, I will use the notation of finite group theory and also elementary properties of groups, such as the Frattini argument, 3-subgroup lemma etc., which are well known to finite group-theorists. Such results can be looked up in any textbook in finite group theory, for example [Asc86].

There are exercises included in each section. Some of them are very elementary, but others are more like little theorems, which are too special to be included in the text. The latter have extensive hints.

There are several ways in which one can make a graduate course out of this book. For example §1–§3 of Chapter I and §1–§4 of Chapter II would make a course on the (elementary) theory of abstract root subgroups. §1, 2 and 4 of Chapter I and §1, 2 and 5 of Chapter II would yield a class on Lie type groups.

Some word about the references. The main Chapters I–III depend only on references available in books. Here journals are quoted only to tell the origin of the quoted results. I have not tried to list all papers connected with the subject of this book, but just those which are either of importance for the results provided or which illustrate applications of the theory developed.

Finally I want to express my gratitude to Mrs. C. Klein for typing the manuscript, to Mrs. B. Brink and Mrs. A. Steinbach for reading parts of the manuscript and to A. Pasini for his careful reading of Chapters I and II.

Chapter I
Rank One Groups

The notion of a rank one group with abelian unipotent subgroups grew out of the necessity to generalize the concept of k-root subgroups of [Tim91] to the demands of quadratic pairs [Tim99]. It is clear that it is not sufficient just to allow SL_2s for this purpose, since other types of "rank one groups", for example 2-dimensional classical groups of Witt index 1, arise in the situation of quadratic pairs.

For the study of arbitrary Lie type groups, i.e. the subgroups generated by two opposite root subgroups, in [Tim00b] it became necessary to allow arbitrary nilpotent unipotent subgroups. Hence the concept of a rank one group became equivalent to what is known as a group with a "split BN-pair of rank one", see (1.3) below. But still the conjugation equation $A^b = B^a$ is, at least for me, the most important tool for the treatment of such groups.

The results of this chapter are, if not mentioned otherwise, from [Tim99] or [Tim00b].

§1 Definition, examples, basic properties

(1.1) Definition

A group X generated by two different nilpotent subgroups A and B satisfying:

$(*)$ For each $a \in A^{\#}$ there exists a $b \in B^{\#}$ satisfying $A^b = B^a$ and vice versa,

will be called a *rank one group*. The conjugates of A (and B) will be called the *unipotent subgroups* of the rank one group X and the conjugates of $H := N_X(A) \cap N_X(B)$ will be called the *diagonal subgroups*.

If A (and B) is abelian, X will be called a rank one group with abelian unipotent subgroups, abbreviated AUS. Moreover, if for each $a \in A^{\#}$ and $b \in B^{\#}$

which satisfy $(*)$ above, also

$$(**) \qquad\qquad a^b = b^{-a}(= (b^{-1})^a)$$

holds, X is called a *special rank one group*.

Before we start to discuss examples of such rank one groups, we state three little lemmas which make the definition more understandable.

(1.2) Lemma

Let $X = \langle A, B \rangle$ be a rank one group with unipotent subgroups A and B. Then the following hold:

(1) $\Omega := A^B \cup \{B\} = B^A \cup \{A\} = A^X$ is X-invariant.

(2) For each $a \in A^{\#}$ the element $b(a) := b \in B^{\#}$ with $A^b = B^a$ is uniquely determined. Moreover the map $a \to b(a)$ is a bijection of $A^{\#}$ to $B^{\#}$.

(3) $N_A(B) = 1$.

Proof. Since for each $a \in A^{\#}$ there exists a $b \in B^{\#}$ with $A^b = B^a$, we have $B^A \subseteq A^B \cup \{B\}$, whence $B^A \cup \{A\} \subseteq \Omega$. Since we require (1.1) $(*)$ symmetrically in A and B, this implies

$$\Omega = A^B \cup \{B\} = B^A \cup \{A\}.$$

As $X = \langle A, B \rangle$, this equation shows that Ω is X-invariant, whence (1) holds.

Let $1 \ne \bar{b} \in N_B(A)$. Then there exists by (1.1) $(*)$ an $\bar{a} \in A^{\#}$ with $A = A^{\bar{b}} = B^{\bar{a}}$, whence $A = B$, a contradiction since we require in (1.1) that $A \ne B$. With symmetry this shows that $1 = N_A(B) = N_B(A)$.

If now for some $a \in A^{\#}$ elements $b_1, b_2 \in B$ exist with $A^{b_1} = B^a = A^{b_2}$, then $b_1 b_2^{-1} \in N_B(A)$ and thus $b_1 = b_2$. Since $b(a)$ is uniquely determined, the map $a \to b(a)$ is injective. Let $b \in B^{\#}$. Then, since we require (1.1)$(*)$ symmetrically in A and B, there exists an $a = a(b)$ with $A^b = B^{a(b)} = B^a$. But then $b = b(a)$ by definition. Hence the map $a \to b(a)$ is also surjective. $\square$

Notice that (1.2)(2) implies that the definition of a special rank one group is also symmetric in A and B (without demanding it in particular). This follows directly by inverting the equation $(**)$ of (1.1). $\square$

We will from now on simply write $A^B \cup B$ instead of the correct, but too complicated, $A^B \cup \{B\}$.

(1.3)   Proposition

The following are equivalent:

(1) $X = \langle A, B \rangle$ is a rank one group with unipotent subgroups A and B.

(2) There exists a group Y which acts doubly transitively on a set Ω with $|\Omega| \geq 3$, such that for some $\alpha \in \Omega$ the group Y_α contains a nilpotent normal subgroup A which is regular on $\Omega - \{\alpha\}$ and $X = \langle A^y \mid y \in Y \rangle$.

Remark. (1.3) shows that the concepts of a rank one group and of a group with a split BN-pair of rank one are equivalent. Indeed if X is a rank one group, then

$$X_\alpha = N_X(A) = AH, A \cap H = 1 \text{ and } H = X_{\alpha,\beta},$$

whence X has a split BN-pair. On the other hand, if the group Y has a split BN-pair of rank one, then Y acts doubly transitively on $\Omega = \{B^y \mid y \in Y\}$. As $B = U \cdot H$ with $U \trianglelefteq B$ nilpotent, $H = B \cap B^y$ for some $y \in Y - B$ and $U \cap H = 1$, the Frattini argument shows that the hypothesis of (1.3)(2) holds. (I.e. as $H = Y_{\alpha,\beta}$, U is regular on $\Omega - \{\alpha\}$).

Tits introduced in [Tit92] the concept of a Moufang set. This is a set Ω with $|\Omega| \geq 3$ and a family $(U_\alpha)_{\alpha \in \Omega}$ of subgroups of $S(\Omega)$ (of the symmetric group on Ω) satisfying:

(i) U_α fixes α and is regular on $\Omega - \{\alpha\}$.

(ii) U_α normalizes $\{U_\beta \mid \beta \in \Omega\}$ (as subgroup of $S(\Omega)$!)

If one requires in addition that the U_α are nilpotent, then this concept is slightly more special. Indeed the group $X = \langle U_\alpha \mid \alpha \in \Omega \rangle$ clearly satisfies (1.3)(2) with $X = Y$, but it also has to act faithfully on Ω. So such Moufang sets are equivalent to rank one groups acting faithfully on Ω.

Notice that for the more elementary properties of rank one groups we do not use the nilpotence of A.

Proof. We first show that (1) implies (2). Suppose (1) holds. Then by (1.2) (1) $\Omega = A^B \cup B = B^A \cup A = A^X$, whence X acts by conjugation on Ω. If $\alpha = A$, then $A \trianglelefteq N_X(A) = X_\alpha$ and as $\Omega - \{\alpha\} = B^A$, A acts also transitively on $\Omega - \{\alpha\}$. As shown in (1.2) $N_A(B) = 1$, which implies that A acts also fixed-point-freely on $\Omega - \{\alpha\}$ whence regularly on $\Omega - \{\alpha\}$. By symmetry B is also regular on $\Omega - \{B\}$. Hence, as $|\Omega| \geq 3$, an easy exercise on permutation groups, (1.13)(1), shows that X is doubly transitive on Ω. Setting $Y = X$, this shows that for the action of Y on Ω, (2) holds.

Now suppose (2) holds. If $y \in Y - Y_\alpha$, then $A^y \cap A = 1$ since A is regular on $\Omega - \{\alpha\}$. Set $B := A^y$. We first show:

$$(*) \qquad\qquad X = \langle A^y \mid y \in Y \rangle = \langle A, B \rangle.$$

Indeed, if $\bar{y} \in Y - Y_\alpha$ with $\alpha^{\bar{y}} \neq \alpha^y$, then, since B is regular on $\Omega - \{\alpha^y\}$, there exists a $b \in B$ with $\alpha^{\bar{y}} = \alpha^b$. Hence $A^{\bar{y}} = A^b \leq \langle A, B \rangle$. If $\alpha^{\bar{y}} = \alpha^y$, then $A^{\bar{y}} = A^y = B \leq \langle A, B \rangle$. This shows

$$X = \langle A^y \mid y \in Y \rangle \leq \langle A, B \rangle,$$

which implies $(*)$.

Now A is regular on $\Omega - \{\alpha\}$ and B is regular on $\Omega - \{\beta\}$, where $\beta = \alpha^y$. Hence, if $a \in A^\#$, then $\beta \neq \beta^a \in \Omega - \{\beta\}$, so that there exists a $b \in B^\#$ with $\alpha^b = \beta^a$. As shown above this implies $A^{ba^{-1}} = B$, whence $A^b = B^a$. With symmetry this shows that $(1.1)(*)$ holds. Hence X is a rank one group with unipotent subgroups A and B. $\qquad\square$

(1.4)   Corollary

Let $X = \langle A, B \rangle$ be a rank one group with unipotent subgroups A and B and $\Omega = A^X$. Then for all $C \neq D \in \Omega$ and $d \in D^\#$ we have

$$X = \langle C, D \rangle = \langle C, d \rangle.$$

Proof. Assume without loss of generality that $C \neq A$ and $D \neq B$. By (1.3) C acts transitively on $\Omega - \{C\}$ and D transitively on $\Omega - \{D\}$. Since $\{A, D\} \subseteq \Omega - \{C\}$ and $\{B, C\} \subseteq \Omega - \{D\}$, there exist $c \in C$ and $\bar{d} \in D$ with $D^c = A$ and $C^{\bar{d}} = B$. Hence

$$\langle C, D \rangle = \langle D^c, C^{\bar{d}} \rangle = \langle A, B \rangle = X.$$

Moreover, by (1.2), $N_D(C) = 1$. Hence $C \neq C^d$ and, applying the same argument to the pair C, C^d we obtain

$$\langle C, d \rangle \geq \langle C, C^d \rangle = X. \qquad\qquad\square$$

(1.5)   Example

Let R be a ring with one-element 1 and $L \subseteq R$ satisfying:

  (1) $1 \in L$ and L is an additive subgroup of R.
  (2) All elements of $L^* = L - \{0\}$ are units in R and L^* is closed under inversion.
  (3) If $t, c \in L$, then $tct \in L$.

Let $a(t) = \left(\begin{smallmatrix} 1 & \\ t & 1 \end{smallmatrix}\right), b(t) = \left(\begin{smallmatrix} 1 & t \\ & 1 \end{smallmatrix}\right), t \in L$ † and $A = \{a(t) \mid t \in L\}, B = \{b(t) \mid t \in L\}$ and $X = \langle A, B \rangle$ (as a subgroup of $GL_2(R)$!) Then the following hold for X:

(a) X is a special rank one group with AUS.

(b) If $a = a(t), b = b(t^{-1}), t \in L^*$, then $a^b = b^{-a}$ and $A^b = B^a$.

(c) Let $n(t) = a(t)b(-t^{-1})a(t), t \in L^*$. Then $a(c)^{n(t)} = b(-t^{-1}ct^{-1}), A^{n(t)} = B$ and $n(t)^2 \in Z(X)$.

(d) Let $h(t) = n(t)n(-1), t \in L^*$. Then $b(c)^{h(t)} = b(tct)$ and $a(c)^{h(t)} = a(t^{-1}ct^{-1})$.

(e) Let $H = N_X(A) \cap N_X(B)$. If $|L| > 3$, then $[B, H] = B$. In particular X is perfect (and thus quasi-simple by (1.10)).

Abusing notation we call such a group $SL_2(L)$.

Proof. (b) We have for $c \in L$

$$(\ast) \qquad a(c)^{ba^{-1}} = \begin{pmatrix} 1 - t^{-1}c & -t^{-1}ct^{-1} \\ c & ct^{-1} + 1 \end{pmatrix}^{a^{-1}} = \begin{pmatrix} 1 & -t^{-1}ct^{-1} \\ & 1 \end{pmatrix}$$
$$= b(-t^{-1}ct^{-1}).$$

As $t^{-1} \in L^*$ by (2) we have $-t^{-1}ct^{-1} \in L$ by (3). Hence $a(c)^{ba^{-1}} \in B$ for each $c \in L$ and thus $A^b \le B^a$.

On the other hand, since $L = \{-t^{-1}ct^{-1} \mid c \in L\}$, $(\ast)$ shows $A^{ba^{-1}} = B$.

Setting $c = t$ $(\ast)$ implies

$$a^{ba^{-1}} = b(-t^{-1}) = b(t^{-1})^{-1} = b^{-1}.$$

This shows (a) and (b). To prove (c) and (d) notice that $n(t) = \left(\begin{smallmatrix} & -t^{-1} \\ t & \end{smallmatrix}\right)$ and $h(t) = \left(\begin{smallmatrix} t^{-1} & \\ & t \end{smallmatrix}\right)$. Now (c) and (d) are easy matrix calculations.

So (e) remains to be proved. Applying (d) we have

$$(+) \qquad [b(c), h(t)] = b(-c)b(c)^{h(t)} = b(tct - c); c \in L, t \in L^*.$$

We show:

$$(\ast\ast) \qquad L - \{1, -1\} \subseteq \{tct - c \mid c \in L, t \in L^*\}.$$

Indeed if $t \in L^*$ with $t^2 = 1$ but $t \ne \pm 1$, then

$$(t + 1)(t - 1) = 0 \text{ and } \{t + 1, t - 1\} \subseteq L^*,$$

---

†We will always omit 0 entries in matrices. So $\left(\begin{smallmatrix} 1 & \\ t & 1 \end{smallmatrix}\right)$ stands for $\left(\begin{smallmatrix} 1 & 0 \\ t & 1 \end{smallmatrix}\right)$

which is impossible since the elements of L^* are units in R. Hence if $a \in L^*, a \neq \pm 1$, then $a^2 \neq 1$ and $\frac{a^2 - 1}{a} = a - a^{-1} \in L^*$ by (2). This implies

$$\frac{a}{a^2 - 1} = \left(\frac{a^2 - 1}{a}\right)^{-1} \in L^* \quad \text{and} \quad a\left(\frac{a}{a^2 - 1}\right)a - \frac{a}{a^2 - 1} = a,$$

which proves $(**)$. Now $(**)$ implies by $(+)$

$$B - \{b(1), b(-1)\} \subseteq \{[b(c), h(t)] \mid c \in L, t \in L^*\} \subseteq [B, H].$$

Since $|B| = |L| \geq 4$ this implies $B = [B, H]$, since a group G with $|G| \geq 5$ contains no proper subgroup U with $|G - U| \leq 2$. (If $|B| = |L| = 4$, then $1 = -1$ and thus $|[B, H]| \geq 3$. This implies $B = [B, H]$ as before!) $\square$

(1.6) Remark

In the following three cases the hypotheses of (1.5) are satisfied:

(1) R is a division ring, σ an anti-automorphism of R and $L = R_\sigma = \{c \in R \mid c^\sigma = c\}$.

(2) K is a non-perfect field of Char 2, $k = K^2$ and $k \subseteq L \subseteq K$ such that L is a vector space over k.

That L satisfies the conditions (1)–(3) of (1.5) in case (1) is obvious. In case (2) we have for $t, c \in L$:

$$tct = t^2 c \in L, \text{ since } t^2 \in K^2 = k \subseteq L$$

and since L is a vector space over k. Moreover, if $c \in L^*$ then $c^2 \in k$ and thus

$$c^{-1} \in k(c) = k + k \cdot c \subseteq L.$$

(3) Let $K(= R)$ be a division ring with anti-automorphism σ satisfying $\sigma^2 = \text{id}$. Set

$$\begin{aligned}\Lambda := \Lambda_{\min} &= \{c + c^\sigma \mid c \in K\}, \\ \Lambda_{\max} &= \{c \in K \mid c = c^\sigma\}.\end{aligned}$$

Let V be a (left) vector space over K endowed with a pseudo-quadratic form $q : V \to K/\Lambda$ with associated $(\sigma, -1)$ hermitian form $f : V \times V \to K$. (compare [Tit74, p. 123]) such that $1 \in \Lambda$.

Set $R := \{v \in \text{Rad}(V, f) \mid q(v) = 0 + \Lambda\}$ and decompose $V = U \perp R$. Then $q \mid_U$ is non-degenerate.

Then the set $L := \{c \in K \mid \text{there exists a } v_c \in \text{Rad}(U, f) \text{ with } q(v_c) = c + \Lambda\}$ satisfies (1)–(3) of (1.5).

Proof. We first show $L \subseteq \Lambda_{\max}$. Denote by $g : V \times V \to K$ the σ-sesquilinear form with $q(x) = g(x, x) + \Lambda, x \in V$. Then for $c \in L$ and $v_c \in \text{Rad}(U, f)$ we have

$$0 = f(v_c, v_c) = g(v_c, v_c) - g(v_c, v_c)^\sigma$$

(see [Tit74, p. 122]).

Hence $g(v_c, v_c) \in \Lambda_{\max}$ and $c \in \Lambda_{\max}$. Thus $\Lambda_{\min} \subseteq L \subseteq \Lambda_{\max}$.

Now for $c \in L$ there exists a unique $v_c \in \text{Rad}(U, f)$ with $q(v_c) = c + \Lambda$. Indeed if $q(v_1) = c + \Lambda = q(v_2)$, then

$$q(v_1 - v_2) = c + c + \Lambda = c + c^\sigma + \Lambda = \Lambda, \text{ since } c = c^\sigma.$$

Hence $v_1 = v_2$ as U is non-degenerate.

If now Char $K \neq 2$, then $\Lambda_{\min} = L = \Lambda_{\max}$ and (1)–(3) are satisfied by example (1). So assume Char $K = 2$. Then

(1) $1 \in \Lambda$ by assumption. If $c, d \in L$ then $q(v_c + v_d) = c + d + \Lambda$ and $c + d \in L$. Hence L is an additive subgroup of K.

(2) Let $c \in L^*$. Then $c^\sigma = c$ and $c^{-\sigma} = c^{-1}$. Hence

$$q(c^{-1}v_c) = c^{-1}q(v_c)c^{-\sigma} = c^{-1}cc^{-1} + \Lambda = c^{-1} + \Lambda.$$

Thus $c^{-1} \in \Lambda$.

(3) Let $c, t \in L$. Then

$$q(tv_c) = tct^\sigma + \Lambda = tct + \Lambda$$

and $tct \in L$.

Notice that example (3) contains example (2), when K is a field of Char 2, $q : V \to K$ being an ordinary quadratic form with $\text{Rad}(V, f) \neq 0$ and $1 \in L := q(\text{Rad}(V, f))$. (Which is always possible to achieve passing to a proportional form.) $\square$

(1.7) Example

Let k be a field, $k \neq GF(2)$, and X the 1-dimensional affine group over k, i.e. the set of all maps of the form

$$x \to ax + b \quad ; \qquad a \in k^*, b \in k$$

from k into k. Then X is doubly transitive on k. Moreover $X_0 = \{x \to ax \mid a \in k^*\} \simeq k^*$ and thence is abelian. Thus, to show that X is a rank one group, it suffices by (1.3) to show that X is generated by the conjugates of X_0. But this is obvious, since by the doubly transitivity of X, X_0 is a maximal subgroup of X.

Hence X is a rank one group with AUS. We show that X is special if and only if $|k| = 3$ or 4. Let

$$A = X_0 = \{x \to ax \mid a \in k^*\} \qquad \text{and}$$
$$B = X_1 = \{x \to ax + b \mid b = 1 - a, a \in k^*\}$$

and pick $1 \neq a \in k^*$ and identify a with $a : x \to ax$. Then $b(a)$ is the unique element in B with $b(a) : 0 \to a$. Hence $b(a) : x \to (1-a)x + a$. Thus $a^{-1} : x \to a^{-1}x$ and $b(a)^{-1} : x \to \frac{1}{1-a}x - \frac{a}{1-a}$. Now an easy computation shows that

$$b(a)^{-1}ab(a) = a^{-1}b(a)^{-1}a$$

if and only if $a = \frac{1}{1-a}$ for $0 \neq a \neq 1$ in k. The latter equation holds if and only if $k = GF(3)$ or $GF(4)$.   □

(1.8)   Example: $SL_2(K)$, K – a Cayley division algebra [Tim94a]

Let in (1.8) K be a division ring with $|K| > 3$ or a Cayley division algebra, see [Pic55, p173], with center k. The latter is an alternative division algebra over the field k (alternative means that each subalgebra generated by 2-elements is associative) satisfying the right and left inverse properties:

(1) $a(a^{-1}b) = b, (ba^{-1})a = b$.

For $t \in K$ and $V = K^2$ let $a(t), b(t)$ be the maps from $V \to V$ defined by:

$$(c, d)^{a(t)} = (c + dt, d),$$
$$(c, d)^{b(t)} = (c, ct + d).$$

Then $a(t), b(t), t \in K$ are bijective linear mappings if one considers V as a vector space over k. Let $A = \{a(t) \mid t \in K\}, B = \{b(t) \mid t \in K\}$. Then A and B are abelian subgroups of $GL_k(V)$ with multiplication given by

(2) $a(t)a(\tau) = a(t + \tau);\;\; b(t)b(\tau) = b(t + \tau)$

Let finally X be the subgroup of $GL_k(V)$ generated by A and B. We will show that X is a special rank one group with AUS. To do this we need to compute some facts about the action of X on V.

For $d \in K$ let

$$V_d = \{(c, cd) \mid c \in K\} \text{ and } W = \{(0, c) \mid c \in K\}.$$

Then

(3) $\mathcal{P} = \{V_d \mid d \in K\} \cup W$ is a partition of V. Further $\{(cd, c) \mid c \in K\} = V_{d^{-1}}$.

Indeed by (1) $(cd, c) = (a, ad^{-1})$ with $a = cd$. This establishes the second equation in (3). That $\mathcal{P}$ is a partition of V follows from (1).

Next we show:

(4) For each $0 \neq t \in K$ one has

$$\begin{aligned} U := V_0 &= C_V(a(t)) = [V, a(t)] := V^{a(t) - \mathrm{id}} \\ W = C_V(b(t)) &= [V, b(t)] := V^{b(t) - \mathrm{id}} \end{aligned}$$

($V^{a(t) - \mathrm{id}}$ is the image of V under the linear map $a(t) - \mathrm{id}$. $[V, a(t)]$ is the group theoretic commutator taken in the semidirect product of V with $GL_k(V)$. It is well-known that both are the same, if one switches from additive to multiplicative notation.)

$U = C_V(a(t))$ and $W = C_V(b(t))$ follow directly from the definition of the action of A and B on V. Moreover

$$\begin{aligned} (c, d)^{a(t) - \mathrm{id}} &= (dt, 0), \\ (c, d)^{b(t) - \mathrm{id}} &= (0, ct), \end{aligned}$$

which establishes (4).

(5) Let $w \in W^{\#}$ and $u \in U^{\#}$. Then

$$\begin{aligned} U = [w, A] &:= \{w^{a(t) - \mathrm{id}} \mid t \in K\}, \\ W = [u, B] &:= \{u^{b(t) - \mathrm{id}} \mid t \in K\}. \end{aligned}$$

Indeed if $w = (0, c), c \neq 0$, then $[w, A] = \{(ct, 0) \mid t \in K\} = U$.

Next, immediately from the definition of $a(t), b(t)$ one obtains:

(6) For each $t \in K^*$ one has

$$U^{b(t)} = V_t \text{ and } W^{a(t)} = V_{t^{-1}}.$$

From these equations we obtain:

(7) The following hold:

(α) B stabilizes W and permutes $\{V_d \mid d \in K\}$ regularly.

(β) A stabilizes U and permutes $\{V_d \mid d \in K^*\} \cup \{W\}$ regularly.

(γ) X respects the partition $\mathcal{P}$.

Indeed by (2) and (6) we have for $t \in K^*$:

$$V_t^{b(\tau)} = U^{b(t)b(\tau)} = U^{b(t+\tau)} = V_{t+\tau},$$

Similarly, if $t + \tau \neq 0$,

$$(V_{t^{-1}})^{a(\tau)} = W^{a(t)a(\tau)} = W^{a(t+\tau)} = V_{(t+\tau)^{-1}}$$

which implies (α) and (β). Now (γ) is a consequence of (α) and (β), since both A and B respect $\mathcal{P}$.

Next we show:

(8) The following hold:

(α) X acts doubly transitively on $\mathcal{P}$.

(β) $\begin{array}{ll} A &= \{x \in X \mid [V,x] \subseteq U \subseteq C_V(x)\}, \\ B &= \{y \in X \mid [V,y] \subseteq W \subseteq C_V(y)\}. \end{array}$

(α) is a consequence of (7), since A is transitive on $\mathcal{P} - \{U\}$ and B is transitive on $\mathcal{P} - \{W\}$. By (4) one has

$$A \subseteq \{x \in X \mid [V,x] \subseteq U \subseteq C_V(x)\}.$$

Pick $x \in X$ with $[V,x] \subseteq U \subseteq C_V(x)$ and $0 \neq w \in W$.

Then there exists by (5) a $t \in K^*$ with

$$[w, xa(t)] = [w,x] + [w,a(t)] = 0.$$

Hence $xa(t) \in C_X(w) \subseteq N_X(W)$, since X respects $\mathcal{P}$. This implies

$$[W, xa(t)] \subseteq W \cap U = \{0\}$$

and thus $xa(t)$ centralizes W and U and thence V. This shows $xa(t) = 1$ and $x = a(-t)$. With symmetry this proves (β).

From (8) we obtain immediately:

(9) X is a rank one group with AUS.

Indeed (8) (β) implies $A \trianglelefteq N_X(U), B \trianglelefteq N_X(W)$. Moreover by (7) A is regular on $\mathcal{P} - \{U\}$ and B is regular on $\mathcal{P} - \{W\}$. Hence (9) is a consequence of (1.3).

(10) For $t \in K^*$ one has

$$a(t)^{b(t^{-1})} = b(-t^{-1})^{a(t)} \text{ and } A^{b(t^{-1})} = B^{a(t)}.$$

In particular X is special.

An easy calculation shows:

$$(c,d)^{b(-t^{-1})a(t)b(t^{-1})} = (dt, -ct^{-1} + 2d)$$
$$= (c,d)^{a(-t)b(-t^{-1})a(t)}.$$

Since $X \subseteq GL_k(V)$ this implies the first equation. The second equation follows from (8) (β) and

$$U^{b(t^{-1})} = V_{t^{-1}} = W^{a(t)}$$

by (6).

(11) Let $H = N_X(A) \cap N_X(B)$. Then $[B, H] = B$. In particular X is perfect.

Indeed for $a \in K^*$ with $a^2 \neq 1$ one has

$$(*) \qquad a\left(\frac{a}{a^2 - 1}\right)a - \frac{a}{a^2 - 1} = a.$$

(For example, since this expression is computed in the subfield $k(a)$!)
Let $n(t) = a(t)b(-t^{-1})a(t)$ and $h(t) = n(t)n(-1)$ for $t \in K^*$. Then one has for arbitrary $c \in K$:

$$(0,c)^{n(t)} = (ct, 0) \text{ and } (c,0)^{n(t)} = (0, -ct^{-1})$$

and $n(t)^2 = -\mathrm{id}_V$. From this we obtain:

$$a(c)^{n(1)} - b(-c),$$
$$(c,0)^{h(t)} = (ct^{-1}, 0),$$
$$(0,c)^{h(t)} = (0, ct).$$

In particular $h(t)^{-1} = h(t^{-1})$. Together with the definition of the action of $b(\tau)$ on V this implies

$$(c,0)^{b(\tau)^{h(t)}} = (ct, 0)^{b(\tau)h(t)} = (ct, (ct)\tau))^{h(t)},$$
$$= (c, [(ct)\tau]t).$$

On the other hand

$$(c,0)^{b(t\tau t)} = (c, c(t\tau t)).$$

Now for all $c, t, \tau \in K^*$ one has

$$[(ct)\tau]t = c(t\tau t).$$

(This is part of the so called Moufang identities, see for example [Pic55, p160]).
Hence we obtain

$$b(t\tau t) = b(\tau)^{h(t)} \; ; \; \tau \in K, t \in K^*.$$

Now $h(t) \in H$ for each $t \in K^*$. Since

$$[b(\tau), h(t)] = b(t\tau t - \tau)$$

the same argument as in the proof of (1.5) (e) implies

$$[B, H] = B, \text{ which proves (11) .} \qquad \square$$

The group X will be called $SL_2(K)$ and the $\mathbb{Z}X$-module V, together with the
partition $\mathcal{P}$, the natural $\mathbb{Z}SL_2(K)$-module. If K is a division ring, then V is a
2-dimensional (left)-K-vector space and $\mathcal{P}$ the set of 1-dimensional subspaces
of V. Hence it is easy to see that this definition coincides with the group X
defined in (1.5), when $L = K$. (With the roles of A and B interchanged.)

(1.9) Example: Classical groups of Witt index 1

Let V be a vector space over a division ring K together with a non-degenerate
(σ, ϵ)-hermitian form f or pseudoquadratic form q of Witt index 1. (In the notation of [Tit74, (8.1),(8.2)]. The reader who is not familiar with this notation
might just take for f an antisymmetric or for q a quadratic form of Witt index
1!) Suppose there exist more than two isotropic (singular) points (1-spaces) in
V and denote by G the group of isometries of V respecting f (resp. q). For
each isotropic (singular) point P of V let

$$U_P = \{\sigma \in G \mid P^{\sigma - \mathrm{id}} = 0, P^{\perp(\sigma - \mathrm{id})} \subseteq P\}.$$

(Notice that $P^{\sigma - \mathrm{id}} = 0$ implies $V^{\sigma - \mathrm{id}} \subseteq P^\perp$!) Then the following hold:

(1) U_P is nilpotent of class at most two.
(2) $R = \langle U_P \mid P \text{ an isotropic (singular) point of } V \rangle$ is a rank one group with
    unipotent subgroups U_P.
(3) Either R is perfect or G is isomorphic to one of the following groups:
    $Sp_2(2), Sp_2(3), U_2(2), U_2(3), U_3(2)$ and $O_3(3)$.

Proof. (1) is an easy consequence of the 3-subgroup lemma. Indeed let

$$T_P = \{\sigma \in G \mid V^{\sigma - \mathrm{id}} \subseteq P, P^{\perp(\sigma - \mathrm{id})} = 0\}.$$

(If $T_P \neq 1$ we say that G is a unitary group.) Then the 3-subgroup lemma applied to the action of U_P on $P^\perp$ and V/P implies $U'_P \leq T_P$. Moreover

$$[V, U_P, T_P] \subseteq [P^\perp, T_P] = 0 = [P, U_P] = [V, T_P, U_P]$$

so that $T_P \leq Z(U_P)$.

To prove (2) let $P = \langle x \rangle$, $Q = \langle y \rangle \neq P$ be another singular point and choose x, y such that $f(x, y) = 1$. Then each singular point R with $P \neq R \neq Q$ is of the form

$$R = \langle cx + z + y \rangle, c \in K \text{ and } z \in x^\perp \cap y^\perp.$$

Hence Q is mapped onto R by a linear map σ corresponding to the matrix:

$$\begin{bmatrix} 1 & & \\ \hline u & I & \\ \hline c & z & 1 \end{bmatrix} ; \quad \text{where } u \in x^\perp \cap y^\perp \text{ is arbitrary.}$$

Hence one can choose u so that σ respects the form f (resp. q), whence $\sigma \in U_P$. This shows that U_P is transitive on the set of singular points different from P. Suppose now $x \in U_P$ fixes the isotropic point $Q \neq P$. Then

$$[Q^\perp \cap P^\perp, x] \leq P \cap Q^\perp = 0 \text{ (since the Witt index of } V \text{ is 1).}$$

Hence x centralizes $P^\perp = P + (Q^\perp \cap P^\perp)$. Since $[Q, x] \leq Q \cap P^\perp = 0$ this implies $x = 1$. This shows

(∗) U_P is regular on the isotropic (singular) points of V different from P.

Clearly, by definition, $U_P \trianglelefteq G_P$ so that (1.3) implies (2).

Since the elements of U_P are Eichler transformations, R is the normal subgroup of G generated by all Eichler transformations on V. Hence (3) follows from [HO89, (6.3.15)]. $\quad\square$

(We will not need "Eichler transformations" elsewhere in this book, but just a special case of these, the "Siegel transformations" in chapter II. Because of this reason I did not include the details, which consist of computations with Eichler transformations, in the treatment of this example!)

(1.10) Lemma

Suppose X is a rank one group with unipotent subgroups A and B and N is a normal subgroup of X. Then either $X = NA$ or $N \leq Z(X)$. In particular perfect rank one groups are quasi-simple. (Here a perfect group G is called *quasi-simple* if $G/Z(G)$ is simple. This notation comes from classification theory of finite simple groups.)

Proof. If $N \not\leq N_X(A)$, then $A \neq A^h$ for some $h \in N$ and by (1.4)

$$X = \langle A, A^h \rangle \leq NA.$$

If $1 \neq a \in A \cap N$, then $X = \langle a, B \rangle \leq NB$, whence $X = NA$ again by (1.4). So we may assume

$$[A, N] \leq A \cap N = 1 = B \cap N = [B, N].$$

But then $N \leq Z(X)$.

Suppose finally $X = NA = X'$. Then, as A is nilpotent $A^{(n)} = 1$ for some $n \in \mathbb{N}$. ($A^{(n)}$ is defined recursively by $A^{(1)} = A'$ and $A^{(n)} = (A^{(n-1)})'$!)

Hence

$$X = X^{(n)} = (NA)^{(n)} \leq NA^{(n)} \leq N.$$

This shows that in a perfect rank one group X each proper normal subgroup N lies in $Z(X)$.   □

(1.11) Lemma

Let $X = \langle A, B \rangle$ be a rank one group with unipotent subgroups A and B and diagonal subgroup $H = N_X(A) \cap N_X(B)$ satisfying $[A, H] = A$. Let $Y = \langle \overline{A}, \overline{B} \rangle$ be another rank one group with unipotent subgroups $\overline{A}$ and $\overline{B}$. Suppose there exists a homomorphism $\sigma : Y \to X$ with $\sigma(\overline{A}) = A, \sigma(\overline{B}) = B$. Then the following holds:

(1) $\ker \sigma \leq Z(Y)$.
(2) If $\overline{H} = N_Y(\overline{A}) \cap N_Y(\overline{B})$, then also $[\overline{A}, \overline{H}] = \overline{A}$. In particular Y is quasi-simple.

Proof. By (1.4) $\overline{A}$ is the only unipotent subgroup of Y contained in $\overline{A} \ker \sigma$. Further arguing as in the proof of (1.10) $\overline{A} \cap \ker \sigma = 1$. Hence $\sigma \mid_{\overline{A}} : \overline{A} \to A$ is injective and, as in (1.10), $\ker \sigma \leq Z(Y)$.

If $y \in Y$ is in the coimage of $N_X(A)$, then $\overline{A}^y \leq \overline{A} \ker \sigma$ and thus $\overline{A}^y = \overline{A}$. This shows that $N_Y(\overline{A})$ is the coimage of $N_X(A)$, $N_Y(\overline{B})$ is the coimage of $N_X(B)$ and $\overline{H}$ is the coimage of H. Let $\overline{A}_0 = [\overline{A}, \overline{H}]$. If $\overline{A}_0 \neq \overline{A}$, then $\sigma(\overline{A}_0) \neq A$. But

$$\sigma(\overline{A}_0) = \sigma([\overline{A}, \overline{H}]) = [\sigma(\overline{A}), \sigma(\overline{H})] = [A, H] = A.$$

This implies (2).   □

Let $X = \langle A, B \rangle$ be a rank one group with unipotent subgroups A and B and let $1 \neq A_0 \leq A$ be $N_X(A)$-invariant. Then we say that X is *relatively special* with respect to A_0, if for each $a \in A_0^{\#}$ we have $a^{b(a)} = b(a)^{-a}$. Then we have:

(1.12)   Lemma

Let $X = \langle A, B \rangle$ be a rank one group and $1 \neq A_0 \leq A$ be $N_X(A)$-invariant. Let $B_0 = A_0^x$, where $x \in X$ with $A^x = B$. Then the following are equivalent:

(1) $X_0 = \langle A_0, B_0 \rangle$ is a special rank one group.

(2) X is relatively special with respect to A.

Proof. First notice that, as A_0 is $N_X(A)$-invariant, we have $B_0 = A_0^y$ for each $y \in X$ with $A^y = B$. If now X_0 is special, then for $a \in A_0^{\#}$ there exists a $b \in B_0^{\#}$ with $a^b = b^{-a}$. Hence $A^b = B^a$ by (1.3) and the uniqueness of $b(a)$ implies $b = b(a)$. This shows that X is relatively special with respect to A_0.

Now suppose X is relatively special with respect to A and let $a \in A_0^{\#}$ and $b = b(a) \in B$ with $A^b = B^a$. Then

$$b^{-1} = a^{ba^{-1}} \in A_0^{ba^{-1}} = B_0$$

by the above. Thus, to show that X_0 is a special rank one group, it remains to show that for each $\beta \in B_0^{\#}$ there exists an $\alpha \in A_0^{\#}$ with $B_0^{\alpha} = A_0^{\beta}$. Now, since X is a rank one group, there exists an $\alpha \in A^{\#}$ with $A^{\beta} = B^{\alpha}$. Hence $A_0^{\beta} = B_0^{\alpha}$ by the above. It remains to show $\alpha \in A_0$.

For this let $w \in X$ interchanging A and B (i.e. $A^w = B, B^w = A$). Then for each $a \in A^{\#}$ we have:

$$A^{a^w} = B^{wa^w} = B^{aw} = A^{b(a)w} = A^{wb(a)^w} = B^{b(a)^w}.$$

Now, as $B_0 = A_0^w, \beta = a^w$ for some $a \in A_0^{\#}$. Hence $\alpha = b(a)^w \in B_0^w = A_0$, since $b(a)^w \alpha^{-1} \in N_A(B) = \{1\}$ by (1.2).   □

(1.13)   Exercises

(1) Suppose G acts on Ω with $|\Omega| \geq 3$, such that for $\alpha \neq \beta \in \Omega$ we have: G_α is transitive on $\Omega - \{\alpha\}$ and G_β is transitive on $\Omega - \{\beta\}$. Show that G is doubly transitive on Ω.

(2) Let X be a rank one group with unipotent subgroups A and B and σ a homomorphism of X with $\sigma(A) \neq \sigma(B)$. Show that $\sigma(X)$ is a rank one group with unipotent subgroups $\sigma(A)$ and $\sigma(B)$.

(3) (a) Show that the 1-dimensional affine group over $GF(3)$ is isomorphic to $SL_2(2)$.

(b) Show that the 1-dimensional affine group over $GF(4)$ is isomorphic to $PSL_2(3)$.

(This explains why such a 1-dimensional affine group is special, if and only if $|k| = 3$ or 4!)

(4) (A. Steinbach) Let $X = \langle A, B \rangle$ be a rank one group with AUS. Show:

   (a) For each $a \in A^{\#}$ there exist unique elements $b_1 \in B^{\#}$ and $a_1 \in A^{\#}$ such that ab_1a_1 interchanges A and B (by conjugation).

   (b) Suppose that $a^{-1}b_1^{-1}a_1^{-1}$ also interchanges A and B. Then $b(a^{-1}) = b(a)^{-1}$

(5) Let $X = \langle U_P, U_Q \rangle$ be a unitary group of Witt index 1 (in the notation of (1.9)), where P and Q are different isotropic points of the underlying vector space. Set $A = U_P, B = U_Q$ and

$$A_0 = T_P \text{ as defined in } (1.9) .$$

Show that X is relatively special with respect to A_0.

(6) Suppose $X = \langle A, B \rangle$ is a special rank one group and $N \trianglelefteq X$ with $A \cap N \neq 1$. Show that $\langle A \cap N, B \cap N \rangle$ is also a special rank one group.

(7) Let K be a Cayley division algebra, $Y = SL_2(K)$ and $A = \{a(c) \mid c \in K\}$, $B = \{b(c) \mid c \in K\}$ in the notation of (1.8).

   (i) Let $\lambda(t) : (x, y) \to (tx, tyt); x, y \in K$ and $t \in K^*$. Show $\lambda(t) \in \text{Aut}(K^2)$ respecting the partition $\mathcal{P}$ of (1.8).

   (ii) $\lambda(t)$ normalizes A and B as subgroups of $\text{Aut}(K^2)$.

   (iii) Let $\widehat{H} = \langle h(t), \lambda(t) \mid t \in K^* \rangle \subseteq \text{Aut}(K^2), h(t)$ as in (1.8). Show that $\widehat{H}$ acts transitively on $A^{\#}$ and $B^{\#}$.

(8) Let K, Y, A and B be as in (7). Let L be a subgroup of $(K, +)$ satisfying (1)–(3) of (1.5) with $|L| > 3$. Show that $X = \langle a(c), b(c) \mid c \in L \rangle$ is a quasi-simple rank one group.

(9) Let $X = \langle A, B \rangle \simeq (P)SL_2(k)$, k a field with Char $k \neq 2$, such that A and B are full unipotent subgroups of X. Let $A_0 \leq A$, $B_0 \leq B$ such that $X_0 = \langle A_0, B_0 \rangle$ is a rank one group. Show $X_0 \simeq (P)SL_2(L)$, L a subfield of k.

   Hint. Let $A = \{a(\lambda) \mid \lambda \in k\}$ and $B = \{b(\lambda) \mid \lambda \in k\}$ as in (1.5) and let $L = \{\lambda \in k \mid a(\lambda) \in A_0\}$ and assume without loss of generality that $1 \in L$. Then $b(1) = b(a(1)) \in B_0$ and $w = n(1) \in X_0$. Hence $B_0 = \{b(\lambda) \mid \lambda \in L\}$. On the other hand $b(a(\lambda)) = b(\lambda^{-1}) \in B_0$ for

$\lambda \in L^*$, so that L is closed under inversion. Hence $h(t) \in X_0$ for $t \in L^*$ and so $t^2 \in L^*$, since $a(1)^{h(t)} = a(t^2)$. Now the equation

$$(\lambda + t)^2 = \lambda^2 + 2t\lambda + t^2$$

implies $t\lambda \in L^*$ if $t, \lambda \in L^*$. Thus L is a field. Now use that $SL_2(L)$ is generated by $a(\lambda), b(\lambda)$ as λ runs over L.

§2 On the structure of rank one groups

We assume in this section that $X = \langle A, B \rangle$ is a rank one group with unipotent subgroups A and B and diagonal subgroup $H = N_X(A) \cap N_X(B)$.

Let $\Omega = A^X$. We will try to obtain information about the structure of X and also conditions which guarantee that, for H-invariant, $A_0 \le Z(A)$ and $B_0 = A^x, x \in X$ with $A^x = B, X_0 = \langle A_0, B_0 \rangle$ is a special rank one group.

(2.1) Lemma

$Z_2(X) = Z(X)$. In particular X is not nilpotent.

Proof. If X is nilpotent, then since $A \ne X$ there exists a proper normal subgroup M with $A \le M \lhd X$. But this is obviously impossible, since by (1.4)

$$X = \langle A, A^x \rangle \text{ for each } x \in X - N(A).$$

So X is not nilpotent. On the other hand $Z_2(X)A$ is nilpotent, since

$$1 \le Z(X) \le Z_2(X) \le Z_2(X)Z(A) \le Z_2(X)Z_2(A) \le \cdots \le Z_2(X)A$$

is an ascending central series. Hence (2.1) is a consequence of (1.10). □

(2.2) Lemma

For $a \in A^{\#}$ the following are equivalent:

(1) $b(a^{-1}) = b(a)^{-1}$.
(2) $a^{b(a)} = b(a)^{-a}$.

In particular both properties hold for each $a \in A^{\#}$, if there exists an automorphism of X which inverts A and B.

Proof. Suppose (1) holds. Then $B^{a^{-1}} = A^{b^{-1}}$ for $b = b(a)$. Hence $A^{b^{-1}a} = B$ and we obtain:

$$A^{a^b} = A^{b^{-1}ab} = B^b = B = A^{b^{-1}a} = A^{a^{-1}b^{-1}a} = A^{(b^{-1})^a}.$$

Now, since $A^{ba^{-1}} = B$, there exists a $\bar{b} \in B$ with $a^b = \bar{b}^a$. Then $A^{a^b} = A^{\bar{b}^a} = A^{(b^{-1})^a}$ and thus, conjugating with b^a:

$$A = A^{\bar{b}^a b^a} = A^{(\bar{b}b)^a} \text{ and } \bar{b}b \in N_B(A).$$

By (1.2) $N_B(A) = \{1\}$ and $\bar{b} = b^{-1}$, which shows that (2) holds.

Next suppose (2) holds. Then $a^b = b^{-a}$ implies

$$(a^{-1})^{b^{-1}} = b(b^{-1})^{a^{-1}} = ba^b a^{-1} = b^{a^{-1}}.$$

From this equation we obtain by (1.3) that $A^{b^{-1}} = B^{a^{-1}}$, since $A^{b^{-1}} \cap B^{a^{-1}} \neq \{1\}$ and since by the regularity of A on $\Omega - \{A\}$ the elements of Ω are trivial intersection subgroups. (I. e. $C = D$ or $C \cap D = \{1\}$ for $C, D \in \Omega$!) Hence by (1.2)

$$b(a^{-1}) = b^{-1} = b(a)^{-1}.$$

Finally assume that $\tau \in \mathrm{Aut}(X)$ inverts A and B. Then

$$A^{b^{-1}} = A^{b^\tau} = A^{b\tau} = B^{a\tau} = B^{a^\tau} = B^{a^{-1}}$$

for each $a \in A^\#$ and $b = b(a)$. Thus $b^{-1} = b(a^{-1})$, by the uniqueness of $b(a^{-1})$ ((1.2)(2)) and (1), and (2) hold. $\square$

(2.3)   Lemma

Suppose $1 \neq A_0 \leq Z(A)$ is H-invariant and $x \in X$ with $A^x = B$. Suppose further that for each $\alpha \in A_0^\#$ there exists no $\beta \in B^\#$ such that $\alpha\beta$ is conjugate to some element of A (in X). Then $X_0 = \langle A_0, B_0 \rangle$ is a special rank one group with AUS.

Proof. Since $A_0 \leq Z(A)$ it follows that A_0 is $AH = N_X(A)$ invariant. Hence by (1.12) it suffices to show that for each $a \in A_0^\#$ and $b = b(a)$ one has $a^b = b^{-a}$.

Because of $a^b \in A^b = B^a$ there exists a $\bar{b} \in B^\#$ with $a^b = \bar{b}^a$. If $\bar{b} = b(a)^{-1}$ for each $a \in A_0^\#$ then (2.3) holds. So we may assume $\bar{b} \neq b(a)^{-1} = b^{-1}$ for some fixed $a \in A_0^\#$. Then $\bar{b} = (a^b)^{a^{-1}} = aa^b a^{-1} = ab^{-1}b^{a^{-1}}$ and

$$a^{b^{-1}} b^{a^{-1}} = bab^{-1}b^{a^{-1}} = b\bar{b} \in B^\#.$$

This implies that $ab^{a^{-1}b} \in B^\#$. Now, as $A \cap B = 1$, $b^{a^{-1}b} \notin A$. Since A is by (1.3) transitive on $\Omega - \{A\}$ there exists a $y \in A$ such that $(b^{a^{-1}b})^y \in B$. We obtain that $ab^{a^{-1}by} = (ab^{a^{-1}b})^y$ is conjugate to some element of $A^\#$, a contradiction to the assumption. $\square$

(2.4) Proposition

Let $1 \neq A_0 \leq Z(A)$ be H-invariant and set $B_0 = A_0^x$ for some $x \in X$ with $A^x = B$. Set $X_0 = \langle A_0, B_0 \rangle$. Suppose there exists a $\mathbb{Z}X$-module N satisfying

(a) $[N, X, X_0] \neq 0$,
(b) $[N, A_0, A] = 0$.

Then X_0 is a special rank one group with AUS.

Proof. We show that the hypothesis of (2.3) is satisfied. By the 3-subgroup lemma $[N, A, A_0] = 0$ so that $[N, A] \leq C_N(A_0)$ and $[N, A_0] \leq C_N(A)$. Further, by conjugation, the same holds for B and B_0. Let

$$N_0 = [N, A] + [N, B_0] \quad \text{and} \quad Z = \bigcap_{x \in X} [N, A]^x.$$

Then, since by (1.4) $X = \langle A, B_0 \rangle$, N_0 is X-invariant. Hence $N_0 = [N, X]$. Moreover by definition Z is X- invariant, $Z \leq C_N(X_0)$ and $Z \leq N_0$. If $Z = [N, A]$, then $[N, A] = [N, B] = [N, X]$ and $[N, X, X_0] = 0$, a contradiction to (a). This shows

(∗) $\qquad\qquad\qquad N_0 \geq [N, A] > Z.$

Moreover by (1.10) $X = C_X(Z)A$ since $X_0 \leq C_X(Z)$. Let $M_0 = N_0/Z$ and $M = [M_0, A] + [M_0, B] = [M_0, X]$. If $M = 0$ then $[N_0, X] \leq Z$ and thus X normalizes $[N, A]$ by (∗). But then $[N, A] = Z$, a contradiction to (∗). Thus $M \neq 0$. Next suppose $Z_1 = [M_0, A] \cap C_M(X) \neq 0$. Then

$$Z_1 \leq \bigcap_{x \in X} [M_0, A]^x$$

and so the coimage of Z_1 in N is contained in Z, a contradiction to $Z_1 \neq 0$. In particular we obtain

(+) $\qquad\qquad [M_0, A] \cap [M_0, B_0] \leq [M_0, A] \cap C_{M_0}(X) = 0$

since $X = \langle B, A_0 \rangle$ and thus

(1) $M = [M_0, A] \oplus [M_0, B_0] = [M_0, B] \oplus [M_0, A_0].$

(Since there exists an element w in X interchanging A and B and also A_0 and B_0!) If now $[M_0, B_0] = 0$, then X normalizes $[M_0, A]$ and thus also $[N, A]$ by (∗) above, since $[N, A, B_0] \leq [N_0, B_0] \leq Z \leq [N, A]$ and $X = \langle A, B_0 \rangle$. But this is again a contradiction to $[N, A] > Z$. This shows that all direct summands in (1) are non-trivial. Next we show:

(2) $[M_0, A] = C_M(A_0) = C_M(a)$ for each $a \in A_0^\#$.

Indeed $[M_0, A] \subseteq C_M(A_0) \subseteq C_M(a)$ for each $a \in A_0^\#$ by hypothesis (b). Suppose $[M_0, A] \subset C_M(a)$ for some $a \in A_0^\#$. Then, as $X = \langle B, a \rangle$ by (1.4),

$$0 \neq Z_2 = C_M(a) \cap [M_0, B_0] \subseteq C_M(X)$$

by (1), a contradiction to $Z_1 = 0$ since A and B are conjugate in X. This implies (2).

We now are in the position to show that the hypothesis of (2.3) is satisfied for X. Namely suppose $d = \alpha\beta, \alpha \in A_0^\#, \beta \in B^\#$ is conjugate to some element of A in X. Then, as

$$0 \neq [M_0, A_0] \subseteq C_M(A),$$

there exists a $1 \neq m \in C_M(d)$. (We now use multiplicative notation for M to apply commutator formulas.) Let by (1) $m = m_1 m_2$ with $m_1 \in [M_0, A]$ and $m_2 \in [M_0, B_0]$. Then

$$
\begin{aligned}
m &= (m_1 m_2)^{\alpha\beta} = m_1^\beta m_2^{\alpha\beta} = m_1[m_1, \beta](m_2[m_2, \alpha])^\beta \\
  &= m_1[m_1, \beta]m_2[m_2, \alpha]^\beta = m_1 m_2[m_1, \beta][m_2, \alpha]^\beta.
\end{aligned}
$$

Hence

$$[m_2, \alpha]^\beta = [m_1, \beta]^{-1} \in [M_0, A_0]^\beta \cap [M_0, B] = 0$$

by (1). Hence $[m_2, \alpha] = 1$ and thus $m_2 = 1$ by (1) and (2). This implies $m = m_1 \in [M_0, A]$. Since this holds for each $m \in C_M(d)$ we obtain $C_M(d) \subseteq [M_0, A]$. Let $d \in D \in \Omega$. Then as $d \notin A$, (1.4) implies $D = B^a$ for some $a \in A$. Let $D_0 = B_0^a$. Then by the above

$$[M_0, D_0] \subseteq C_M(D) \subseteq C_M(d) \subseteq [M_0, A]$$

and thus also

$$[M_0, B_0] = [M_0, D_0]^{a^{-1}} \subseteq [M_0, A],$$

a contradiction to (1). This shows that no such d exists. Now (2.4) is a consequence of (2.3). □

(2.5) Corollary

Suppose X has abelian unipotent subgroups. Then X is special if there exists a $\mathbb{Z}X$-module N with

(a) $[N, X, X] \neq 0$,
(b) $[N, A, A] = 0$.

(2.6) Examples

(1) Let X be a unitary group of Witt index 1, $A = U_P$ and $A_0 = T_P$ in the notation of (1.9) and (1.13)(5). Then, if V is the natural $\mathbb{Z}X$-module, then $[V, A_0] \leq P$ whence $[V, A_0, A] = 0$. Thus X and V satisfy the hypothesis of (2.4). This gives an abstract (non-computational) solution of Exercise (1.13)(5).

(2) Let $X = Sz(2^{2n+1}), n \in \mathbb{N}$ be a Suzuki group ($^2B_2(2^{2n+1})$ in Lie notation.) Let A and B be different 2-Sylow subgroups of X and $A_0 = Z(A) = A'$, $B_0 = Z(B) = B'$. Then it follows from (1.3) and [Suz62] that X is a rank one group with unipotent subgroups A and B. Moreover by [Suz62] A_0 and B_0 are elementary abelian 2-groups.

Suppose now for some $a \in A_0^{\#}$ and $b = b(a) \in B$ we have $a^b = b^{-a}$. Then, since a is an involution also, $b^{-1} = a^{ba^{-1}}$ is an involution, whence

$$aba = b^a = (b^{-1})^a = a^b = bab.$$

This implies $o(ab) = 3$. But $3 \nmid |X|$. This shows that X admits no non-trivial $\mathbb{Z}X$-module V with $[V, A_0, A] = 0$. (Or $[V, A, A_0] = 0$!)

(3) Let V be a vector space over the field k together with a non-degenerate quadratic form $q : V \to k$ of Witt index 1, such that there exist more than two singular points in V. Let

$$R = \langle U_P \mid P \text{ a singular point of } V \rangle \leq O(V, q)$$

in the notation of (1.9). Then by (1.9) R is a rank one group with unipotent subgroups U_P. Moreover U_P is abelian.

Let W be the so-called "spin-module" for $O(V, q)$. Then it is known that $[W, U_P, U_P] = 0$. Hence by (2.5) R is a special rank one group. Indeed the group $SL_2(K)$, K a Cayley division algebra, see (1.8), is a special case of this example, since this group can be considered as a 10-dimensional orthogonal group of Witt index 1 over $k = Z(K)$, see [Tit74, (5.12)].

We now start to discuss (quasi) simplicity of rank one groups. Clearly the groups of example (1.7) are solvable. So we concentrate on special (or relatively special) rank one groups. We first show:

(2.7) Lemma

Suppose X is special with AUS. Let K be an additive group coordinatizing A, i.e. $A = \{a(\lambda) \mid \lambda \in K\}$ with $a(\lambda)a(\mu) = a(\lambda + \mu)$. For $\lambda \in K^*$ set $n(\lambda) = a(\lambda)b(a(-\lambda))a(\lambda)$. Then the following holds:

(1) $b(a(-\lambda)) = b(a(\lambda))^{-1}, n(-\lambda) = n(\lambda)^{-1}$,

(2) $A^{n(\lambda)} = B, B^{n(\lambda)} = A$,

(3) $a(\lambda)^{n(\lambda)} = b(a(-\lambda)), b(a(\lambda))^{n(\lambda)} = a(-\lambda)$.

Proof. Since X is special we have by (2.2)

$$b(a(\lambda))^{-1} = b(a(\lambda)^{-1}) = b(a(-\lambda)), \quad \lambda \in K.$$

Hence by definition of $n(\lambda)$:

$$
\begin{aligned}
n(-\lambda) &= a(-\lambda)b(a(\lambda))a(-\lambda) = a(\lambda)^{-1}b(a(-\lambda))^{-1}a(\lambda)^{-1} \\
&= (a(\lambda)b(a(-\lambda))a(\lambda))^{-1} = n(\lambda)^{-1}.
\end{aligned}
$$

Now

$$A^{n(\lambda)} = A^{b(a(-\lambda))a(\lambda)} = B^{a(-\lambda)a(\lambda)} = B$$

and

$$
\begin{aligned}
B^{n(\lambda)} &= B^{a(\lambda)b(a(-\lambda))a(\lambda)} = A^{b(a(\lambda))b(a(-\lambda))a(\lambda)} \\
&= A^{a(\lambda)} = A.
\end{aligned}
$$

Similarly

$$
\begin{aligned}
a(-\lambda)^{n(\lambda)} &= a(-\lambda)^{b(a(-\lambda))a(\lambda)} = (b(a(-\lambda))^{-1})^{a(-\lambda)a(\lambda)} \\
&= b(a(\lambda)).
\end{aligned}
$$

Taking the inverse this implies

$$a(\lambda)^{n(\lambda)} = b(a(\lambda))^{-1} = b(a(-\lambda)) \text{ by } (1).$$

Moreover

$$
\begin{aligned}
b(a(-\lambda))^{n(\lambda)} &= (b(a(\lambda))^{-1})^{a(\lambda)b(a(-\lambda))a(\lambda)} \\
&= a(\lambda)^{b(a(\lambda))b(a(-\lambda))a(\lambda)} = a(\lambda)
\end{aligned}
$$

again by (1).   □

(2.8) Proposition

Suppose X is special with AUS. Then one of the following holds:

(1) $[A, H] \neq 1$

(2) $X \simeq (P)SL_2(2)$ or $(P)SL_2(3)$.

(For abbreviation we write $(P)SL$ if the group is an image of SL with kernel in the center.)

Proof. We use the notation of (2.7). Let $\overline{X} = X/Z(X)$. Then by (2.1) $Z(\overline{X}) = 1$. Hence if $\overline{X} \simeq PSL_2(2)$ or $PSL_2(3)$ then clearly (2) holds. By (1.4) A is the unique element of Ω in $AZ(X)$. Hence $N_X(A)$ is the coimage of $N_{\overline{X}}(\overline{A})$ in X. This implies that $[A, N_X(A)] \neq 1$ if $[\overline{A}, N_{\overline{X}}(\overline{A})] \neq 1$. Since $N_X(A) = AH$ this shows that (2.8)(1) also holds for X if it holds for $\overline{X}$.

So we may assume $Z(X) = 1$. We will assume for the rest of the proof that $[A, H] = 1$ and then show that (2) holds. Clearly H centralizes with each $a(\lambda) \in A$ also the element $b(a(\lambda)) \in B$. Hence $H \leq Z(X) = 1$. Since $N_X(A) = AH$ this implies $N_X(A) = A$. Now by (2.7) the elements $n(\lambda), \lambda \in K^*$ interchange A and B. Hence we obtain:

(∗) $n(\lambda)^2 = 1, n(\lambda) = n(\mu)$ for all $\lambda, \mu \in K^*$.

Pick $\lambda \in K$ with $2\lambda \neq 0$. Then by (2.7) and (∗)

(+)
$$b(a(\lambda))^2 = (a(-\lambda)^{n(\lambda)})^2 = a(-2\lambda)^{n(\lambda)} = a(-2\lambda)^{n(2\lambda)}$$
$$= b(a(2\lambda))$$

and

$$
\begin{aligned}
B^{a(-\lambda)} &= B^{a(\lambda)a(-\lambda)b(a(\lambda))a(-\lambda)} = B^{a(\lambda)n(-\lambda)} \\
&= B^{a(\lambda)n(\lambda)} = B^{a(\lambda)a(\lambda)b(a(-\lambda))a(\lambda)} \\
&= B^{a(2\lambda)b(a(-\lambda))a(\lambda)} = A^{b(a(2\lambda))b(a(-\lambda))a(\lambda)} \\
&= A^{b(a(\lambda))a(\lambda)} = B^{a(\lambda)a(\lambda)} = B^{a(2\lambda)}.
\end{aligned}
$$

Since $N_A(B) = 1$ by (1.3) this implies $2\lambda = -\lambda$ or $3\lambda = 0$. We have shown

(∗∗) $2\lambda = 0$ or $3\lambda = 0$ for each $\lambda \in K$.

Hence each element of $A^{\#}$ has order 2 or 3. Suppose now $\lambda, \mu \in K^*$ with $2\lambda = 0 = 2\mu$. Then by (∗)

$$b(a(-\mu))^{a(\mu)} = n(\mu) = n(\lambda) = b(a(-\lambda))^{a(\lambda)}.$$

Since the elements of Ω are trivial intersection subgroups this implies

$$a(\mu)a(-\lambda) \in N_A(B) = 1 \text{ and thus } a(\mu) = a(\lambda).$$

We obtain: Either $K = \mathbb{Z}_2$ or each element of K^* has order 3.

In the first case by (1.3) $|\Omega| = 3$. Hence easily $X \simeq \Sigma_3 \simeq PSL_2(2)$. ($\Sigma_n$ denotes the symmetric group on n-letters.) So it remains to show that $X \simeq (P)SL_2(3)$

in the second case. Fix in this case $\lambda \in K^*$ and let $\Delta = \{A, B, B^{a(\lambda)}, B^{a(2\lambda)}\} \subseteq \Omega$. We have

$$B^{a(2\lambda)n(\lambda)} = B^{b(a(-\lambda))a(\lambda)} = B^{a(\lambda)}$$

since $3\lambda = 0$. As $n(\lambda)^2 = 1$ this implies that $n(\lambda)$ interchanges $B^{a(\lambda)}$ and $B^{a(2\lambda)}$ and fixes Δ. Now

$$B^{n(\lambda)a(\lambda)} = B^{b(a(-\lambda))a(2\lambda)} = B^{a(2\lambda)}$$

and

$$A^{n(\lambda)a(\lambda)} = A^{n(\lambda)a(\lambda)} = B^{a(\lambda)}.$$

Thus also $n(\lambda)^{a(\lambda)}$ fixes Δ and acts fixed-point-freely on Δ. Since both act as elements of the Klein 4-group on Δ, the commutator fixes all elements of Δ. Hence

$$[n(\lambda), n(\lambda)^{a(\lambda)}] \in H = \{1\}.$$

As $n(\lambda) = n(\mu)$ for $\lambda, \mu \in K^*$ by $(*)$ this shows

$$[n(\lambda), n(\lambda)^{a(\mu)}] = [n(\mu), n(\mu)^{a(\mu)}] = 1,$$

which implies that $N = \langle n(\lambda)^A \rangle$ is an elementary abelian 2-group. Now $X = \langle n(\lambda), A \rangle = NA$ and A is an elementary abelian 3-group. As A is regular on $A^N - \{A\}$, A acts also regularly on $N^\#$. Therefore, if $|A| = 3^r$ and $|N| = 2^s$, we get $3^r = 2^s - 1$, which implies $r = 1$ and $s = 2$. Hence

$$X \simeq A_4 \simeq PSL_2(3). \qquad \square$$

(2.9)   Lemma

Suppose X is special with AUS and $|A| > 3$. Then $|[A, H]| > 3$.

Proof. Let $A_0 = [A, H]$ and assume $|A_0| \le 3$. Then by (2.8) $|A_0| = 2$ or 3. This leads to a contradiction. For each prime p let

$$A_p = \langle x \in A \mid o(x) = p^n, n \in \mathbb{N} \rangle \text{ and similarly } B_p.$$

Then $A_p = \{x \in A \mid o(x) = p^n, n \in \mathbb{N} \cup \{0\}\}$. Further if $a \in A_p^\#$, then $o(b(a))$ is also a p-power, since $b(a)^{-1} = a^{b(a)a^{-1}}$. Hence $b(a) \in B_p$ for each $a \in A_p^\#$. Moreover since $A_p^{b(a)a^{-1}} \subseteq B$, we obtain $A_p^{b(a)} \subseteq B_p^a$ and thus, with symmetry, $A_p^{b(a)} = B_p^a$. This shows that $X_p = \langle A_p, B_p \rangle$ is a special rank one group for each prime p for which $A_p \ne 1$.

Suppose now $A_p \ne 1$ for some prime $p \ge 5$. Then by (2.8)

$$|A_0| \ge |A_0 \cap A_p| \ge p > 3,$$

a contradiction. Thus $A_p = 1$ for each $p \ge 5$. Next, if $|A_2 \cap A_0| \ge 2$ and $|A_3 \cap A_0| \ge 3$, then also $|A_0| > 3$. Hence by (2.8) one of the following holds:

(α) There exists a subgroup $A_0 \le A_1 \le A$ with $|A_1| = 6$,

(β) $A_2 = 1$,

or

(γ) $A_3 = 1$.

From these possibilities we obtain easily the following possibilities:

(1) there exists a torsion-free subgroup A^∞ of A with $A = A_0 \times A^\infty$

or

(2) there exists a subgroup A_1 of A with $A_0 \le A_1$ and $|A_1| = 4, 6$ or 9.

In case (1) $A^2 = \{a^2 \mid a \in A\} \le C_A(H)$ resp. $A^3 \le C_A(H)$, since $A^2 \cap A_0 = 1$ resp. $A^3 \cap A_0 = 1$. Thus setting $B^n = (A^n)^x$, $n = 2$ or 3 resp. $B_1 = A_1^x$, where $x \in X$ with $A^x = B$, then (1.12) implies that $X^n = \langle A^n, B^n \rangle$ resp. $X_1 = \langle A_1, B_1 \rangle$ is a special rank one group. In the first case by (2.8) $|[A, H]| \ge |[A^n, H]| > 3$, since A^n is torsion-free.

In the second case $X_1 / Z(X_1)$ is a doubly transitive permutation group of degree 5, 7 or 10. Hence it follows easily, see exercise (2.13)(9), that X_1 is isomorphic to $(P)SL_2(4), (P)SL_2(9)$ or to a Frobenius group of order $5 \cdot 4$ or $7 \cdot 6$. The first two cases contradict $|A_0| \le 3$, while the last two cases contradict (2.8) (or (1.7)!). $\qquad\square$

(2.10) Theorem

Suppose X is special with AUS. Then one of the following holds:

(1) $|A| \le 3$ and $X \simeq SL_2(2)$ or $(P)SL_2(3)$.

(2) $X = X'A$, X' quasi-simple and $|[A, H]| > 3$. ($[A, H] \le X' \cap A!$)

Proof. By (2.8) we may assume $|A| > 3$, since otherwise (1) holds. Hence by (2.9) $|[A, H]| > 3$. By (1.10) $X = X'A$. So X' quasi-simple remains to be shown. Applying (1.13)(6) and (2.8) on $Y = \langle X' \cap A, X' \cap B \rangle$ we obtain $Y' \cap A \ne 1$, whence $Y' \not\le Z(Y)$ and thus $X'' \not\le Z(X)$. Now again by (1.10) $X = X''A$ and thus $X' = X''$, since A is abelian. By the same argument $X' = \langle C \cap X' \mid C \in \Omega \rangle$, since the latter is a normal subgroup of X.

Let now $N \lhd X'$. We claim $N \le Z(X')$. Suppose first $A \cap N = 1$. Then $C \cap N = 1$ for each $C \in \Omega$, since by (1.13)(6) $A \cap X'$ and $C \cap X'$ are conjugate in X'. If N normalizes $A \cap X'$, then

$$[N, C \cap X'] \le C \cap N = 1 \text{ for each } C \in \Omega,$$

whence $N \le Z(X')$ which was to be shown. So we may assume that there exists an $n \in N$ with $A \cap X' \ne (A \cap X')^n$. Thus by (1.4)

$$Y = \langle A \cap X', (A \cap X')^n \rangle \le N(A \cap X').$$

But then by (2.8) $A \cap N \ne 1$, since $|A \cap X'| > 3$.

This shows $A \cap N \neq 1$. Let now $N_0 = \langle C \cap N | C \in \Omega \rangle$ and $N_1 = \bigcap_{\alpha \in A} N_0^\alpha$. Then $N_0 \triangleleft X'$, whence also $N_1 \triangleleft X'$.

Moreover, since $X = X'A$ and since N_1 is A-invariant, we obtain $N_1 \triangleleft X$. Clearly $N_1 \not\leq Z(X)$ as $N_1 \cap A \neq 1$. Hence by (1.10) $X = N_1 A$ and thus $X' \leq N_1$, a contradiction to $N_1 \leq N < X'$.

This shows that each proper normal subgroup of X' lies in $Z(X')$, whence X' is quasi-simple. $\qquad\square$

Remark. I believe that each special rank one group with AUS is either quasi-simple or isomorphic to $SL_2(2)$ or $(P)SL_2(3)$. If one could prove this, it would quite simplify the proofs of simplicity for classical and Lie type groups.

(2.11) Corollary

Suppose X is relatively special with respect to $1 \neq A_0 \leq Z(A)$. Let $B_0 = A_0^x, x \in X$ with $A^x = B$ and $X_0 = \langle A_0, B_0 \rangle$. Then one of the following holds:

(1) $X_0 \simeq SL_2(2)$ or $(P)SL_2(3)$.
(2) Let $n \in \mathbb{N}$ with $A^{(n)} = 1$. Then $X = X^{(n)}A, X^{(n)}$ is quasi-simple and $|A_0 \cap X^{(n)}| > 3$.

Proof. First remember that $B_0 = A_0^y$ for each $y \in X$ with $A^y = B$, since by the definition of "relatively special" A_0 must be $N_X(A)$ invariant. Now by (1.12) X_0 is a special rank one group with AUS. Hence we may apply (2.10) to X_0. Thus either (1) holds or $X_0 = X_0'A_0, X_0'$ is quasi-simple and $|A_0 \cap X_0'| > 3$. So assume from now on the latter. We will show that in this case (2) holds.

Let $R = \langle (A_0 \cap X_0')^X \rangle$. Then $R \triangleleft X$ and $R \not\leq Z(X)$ since $A_0 \cap R \neq 1$. Hence (1.10) implies $X = RA$. In particular

$$X^{(n)} = (RA)^{(n)} \leq RA^{(n)} = R.$$

On the other hand $X_0' = X_0'' \leq R'$. Hence $A_0 \cap X_0' \leq R'$ and thus by definition $R = R'$. This shows that $R = X^{(n)}$ is perfect. So it remains to show that each proper normal subgroup of R lies in $Z(R)$.

Let $N \triangleleft R$ with $N \not\leq Z(R)$. For $C \in \Omega$ let $C_0 = A_0^g, g \in X$ with $A^g = C$ and $C_1 = C_0 \cap R$. Then $A_0 \cap X_0' \leq A_1$ so that $R = \langle C_1 | C \in \Omega \rangle$. If $A_1 \cap N \neq 1$ then, since R is already transitive on Ω, $C_1 \cap N \neq 1$ for each $C \in \Omega$. If now $C_1 \neq A_1$, then by (1.13)(6) $\langle C_1, A_1 \rangle$ is a rank one group so that by (1.4)

$$C_1 \leq \langle C_1 \cap N, A_1 \rangle \leq NA_1.$$

This implies $R \leq NA_1$, whence $R = R' \leq N$ which was to be shown.

So we may assume $A_1 \cap N = 1$. Suppose $A_1 \neq A_1^n$ for some $n \in N$. Then by (1.12) $F = \langle A_1, A_1^n \rangle$ is a special rank one group with AUS. Since $F \leq NA_1$ and since $|A_1| > 3$, we obtain by (2.8) applied to F:

$$1 \neq A_1 \cap F' \leq A_1 \cap N,$$

a contradiction to $A_1 \cap N = 1$. $\square$

If L is as in (1.5) or L is a Cayley division algebra we denote by $(P)SL_2(L)$ any center-factor group of $SL_2(L)$. Remember that by (1.5) and (1.8) $(P)SL_2(L)$ is quasi-simple. With this notation we have the following more special simplicity criterion.

(2.12) Lemma

Suppose $X = \langle A, B \rangle$ is a rank one group with unipotent subgroups A and B and there exists a homomorphism

$$\sigma : X \to Y = (P)SL_2(L), L \text{ as in (1.5) or (1.13)(8)}$$

mapping A and B into different unipotent subgroups of $(P)SL_2(L)$. Then one of the following holds:

(1) $X \simeq SL_2(2)$ or $(P)SL_2(3)$.

(2) X is special and quasi-simple with AUS.

Moreover in case (2) there exist subgroups $A_0 \leq A$ and $B_0 = \{b(a) \mid a \in A_0\} \leq B$ such that $X_0 = \langle A_0, B_0 \rangle$ is a perfect central extension of $PSL_2(k)$, k a subfield of L with $|k| \geq 4$.

Proof. Being slightly incorrect in notation, assume that $\widehat{A} = \{a(c) \mid c \in L\}$ and $\widehat{B} = \{b(c) \mid c \in L\}$, in the notation of (1.5) (resp. (1.13)(8)), are the unipotent subgroups of Y into which A and B are mapped by σ.
For $x \in X$ write $\overline{x}$ for $\sigma(x)$. Conjugating with a diagonal automorphism of Y, see exercise (1.13)(7), we may assume $a(1) \in \overline{A} \leq \widehat{A}$, whence $b(1) = b(a(1)) \in \overline{B}$ and thus

$$w := a(1)b(-1)a(1) \in \overline{X}.$$

Then $w = n(1)$ in the notation of (1.8) (resp. $w = \begin{pmatrix} & 1 \\ -1 & \end{pmatrix}$ in (1.5)). Hence $w^2 = -\mathrm{id}_V$, V the natural Y-module and $a(c)^w = b(-c)$ for each $c \in L$. Let F be an additive subgroup of L with $\overline{A} = \{a(f) \mid f \in F\}$. Then $1 \in F$ and $\overline{B} = \{b(f) \mid f \in F\}$. Now for $f \in F$ we have $b(a(f)) = b(f^{-1}) \in \overline{B}$ (See (1.8)(10)) and thus $F = \{f^{-1} \mid f \in F\}$. Now for each $f \in F$ also

$$n(f) = a(f)b(-f^{-1})a(f) \in \overline{X}$$

and thus also $h(f) = n(f)n(-1) \in \overline{X}$. Now $h(f)$ normalizes $\overline{A}$ and $\overline{B}$ and acts on $\overline{B}$ by

$$b(c)^{h(f)} = b(fcf)\,;\; c \in F, f \in F^*$$

(see (1.8)(11)!)

Hence F is a subgroup of L satisfying:

(a) $1 \in F$,
(b) If $c \in F^*$, then $c^{-1} \in F^*$,
(c) If $\lambda \in F^*, c \in F$, then $\lambda c \lambda \in F$.

In particular in the situation of (1.5) F is a subgroup of L satisfying the same properties as L. In any case either (1.5) or the proof of (1.8)(11) show that, if we set $\overline{H} = N_{\overline{X}}(\overline{A}) \cap N_{\overline{X}}(\overline{B})$, then either

(i) $|F| \leq 3$ or
(ii) $[\overline{A}, \overline{H}] = \overline{A}$. In particular $\overline{X}$ is quasi-simple by (1.10).

Now in case (ii), $\ker \sigma \leq Z(X)$ and $[A, H] = A$ for $H = N_X(A) \cap N_X(B)$ by (1.11). In particular X is quasi-simple with AUS. That X is also special follows from the fact that $\overline{X}$ is special (since Y is special) and since $\sigma|_A$ and $\sigma|_B$ are isomorphisms. By (a) and (b) above the prime field k of L is contained in F. Further if $|k| \leq 3$ and $\mu \in F - k$, then the field extension $k(\mu) \subseteq F$. Setting $A_0 = \{a(\lambda) \mid \lambda \in k \text{ resp. } \lambda \in k(\mu)\}$ this shows that (2) holds in case (ii).

The proof of (1) in case (i) is an exercise.   □

(2.13)   Exercises

(1) Let G be nilpotent of class n and $U \leq G$. Set $U = U^0$ and define recursively $U^i = N(U^{i-1})$. Show $U^n = G$. In particular each proper subgroup of G is contained in a proper normal subgroup.

(2) Let $U \leq V \leq G$. Then U is *strongly closed* in V with respect to G, if for $u \in U^\#$ and $g \in G$ with $u^g \in V$ it follows that $u^g \in U$. Now suppose $X = \langle A, B \rangle$ is a rank one group. Show that A is strongly closed in each proper subgroup of X containing A.

(3) Suppose X is a rank one group with unipotent subgroups A and B. Show

   (i) If $|A| = 2$, then $X \simeq SL_2(2)$,
   (ii) If $|A| = 3$, then $X \simeq (P)SL_2(3)$.

(4) Suppose X is a finite rank one group, which satisfies case (1) of (2.11) with $X_0 \simeq SL_2(2)$. Let $1 \neq d \in A_0$ and $D = d^X$. Show:

(i) $o(ef) = 3$ for all $e \neq f \in D$.

(ii) Let $Y = \langle D \rangle$. Then $Y = Q\langle d \rangle$ with $Q = O_3(Y)$. Further $X = QA$.
($Q \not\leq Z(X)$ since $N_X(A_0) \leq N_X(A)$!)
Hint. Use the $\mathbb{Z}^*$-theorem of Glauberman [Asc86, p261].

(iii) $C_Q(A_0) = C_Q(A) = C_Q(a) = Z(X)$ for each $a \in A^{\#}$.

(iv) Let $\overline{X} = X/Z(X)$. Then $\overline{X} \simeq (\mathbb{Z}_3 \times \mathbb{Z}_3)Q_8$ or $\overline{X} \simeq F_q \cdot F_q^*$, where $q = 3^m$.

Hint. By (iii) $\overline{X}$ is a doubly transitive Frobenius group with kernel $\overline{Q}$ an elementary abelian 3-group and nilpotent complement $\overline{A}$. Hence all elements of odd order of $\overline{A}$ are in $Z(\overline{A})$. Now $Z(\overline{A}) \subseteq \mathrm{Hom}_{\overline{A}}(\overline{Q}, \overline{Q}) = F = GF(3^{\alpha})$ for some $\alpha \geq 1$ and $\overline{Q}$ is an F-vector space, say of dimension n. Thus

$$\frac{3^{n\alpha} - 1}{3^{\alpha} - 1} = \frac{|\overline{Q}| - 1}{|F| - 1} = \frac{|\overline{A}|}{|F^*|} \mid |\overline{A} : Z(\overline{A})| = 2 - \text{power}.$$

Hence $n = 1$ or $\alpha = 1$ and $n = 2$.

(5) Let X be a solvable rank one group with unipotent subgroup A and $\overline{X} = X/Z(X)$. Show:

(i) There exists a unique minimal normal subgroup $1 \neq \overline{N}$ of $\overline{X}$.

(ii) $X = N \cdot A$, $N \cap A = 1$.

(iii) Either $\overline{N}$ is torsion-free or $\overline{N}$ is an elementary abelian p-group. (I. e. each element of $\overline{N}^{\#}$ has order p.)

(iv) Each element of $A^{\#}$ acts fixed-point-freely on $\overline{N}$.

(v) A acts transitively on $\overline{N}^{\#}$.

(6) Construct an example of a finite solvable rank one group with unipotent subgroup A, which is neither abelian nor of prime power order.

(7) Suppose the group G is generated by conjugates of the subgroup U. Then, if X and Y are subgroups of G with

$$U^g \leq X \text{ or } U^g \leq Y \text{ for each } g \in G,$$

either $G = X$ or $G = Y$.
Hint. Let $\Sigma = U^G$. Then $\Sigma - (\Sigma \cap X) \subseteq Y$.

(8) Let G and U be as in (7) and let $V = U^g \neq U$. Then there exists a conjugate U^h of U, which neither normalizes U nor V.

(9) Determine the doubly transitive permutation groups of degree $5, 7$ and 10, where the stabilizer of a point has an abelian regular normal subgroup (on the remaining points!).

Hering, Kantor and Seitz have in [HKS72] determined all such permutation groups, with no restriction on the degree.

(10) Let $X = \langle A, B \rangle$ be a rank one group and $1 \neq a \in A$. Then $\langle a^X \rangle$ is not nilpotent.

Hint. Assume without loss of generality $Z(X) = 1$ and use (1.10).

§3 Quadratic modules

Let X be a rank one group with unipotent subgroup A. A faithful $\mathbb{Z}X$-module V (i. e. an abelian group V upon which X acts faithfully) is called a quadratic X-module, if $[V, A, A] = 0$. Notice that, by the 3-subgroup-lemma, A is abelian, if X admits such a quadratic module. Hence by (2.5) X is special. On the other hand, as example (2.6)(3) shows, all orthogonal groups of Witt index 1 are special rank one groups admitting a quadratic module. Hence the determination of such rank one groups seems to be difficult. In the main result of this section, (3.2), we give a partial solution to this problem, i.e. we determine $SL_2(K)$, K a division ring or a Cayley division algebra with its quadratic action.

Let in this whole section X be again a rank one group with unipotent subgroups A and B and diagonal subgroup H. We first show:

(3.1) Lemma

Suppose X admits a quadratic module V. Let $V_1 = [V, X], V_2 = V_1 \cap C_V(X)$ and $\overline{V} = V_1/V_2$. Then the following holds:

(1) $\overline{V} = [\overline{V}, A] \oplus [\overline{V}, B]$. Further

$$[\overline{V}, A] = [\overline{V}, a] = C_{\overline{V}}(a) = C_{\overline{V}}(A) \text{ for each } a \in A^{\#}.$$

(2) Let K be the kernel of the action of X on $\overline{V}$. Then $K \leq Z(X)$.

(3) Either A is torsion-free or an elementary abelian p-group (for some prime p).

Proof. Since $V_2 = [V, A] \cap [V, B]$ we have

$$\overline{V} = [\overline{V, A}] \oplus [\overline{V, B}]$$

by definition of $\overline{V}$. As $[\overline{V}, A] = [\overline{V, A}]$ this shows the first part of (1). Now, since $\overline{V}$ is a quadratic module, we have

$(*)$ $\qquad\qquad [\overline{V}, a] \subseteq [\overline{V}, A] \subseteq C_{\overline{V}}(A) \subseteq C_{\overline{V}}(a)$ for each $a \in A^{\#}$.

Since $X = \langle a, B \rangle$ by (1.4) we also have

$$\overline{V} = [\overline{V}, a] \oplus [\overline{V}, B],$$

since the latter is X-invariant and thus contains $[\overline{V}, A]$. If now $[\overline{V}, a] \neq C_{\overline{V}}(a)$, then $0 \neq C_{\overline{V}}(a) \cap [\overline{V}, B] \subseteq C_{\overline{V}}(X)$ and thence also $[\overline{V}, A] \cap [\overline{V}, B] \neq 0$ a contradiction. This shows that we have equality in $(*)$, which proves (1).

Clearly by (1) $X \neq KA$. Hence (1.10) implies (2). Finally suppose $a \in A^{\#}$ is of order p, p a prime. Then for each $\overline{v} \in \overline{V}$ we have

$$[\overline{v}, a]^p = [\overline{v}, a^p] = [\overline{v}, 1] = 0.$$

Now (1) implies that $\overline{V}$ is an elementary abelian p-group. But then for each $\alpha \in A^{\#}$ and $\overline{v} \in \overline{V}$:

$$0 = [\overline{v}, \alpha]^p = [\overline{v}, \alpha^p]$$

and whence $\alpha^p \in C_X(\overline{V}) = K$. Since $A \cap Z(X) = 1$ now (2) shows $\alpha^p = 1$ and A is an elementary abelian p-group. This proves (3). $\qquad\qquad \square$

The following theorem, which is theorem 1 of [Tim94a], gives a classification of $\overline{V}$ and X under an additional condition.

(3.2) Theorem

Suppose V is a quadratic module for X satisfying:

  (a) $V = [V, X]$ and $C_V(X) = 0$.
  (b) $[V, A] = [v, A]$ for each $v \in V - [V, A]$.

Then there exists a division ring or a Cayley division algebra K such that $X \simeq SL_2(K)$ and V is the natural $\mathbb{Z}X$-module.

Proof. Set $U = [V, A]$ and $W = [V, B]$. The proof consists of a simultaneous coordinatization of U, W and A, B, which we do in a series of intermediate steps. By (3.1) we have:

  (1) $U = [V, a] = C_V(a) = C_V(A)$ for each $a \in A^{\#}$,
     $W = [V, b] = C_V(b) = C_V(B)$ for each $b \in B^{\#}$
     and $V = U \oplus W$.

The additional hypothesis (b) implies

  (2) $U = [w, A]$ and $W = [u, B]$ for each $w \in W^{\#}, u \in U^{\#}$,

since there exists a $\omega \in X$ interchanging A and B. Next we show:

(3) $\mathcal{P} = \{U\} \cup W^A = U^B \cup \{W\}$ is an X-invariant partition of V.

$\Omega = A^B \cup \{B\} = B^A \cup \{A\}$ and (1) imply that $\mathcal{P} = \{U\} \cup W^A = U^B \cup \{W\}$ is X-invariant. As $U \cap W = 0$ also $U \cap W^a = 0$ for each $a \in A$. Suppose $W \cap W^a \neq 0$ for some $a \in A^{\#}$. Then $w^a \in W$ for some $w \in W^{\#}$ and thus $w^{-1}w^a = [w,a] \in U \cap W = 0$, which implies $w = w^a$. But this is a contradiction to $U = C_V(a)$. From this we obtain that

$$W^{a_1} \cap W^{a_2} = 0 \text{ for } a_1 \neq a_2 \in A,$$

which implies that the elements of $\mathcal{P}$ have pairwise trivial intersection.

So it remains to show that each element of V lies in some element of $\mathcal{P}$. So let $v \in V$ and assume without loss of generality that $v \notin U$ and $v \notin W$. Then by (1) $v = u + w, u \in U^{\#}, w \in W^{\#}$. By (2) there exists an $a \in A$ with

$$w^a - w = [w,a] = u.$$

Hence $v = u + w = w^a \in W^a \in \mathcal{P}$.

(4) Let $a \in A^{\#}$ and $b = b(a) \in B^{\#}$. Then $w = [w,a,b]$ for each $w \in W^{\#}$.

As $[w,a] \in U^{\#}$ there exists by (2) a $b_1 = b_1(w)$ with $w = [w,a,b_1]$ (b_1 depends on w!). Set $u = [w,a]$. Then $u^{b_1} = w + u$ and thus $u^{b_1^{-1}} = u - w$. This implies

$$w^{a b_1^{-1}} = w^{ab_1^{-1}} = (w+u)^{b_1^{-1}} = w + (u - w) = u.$$

Now we obtain from (3):

$$W^{a b_1^{-1}} = U \text{ and thus } W^a = U^{b_1}.$$

On the other hand, as $A^b = B^a$ (1) implies that also $W^a = U^b$, whence $U^b = U^{b_1}$ and $b_1 b^{-1} \in N_B(U)$. But then, again by (1), $b = b_1$ is independent of w.

(5) A, B, U and W are isomorphic groups, which are either all torsion-free or elementary abelian p-groups.

(3) implies that U and W and (2) that U and A and W and B are isomorphic, since the commutator maps are always isomorphisms. Now (5) follows from (3.1).

Now we start labelling the groups A, B, U and W. Let K be an additive group isomorphic to A, $K^* = K - \{0\}$ and fix an element $1 \in K^*$. Let

$$A = \{a(\lambda) \mid \lambda \in K\} \text{ with } a(\lambda)a(\mu) = a(\lambda + \mu)$$

and set $b(1) := b(a(1))$ and $b(-1) := b(a(-1))$. Then, since X is special $b(-1) = b(1)^{-1}$ by (2.2). Fix an element $w = w(1) \in W^{\#}$ and set

$$
\begin{aligned}
n(t) &:= a(t)b(a(-t))a(t), t \in K^* \\
\omega &:= n(1) \\
u(\lambda) &:= [w(1), a(\lambda)], \lambda \in K.
\end{aligned}
$$

(Notice that again by (2.2) $b(a(-t)) = b(a(t)^{-1}) = b(a(t))^{-1}$!) Then we have:

(6) The following holds:

(a) $U = \{u(\lambda) \mid \lambda \in K\}$ with $u(\lambda) + u(\mu) = u(\lambda + \mu)$.

(b) $w(1)^{n(t)} = u(t)$ and $u(t)^{n(t)} = -w(1) =: w(-1)$.

(c) $n(t)^2 = -\mathrm{id}_V$ independent of t.

Proof. By (2) the commutator map

$$
a(\lambda) \to u(\lambda) = [w(1), a(\lambda)]
$$

is an isomorphism of A onto U. This implies (a). Now by (4)

$$
u(t)^{b(a(t))} - u(t) = [u(t), b(a(t))] = [w(1), a(t), b(a(t))] = w(1).
$$

Hence

$$
u(t)^{b(a(t))} = w + u(t) = w^{a(t)}, w = w(1).
$$

So we obtain

$$
w^{n(t)} = w^{a(t)b(a(t))^{-1}a(t)} = u(t)^{a(t)} = u(t).
$$

Conjugating the equation before by $b(a(t))^{-1}$ we also have $u(t)^{b(a(t))^{-1}} = u(t) - w$. Hence we obtain

$$
u(t)^{n(t)} = (u(t) - w)^{a(t)} = u(t), (u(t) + w) = -w = w(-1)
$$

which proves (b).

Now by (b) $u(t)^{n(t)^2} = -u(t)$ and $w^{n(t)^2} = -w$. Because of

$$
[-w, a(\lambda)] = -[w, a(\lambda)] = -u(\lambda) = u(-\lambda)
$$

this shows by (2.7) that $n(t)^2$ centralizes A and normalizes B. But then $n(t)^2$ centralizes with each element $a \in A^{\#}$ also the unique $b(a) \in B^{\#}$, whence $n(t)^2 \in Z(X)$. Since $n(t)^2$ is $-\mathrm{id}$ on U this proves (c).

Now we set:

$$
w(\lambda) := u(-\lambda)^{\omega} \text{ and } b(\lambda) := a(-\lambda)^{\omega}, \lambda \in K \ (\omega = n(1)!)
$$

Then by (6)(b) $u(-1)^\omega = w = w(1)$. Further

$$u(1)^{b(-1)} = u(1)^{n(1)a(-1)} = -w(1)^{a(-1)} = -(u(-1) + w(1)) = u(1) - w(1)$$

and

$$u(1)^{\omega^{-1}a(1)\omega} = w(1)^{a(1)\omega} = (u(1) + w(1))^\omega = u(1) - w(1).$$

Hence

$$[u(1), b(-1)] = -w(1) = [u(1), a(1)^\omega].$$

This shows that $b(1)a(1)^\omega$ centralizes W and $u(1)$ and thus $b(-1) = a(1)^\omega$ by (2). So the notation $b(\lambda) := a(-\lambda)^\omega$ is consistent with the earlier definition of $b(-1)$.

Now for $\lambda, \mu \in K$ set: $\lambda \cdot \mu$ is the unique element of K satisfying $u(\lambda \cdot \mu) = [w(\lambda), a(\mu)]$.

Then we obtain:

(7) The following hold:

    (a) $K^* = \{\lambda \cdot \mu \mid \lambda \in K^*\} = \{\nu \cdot \lambda \mid \lambda \in K^*\}$ for all fixed $\mu, \nu \in K^*$.

    (b) $\lambda \cdot 1 = \lambda$ for all $\lambda \in K^*$.

    (c) $1 \cdot \lambda = \lambda$ for all $\lambda \in K^*$.

Proof. (a) follows from

$$\begin{aligned} U^\# &= [W^\#, a(\mu)] = \{[w(\lambda), a(\mu)] \mid \lambda \in K^*\} \quad \text{and} \\ U^\# &= [w(\nu), A^\#] = \{[w(\nu), a(\lambda)] \mid \lambda \in K^*\} \end{aligned}$$

by (1) and (2). (c) is a part of the definition of the multiplication on K. So (b) remains to be proved. Because of $\omega^2 = -\mathrm{id}_V$ we have $w(\lambda)^\omega = u(\lambda)$. Hence

$$(*) \qquad\qquad w(\lambda)^{a(1)} = u(\lambda)^{b(1)}.$$

Now by (4), using additive notation:

$$(w(\lambda)^{a(1)} - w(\lambda))^{b(1)} - (w(\lambda)^{a(1)} - w(\lambda)) = w(\lambda).$$

Hence

$$w(\lambda)^{a(1)b(1)} - w(\lambda)^{a(1)} = w(\lambda).$$

Together with $(*)$ this implies

$$u(\lambda)^{b(1)^2} - u(\lambda)^{b(1)} = (u(\lambda)^{b(1)} - u(\lambda))^{b(1)} = w(\lambda)$$

and thus

$$u(\lambda)^{b(1)} - u(\lambda) = w(\lambda).$$

Now again by (∗)

$$u(\lambda) = w(\lambda)^{a(1)} - w(\lambda) = [w(\lambda), a(1)]$$

which proves (b).

Next we show:

(8) The following hold:

(a) The multiplication on K satisfies both distributive laws.

(b) For $x, y \in K^*$ there exist unique $t, \tau \in K^*$ satisfying $x \cdot t = y = \tau \cdot x$.

(a) is equivalent to the bilinearity of the commutator map, which follows from quadratic action. (b) follows from (7)(a) and the fact that the commutator maps are isomorphisms.

(9) For all $\mu, \lambda \in K$ we have $w(\mu \cdot \lambda) = [u(\mu), b(\lambda)]$.

Namely by definition of $b(\lambda)$ and $w(\mu)$ we have

$$
\begin{aligned}
[u(\mu), b(\lambda)] &= [u(\mu), a(-\lambda)^\omega] = [w(\mu)^\omega, a(-\lambda)^\omega] = [w(\mu), a(-\lambda)]^\omega \\
&= u(\mu(-\lambda))^\omega = u(-\lambda\mu)^\omega = w(\mu\lambda),
\end{aligned}
$$

since $\mu(-\lambda) + \mu\lambda = \mu(-\lambda + \lambda) = 0$ by (8)(a).

(10) For $t \in K^*$ let t^{-1} be the element of K^* satisfying $t \cdot t^{-1} = 1$. Then the following hold:

(a) $b(t^{-1}) = b(a(t))$.

(b) $(xt)t^{-1} = x$ for all $x \in K$.

(c) $t^{-1}t = 1$.

Proof. By (4) we have

$$[w(1), a(t), b(a(t))] = w(1).$$

On the other hand $[w(1), a(t)] = u(t)$, so that by (9) $b(t^{-1})$ satisfies the same equation. Since, as shown in (4), $b(a(t))$ is the unique element in B with this property this shows (a).

Now by (4), (9) and (a)

$$
\begin{aligned}
w((xt)t^{-1}) &= [u(xt), b(t^{-1})] = [u(xt), b(a(t))] \\
&= [w(x), a(t), b(a(t))] = w(x),
\end{aligned}
$$

which proves (b). Finally by (4), (9), (a) and (b):

$$w((t^{-1}t)t^{-1}) = [u(t^{-1}t), b(t^{-1})] = [w(t^{-1}), a(t), b(a(t))]$$
$$= w(t^{-1}).$$

Hence $(t^{-1}t)t^{-1} = t^{-1}$. On the other hand by (7) $1 \cdot t^{-1} = t^{-1}$ so that by (8)(b) $t^{-1}t = 1$.

Now by (7)–(10) K is a (non-necessary associative) division ring with right inverse property in the notation of §6 of [Pic55]. Hence by the theorem of Skornyakov-San Souci [HP73, (6.16)] K is an alternative division ring (i. e. the associative law holds for all subrings generated by 2 elements.) Now the theorem of Bruck-Kleinfeld [VM98, Ap. B] shows that

(11) Either K is an (associative) division ring or a Cayley division algebra.

Let now $\varphi : V \rightarrow K^2$ given by $\varphi : u(\lambda) + w(\mu) \rightarrow (\lambda, \mu)$. Then φ is an isomorphism of abelian groups. Further φ maps the partition $\mathcal{P}$ of (3) onto the partition $\mathcal{P}$ of K^2 given in (1.8)(3). Let $\sigma : x \rightarrow \varphi^{-1}x\varphi$. Then σ is an injective homomorphism of X into $\operatorname{Aut}(K^2)$. Now for $\lambda, \nu, \mu \in K$ we have

$$(\lambda, \nu)^{\sigma(a(\mu))} = (u(\lambda) + w(\nu))^{a(\mu)\varphi} = (u(\lambda) + w(\nu)^{a(\mu)})^\varphi$$
$$= (u(\lambda) + u(\nu \cdot \mu) + w(\nu))^\varphi = (\lambda + \nu\mu, \nu)$$

by definition of the multiplication in K. Further

$$(\lambda, \nu)^{\sigma(b(\mu))} = (u(\lambda) + w(\nu))^{b(\mu)\varphi} = (u(\lambda) + w(\lambda\mu) + w(\nu))^\varphi$$
$$= (\lambda, \lambda\mu + \nu).$$

Hence the images of $a(\mu), b(\mu), \mu \in K$ under σ are exactly the automorphisms of K^2 given in (1.8). Since $SL_2(K)$ was defined as the subgroup of $\operatorname{Aut}(K^2)$ generated by these automorphisms, this shows that σ is an isomorphism of X onto $SL_2(K)$. Now, as $x\varphi = \varphi x^\sigma$ by definition of σ, this shows that V and K^2 are equivalent $\mathbb{Z}SL_2(K)$-modules (identifying X with $SL_2(K)$ via σ). □

(3.3) **Remark**

(3.2) can be used for a completely group theoretic approach to Moufang-planes. Namely let $\mathcal{P}$ be a translation plane for two distinct points P and Q (resp. distinct lines.) Let ℓ be the line through P and Q and U be the set of all elations corresponding to (P, ℓ) and W the set of all elations corresponding to (Q, ℓ). Then by a known theorem [HP73, (4.14)] $V = U \times W$ is an abelian group. Let now $g \neq \ell$ be another line through P and $h \neq \ell$ another line through Q and A the set of all elations corresponding to (P, g) and B the set of all elations corresponding to (Q, h). Then it is easy to show that, for the action of the group $X = \langle A, B \rangle$ on V, the hypothesis of (3.2) is satisfied. This implies then that by (3.2) $\mathcal{P}$ is uniquely determined by some division ring or Cayley division algebra K. For details see [Tim94a] (or the exercises to this section).

Finally we prove in this section a theorem which shows that the section $\overline{V}$ of (3.1) is indeed a direct sum of natural $\mathbb{Z}X$-modules, if $X \simeq (P)SL_2(k)$, k a commutative field. This theorem was proved first as a proposition [Tim90a, (2.7)], but only in case k is perfect, if Char $k = 2$. The present proof is from [Stb92].

(3.4) Theorem

Let $X \simeq (P)SL_2(k)$, k a commutative field and V a quadratic X-module with $C_V(X) = 0$. Let $0 \neq v \in C_V(A)$ and $W = \langle v^X \rangle$. Then there exists a scalar action of k on W such that W is the natural kX-module. In particular $X \simeq SL_2(k)$.

Before we start with the proof of (3.4) we state a subsidiary lemma for which we use the notation of (1.5); i. e. we fix the elements $a(c), b(c), n(c)$ and $h(c)$, $c \in k$ resp. $c \in k^*$ as in (1.5) and set $\omega = n(1)$. With this notation we have:

(3.5) Lemma

Let V and v be as in (3.4). Then the following hold:

(1) $v^{\omega h(c)} = v^{b(-c)a(c^{-1})} = v^{b(-c)} - v$ for $c \in k^*$.
(2) $v^{\omega^2} = -v$.
(3) $v^{h(c^{-1})} = v^{\omega^{-1}(a(c)-\mathrm{id})}$ for $c \in k^*$.

Proof. We have

$$a(c^{-1})b(-c)a(c^{-1}) = n(c^{-1}) = h(c^{-1})\omega = \omega h(c),$$

since by (1.5)(c) $\omega^2 \in Z(X)$ and thus $\omega^4 = 1$ and since ω inverts $h(c)$. Now set

$$w := v - v^{b(-c)} + v^{b(-c)a(c^{-1})}.$$

Then

$$\begin{aligned} w &= v + [v^{b(-c)}, a(c^{-1})] \in C_V(A) \\ &= -[v, b(-c)] + v^{b(-c)a(c^{-1})} = -[v, b(-c)] + v^{\omega h(c)}. \end{aligned}$$

As ω interchanges A and B this implies

$$w \in C_V(A) \cap C_V(B) = \{0\}.$$

From this we obtain

$$v^{\omega h(c)} = v^{b(-c)a(c^{-1})} = v^{b(-c)} - v.$$

Now applying (1) for $c = 1$ we have:

$$\begin{aligned}
v^{\omega^2} &= v^{\omega h(1)\omega} = (v^{b(-1)} - v)^\omega = v^{\omega a(1)} - v^\omega \\
&= v^{b(-1)a(1)} - v^{a(1)} - v^\omega = -v^{a(1)} = -v
\end{aligned}$$

since by (1.5)(c) we have

$$b(-c)^\omega = a(c)^{\omega^2} = a(c) \text{ for } c \in k^*.$$

Conjugating (1) with ω^{-1} we obtain:

$$\begin{aligned}
v^{h(c^{-1})} &= v^{\omega h(c)\omega^{-1}} = (v^{b(-c)} - v)^{\omega^{-1}} = v^{\omega^{-1}a(c)} - v^{\omega^{-1}} \\
&= v^{\omega^{-1}(a(c)-\mathrm{id})}.
\end{aligned}$$

$\square$

Proof of (3.4). For $c \in k$ let

$$c \cdot v := v^{\omega^{-1}(a(c)-\mathrm{id})} = -v^{\omega(a(c)-\mathrm{id})}$$

and set $U = \{c \cdot v \mid c \in k\}$. Since the commutator map

$$a(c) \rightarrow [v^{\omega^{-1}}, a(c)] = v^{\omega^{-1}(a(c)-\mathrm{id})}$$

is an injective homomorphism from A into V with image U, U is a subgroup of $C_V(A)$. (I. e. we have $c \cdot v + d \cdot v = (c+d) \cdot v$!) Now for $d \in k$ and

$$c \cdot v \in U, \quad \text{set} \quad d \cdot (c \cdot v) := (dc) \cdot v.$$

Then we have defined a scalar multiplication on U, which satisfies $1 \cdot v = v$ by (3.5)(3). We compute the distributive law:

$$\begin{aligned}
(d_1 + d_2)(cv) &= (d_1c + d_2c) \cdot v = -v^{\omega(a(d_1c+d_2c)-\mathrm{id})} \\
&= -v^{\omega(a(d_1c)a(d_2c)-\mathrm{id})} \\
&= -v^{\omega(a(d_1c)-\mathrm{id})a(d_2c)} - v^{\omega(a(d_2c)-\mathrm{id})} \\
&= -v^{\omega(a(d_1c)-\mathrm{id})} - v^{\omega(a(d_2c)-\mathrm{id})}, \quad \text{since } [V, A] \leq C_V(A) \\
&= (d_1c) \cdot v + (d_2c)v = d_1(cv) + d_2(c \cdot v).
\end{aligned}$$

From this one immediately obtains:

$$\begin{aligned}
(d_1 + d_2)v &= d_1v + d_2v \\
d(c_1v + c_2v) &= d(c_1v) + d(c_2v) \\
(d_1d_2)(cv) &= d_1(d_2(cv)) \\
1(cv) &= cv.
\end{aligned}$$

Hence U is a 1-dimensional vector space over k. Now we extend this scalar-multiplication to U^ω by setting

$$c \cdot v^\omega := (c \cdot v)^\omega$$

and

$$d(c \cdot v^\omega) := (dc) \cdot v^\omega.$$

Then also U^ω is a 1-dimensional vector space over k and thus $W = U \oplus U^\omega$ is a 2-dimensional vector space over k, if we extend this scalar multiplication linearly ($U \cap U^\omega = \{0\}$, since $U^\omega \subseteq C_V(B)$ and $C_V(A) \cap C_V(B) = \{0\}$!)

We next compute that W is X-invariant. Since $\omega^2 = -\mathrm{id}$ on W by (3.5)(2), W is clearly ω-invariant. We compute

$$
\begin{aligned}
(c \cdot v^\omega)^{(a(d)-\mathrm{id})} &= v^{h(c^{-1})\omega(a(d)-\mathrm{id})} = -v^{h(c^{-1})\omega^{-1}(a(d)-\mathrm{id})} \\
&= -v^{h(c^{-1})h(d^{-1})} = -v^{h((cd)^{-1})} \\
&= -(cd)v = c(-d \cdot v) = cv^{\omega(a(d)-\mathrm{id})}
\end{aligned}
$$

by (3.5)(3) applied to $-v^{h(c^{-1})} \in C_V(A)$. This shows that $[U^\omega, A] \le U$ and thus that W is $\langle \omega, A \rangle = X$-invariant. Moreover, it also shows that $a(d)$ is $-\mathrm{id}$ and whence also $a(d)$ respects the scalar multiplication on W. Since by definition and $\omega^2 = \mathrm{id}$, also ω respects the scalar multiplication, this shows that W is a 2-dimensional kX-module.

Now with respect to the basis $(v, -v^\omega)$ of W, $a(c)$ is represented by the matrix $\begin{pmatrix} 1 & \\ c & 1 \end{pmatrix}$ and ω by the matrix $\begin{pmatrix} & -1 \\ 1 & \end{pmatrix}$. Hence W is equivalent to the natural $kSL_2(k)$-module. (In particular $X \simeq SL_2(k)$, although we did not assume this in the hypothesis of (3.4).) $\square$

(3.6) Remark

Suppose that the hypothesis of (3.4) is satisfied and V is a k-vector space. Then (3.4) does not say that V is the direct sum of 2-dimensional subspaces, which are all equivalent to natural $kSL_2(k)$-modules. Indeed in the proof of (3.4) the scalar action constructed depends on the action of X and thus may be different on each direct summand. An example of where this happens is when the elements $D \in SL_2(k)$ act as a matrix

$$
\begin{bmatrix}
D^{\alpha_1} & & & \\
& D^{\alpha_2} & & \\
& & \ddots & \\
& & & D^{\alpha_n}
\end{bmatrix}, \alpha_i \in \mathrm{Aut}(k).
$$

(3.7) Corollary

Assume the same hypothesis as in (3.4) and in addition $V = [V, X]$ if char $k = 2$. Then there exists an index set I such that

$$V = \oplus_{i \in I} V_i \text{ with natural } \mathbb{Z}X - \text{modules } V_i.$$

Proof. We first show that in any case $V = [V, X]$. Indeed by (3.4) for $0 \neq v \in C_V(A)$ we know that $\langle v^X \rangle$ is the natural kX-module for some scalar action of k. In particular $X \simeq SL_2(k)$ and there exists a central involution t if Char $k \neq 2$. If $V_+ = C_V(t) \neq 0$ then $C_{V_+}(A) \neq 0$, a contradiction to the fact that $\langle w^X \rangle \subseteq V_+$ and is a natural $\mathbb{Z}X$-module for $0 \neq w \in C_{V_+}(A)$. By the same argument $W = [V, t] = [V, X]$ is inverted by t. If $V \neq W$ then we find a $v \in V - W$ with $v^t = v + w$, $w \in C_W(A)$. Since $\langle w^X \rangle$ is a natural $\mathbb{Z}X$-module, there exists a $\overline{w} \in W$ with $2\overline{w} = w$. Hence $v + \overline{w} \in C_V(t)$, a contradiction to $V_+ = 0$. Thus $V = W = [V, X]$.

Now $V = [V, X]$ immediately implies $V = [V, A] \oplus [V, B]$ and $[V, A] = C_V(A)$. By the axiom of choice and (3.4) we may for each $0 \neq v \in C_V(A)$ pick an index $j(v)$ such that $\langle v^X \rangle = V_{j(v)}$ is a natural $\mathbb{Z}X$-module. Let $J = \{j(v) \mid 0 \neq v \in C_V(A)\}$. Then, also by (3.4) $\{V_j \mid j \in J\}$, is the set of all natural $\mathbb{Z}X$-submodules of V. Clearly

$$V = \langle V_j \mid j \in J \rangle,$$

since $C_V(A) \subseteq \langle V_j \mid j \in J \rangle$. We need to show:
There exists a subset $I \subseteq J$ such that $V = \oplus_{i \in I} V_i$.

For this let $\mathcal{J}$ be the set of all subsets I of J satisfying

$$V_I := \langle V_i \mid i \in I \rangle = \oplus_{i \in I} V_i.$$

Then clearly the union of an ascending chain of elements of $\mathcal{J}$ also belongs to $\mathcal{J}$. Hence, by Zorn's lemma, $\mathcal{J}$ contains a maximal element, say I_1. Now V_{I_1} is X-invariant. Hence either $V_j \subseteq V_{I_1}$ or $V_j \cap V_{I_1} = \{0\}$ for each $j \in J$. Thus the maximality of I_1 implies $V = V_{I_1} = \oplus_{i \in I_1} V_i$, which was to be shown. $\quad\square$

(3.8) Exercises

(1) Let $X = (P)SL_2(k)$, k a commutative field with Char $k \neq 2$ and $k \neq GF(3)$ and V a quadratic X-module. Show $V = C_V(X) \oplus [V, X]$ with $[V, X]$ the direct sum of natural $\mathbb{Z}SL_2(k)$-modules.

   Hint: By (3.4) $X = SL_2(k)$. Let σ be the central involution of X. Then by (3.1) and (3.4) $W = [V, \sigma] = [V, X]$ is inverted by σ and $C_W(X) = 0$, since otherwise by (3.1) $W^2 = \{2w \mid w \in W\} \neq W$, a contradiction as

$[W, X] = W$ and X can not act non-trivially on W/W^2. Hence W is by (3.7) the direct sum of natural $\mathbb{Z}X$-modules and thus for each $w \in W^{\#}$ there exists an element $\frac{1}{2}w \in W$. Thus, if $v \in V - W$ and $v^\sigma = v + w$, then $v + \frac{1}{2}w \in C_V(\sigma)$, whence $V = C_V(\sigma) \oplus W$ and $C_V(\sigma) = C_V(X)$.

(2) Let K be a division ring or a Cayley division algebra, L a subgroup of K satisfying (1)–(3) of (1.5) and suppose Char $K \neq 2$. Assume that $X = (P)SL_2(L)$ and V is a quadratic X-module. Show that $V = C_V(X) \oplus [V, X]$.

Hint: Pick a subfield $k \subseteq L$ of K and $(P)SL_2(k) = X_0 \leq X$. Then, applying (1) to the action of X_0 on V, we have $\mathrm{id}_V \neq n(1)^2$. Hence by (1.5)(c) $n(1)^2$ is a central involution of X with $[V, n(1)^2] = [V, X_0]$.

(3) Use the notation of (3.3). For a point R on ℓ let $E_{R,\ell}$ be the elation group corresponding to the point R and line ℓ. Show that V is partitioned by $\{E_{R,\ell} \mid R \text{ is a point on } \ell\} = \mathcal{P}$ and that A is regular on $\mathcal{P} - \{U\}$ and B is regular on $\mathcal{P} - \{W\}$, whence X is doubly transitive on $\mathcal{P}$.

(4) Let X be as in (3). Show that X satisfies the hypothesis of (1.3)(2) and thus is a rank one group. Show further that for $v \in V$ and $\alpha \in A^{\#}$ the commutator $[v, \alpha] = v^{-1}v^\alpha$ fixes all lines through P so that

$$U = C_V(A) = [V, \alpha] \text{ for each } \alpha \in A^{\#}.$$

Hence V is a quadratic module for X.

(5) Let $X = \langle A, B \rangle$ be a special rank one group admitting a quadratic module. Show $AB \cap B^a = 1$ for each $a \in A^{\#}$.

Hint. Argue as in the proof of (2.4).

§ 4 Rank one groups and buildings

In this section we will discuss the action of root groups on spherical buildings. For this purpose we very briefly repeat the definition of a building in a way appropriate for us and state some of the results necessary for us. It is not possible to do this in sufficient detail, since buildings are not the topic of this book. For more details, the interested reader is referred to [Ron89],[Sch95], or the original work of [Tit74].

The main result of this section is (4.12), which shows that on a Moufang building the group generated by two opposite root groups is a rank one group.

A *chamber system* $\mathcal{C}$ over some index set I is a set $\mathcal{C}$ of "chambers" together with partitions $\mathcal{P}_i, i \in I$ of $\mathcal{C}$. If $J \subseteq I$ we will denote by $\mathcal{P}_J$ the "smallest" partition of $\mathcal{C}$ "containing" all $\mathcal{P}_j, j \in J$. If $J \subseteq I$ and $c \in \mathcal{C}$, we denote by $\Delta_J(c)$ the element of $\mathcal{P}_J$ containing c. It is called the *residue* of type J containing c. The chambers $d, e \in \mathcal{C}$ are J-adjacent if and only if they are contained in some $\Delta_J(c)$. $\mathcal{C}$ is *thick* if $|\Delta_i(c)| \geq 3$ for all $i \in I$ and *connected* if $\Delta_I(c) = \mathcal{C}$ for all $c \in \mathcal{C}$. In this book, we will always require that $|I| < \infty$. The most important examples for our purpose are the chamber systems obtained from groups. Indeed, if G is a group, B a subgroup of G and $(P_i)_{i \in I}$ a system of subgroups of G containing B, then $\mathcal{C} = \mathcal{C}(G, B, (P_i)_{i \in I})$ has as chambers the (right) cosets of B, two such cosets Bg and Bh being i-adjacent if and only if $P_i g = P_i h$.

Let $\mathcal{C}$ and $\mathcal{D}$ be two chamber systems over the same index set I. Then a bijective map α from $\mathcal{C}$ onto $\mathcal{D}$ is an *isomorphism* if it respects i-adjacency; i.e. c and $c' \in \mathcal{C}$ are i-adjacent in $\mathcal{C}$, if and only if c^α and $(c')^\alpha$ are i-adjacent in $\mathcal{D}$. If $\mathcal{C} = \mathcal{D}$ such an α is called an *automorphism* of $\mathcal{C}$. Notice that usually such isomorphisms or automorphisms are called type-preserving. But since it is not necessary for us to consider any other, we will omit the word type preserving.

It is clear that the elements of G act via right multiplication as automorphisms on $\mathcal{C} = \mathcal{C}(G, B, (P_i)_{i \in I})$. Further, if $B_G = \cap_{g \in G} B^g$, then B_G is the kernel of the action of G on $\mathcal{C}$ and $G/B_G \leq \mathrm{Aut}(\mathcal{C})$.

A rank one residue $\Delta_i(c)$ of $\mathcal{C}$ is called a *panel* of $\mathcal{C}$.

(4.1) Lemma

Let $\mathcal{C} = \mathcal{C}(G, B, (P_i)_{i \in I})$. Then $\mathcal{C}$ is connected if and only if $G = \langle P_i \mid i \in I \rangle$.

(4.2) Lemma

Suppose G acts chamber-transitively on $\mathcal{C}$. Let B be the stabilizer of $c \in \mathcal{C}$ and P_i be the stabilizer of $\Delta_i(c), i \in I$ in G. Then

$$\mathcal{C} \simeq \mathcal{C}(G, B, (P_i)_{i \in I}).$$

Proof. Exercise.

The *rank* of $\mathcal{C}$ is simply $|I|$. A *gallery* of type $J \subseteq I$ is a finite sequence

$$(c_0, \ldots, c_k)$$

of chambers, such that $c_{i-1} \overset{j_i}{\sim} c_i$ (i. e. c_{i-1}, c_i are j_i-adjacent) for $1 \leq i \leq k$ and $j_i \in J$.

(4.3)   Example

Let W be a Coxeter group of type $M = M(I)$ over the index set I. That is, $M(I) = (m_{ij})$ is a symmetric matrix with coefficients in $\mathbb{N} \cup \{\infty\}$, satisfying $m_{ii} = 1$ for $i \in I$ and $m_{ij} \geq 2$ for $i \neq j$, and $W = W(I)$ is the group given by the presentation

$$W = \langle w_i \mid (w_i w_j)^{m_{ij}} = 1 \text{ for } i, j \in I \rangle.$$

Then a Coxeter (chamber) system of type M over I is any chamber system isomorphic to

$$\mathcal{C} = \mathcal{C}(W, \{1\}, \langle w_i \rangle_{i \in I}).$$

It is clear that such Coxeter systems are connected, thin (i. e. $|\Delta_i(c)| = 2$ for each $i \in I$) and that W acts via right-multiplication as the full automorphism group on $\mathcal{C}$. (Exercise!)

The *diagram* $\Delta(I)$ of such a Coxeter system is the marked graph with vertex set I, two different vertices being connected by an edge of strength $m_{ij} - 2$. (I. e. if $m_{ij} = 2$, then i and j are not connected!)

Finally $\mathcal{C}$ is *spherical* iff $|W| < \infty$. For the list of spherical Coxeter groups with connected $\Delta(I)$ see [Bou81].

(4.4)   Definition

A *building* $\mathcal{B}$ over I of type $M = M(I)$ is a connected chamber system over I together with a family $\mathcal{F}$ of subsystems of $\mathcal{B}$ (called *apartments*) satisfying:

(B1) Each $\mathcal{A} \in \mathcal{F}$ is a Coxeter system of type M over I.
(B2) For each pair c, d of chambers of $\mathcal{B}$, there exists an apartment $\mathcal{A} \in \mathcal{F}$ containing $\{c, d\}$.

Let now $\mathcal{A} \in \mathcal{F}$ be an apartment of $\mathcal{B}$ and $\Delta_i(c)$ a panel of $\mathcal{B}$. Then we say the panel $\Delta_i(c)$ *lies in* $\mathcal{A}$ iff $\Delta_i(c) \cap \mathcal{A} \neq \emptyset$. This is equivalent to demanding that $\Delta_i(c) \cap \mathcal{A}$ is a panel of $\mathcal{A}$.

(B3) Let x and y be chambers of $\mathcal{B}$, resp. x a chamber and $\Delta_i(y)$ a panel of $\mathcal{B}$, lying in apartments $\mathcal{A}$ and $\mathcal{A}'$ of $\mathcal{B}$. Then there exists an isomorphism $\sigma : \mathcal{A} \to \mathcal{A}'$ stabilizing x and y resp. x and $\Delta_i(y)$.

Notice that our axiom (B3) is equivalent to the axiom (B3 Tits) of [Tit74], restated in the language of chamber systems:

(B3 Tits) If x is a chamber of $\mathcal{B}$ and Δ is any residue of $\mathcal{B}$ and if $\mathcal{A}$ and $\mathcal{A}'$ are apartments containing x such that $\mathcal{A} \cap \Delta \neq \emptyset \neq \mathcal{A}' \cap \Delta$, then there exists an isomorphism $\sigma : \mathcal{A} \to \mathcal{A}'$ stabilizing x and Δ.

Indeed, if $y \in \mathcal{A} \cap \Delta$ and $y' \in \mathcal{A}' \cap \Delta$, take a gallery $y = y_0 \overset{j_1}{\sim} y_1 \sim \cdots \overset{j_k}{\sim} y_k = y'$ inside Δ and let by (B2) $\mathcal{A}_i$ be an apartment containing x and y_i with $\mathcal{A}_0 = \mathcal{A}$ and $\mathcal{A}_k = \mathcal{A}'$. Then there exist by (B3) isomorphisms $\sigma_i = \mathcal{A}_i \to \mathcal{A}_{i+1}$ stabilizing x and $\Delta_{j_i}(y_i)$. Hence $\sigma = \sigma_{k-1} \ldots \sigma_0$ is the required isomorphism.

A building is called *spherical*, if its apartments are spherical (as Coxeter systems). Such a family $\mathcal{F}$ of subsystems of $\mathcal{B}$ satisfying (B1)–(B3) is called an *apartment-system* of $\mathcal{B}$. Although such an apartment-system is not necessarily unique, there exists a unique maximal apartment-system of $\mathcal{B}$ (if $\mathcal{B}$ is a building).

(4.5) Example: Generalized m-gons

Let $m \geq 2$ be an integer. Then a *generalized m-gon* is a connected, bipartite graph Γ of diameter m and girth $2m$ such that each vertex lies on at least two edges. (Here a graph Γ is *bipartite* if the set of vertices of Γ is partioned into two subsets, such that no pair of vertices in the same subset lies on a common edge. The *diameter* of a graph is the maximal distance between two vertices and the *girth* is the minimal length of a circuit. Notice that any connected graph admits at most one partition into two subsets, such that no pair of vertices in the same subset lies on a common edge.)

If now Γ is such a generalized m-gon and if the vertex set $V(\Gamma) = \mathcal{P} \dot{\cup} \mathcal{L}$ such that no pair of vertices of $\mathcal{P}$ and of $\mathcal{L}$ lie on a common edge, call the elements of $\mathcal{P}$ points and the elements of $\mathcal{L}$ lines, a point $P \in \mathcal{P}$ lying on (being incident with) a line $\ell \in \mathcal{L}$ if and only if the pair (P, ℓ) has a common edge. Let $\mathcal{C} = \mathcal{C}(\Gamma)$ be the *flag complex* of the incidence-geometry $(\mathcal{P}, \mathcal{L}, I)$; i. e., the chambers of $\mathcal{C}$ are the incident point-line pairs, two such chambers being 1-adjacent if and only if they have a common point and 2-adjacent if and only if they have a common line. We show that $\mathcal{C}$ is a building over $I = \{1, 2\}$ of type $M = \left(\begin{smallmatrix} 1 & m \\ m & 1 \end{smallmatrix} \right)$. Indeed, let $\mathcal{F}'$ be the set of ordinary m-gons in $(\mathcal{P}, \mathcal{L}, I)$ and for $\mathcal{A}' \in \mathcal{F}'$ let $\mathcal{A} := \mathcal{C}(\mathcal{A}')$, the flag complex of the incidence geometry $\mathcal{A}'$ as defined above. Let $\mathcal{F} = \{\mathcal{A} \mid \mathcal{A}' \in \mathcal{F}'\}$. Now it is easy to show (Exercise) that:

(B1) If $\mathcal{A} \in \mathcal{F}$, then $\mathcal{A}$ is isomorphic to the Coxeter complex of a dihedral group of order $2m$ (i.e. presented by $\langle w_1, w_2 \mid w_i^2 = (w_1 w_2)^m = 1 \rangle$).

Now let $v, w \in V(\Gamma)$ be of distance m. Then, if m is even, both v and w are points (resp. lines) and if m is odd, one is a point and the other a line. From this it follows immediately that, if $v_1 \in V(\Gamma)$ is incident to v and w_1 incident to w, then $\{v, v_1, w_1, w\}$ is contained in an ordinary m-gon of $(\mathcal{P}, \mathcal{B}, I)$. This implies (B2).

Let now c, d be chambers of $\mathcal{C}$ of maximal distance (resp. c a chamber and y a panel of maximal distance). Then it follows easily from the definition of a generalized m-gon that there exists a unique apartment containing both. Hence the identity is the desired isomorphism.

Let now c, d be any pair of chambers of $\mathcal{C}$ (resp. c a chamber and y a panel) and let $\mathcal{A}$ and $\mathcal{A}'$ be apartments of $\mathcal{C}$ containing c and d resp. c and y. (Remember that we say y lies in $\mathcal{A}$, if $\mathcal{A} \cap y \neq \emptyset$!) Let $e \in \mathcal{A}$ and $f \in \mathcal{A}'$ be of maximal distance (i.e. m) from c. Then, by the above, $\mathcal{A}$ is the unique apartment of $\mathcal{C}$ containing c and e and $\mathcal{A}'$ is the unique apartment containing c and f. Hence the map $\sigma : c \to c,\ e \to f$ extends to an isomorphism from $\mathcal{A}$ onto $\mathcal{A}'$ stabilizing d (resp. y). This proves (B3).

On the other hand, if $\mathcal{B}$ is a building of type $M = \left(\begin{smallmatrix} 1 & m \\ m & 1 \end{smallmatrix} \right)$, call the residues of type 1 points and the residues of type 2 lines, a point lying on a line if and only if they have some chambers in common. Let $\mathcal{G}$ be the corresponding point-line geometry. Then it is easy to show that the incidence graph Γ of $\mathcal{G}$ is a generalized m-gon in the original definition.

(4.6) Definitions

Let $I = \{1, \ldots, \ell\}$. The *diagram* $\Delta(I)$ of a building $\mathcal{B}$ over I is just the diagram of any of its apartments. $\mathcal{B}$ is *irreducible* if the diagram $\Delta(I)$ is connected.

If x, y are chambers of $\mathcal{B}$, then $d(x, y)$ is the minimal length of a gallery from x to y. If $\mathcal{B}$ is spherical, then $d(x, y)$ is restricted, since x and y are by (B2) contained in an apartment of $\mathcal{B}$. In that case we call x and y *opposite* if $d(x, y)$ is maximal. The following facts are well-known properties of finite Coxeter groups (and for example contained in [Hum90]).

Suppose x and y are chambers in some apartment $\mathcal{A}$ of the spherical building $\mathcal{B}$. Identify $\mathcal{A}$ with $\mathcal{C}(W, \{1\}, \langle w_i \rangle_{i \in I})$ as in (4.3). Then we have

(1) x and y are opposite if and only if $y = w_0 x$, w_0 being the longest word in W. In particular there exists exactly one chamber opposite to x in $\mathcal{A}$.

(2) If x and y are opposite in $\mathcal{A}$, then each chamber of $\mathcal{A}$ lies on some minimal gallery from x to y.

Call a subset $\mathcal{W}$ of $\mathcal{B}$ *convex*, if for each pair c, d of chambers of $\mathcal{W}$ all chambers on a minimal gallery from x to y lie in $\mathcal{W}$. Then we have

(3) Residues of $\mathcal{A}$ are convex. (As subsets of $\mathcal{A}$!)

A *reflection* w_r of $\mathcal{A}$ is any conjugate of one of the generators w_i, $i \in I$ of W. The *reflecting wall* M_r (often also denoted by ∂r) is the set of proper residues of $\mathcal{A}$ invariant under w_r. It can be shown that $\mathcal{A}$ is split into two disjoint subsets r and $-r$, such that $\mathcal{A} = r \,\dot\cup\, (-r)$ and $r^{w_r} = -r$. These two subsets of $\mathcal{A}$ (together with the residues of $\mathcal{A}$ contained in these) are called the *roots* or *half-apartments* of $\mathcal{A}$ determined by w_r (or M_r). The two roots r and $-r$ are called *opposite*. We have:

(4) Roots of $\mathcal{A}$ are convex. (As subsets of $\mathcal{A}$.) Moreover, given chambers x and y of $\mathcal{A}$, a chamber z lies on a minimal gallery from x to y, if and only if it lies in any root of $\mathcal{A}$ containing x and y.

Finally, since there is exactly one minimal gallery in $\mathcal{B}$ of "given type" from x to y, see [Ron89, 3.1], it follows from (2) that:

(5) If x and y are opposite in $\mathcal{B}$, then there exists exactly one apartment of $\mathcal{B}$ containing x and y. Moreover apartments are convex. (as subsets of $\mathcal{B}$).

Notice that by [Hum90] our Coxeter-group W acts faithfully on some ℓ-dimensional $\mathbb{R}$-vector space V and respects some (not necessarily positive definite) scalar-product $(,)$, such that the w_i, $i \in I$ act as reflections (corresponding to a hyperplane M_i) on V. This representation of W is often called the geometric realization of W. In this geometric realization of W of course the conjugates w_r of the w_i are also reflections on V and the reflecting wall M_r can be identified with the corresponding hyperplane of V.

Now W is spherical if and only if $(,)$ is positive definit. Suppose from now on that W is spherical. In that case the roots r and $-r$ can be identified with the two vectors of length 1 perpendicular to M_r. Let now Φ be the set of such roots r, considered as vectors of V. Then there exists a subset $\Pi = \{r_1, \ldots, r_\ell\}$ of Φ, which is a basis of V, such that each root of Φ is a linear combination of roots in Π with non-negative resp. non-positive coefficients and such that $w_i = w_{r_i}$ for $i = 1, \ldots, \ell$. One calls

$$\Phi^+ := \{r \in \Phi \mid r = \Sigma \lambda_i r_i, r_i \in \Pi \text{ and } \lambda_i \geq 0\}-$$

the set of *positive roots* and

$$-(\Phi^+) =: \Phi^-$$

the set of *negative roots*.Moreover Π is called a *fundamental root system*.

The longest word w_0 of W is the unique element of W satisfying $w_0(\Phi^+) = \Phi^-$ and $w_0(\Pi) = -\Pi = \{-r \mid r \in \Pi\}$. It is clear that $w_0^2 = 1$ and that $\langle w_0 \rangle = Z(W)$, if $Z(W) \neq 1$. Moreover, since for $r_i \in \Pi$ we have $w_0(r_i) = -r_j$, $r_j \in \Pi$, it follows that $w_i^{w_0} = w_j$. All these facts are well known and for example contained in [Hum90]

(4.7) Lemma

Let $\mathcal{B}$ be a thick spherical building, r a root in some apartment $\mathcal{A}$ with wall M_r and $\Pi = \{x, y\}$ a panel of $\mathcal{A}$ in M_r. Suppose $x \in r$, $y \in -r$ and let x' be opposite to x and y' be opposite to y in $\mathcal{A}$. Then the following hold

(a) $d(x, y') = d - 1 = d(y, x')$ $(d = d(x, x'))$.

(b) $\Pi' = \{x', y'\}$ is a panel of $\mathcal{A}$ in M_r, with $y' \in r$ and $x' \in -r$.

(c) Let $\Delta(x)$ be the rank one residue of $\mathcal{B}$ containing Π and $z \in \Delta(x) - \{x\}$. Then there exists a unique apartment of $\mathcal{B}$ containing r and z.

(d) r is the only root of $\mathcal{B}$ containing x and y'.

Proof. For the proof of (a) and (b), identify $\mathcal{A}$ with the Coxeter complex of W. Then by (4.6) (1) $x' = w_0 x$, $y' = w_0 y$. Suppose y is i-adjacent to x. Then $y = w_i x$ and

$$y' = w_0 w_i x = w_0 w_i w_0^{-1} x' = w_i^{w_0} x'.$$

Now the longest element of a spherical Coxeter group conjugates each fundamental reflection w_i onto some other w_j. (Since $\Pi^{w_0} = -\Pi$, Π a fundamental root system.) Hence $y' = w_j x'$ and x' and y' are j-adjacent. On the other hand, if w_r is the reflection corresponding to r then $x w_r = y$ and thence $x' w_r = y'$.

Since $d(x, x') = d = d(y, y')$ this implies (a) and that Π' is a panel in M_r. Since opposite chambers cannot lie in a root by (4.6)(2) and (4), this also implies (b).

Now by (a),(b) and (4.6)(5) we have $d(z, y') = d$. Hence z and y' are opposite and thus by (4.6)(5) contained in a unique apartment $\mathcal{A}'$ of $\mathcal{B}$. Now by (4.6)(4) every chamber of r lies on a minimal gallery from x to y', whence on a minimal gallery from z to y' of the form

$$z \overset{i}{\sim} x \sim \cdots \sim y'.$$

Hence (4.6)(5) implies $r \subseteq \mathcal{A}'$ which proves (c). For (d), we show first that r is the only root of $\mathcal{A}$ containing x and y'. Indeed we have

$$d = \ell(w_0) = \frac{1}{2}(\# \text{ of roots of } \mathcal{A}),$$

since w_0 maps each positive root on a negative root. Hence also

$$d = \# \text{ of roots of } \mathcal{A} \text{ containing } x.$$

Now, if we identify $\mathcal{A}$ with W and set without loss of generality $x = 1$, then $d - 1 = d(x, y') = \ell(y')$ is the number of positive roots mapped by y' onto negative roots. Hence there remains exactly one root of $\mathcal{A}$ containing x and y'.

Let now $z \in r$ be a chamber different from x and y'. Then by the above:

$$
\begin{aligned}
d(x, z) &= \# \text{ roots of } \mathcal{A} \text{ containing } x \text{ but not } z, \\
d(z, y') &= \# \text{ roots of } \mathcal{A} \text{ containing } z \text{ but not } y', \\
&= (\# \text{ roots of } \mathcal{A} \text{ containing } z \text{ and } x) - 1,
\end{aligned}
$$

since $-r$ is the only root of $\mathcal{A}$ which neither contains x nor y'. Hence we obtain

$$
d(x, z) + d(z, y') = (\# \text{ roots of } \mathcal{A} \text{ containing } x) - 1 = d - 1 = d(x, y').
$$

We have shown: Each chamber of r lies on a minimal gallery from x to y'.

Suppose now r' is another root containing x and y' in some apartment $\mathcal{A}'$. Then, since the above also holds for r' and since apartments are convex, we have $r' \subseteq \mathcal{A}$. Hence as shown, $r = r'$ since by (B3) r' is a root in every apartment containing it. $\qquad\square$

(4.8) Proposition

Let $\mathcal{B}$ be a thick, spherical building over I, c and b opposite chambers of $\mathcal{B}$ and σ an automorphism of $\mathcal{B}$ fixing b and all chambers in

$$
\Delta(c) := \bigcup_{i \in I} \Delta_i(c).
$$

Then $\sigma = \mathrm{id}$.

Proof. We show first:

(1) σ fixes all chambers in $\Delta(b)$.

To prove (1) let $\mathcal{A}$ be the apartment of $\mathcal{B}$ containing c and b. Then $\sigma|_{\mathcal{A}} = \mathrm{id}$ by (4.3). Hence σ fixes each $b_1 \in \Delta(b) \cap \mathcal{A}$. Let $b_1 \in \Delta_i(b) - \mathcal{A}$. Then $d(c, b_1) = d = d(c, b)$, since if $d(c, b_1) = d - 1$, b_1 must lie on a minimal gallery from c to b and thus $b_1 \in \mathcal{A}$ by (4.6). Let $b \neq b' \in \Delta_i(b) \cap \mathcal{A}$. Then $d(c, b') = d - 1$. Let $c' \in \mathcal{A}$ be opposite to b' and $j \in I$ with $c' \in \Delta_j(c)$. Then there exists a $c_1 \in \Delta_j(c)$ with $d(b_1, c_1) = d - 1$. (Indeed embed c and b_1 in an apartment $\mathcal{A}'$. Then $|\mathcal{A}' \cap \Delta_j(c)| = 2$ and there exists a $c \neq c_1 \in \mathcal{A}' \cap \Delta_j(c)$. Hence $d(b_1, c_1) = d - 1$.)
Now $d(c_1, b) = d$, since if $d(c_1, b) = d - 1$, then $c_1 \in \mathcal{A}$, a contradiction to $\Delta_j(c) \cap \mathcal{A} = \{c, c'\}$. Hence b_1 lies in the unique apartment of $\mathcal{B}$ containing c_1 and b ((4.6)(5)!) and thus $\sigma(b_1) = b_1$.

Next we show:

(2) Let $b' \in \Delta(b)$. Then there exists a $c' \in \Delta(c)$ opposite to b', such that σ fixes all chambers in $\Delta(c')$.

Indeed, if $d(b', c) = d$, set $c' = c$. So we may assume $d(b', c) = d - 1$. Now, as shown in (1), we have for $c' \in \Delta(c) - \mathcal{A}$, that $d(c', b') = d = d(c', b)$. Hence by (1) applied to (b, c'), σ fixes all chambers in $\Delta(c')$.

Now it follows from (2) by induction on $d(e, b)$, that σ fixes all chambers e of $\mathcal{B}$. Indeed if $d(e, b) = 2$, pick $b' \in \Delta(e) \cap \Delta(b)$. Then there exists by (2) a c' opposite to b' such that σ fixes all chambers of $\Delta(c')$. Now, applying (1) to (c', b'), σ fixes e. $\qquad\square$

Now for a root r of some apartment $\mathcal{A}$ of $\mathcal{B}$ let $\mathcal{W}(r)$ be the set of apartments of $\mathcal{B}$ containing r, and

$$A_r := \{\sigma \in \mathrm{Aut}(\mathcal{B}) \mid c^\sigma = c \text{ for each } c \in \mathcal{B} \text{ with } \Delta_i(c) \cap \mathcal{A} \text{ a panel of } \mathcal{A}$$
$$\text{contained in } r \text{ for some } i \in I\}.$$

A_r is called the *root subgroup* of $\mathrm{Aut}(\mathcal{B})$ corresponding to the root r.

(4.9) Corollary

Let $\mathcal{B}$ be a thick, irreducible, spherical building over I, $|I| \geq 2$ and r a root of some apartment $\mathcal{A}$ of $\mathcal{B}$. Then A_r acts fixed-point-freely on $\mathcal{W}(r)$.

Proof. We show first:

$(*)$ $\qquad$ There exists a chamber $c \in r$ with $\Delta(c) \cap \mathcal{A} \subseteq r$.

To prove $(*)$ identify $\mathcal{A}$ with $\mathcal{C}(W, \{1\}, \langle w_i \rangle_{i \in I})$ as in (4.3). Pick $d \in r$ with $\Delta_i(d) \cap \mathcal{A} \subseteq M_r$ and let $c \in \Delta_j(d) \cap r$ with $m_{ij} > 2$. (Exists!)

As $\Delta_i(d) \cap \mathcal{A} \subseteq M_r$ we have $\Delta_i(d) \cap \mathcal{A} = \{d, dw_r\}$, i.e. $dw_r = w_i d$. Since c is j-connected to d, we have $c = w_j d$. Assume $e \in \Delta_k(c) \cap (-r)$ for some $k \in I$. Then $e = w_k c$. Since the gallery $c \overset{k}{\sim} e$ crosses the wall M_r, we have $\{c, e\} \subseteq M_r$ and $e = cw_r$ (This is the definition of "crossing the wall". See [Ron89, p13]. Hence we obtain $w_k c = cw_r$ and

$$w_j w_i d = w_j dw_r = cw_r = e = w_k c = w_k w_j d.$$

Hence $w_j w_i = w_k w_j$, which is impossible since $o(w_i w_j) = m_{ij} > 2$ and since there don't hold any relations in W except consequences of the defining relations. This proves $(*)$.

Now suppose $\mathcal{A} \in \mathcal{W}(r)$ is an apartment fixed by $\alpha \in A_r$. Let by $(*)$ $c \in r$ with $\Delta_k(c) \cap \mathcal{A} \subseteq r$ for all $k \in I$. Then α fixes c and the opposite chamber b of $\mathcal{A}$. Hence it suffices by (4.8) to show that α fixes all chambers in $\Delta(c)$, to show that $\alpha = 1$. If now $d \in \Delta_k(c)$, then $\Delta_k(d) \cap \mathcal{A} = \Delta_k(c) \cap \mathcal{A}$ is a panel of $\mathcal{A}$ in r. Thus $d^\alpha = d$ by definition of A_r. $\qquad\square$

Without proof we state the following important theorem of J. Tits [Tit77, Satz 1].

(4.10)   Theorem

Let $\mathcal{B}$ be a thick, irreducible, spherical building over I with $|I| \geq 3$ and r a root of some apartment of $\mathcal{B}$. Then A_r acts transitively on $\mathcal{W}(r)$.

Together with (4.9) this shows that A_r acts regularly on $\mathcal{W}(r)$. A thick irreducible spherical building satisfying (4.10) for each root r of some apartment $\mathcal{A}$ of $\mathcal{B}$ is called a *Moufang building*. (4.10) shows that all thick, irreducible, spherical buildings of rank ≥ 3 are Moufang buildings. It is well known that this is no longer true in rank two, since for example there exist projective planes, which do not satisfy the Moufang property.

The proof of (4.10) is one of the most complicated parts of the theory of spherical buildings. There exists still only the original proof of Tits, which is given in [Tit77] and which is based on theorem (4.12) of [Tit74]. All books on buildings, even if they treat the Moufang property as in [Ron89] and [Sch95], leave out the critical proof of (4.10), a situation I find somewhat unsatisfactory.

(4.11)   Lemma

Let $\mathcal{B}$ be a building of type M over I, $J \subseteq I$ and Δ_J a J-residue of $\mathcal{B}$. Then the following hold:

(1) Δ_J is a building of type M_J (i.e. $M|_J$) over J.
(2) If $\mathcal{B}$ is a Moufang building, $|J| \geq 2$ and Δ_J is irreducible, then Δ_J is also a Moufang building.

Proof. Let $\mathcal{F}$ be a set of apartments of $\mathcal{B}$ satisfying (B1)–(B3) and

$$(*) \qquad \mathcal{F}_J = \{\mathcal{A} \cap \Delta_J \mid \mathcal{A} \in \mathcal{F} \text{ and } \mathcal{A} \cap \Delta_J \text{ a } J\text{-residue of } \mathcal{A}\}.$$

Then it is easy to see that $\mathcal{F}_J$ also satisfies (B1)–(B3) in Δ_J, whence Δ_J is a building of type M_J. ($\mathcal{A} \cap \Delta_J \simeq \mathcal{C}(W_J, \{1\}, \langle w_j \rangle_{j \in J})$!)

To prove (2) let w_{r_0} be a reflection of $\mathcal{A} \cap \Delta_J$ with roots r_0, $-r_0$. Then, since W_J is a Coxeter-subgroup of W, w_{r_0} is also a reflection of $\mathcal{A}$. Further, if r and $-r$ are the roots of $\mathcal{A}$ corresponding to this reflection, then $r_0 = r \cap \Delta_J$ and $-r_0 = -r \cap \Delta_J$. Now the root subgroup A_r (on $\mathcal{B}$!) fixes each chamber in r and thus each chamber in r_0 and thence acts on Δ_J.

Suppose $1 \neq \alpha \in A_r$ induces the identity on Δ_J. Then α fixes some panel Π of $\mathcal{A}$ in $\partial r_0 \subset \partial r$. Hence by (4.7) there exists a unique apartment of $\mathcal{B}$ containing Π and r, which is then fixed by α, a contradiction to (4.9).

This shows that A_r acts faithfully on Δ_J and thus $A_r \leq A_{r_0}$ (A_{r_0} the root group corresponding to r_0 on Δ_J) by definition of root groups. On the other hand by $(*)$ and $(4.7)(c)$ each apartment $\mathcal{A}'_0$ of Δ_J containing r_0 is of the form $\mathcal{A}'_0 = \mathcal{A}' \cap \Delta_J$, $\mathcal{A}'$ an apartment of $\mathcal{B}$ containing r. This implies that A_r and whence A_{r_0} acts transitively on the set of these apartments and thus Δ_J is a Moufang building.   □

In fact (4.9) applied to Δ_J and the Frattini argument imply $A_r = A_{r_0}$, a fact we will need for the proof of the main result of this section.

(4.12) Proposition

Let $\mathcal{B}$ be a thick irreducible Moufang building, $\mathcal{A}$ an apartment of $\mathcal{B}$ and $r, -r$ opposite roots of $\mathcal{A}$. Pick a panel $\Pi = \{x, y\}$ of $\mathcal{A}$ contained in the wall M_r, with $x \in r$ and $y \in -r$ and let Δ be the rank one residue of $\mathcal{B}$ containing Π. Finally, let x' (resp. y') be opposite to x (resp. y) in $\mathcal{A}$. Then the following hold:

(1) $X_r = \langle A_r, A_{-r} \rangle$ acts doubly transitively on Δ with $A_r \trianglelefteq (X_r)_x$ acting regularly on $\Delta - \{x\}$. (And $A_{-r} \trianglelefteq (X_r)_y$ regular on $\Delta - \{y\}$.)

(2) Let Δ' be the rank one residue of $\mathcal{B}$ containing the panel $\Pi' = \{y', x'\}$ and
$$\mathcal{W} = \{\alpha \mid \alpha \text{ a root of } \mathcal{B} \text{ with } \mid \Delta \cap \alpha \mid = 1 = \mid \Delta' \cap \alpha \mid\}.$$

Then X_r acts doubly transitively on $\mathcal{W}$ with $A_r \trianglelefteq (X_r)_r$ acting regularly on $\mathcal{W} - \{r\}$.

(3) X_r is a rank one group with unipotent subgroups A_r and A_{-r} of nilpotency class at most 2.

Moreover, the permutation actions of X_r in (1), (2) and (3) on $\Omega = A_r^{X_r}$ are equivalent.

Proof. (1) Pick $z \in \Delta - \Pi$ (Exists because of thickness of $\mathcal{B}$!) Then there exists by $(4.7)(c)$ a unique apartment $\mathcal{A}'$ containing r and z. Hence, by the Moufang property, there exists an $a \in A_r$ with $\mathcal{A}^a = \mathcal{A}'$. This implies $y^a = z$, which shows that A_r acts transitively and whence by $(4.9)(1)$ regularly on $\Delta - \{x\}$. By the same reason, A_{-r} acts regularly on $\Delta - \{y\}$, so that by exercise $(1.13)(1)$ X_r acts doubly transitively on Δ.

By the same reason X_r acts doubly transitively on Δ'. Now by $(4.7)(a)$ y' is the only chamber of Δ' with $d(x, y') = d - 1$. Hence $(X_r)_x \leq (X_r)_{y'}$ and thus by $(4.7)(d)$ $(X_r)_x$ fixes r and whence normalizes A_r.

Now for $z \in \Delta$ let $u(z)$ be the unique chamber of Δ' with $d(z, u(z)) = d - 1$ and let by $(4.7)(d)$ $r(z)$ be the unique root of $\mathcal{B}$ containing $\{z, u(z)\}$. Then the map $z \to r(z)$ is a bijection of Δ onto $\mathcal{W}$, which is respected by the action of X_r. This proves (1) and (2).

For (3) we have by (1.3) to show that A_r is nilpotent of class ≤ 2. For this let Ω be an irreducible rank two residue of $\mathcal{B}$ containing Δ. Then X_r fixes Ω and A_r, A_{-r} act by (4.11) (i.e. comment after the proof) as root subgroups corresponding to $r \cap \Omega$ and $-r \cap \Omega$ on Ω. In particular Ω is a Moufang n-gon, $n \geq 3$. Hence by the commutator relations of root groups on Moufang n-gons by [Tit94] (see also [VM98, section 5.4]), A_r is nilpotent of class at most 2.

Finally, the equivalence or permutation actions is obvious from the construction.   □

(4.13)   Example

Let $\mathcal{B}$ be a Moufang hexagon. Then an apartment $\mathcal{A}$ of $\mathcal{B}$ is by (4.5) an ordinary hexagon. We visualize $\mathcal{A}$ as a point-line set:

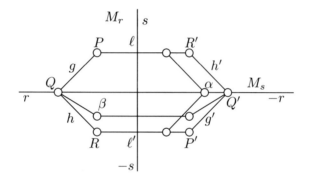

Then
A_r is transitive on the points of ℓ different from P,
A_r is transitive on the points of ℓ' different from R,
X_r is doubly transitive on the points of ℓ and ℓ',
X_r is doubly transitive on the set of roots α of $\mathcal{B}$ with $\alpha \cap \ell$ and $\alpha \cap \ell'$ points,
A_s is transitive on the lines through Q different from g,
A_s is transitive on the lines through Q' different from h',
X_s is doubly transitive on the lines through Q and Q',
X_s is doubly transitive on the set of roots β of $\mathcal{B}$ with $\beta \cap Q, \beta \cap Q'$ points.

(4.14)   Exercises

(1) Prove (4.1).

(2) Prove (4.2).

(3) Let $\mathcal{C}$ be a Coxeter system isomorphic to $\mathcal{C}(W, \{1\}, \langle w_i \rangle_{i \in I})$. Show that $\mathrm{Aut}(\mathcal{C}) \simeq W$.

(4) Let K be a division ring, V a 3-dimensional vector space over K, $\mathcal{P}$ the set of 1-spaces of V and $\mathcal{B}$ the set of 2-spaces of V.

    (a) Show that the incidence graph Γ of the incidence geometry $(\mathcal{P}, \mathcal{B}, \subseteq)$ is a generalized 3-gon in the sense of (4.5).

    (b) Consider the building $\mathcal{C}(\Gamma)$ defined in (4.5). Show that for any given basis $B = (v_1, v_2, v_3)$ of V, the set of all maximal flags of $\mathcal{C}(\Gamma)$ obtained from B is an apartment $\mathcal{A}$ of $\mathcal{C}(\Gamma)$.

    (c) Show that the roots of $\mathcal{A}$ are in 1-1 correspondence to point-line pairs of the form $(\langle v_i \rangle, \langle v_i, v_j \rangle)$, $1 \leq i \neq j \leq 3$.

    (d) Show, that if $r \leftrightarrow (\langle v_i \rangle, \langle v_i, v_j \rangle)$, then A_r is the transvection group corresponding to point-hyperplane pair $(\langle v_i \rangle, \langle v_i, v_j \rangle)$.

    (e) Show by elementary matrix-calculations that

$$\langle A_r \mid r \text{ a root of } \mathcal{A} \rangle = SL_3(K).$$

(5) Let V be an $(n+1)$-dimensional vector space over some division ring K. Then a chamber c of V is a set

$$c = (V_1, \ldots, V_n) \text{ of subspaces } V_i \text{ of } V \text{ with } \dim V_i = i \text{ and } V_i \subset V_{i+1} \text{ for}$$
$$i = 1, \ldots, n-1.$$

Let $\mathcal{C}(V)$ be the set of all such chambers of V. For $i \in I = \{1, \ldots, n\}$ we call the chambers $c = (V_1, \ldots, V_n)$ and $d = (W_1, \ldots, W_n)$ i-adjacent, if and only if $V_j = W_j$ for all $j \neq i$.

    (a) Show that $\mathcal{C}(V)$ is a connected chamber system over I.

For some ordered basis $\mathcal{B} = (v_1, \ldots, v_{n+1})$ of V let $\mathcal{C}(\mathcal{B})$ be the set of all chambers $c = (V_1, \ldots, V_n)$ of $\mathcal{C}(V)$ such that each V_i is spanned by vectors in $\mathcal{B}$. Let

$$M = \begin{bmatrix} 1 & 3 & & & & \\ 3 & \ddots & \ddots & & & \\ & \ddots & \ddots & \ddots & & \\ & & \ddots & \ddots & 3 \\ & & & 3 & 1 \end{bmatrix} \quad \text{be an } n \times n \text{ matrix.}$$

    (b) Show that $\mathcal{C}(\mathcal{B})$ is a Coxeter system of type M.

Now let $\mathcal{F} = \{\mathcal{C}(\mathcal{B}) \mid \mathcal{B} \text{ an ordered basis of } V\}$.

    (c) Show that $\mathcal{C}(V)$ together with $\mathcal{F}$ is a building of type M over I.

§5 Structure and embeddings of special rank one groups

As the large list of examples in §1 shows, it seems unlikely that one can determine the isomorphism type, even of special rank one groups with AUS. For example in the situation of (1.5) this would mean that one has to determine the type of the (division)ring k first and from that the additive subgroup L with properties (1)–(3) of (1.5).

In this section we will determine the minimal rank one groups [Tim00a] and also discuss embeddings from one special rank one group in another [Tim94a, 6.2].

For the whole Section 5 let $X = \langle A, B \rangle$ be a special rank one group with AUS with unipotent subgroups A and B. For $a \in A^{\#}$ and $b \in B^{\#}$ we write

$$\chi : a \to b(a), b \to a(b)$$

where $b(a)$ resp. $a(b)$ are the elements introduced in (1.2). Then by (1.2) and the definition of special, χ is a bijection of $A^{\#}$ onto $B^{\#}$, $B^{\#}$ onto $A^{\#}$ with $\chi^2 = \mathrm{id}$.

We first discuss an additional example which then will appear in the conclusion of one of the theorems.

(5.1) Example

Universal perfect central extension of $SL_2(k)$, k a field with $|k| > 4$ and $|k| \neq 9$ [Stg67].

By Theorem 10 of [Stg67] this universal perfect central extension is the group X generated by symbols

$$a(t), b(t), t \in k$$

subject to the relations:

(A) $a(t)a(\tau) = a(t + \tau), b(t)b(\tau) = b(t + \tau)$; $\qquad t, \tau \in k$.
(B) $a(u)^{n(t)} = b(-t^{-2}u); u \in k$ and $t \in k^*$ where $n(t) = a(-t)b(t^{-1})a(-t)$.

($n(-t) = n(t)^{-1}$, so (B) also gives a relation conjugating the elements of B onto A. $n(t)$ is defined slightly differently from §6 of [Stg67]. This is necessary since we conjugate in the usual group theoretic fashion, i. e. $x^y = y^{-1}xy$.)
Now it is easy to see that the relations $(A) + (B)$ are equivalent to $(A) + (B')$ where

(B') $a(u)^{b(t^{-1})} = b(-t^{-2}u)^{a(t)}$; $\qquad u \in k, t \in k^*$.

If now $t \in k^*$ is fixed, then $k = \{-t^{-2}u \mid u \in k\}$, whence

$$A^{b(t^{-1})} = B^{a(t)} \text{ for } A = \{a(u) \mid u \in k\}, B = \{b(u) \mid u \in k\}.$$

Further

$$a(t)^{b(t^{-1})} = b(-t^{-1})^{a(t)} = (b(t^{-1})^{-1})^{a(t)}.$$

Hence, setting $b(a(t)) := b(t^{-1})$, it follows that X is a special rank one group with AUS.

Notice that if $|k| = \infty$, usually X is different from $SL_2(k)$, see §7 of [Stg67].

Before we can start with the proof of the first theorem, a further piece of notation: If for some $n \in \mathbb{N}$ and $a \in A^{\#}$ (resp. $b \in B^{\#}$) there exists a unique $\bar{a} \in A$ with $\bar{a}^n = a$, we write $\bar{a} = a^{1/n}$ (resp. $b^{1/n}$). So if the notation $a^{1/n}$ (resp. $b^{1/n}$) appears in the next theorem, this means always the existence and uniqueness of such an element!

(5.2)   Theorem

Let $X = \langle A, B \rangle$ be a special rank one group with AUS. Then the following hold:

(a) Either

  (i) A is an elementary abelian p-group for some prime p,

  or (ii) A is torsion-free and divisible.

(b) For all $a \in A^{\#}$ and $b \in B^{\#}$ we have:

$$a^{1/n} = \chi(\chi(a)^n), b^{1/n} = \chi(\chi(b)^n)$$

  where in case (i) $n \in \mathbb{N}$ with $(p, n) = 1$, while in case (ii) $n \in \mathbb{N}$ is arbitrary.

We will prove (5.2) in three steps. Before starting with the first, another piece of notation. For $n \in \mathbb{N}$ let $A^n = \{a^n \mid a \in A\}$ and $A_n = \{a \in A \mid a^n = 1\}$ and similarly B^n, B_n. We have:

(5.3)   Lemma

Suppose there exists an $a \in A$ with $a^2 \neq 1$. Then the following hold:

(a) $A_2 = 1$ and $A = A^2$.

(b) For each $a \in A^{\#}$ and $b \in B^{\#}$ we have:

$$a^{1/2} = \chi(\chi(a)^2), b^{1/2} = \chi(\chi(b)^2).$$

((a) also holds for B, since A and B are conjugate!)

Proof. It suffices to prove (b) just for $a \in A^{\#}$, since it then also holds for $b \in B^{\#}$ by symmetry.

Pick $a \in A$ with $a^2 \neq 1$ and set $b = \chi(a)$. Then, as $a^b = b^{-a}$ we have $o(b) = o(b^{-1}) = o(a) \neq 2$. Hence there exists a unique $\bar{a} \in A$ with $A^{b^2} = B^{\bar{a}}$. This implies $b^2 = \chi(\bar{a})$ and, since X is special

$$\bar{a}^{b^2} = (b^2)^{-\bar{a}}.$$

Further by (2.2):

$$b^{-2} = (b^2)^{-1} = \chi(\bar{a})^{-1} = \chi(\bar{a}^{-1})$$

so that $B^{\bar{a}^{-1}} = A^{b^{-2}}$. Now

$$
\begin{aligned}
B^{\bar{a}} &= (A^b)^b = (B^a)^b = B^{a^b} = B^{(b^{-1})^a} = (B^{a^{-1}})^{b^{-1}a} \\
&= (A^{b^{-1}})^{b^{-1}a} = A^{b^{-2}a} = B^{\bar{a}^{-1}a},
\end{aligned}
$$

since by (2.2) $b^{-1} = \chi(a^{-1})$. We obtain $B^{\bar{a}^2 a^{-1}} = B$. Hence $\bar{a}^2 a^{-1} \in N_A(B)$ and thus $\bar{a}^2 = a$ as $N_A(B) = 1$. Since $\bar{a} = \chi(b^2)$ (as $\chi^2 = \mathrm{id}$!) we obtain the equation

(*) $a = \chi(\chi(a)^2)^2$ for each $a \in A^{\#}$ with $a^2 \neq 1$.

Now (*) shows that each element of A with $a^2 \neq 1$ is a square in A. This implies $A = A_2 \cup A^2$. Since no group is the union of two proper subgroups this shows $A = A^2$.

Suppose $\tilde{a} \in A^{\#}$ has even order. If $o(\tilde{a}) \neq 2$, then there exists by (*) an $\bar{a} \in A$ with $\bar{a}^2 = \tilde{a}$ and $\bar{a} = \chi(\chi(\tilde{a})^2)$. Since the elements a and $\chi(a)^{-1}$ are conjugate in X by definition of χ, this implies

$$o(\bar{a}) = o(\chi(\tilde{a})^2) = o(\tilde{a})^2,$$

which obviously contradicts $\bar{a}^2 = \tilde{a}$.

This shows that each element of even order in $A^{\#}$ has order 2. But as $A^2 = A$, this implies that there exists no element of order 2 in A, whence $A_2 = 1$ which proves (a). Now (a) and (*) imply that $\hat{a} = \chi((a)^2)$ is the unique element of A with $\hat{a}^2 = a$. Hence by definition $\hat{a} = a^{1/2} = \chi(\chi(a)^2)$, which proves (5.3). $\square$

Next we show:

(5.4) Lemma

Suppose A is an elementary abelian q-group for some prime q. Then we have for all $m \in \mathbb{N}$ with $(m,q) = 1$ and for all $a \in A^{\#}, b \in B^{\#}$:

$$a^{1/m} = \chi(\chi(a)^m), b^{1/m} = \chi(\chi(b)^m).$$

Proof. We first show that it suffices to prove (5.4) for $m \leq q - 1$. Indeed let $m = n \cdot q + r, r \leq q - 1$. Then, since A and B are elementary abelian q-groups, we have $\chi(a)^m = \chi(a)^r$ and if $\chi(\chi(a)^r)^r = a$, then also $\chi(\chi(a)^m)^m = a$. Hence (5.4) holds for m if it holds for r. We now prove (5.4) for $m \leq q-1$ by induction on m, the induction assumption $m = 2$ being (5.3). Suppose that (5.4) holds for $n < m$. Pick $a \in A^{\#}$ and let $\bar{a} = \chi(\chi(a)^m)$. Then with $b = \chi(a)$ we have:

$$\begin{aligned} B^{\bar{a}} &= A^{b^m} = A^{b^{m-1}b} = B^{\chi(b^{m-1})b} = B^{a^{1/m-1}b} \\ &= B^{(a^{1/m-1})^b} \quad \text{since (5.4) holds for } m-1. \end{aligned}$$

Now, as $a^{ba^{-1}} = b^{-1}$, we have $(a^{1/m-1})^{ba^{-1}} = (b^{-1})^{1/m-1}$, whence $(a^{1/m-1})^b = ((b^{-1})^{1/m-1})^a$. This implies

$$\begin{aligned} B^{\bar{a}} &= B^{(a^{1/m-1})^b} = B^{((b^{-1})^{1/m-1})^a} = B^{a^{-1}(b^{-1})^{1/m-1}a} \\ &= A^{b^{-1}(b^{-1})^{1/m-1}a} = A^{(b^{-m})^{1/m-1}a} \end{aligned}$$

by (2.2) and since

$$b^{-1}(b^{-1})^{1/m-1} = (b^{-1})^{1+1/m-1} = (b^{-1})^{m/m-1} = (b^{-m})^{1/m-1}.$$

Now, since $\bar{a} = \chi(b^m)$, (2.2) implies:

$$\bar{a}^{-1} = \chi(b^m)^{-1} = \chi(b^{-m})$$

and thus applying χ to this equation

$$b^{-m} = \chi(\bar{a}^{-1}).$$

We obtain:

$$(b^{-m})^{1/m-1} = \chi(\bar{a}^{-1})^{1/m-1} = \chi((\bar{a}^{-1})^{m-1})$$

by induction assumption applied to $\tilde{b} = \chi(\bar{a}^{-1})$ and since $\chi^2 = \text{id}$. Substituting this in the above equation, we obtain:

$$B^{\bar{a}a^{-1}} = A^{(b^{-m})^{1/m-1}} = A^{\chi((\bar{a}^{-1})^{m-1})} = B^{(\bar{a}^{-1})^{m-1}}$$

and thus $B^{\bar{a}^m a^{-1}} = B$. Hence, as in (5.3), $\bar{a}^m = a$. This implies $\bar{a} = a^{1/m}$, which proves (5.4) by definition of $\bar{a}$. $\qquad\square$

(5.4) shows that theorem (5.2) holds, if A is an elementary abelian q-group for some prime q. So we assume from now on that this is not the case. We show next:

(5.5) Lemma

Let p be a prime and $a \in A$ with $a^p \neq 1$. Then the following hold:

(i) $A_p = 1$ and $A = A^p$, $B_p = 1$ and $B = B^p$.

(ii) For each $a \in A^\#$ and $b \in B^\#$ we have:

$$a^{1/p} = \chi(\chi(a)^p), b^{1/p} = \chi(\chi(b)^p).$$

Proof. If $p = 2$ (5.5) is (5.3). Proceeding by induction assume that p is the smallest prime for which (5.5) is false. Then, since it holds for all primes $q < p$, we obtain:

(1) If $q < p$ is a prime, then $q \nmid o(a)$ for all $a \in A^\#$.

Indeed if $q \mid o(a)$ for some $a \in A$, then some power of a has order q. But then $A = A_q$, since we assume (5.5) holds for q, which contradicts the assumption we made for the rest of the proof of (5.2).
From (1) we obtain:

(2) If $n \leq p - 1$, then the following hold:

    (i) $A_n = 1$ and $A = A^n$.

    (ii) $a^{1/n} = \chi(\chi(a)^n), b^{1/n} = \chi(\chi(b)^n)$ for all $a \in A^\#, b \in B^\#$.

Indeed (2) holds for each prime $q \mid n$. Hence immediately $A = A^n$ and $A_n = 1$. To prove (ii) let $n = q \cdot r$ with $(q,r) = 1$ and $q > 1, r > 1$ and, proceeding by induction, we may assume that (ii) holds for q and r. Pick $a \in A^\#$ and let $a_1 = a^{1/r}, a_2 = a_1^{1/q}$. Then

$$a_2^n = a_2^{qr} = a_1^r = a.$$

Further, by induction assumption:

$$a_1 = \chi(\chi(a)^r) \text{ and } a_2 = a_1^{1/q} = \chi(\chi(a_1)^q).$$

This implies

$$
\begin{aligned}
a^{1/n} &= a_1^{1/q} = \chi(\chi(a_1)^q) = \chi(\chi(a^{1/r})^q) \\
&= \chi((\chi(a)^r)^q) = \chi(\chi(a)^{rq}) = \chi(\chi(a)^n)
\end{aligned}
$$

since $\chi^2 = \mathrm{id}$.
We now lead the existence of such a prime p to a contradiction. Pick $a \in A$ with $a^p \neq 1$. Then by (2) $a^{p-1} \neq 1$ and $a^{1/p-1} = \chi(\chi(a)^{p-1})$. We now argue

as in the proof of (5.4). Let $\bar{a} = \chi(\chi(a)^p)$. ($\chi(a)^p \neq 1$ since a and $\chi(a)^{-1}$ are conjugate!). Then we have with $b = \chi(a)$:

$$\begin{aligned} B^{\bar{a}} &= A^{b^p} = A^{b^{p-1}b} = B^{a^{1/p-1}b} = B^{(a^{1/p-1})^b} \\ &= B^{(b^{-1/p-1})^a} = (B^{a^{-1}})^{b^{-1/p-1}a} = A^{b^{-1}b^{-1/p-1}a} \\ &= A^{b^{-p/p-1}a} = A^{(b^{-p})^{1/p-1}a}. \end{aligned}$$

Hence $B^{\bar{a}a^{-1}} = A^{(b^{-p})^{1/p-1}}$. Now arguing as in (5.4) $\bar{a} = \chi(b^p)$ implies by (2.2):

$$\bar{a}^{-1} = \chi(b^p)^{-1} = \chi(b^{-p})$$

and thus, since $\chi^2 = \mathrm{id}$:

$$\chi(\bar{a}^{-1}) = b^{-p}.$$

Now by (2)(ii) applied to $\chi(\bar{a}^{-1})$ we obtain:

$$(b^{-p})^{1/p-1} = \chi(\bar{a}^{-1})^{1/p-1} = \chi((\bar{a}^{-1})^{p-1}).$$

Substituting this in the above equation we get:

$$B^{\bar{a}a^{-1}} = A^{\chi((\bar{a}^{-1})^{p-1})} = B^{(\bar{a}^{-1})^{p-1}}.$$

Hence $B^{\bar{a}^p a^{-1}} = B$ and $\bar{a}^p = a$ as before. This shows that we have:

(*) $\qquad \chi(\chi(a)^p)^p = a$ for all $a \in A^{\#}$ with $a^p \neq 1$.

Next we show, as in the proof of (5.3), that if $p \mid o(a)$ for some $a \in A^{\#}$, then $o(a) = p$. Indeed if $o(a) \neq p$ then (*) holds for a. But since a and $\chi(a)^{-1}$ are conjugate we have $o(a) = o(\chi(a))$ and thus by the same argument

$$o(\chi(\chi(a)^p)) = o(\chi(a)^p) = \frac{o(a)}{p},$$

which obviously contradicts (*).

This shows that each $a \in A^{\#}$ with $o(a) \neq p$ satisfies $(o(a), p) = 1$. Now by (*) $A = A_p \cup A^p$ and so, as in (5.3) $A = A^p$. If now $\tilde{a} \in A$ has order p, then since $A = A^p, \tilde{a}$ is a p-power. But this is impossible, since each element whose order is divisible by p has order p. Thus $A_p = 1$ and (5.5) (i) holds. But then $a^{1/p}$ exists for each $a \in A^{\#}$ and (*) implies $a^{1/p} = \chi(\chi(a)^p)$, which proves (5.5). $\square$

Now (5.4) and (5.5) imply that $A_p = 1$ and $A = A^p$ for each prime p or Theorem (5.2) holds. Indeed if $A_p \neq 1$, then $A = A_p$ and Theorem (5.2) holds by (5.4). This shows that A is torsion free and divisible by each prime p, whence it is divisible. Now it follows from (5.5) (ii) with the same argument as in the proof of (2)(ii) that

$$a^{1/n} = \chi(\chi(a)^n) \text{ for each } a \in A^{\#} \text{ and } n \in \mathbb{N}.$$

This proves Theorem (5.2). $\qquad\qquad\qquad\qquad\qquad\qquad\qquad\qquad\qquad\qquad\square$

As consequence of (5.2) we have

(5.6)   Theorem

Let $X = \langle A, B \rangle$ be a special rank one group with AUS. Then one of the following holds:

(a) If A is an elementary abelian p-group, then

$$\langle a, \chi(a) \rangle \simeq (P)SL_2(p) \text{ for each } a \in A^{\#}.$$

   (And of course also $\langle b, \chi(b) \rangle \simeq (P)SL_2(p), b \in B^{\#}$.)

(b) If A is torsion-free and divisible, $a \in A^{\#}$ and $b = \chi(a) \in B^{\#}$ set

   $A(a) = \{a^{m/n} \mid m, n \in \mathbb{Z}, n \neq 0, a^0 = 1\}$.
   $B(b) = \{b^{m/n} \mid m, n \in \mathbb{Z}, n \neq 0, b^0 = 1\}$.
   ($a^{m/n}$ is well-defined by (5.2) and $a^{m/n} = (a^{1/n})^m$!)
   Then $A(a) \simeq (\mathbb{Q}, +) \simeq B(b)$ and $X(a) = \langle A(a), B(b) \rangle$ is a factor group of the universal perfect central extension of $SL_2(\mathbb{Q})$.

Proof. Pick $a \in A^{\#}$ and set $b = \chi(a)$. If A is an elementary abelian p-group, we set, to be able to prove part (a) and (b) together:

$$A_0 = \{a^m \mid m \leq p\}, B_0 = \{b^m \mid m \leq p\}$$

while in case A is torsion-free and divisible we set

$$A_0 = A(a) \text{ and } B_0 = B(b).$$

We now treat both cases (a) and (b) of Theorem (5.6) together, with the convention that, if A is an elementary abelian p-group, all exponents m, n, ℓ, k occurring in the proof are elements of $\mathbb{Z}_p$. Then by definition of $(\bar{a})^{1/n}, \bar{a} \in A^{\#}$ we have:
$$(a^m)^{1/n} = a^{m/n} = (a^{1/n})^m, n \neq 0.$$

Hence
$$(a^{\ell/m} a^{k/n})^{mn} = a^{\ell n} \cdot a^{km} = a^{\ell n + km}$$

for $m \neq 0 \neq n$ and thus:

$$a^{\ell/m} \cdot a^{k/n} = (a^{\ell n + km})^{1/mn} = a^{\frac{\ell n + km}{mn}} = a^{\ell/m + k/n}.$$

This implies that the map $\sigma : \ell/m \to a^{\ell/m}$ is an isomorphism of $(\mathbb{Q}, +)$ (resp. $(\mathbb{Z}_p, +)$) onto A_0. We next show that:

(*) $\chi(a^{m/n}) = b^{n/m}$ for all $m \neq 0 \neq n$.

Now to prove $(*)$ it suffices to prove

$(+)$
$$\chi(\overline{a}^m) = \chi(\overline{a})^{1/m}$$
$$\chi(\overline{a}^{1/m}) = \chi(\overline{a})^m$$
for all $\overline{a} \in A^{\#}$ and $m \neq 0$.

Indeed if these equations hold, then

$$
\begin{aligned}
\chi(a^{m/n}) &= \chi((a^{1/n})^m) = \chi(a^{1/n})^{1/m} = (\chi(a)^n)^{1/m} \\
&= \chi(a)^{n/m} = b^{n/m}.
\end{aligned}
$$

Now, as $\chi^2 = \mathrm{id}$, the second equation in $(+)$ is a consequence of part (b) of Theorem (5.2). Let $\overline{b} = \chi(\overline{a})$. Then also by (5.2)

$$\chi(\overline{a})^{1/m} = \overline{b}^{1/m} = \chi(\chi(\overline{b})^m) = \chi(\overline{a}^m),$$

which proves $(+)$ and $(*)$. $(*)$ shows that χ induces a bijection of $A_0^{\#}$ onto $B_0^{\#}$ and $B_0^{\#}$ onto $A_0^{\#}$. Now for $\lambda = m/n, n \neq 0$ set $a(\lambda) = a^{m/n}$ and $b(\lambda) = b^{m/n}$. Then the group $X_0 = \langle A_0, B_0 \rangle$ is generated by elements $a(\lambda), b(\lambda), \lambda \in \mathbb{Q}$ resp. $\lambda \in \mathbb{Z}_p$.

Further, since σ is an isomorphism, the relations (A) of (5.1) are satisfied. Since the universal central extension of $SL_2(p)$ is $SL_2(p)$ itself (see [Stg67, §7]) it suffices to prove Theorem (5.6), to show that also the relations (B') of (5.1) are satisfied. (We may assume $p > 3$ since otherwise $\{A_0\} \cup B_0^{A_0}$ is X_0-invariant and thus (5.6) holds.)

Now $(*)$ can be expressed as

$(**)$
$$\chi(a(\lambda)) = b(\lambda^{-1}), \lambda \in \mathbb{Q}^*(\text{ resp. } \mathbb{Z}_p^*).$$

Hence we have

$$a(\lambda)^{b(\lambda^{-1})} = a(\lambda)^{\chi(a(\lambda))} = b(\lambda^{-1})^{-a(\lambda)} \text{ for all } \lambda \neq 0.$$

Now let $\lambda = n/m$ and $\delta = r/s$ with $m \neq 0 \neq n$ and $r \neq 0 \neq s$. Then $a(\lambda) = a(\delta)^{sn/rm}$. Hence we obtain:

$$
\begin{aligned}
a(\lambda)^{b(\delta^{-1})} &= (a(\delta)^{sn/rm})^{b(\delta^{-1})} = (b(\delta^{-1})^{-a(\delta)})^{sn/rm} \\
&= (b(-\delta^{-1})^{sn/rm})^{a(\delta)} = ((b^{-1})^{s^2 n/r^2 m})^{a(\delta)} \\
&= b(-\frac{\lambda}{\delta^2})^{a(\delta)},
\end{aligned}
$$

since as shown in the proof of (5.4) we have for all $\overline{a} \in A^{\#}$:

$$(\overline{a}^{m/n})^{b(\overline{a})} = (\overline{a}^{b(\overline{a})})^{m/n}, m \neq 0 \neq n.$$

This shows that the relations (B') are also satisfied and thus (5.6) holds by (5.1). □

In view of (1.10), (2.10) and (5.6) it seems natural to conjecture that special rank one groups with AUS are either quasi-simple or isomorphic to $SL_2(2)$ or $(P)SL_2(3)$. As will be seen in the next chapter, the proof of this conjecture would simplify the existing simplicity proofs of classical and Lie type groups, which are not defined over $\mathbb{Z}_2$ or $\mathbb{Z}_3$.

At the end of §5 we discuss embeddings of $(P)SL_2(K)$ into $(P)SL_2(L)$, where both L and K are division rings. The group of diagonal automorphisms of such a group is the subgroup of the automorphism group induced by all diagonal matrices of $GL_2(K)$. It is well known and easy to compute that the group $\widehat{H}$ of diagonal automorphisms of $SL_2(K)$ acts transitively on $A^{\#}$ and $B^{\#}$, where $A = \{a(t) \mid t \in K\}, B = \{b(t) \mid t \in K\}$ and $a(t) = \left(\begin{smallmatrix} 1 & \\ t & 1 \end{smallmatrix}\right), b(t) = \left(\begin{smallmatrix} 1 & t \\ & 1 \end{smallmatrix}\right)$.

We have:

(5.7)   Theorem

Let $X = (P)SL_2(K)$ and $Y = (P)SL_2(L)$, where K and L are division rings and suppose $\sigma : X \to Y$ is an embedding, mapping unipotent subgroups of X into unipotent subgroups of Y. Then there exists an inner diagonal automorphism φ of Y, such that $\varphi\sigma$ is induced by some embedding or antiembedding $\alpha : K \to L$. In particular if σ is an isomorphism, then K and L are isomorphic or antiisomorphic division rings.

Proof. For the whole proof let A, B (resp. A', B') be different unipotent subgroups of X (resp. Y) with $A = \{a(\lambda) \mid \lambda \in K\}, B = \{b(\lambda) \mid \lambda \in K\}$ with $a(\lambda)a(\mu) = a(\lambda + \mu), b(\lambda)b(\mu) = b(\lambda + \mu)$ (resp. $A' = \{a'(\lambda) \mid \lambda \in L\}, B' = \{b'(\lambda) \mid \lambda \in L\}$ with the same multiplication). Define elements $n(t), w = n(1)$ and $h(t), t \in K^*$ (resp. $n'(t), w'$ and $h'(t), t \in L^*$) as in (1.5). Then we have by (1.5) (c) and (d) the equations:

(1)   (i) $a(t)^{b(t^{-1})} = b(-t^{-1})a(t), A^{b(t^{-1})} = B^{a(t)}, t \in K$.

(ii) $a(c)^w = b(-c), c \in K$.

(iii) $b(c)^{h(t)} = b(tct), a(c)^{h(t)} = a(t^{-1}ct^{-1}), t \in K^*, c \in K$

and similarly for the $'$-elements.

We will prove the theorem with a series of reductions.

(2) Without loss of generality, $\sigma(A) \subseteq A', \sigma(B) \subseteq B'$.

This is a consequence of the fact that Y acts doubly transitively on its unipotent subgroups. We now define a map $\alpha : K \to L$ by

$$a'(\alpha(\lambda)) := \sigma(a(\lambda)), \lambda \in K.$$

Then we have:

(3) α is an embedding of $(K, +)$ into $(L, +)$ which satisfies without loss of generality $\alpha(1_K) = 1_L$.

Since

$$\sigma(a(\lambda + \mu)) = a'(\alpha(\lambda))a'(\alpha(\mu)) = a'(\alpha(\lambda) + \alpha(\mu))$$

the first part of (3) is obvious. The second part follows from the fact that the full diagonal subgroup of $\mathrm{Aut}(Y)$ acts transitively on $(A')^{\#}$.

Now applying (1)(i) on both groups X and Y we obtain:

$$\sigma(b(1)) = \sigma(b(a(1))) = b'(a'(1)) = b'(1)$$

with self-explanatory notation, since σ respects the equation (1)(i). Since

$$w = n(1) = a(1)b(-1)a(1)$$

this implies $\sigma(w) = w'$. Now we obtain with (1)(ii):

(4) $\sigma(b(\lambda)) = b'(\alpha(\lambda)), \lambda \in K$.

Now again by (1)(i) we have

$$b(a(\lambda)) = b(\lambda^{-1}), b'(a'(\lambda)) = b'(\lambda^{-1}), \lambda \neq 0.$$

Hence

$$\begin{aligned} \sigma(b(\lambda^{-1})) &= \sigma(b(a(\lambda))) = b'(\sigma(a(\lambda))) = b'(a'(\alpha(\lambda))) \\ &= b'(\alpha(\lambda)^{-1}) \end{aligned}$$

and thus

(5) $\alpha(\lambda^{-1}) = \alpha(\lambda)^{-1}$ for $\lambda \in K^*$.

Now by definition of $h(t), h'(t)$ in (1.5):

$$\sigma(h(t)) = h'(\alpha(t)), t \in K^*.$$

Hence by (1)(ii) and (iii):

$$\begin{aligned} b'(\alpha(t\lambda t)) &= \sigma(b(t\lambda t)) = \sigma(b(\lambda)^{h(t)}) = b'(\alpha(\lambda))^{h'(\alpha(t))} \\ &= b'(\alpha(t)\alpha(\lambda)\alpha(t)). \end{aligned}$$

This implies:

(6) $\alpha(t\lambda t) = \alpha(t)\alpha(\lambda)\alpha(t); \lambda \in K, t \in K^*$.

In particular $\alpha(t^2) = \alpha(t)^2$ for $t \in K$. Setting now $t = \lambda + \mu$ and using (3) we obtain:

(7) $\alpha(\lambda\mu) + \alpha(\mu\lambda) = \alpha(\lambda)\alpha(\mu) + \alpha(\mu)\alpha(\lambda)$.

On the other hand we obtain from (5) and (6):

(8) $\alpha(\mu\lambda) = \alpha(\lambda\mu(\lambda\mu)^{-1}\mu\lambda) = \alpha(\lambda)(\alpha(\mu)\alpha(\lambda\mu)^{-1}\alpha(\mu))\alpha(\lambda)$.

Now by (7) and (8):

$$[1 - \alpha(\lambda)\alpha(\mu)\alpha(\lambda\mu)^{-1}](\alpha(\lambda\mu) - \alpha(\mu)\alpha(\lambda))$$
$$= \alpha(\lambda\mu) - \alpha(\lambda)\alpha(\mu) - \alpha(\mu)\alpha(\lambda) + \alpha(\lambda)(\alpha(\mu)\alpha(\lambda\mu)^{-1}\alpha(\mu))\alpha(\lambda)$$
$$= \alpha(\lambda\mu) - \alpha(\lambda)\alpha(\mu) - \alpha(\mu)\alpha(\lambda) + \alpha(\mu\lambda) = 0.$$

Hence

(9) For all $\lambda, \mu \in K^*$ either $\alpha(\lambda\mu) = \alpha(\lambda)\alpha(\mu)$ or $\alpha(\lambda\mu) = \alpha(\mu)\alpha(\lambda)$.

For $\lambda \in K$ let D_λ (resp. A_λ) be the set of all $\mu \in K$ satisfying $\alpha(\lambda)\alpha(\mu) = \alpha(\lambda\mu)$ (resp. $\alpha(\lambda\mu) = \alpha(\mu)\alpha(\lambda)$.) Then by (3) D_λ and A_λ are additive subgroups of K. Further by (9) $K = D_\lambda \cup A_\lambda$. Hence either $K = D_\lambda$ or $K = A_\lambda$, since no group is the union of two proper subgroups. Similarly, if $D = \{\lambda \in K \mid K = D_\lambda\}$ and $A = \{\lambda \in K \mid K = A_\lambda\}$, then D and A are additive subgroups of K with $K = D \cup A$. Hence either $K = D$ or $K = A$ and we obtain:

(10) α is an embedding or antiembedding of K into L.

If now σ is an isomorphism, then clearly α is surjective. Hence α is an isomorphism or antiisomorphism. This proves (5.7). □

As an immediate Corollary we obtain:

(5.8) Corollary

Let A be the group of all automorphisms of $(P)SL_2(K)$, K a division ring, mapping unipotent subgroups onto unipotent subgroups. Then A contains a normal subgroup A_0 of index at most two, which is an extension of inner by diagonal and field automorphisms. Further each $\sigma \in A - A_0$ induces an antiautomorphism on K.

If K is a division ring with antiautomorphism α, then $\widetilde{\alpha}^{-1}\varphi \in A - A_0$, where φ is the transpose inverse map and $\widetilde{\alpha} : (a_{ij}) \rightarrow (a_{ij}^\alpha)$, where $(a_{ij}) \in SL_2(K)$. The method of proof of (5.7) goes back to [Hua49].

(5.9) Exercises

(1) Let $X = \langle A, B \rangle$ be a special rank one group with AUS and A an elementary abelian 2-group. Show

(a) $o(ab) = \infty$ or odd for all $a \in A^\#, b \in B^\#$.

(b) X is quasi-simple or isomorphic to $SL_2(2)$, if there exists no $a \in A^\#, b \in B^\#$ with $o(ab) = \infty$.

A Zassenhaus group is a finite doubly transitive permutation group G on a set Ω, satisfying $G_{\alpha,\beta,\gamma} = 1$ for different $\alpha, \beta, \gamma \in \Omega$.

(2) Let G be a Zassenhaus group on Ω. Show:

    (a) G_α is a Frobenius group with kernel A and complement $H = G_{\alpha,\beta}$ acting fixed point freely on A.

    (b) $G_0 = \langle A^G \rangle$ is a rank one group.

(3) Suppose G is as in (2), $|G_\alpha|$ is even and H acts irreducibly on A. Show:

    (a) A is an elementary abelian p-group for some prime p.

    (b) G_0 is a special rank one group with AUS.

Hint: If $p = 2$ (b) holds anyway by (2.2). If p is odd, there exists an involution in H inverting A and B.

The following exercises show how the coordinatization-procedure of this section may be used for an alternative classification of certain Zassenhaus-groups.

(4) Let G be as in (3) and assume $G = G'$. Coordinize A by an abelian additive group K, such that $A = \{a(\lambda) \mid \lambda \in K\}$ with $a(\lambda)a(\mu) = a(\lambda + \mu)$. For $\lambda \in K$ let $n(\lambda) = a(\lambda)b(a(-\lambda))a(\lambda)$ as in (2.7). Show:

    (a) $n(\lambda)^2 = 1$ for each $\lambda \in K$.

    (b) $C_H(n(\lambda)) = 1$ or $\langle z \rangle$, where z is the central involution in H inverting A if $|H|$ is even.

Hint. Show as in [Gor68, 13.1.4] that $|C_H(n(\lambda))|$ is even. Let $S \in \mathrm{Syl}_2(H)$ normalized by $n(\lambda)$ such that $C_S(n(\lambda)) = C_H(n(\lambda))$.
Show that $S\langle n(\lambda)\rangle \in \mathrm{Syl}_2(G)$. Hence by the Thompson transfer lemma, $n(\lambda)$ fuses to z. But if $|C_S(n(\lambda))| > 2$ this is impossible, since z is a square in $C(n(\lambda))$, while each involution different from z in $C(z) = H\langle n(\lambda)\rangle$ is not a square in $C(z)$.

(5) Let G be as in (4). Show that $n(\lambda)$ inverts H, whence H is abelian.
Hint. Let $a(\lambda)^H = \{a(\mu) \mid \mu \in L \subseteq K\}$. Then $|L| = |H|$. For each $\lambda \neq \mu \in L, n(\lambda)n(\mu) \in H$ is inverted by $n(\lambda)$. Show, that if $n(\lambda)n(\mu) = n(\lambda)n(\nu)$ then $\nu = \pm\mu$. Namely there exists an $h \in H$ with $a(\mu)^h = a(\nu)$. Hence
$$n(\mu)^h = n(\nu) = n(\mu) \text{ and so } h \in C_H(n(\mu)) = \langle z \rangle.$$

This implies that $n(\lambda)$ inverts at least $|H|/2$ elements of H. Moreover the same holds for each $n(\lambda)$ invariant subgroup of H.

If $n(\lambda)$ does not invert S, then the set T of elements of S inverted by $n(\lambda)$ is a subgroup of index 2 and all involutions of $S\langle n(\lambda)\rangle$ lie in $T\langle n(\lambda)\rangle$. But again by Thompson transfer an element x of minimal order in $S\langle n(\lambda)\rangle - T\langle n(\lambda)\rangle$ must fuse into $T\langle n(\lambda)\rangle$, which is impossible since z is a power of x.

Hence $n(\lambda)$ inverts S and thus H.

(6) Let G be as in (2) and assume in addition $G = G'$, $|G_\alpha|$ is odd and $|H| \geq \frac{|A|-1}{2}$. Show

   (a) H is abelian and inverted by an involution w interchanging α and β.

   (b) $|H| = \frac{|A|-1}{2}$.

   (c) There exists an $a \in A^{\#}$ with $awa = wa^{-1}w$, whence $wa^{-1}w = b(a)$ and $b(a^{-1}) = b(a)^{-1}$.

   Hint. See the proof of [Gor68, 13.3.3i].

   (d) $b(a^{-1}) = b(a)^{-1}$ for each $a \in A^{\#}$ and G is special. (As $A^{\#} = a^H \cup (a^{-1})^H$!)

Remark. In the situation of (5) and (6), when $|H| \geq \frac{|A|-1}{2}$, it is possible to show $G \simeq PSL_2(K)$, where $K = \operatorname{Hom}_H(A, A)$, by the coordinatization procedure of (5.6).

Chapter II
Abstract Root Subgroups

In this chapter we introduce the notion of a group generated by a class of abstract root subgroups together with certain specializations of it, namely k-root subgroups and abstract transvection groups. After discussing some examples, i. e. classical groups, we start to develop the elementary general theory of groups generated by a class of abstract root subgroups. Finally in §5 we show that each simple group of Lie type (in a general sense) is generated by a class of abstract root subgroups. The results of this chapter are, if not mentioned otherwise, from [Tim90a] ,[Tim91], [Tim99] or [Tim00b].

§1   Definitions and examples

(1.1)   Definition

A set Σ of abelian non-identity subgroups of the group G is called a set of *abstract root subgroups* of G, if it satisfies.

  I $G = \langle \Sigma \rangle$ and $\Sigma^g \subseteq \Sigma$ for each $g \in G$.
  II For each pair $A, B \in \Sigma$ one of the following holds:

   (α) $[A, B] = 1$.
   (β) $X = \langle A, B \rangle$ is a rank one group with unipotent subgroups A and B.
   (γ) $Z(\langle A, B \rangle) \geq [A, B] = [a, B] = [A, b] \in \Sigma$ for each $a \in A^{\#}$ and $b \in B^{\#}$.

(Notice that (γ) implies that $\langle A, B \rangle$ is nilpotent of class 2, since it is abelian $\mathrm{mod}[A, B]$.)

If Σ is a conjugacy class and a set of abstract root subgroups of G, we call it a *class* of *abstract root subgroups* of G. If in case II (β) always $X \simeq (P)SL_2(k)$, k a fixed commutative field with A and B (images) of unipotent subgroups of $SL_2(k)$, we call Σ a set of *k-root subgroups* of G (resp. class of k-root subgroups.) If possibility II (γ) never occurs we call Σ a set (resp. class) of *abstract transvection groups* (resp. *k-transvection groups*) or a *degenerate* set of abstract root subgroups of G.

If all possibilities (α)–(γ) of II occur, we call Σ a *non-degenerate* set (class) of abstract root subgroups of G.

(Notice that if (γ) occurs, then also (α) occurs. The two "very degenerate" cases, when (β) (resp. (α)) never occur, are included in our definition of abstract transvection groups. In the first case G is abelian, whence uninteresting. In the second case $\langle A, B \rangle$ is a rank one group for all $A \neq B \in \Sigma$. This case, which corresponds to classical groups of Witt index 1, is contained in our general treatment. But unfortunately we can not give a classification here. Notice further that if (α) and (γ) occur but never (β), we cannot say very much about G, even not that G is nilpotent. So this case will be mostly excluded by some additional hypothesis.)

Suppose now Σ is a set of abstract root subgroups of G, $\Lambda \subseteq \Sigma$ and $U \leq G$. We then fix for the rest of this book the following notation:

$U \cap \Lambda := \{A \in \Lambda \mid A \leq U\}$

U is a Λ-*subgroup*, if $U = \langle U \cap \Lambda \rangle$

$N_\Lambda(U) := \{A \in \Lambda \mid A \leq N(U)\}$ and similar $C_\Lambda(U)$

$D(\Lambda) := \{a \in A^{\#} \mid A \in \Lambda\}$

Λ is *abelian*, if $\langle \Lambda \rangle' = 1$.

For $A \in \Sigma$ let:

$$\begin{aligned}
\Sigma_A &:= C_\Sigma(A) - \{A\} \\
\Lambda_A &:= \{B \in \Sigma_A \mid \Sigma \cap AB \text{ is a partition of } AB\} \\
\Psi_A &:= \{B \in \Sigma \mid [A, B] \in \Sigma\} \\
\Omega_A &:= \{B \in \Sigma \mid \langle A, B \rangle \text{ is a rank one group }\}.
\end{aligned}$$

Notice that, since rank one groups are by I (2.1) not nilpotent, $\Psi_A \cap \Omega_A = \emptyset$. Hence we have

$$\Sigma = \{A\} \dot\cup \Sigma_A \dot\cup \Psi_A \dot\cup \Omega_A.$$

If Σ is non-degenerate let

$$M_A := \langle A, \Lambda_A \rangle. \text{ Then } M_A \trianglelefteq N(A).$$

We often use the following pictorial notation:

$$
\begin{array}{cc}
A & B \\
\circ & \circ
\end{array}
\qquad \text{if and only if } B \in \Sigma_A
$$

$$
\begin{array}{cc}
A & B \\
\circ \;\sim\; \circ
\end{array}
\qquad \text{if and only if } B \in \Lambda_A
$$

$$
\begin{array}{cc}
A & B \\
\circ\!\!-\!\!-\!\!\circ
\end{array}
\qquad \text{if and only if } B \in \Omega_A
$$

$$
\begin{array}{cc}
A & B \\
\circ\!\!=\!\!=\!\!\circ
\end{array}
\qquad \text{if and only if } B \in \Psi_A.
$$

If $\Lambda \subseteq \Sigma$ there are three graphs, all with vertex set Λ, which will be important for us.

$\mathcal{D}(\Lambda)$ has as edges (A, B); $A \neq B \in \Lambda$ and $[A, B] = 1$
$\mathcal{E}(\Lambda)$ has as edges (A, B); $A \in \Lambda$ and $B \in \Lambda \cap \Lambda_A$
$\mathcal{F}(\Lambda)$ has as edges (A, B); $A \in \Lambda$ and $B \in \Lambda \cap \Omega_A$.

Before we start to discuss examples, we state an easy lemma, which makes the verification of condition II (γ) easier.

(1.2) Lemma

The following hold:

(1) If A, B, C, D are abelian subgroups of some group G, satisfying:

(i) $C \cap D = 1 = [C, D] = [A, C] = [B, D]$,
(ii) $[A, D] = C$, $\qquad [B, C] = D$,
(iii) $X = \langle A, B \rangle$ is a rank one group,

then $[a, D] = C$ and $[b, C] = D$ for all $a \in A^{\#}, b \in B^{\#}$.

(2) Suppose Φ is a root system of type A_2 and $\{A_\alpha \mid \alpha \in \Phi\}$ is a set of abelian subgroups, generating the group G and satisfying:

(i) $[A_\alpha, A_\beta] = 1$ if $\beta \neq \pm\alpha$ and $\alpha + \beta \notin \Phi$.
(ii) $[A_\alpha, A_\beta] \leq A_{\alpha+\beta}$ if $\alpha + \beta \in \Phi$.
(iii) $X_\alpha = \langle A_\alpha, A_{-\alpha} \rangle$ is a rank one group.

Then either

(a) $G = X_r * X_s * X_{r+s}$ (central product) for $\Phi = \{\pm r, \pm s, \pm(r+s)\}$ or
(b) $[a, A_\beta] = A_{\alpha+\beta} = [A_\alpha, b]$ for all $a \in A_\alpha^{\#}, b \in A_\beta^{\#}$ and all $\alpha, \beta \in \Phi$ with $\alpha + \beta \in \Phi$.

Proof. (1) Since X is a rank one group we have $X = \langle a, B \rangle = \langle A, b \rangle$ for all $a \in A^{\#}, b \in B^{\#}$ by I(1.4). Suppose $C_0 = [a, D] < C$ for some $a \in A^{\#}$. Then $C_0 D$ is X-invariant, but $C_0 D < CD$, a contradiction to

$$C = [A, D] \leq [C_0 D, X] \leq C_0 D.$$

(2) Note first that for $\alpha \neq \beta \in \Phi$ we have $A_\alpha \cap A_\beta = 1$. Namely if $1 \neq x \in A_\alpha \cap A_\beta$, then $X_\alpha = \langle x, A_{-\alpha} \rangle = \langle A_\alpha, A_{-\alpha} \rangle$ is a rank one group but on the other hand

$$\langle x, A_{-\alpha} \rangle \leq \langle A_\beta, A_{-\alpha} \rangle$$

is nilpotent, a contradiction to I(2.1).

Now X_s acts on $N = A_r \times A_{r+s}$ and $M = A_{-s} \times A_{-r-s}$. If now $C_N(A_s) = A_{r+s}$ and $C_N(A_{-s}) = A_r$, then an element w of X_s interchanging A_s and A_{-s} also interchanges A_{r+s} and A_r. If now $[A_r, A_s] < A_{r+s}$, then as in (1) $A_r[A_r, A_s] < N$ and is X_s-invariant, a contradiction to $A_r = A_{r+s}^w \leq N$.

Thus either $[A_r, A_s] = A_{r+s}, [A_{r+s}, A_{-s}] = A_r$ and whence by (1) also (2) holds for the pair r, s or there exists a pair $r, s \in \Phi$ with $r + s \in \Phi$ and $C_N(A_s) > A_{r+s}$, where $N = A_r \times A_{r+s}$. We will show that in this case G is the central product of X_r, X_s and X_{r+s}. Namely $C_N(A_s) > A_{r+s}$ implies that there exists a $1 \neq y \in C_{A_r}(A_s)$. Hence $X_r = \langle y, A_{-r} \rangle \leq C(A_s)$. Since X_r acts on $A_s \times A_{s+r}$ and A_r centralizes it, it follows that $X_r \leq C(A_{s+r})$. (A_r and A_{-r} are conjugate in X_r!) Thus $A_r \leq C(A_s A_{s+r})$, which implies in turn $A_s \leq C(A_r A_{s+r})$. Since $A_{-s} \leq N(A_r A_{s+r})$ we obtain $X_s \leq C(A_r A_{s+r})$. Thus $\langle X_r, X_s \rangle \leq C(A_{s+r})$, whence

$$[A_{-r}, A_{-s}] \leq A_{-r-s} \cap C(A_{s+r}) = 1,$$

since X_{s+r} is a rank one group. As $[A_r, A_s] = 1$ this implies $[X_r, X_s] = 1$. Finally the rank one group X_{s+r} acts on $A_r A_{-s}$ and $A_{-r} A_s$ with A_{s+r} centralizing both. Hence X_{s+r} centralizes both groups and thus $X_{s+r} \leq C(X_r * X_s)$. □

The proof of (2) is extracted from the proof of Proposition (6.3) of [Tim98].

(1.3)   Example

The groups $T(V, W^*)$ [CH91] and [Tim99, §11]

Let K be a division ring and V a left vector space over K. Then the dual space V^* is a right vector space. If $W \subseteq V$ let $W^0 = \{\varphi \in V^* \mid W\varphi = 0\}$. For $0 \neq \varphi \in V^*$ and $\Phi = \varphi K$ let Kern $\Phi := $ Kern φ. If $0 \neq p \in$ Kern Φ, $P = Kp$ and $\lambda \in K$ let

$$t(\lambda) \quad : \quad v \to v + (v\varphi)\lambda p$$
$$V \to V$$

and $T_{P,\Phi} = \{t(\lambda) \mid \lambda \in K\}$ – the transvection group corresponding to the points P of V and Φ of V^*. Then $t(\lambda)t(\mu) = t(\lambda+\mu)$ and the $t(\lambda)$ are bijective linear maps. Hence we have:

(1) The following hold:

    (a) $T_{P,\Phi} \leq GL(V)$.

    (b) $T_{P,\Phi} \simeq K^+$.

    (c) $T_{P,\Phi}$ is independent of $0 \neq p \in P$ and $0 \neq \varphi \in \Phi$.

(c) is also clear, since the transvection $t(\lambda)$ defined with respect to the vector cp is equal to $t(\lambda c)$ defined with respect to p.

Next we show

(2) Let P, Q be points of V and Φ, Ψ be points of V^* with $P \subseteq \mathrm{Kern}\ \Phi$, $Q \subseteq \mathrm{Kern}\ \Psi$. Then the following hold:

    (a) $[T_{P,\Phi}, T_{Q,\Psi}] = 1$ if $P \subseteq \mathrm{Kern}\ \Psi, Q \subseteq \mathrm{Kern}\ \Phi$.

    (b) $[t(\lambda), T_{Q,\Psi}] = [T_{P,\Phi}, \tau(\mu)] = [T_{P,\Phi}, T_{Q,\Psi}] = T_{P,\Psi}$ for each $1 \neq t(\lambda) \in T_{P,\Phi}$ and $1 \neq \tau(\mu) \in T_{Q,\Psi}$ if $P \subseteq \mathrm{Kern}\ \Psi, Q \not\subseteq \mathrm{Kern}\ \Phi$.

    (c) $\langle T_{P,\Phi}, T_{Q,\Psi} \rangle \simeq SL_2(K)$ acting naturally on $P + Q$ and $\Phi + \Psi$ and centralizing $\mathrm{Kern}\ \Phi \cap \mathrm{Kern}\ \Psi$ resp. $P^0 \cap Q^0$, if $P \not\subseteq \mathrm{Kern}\ \Psi$ and $Q \not\subseteq \mathrm{Kern}\ \Phi$.

Proof. Notice that by definition of $T_{P,\Psi}$:

$$(*) \qquad\qquad [V, T_{P,\Phi}] = P \text{ and } C_V(T_{P,\Phi}) = \mathrm{Kern}\ \Phi.$$

Hence (a) is a consequence of the 3-subgroup lemma. Also (c) is immediate, since $T_{P,\Phi}$ induces all transvections corresponding to P on $P + Q$ and $T_{Q,\Psi}$ induces all transvections corresponding to Q. (And since $SL_2(K) = \langle \begin{pmatrix} 1 & \\ 1 & 1 \end{pmatrix}, \begin{pmatrix} 1 & \mu \\ & 1 \end{pmatrix} \mid \lambda, \mu \in K \rangle$!)

To prove (b) let $t(\lambda) \in T_{P,\Phi}$ and $\tau(\mu) \in T_{Q,\Psi}$ and $\Psi = \varphi K$, $Q = Kq$. Then an easy computation with the definition of $t(\lambda), \tau(\mu)$ shows that

$$\tau(-\mu)t(-\lambda)\tau(\mu)t(\lambda) : v \to v - (v\psi)\mu(q\varphi)\lambda p$$

for arbitrary $v \in V$. Hence

$$\tau(-\mu)t(-\lambda)\tau(\mu)t(\lambda) = \rho(\nu) \in T_{P,\Psi}$$

with $\nu = -\mu(q\varphi)\lambda$. Since $q\varphi \neq 0$, it is obvious that ν runs for fixed μ (resp. λ) with λ (resp. with μ) over all elements of K. This proves (b) $\qquad\square$

For a subspace W^* of V^* (W^* is arbitrary not just the dual of a subspace of V) with $\operatorname{Ann}_V(W^*) := \{v \in V \mid v\varphi = 0 \text{ for all } \varphi \in W^*\} = 0$, let $\Sigma(W^*) := \{T_{P,\Phi} \mid \Phi \subseteq W^* \text{ and } P \subseteq \operatorname{Kern} \Phi\}$ and

$$T(V, W^*) := \langle T_{P,\Phi} \mid T_{P,\Phi} \in \Sigma(W^*)\rangle \subseteq GL(V).$$

Let $SL(V) := T(V, V^*)$. If $\dim V < \infty$, then V^* is the only subspace of V^* with $\operatorname{Ann}_V(V^*) = 0$. So in this case there is only one such group, namely $SL(V)$. Notice that $T(V, W^*)$ leaves W^* invariant in its natural action on V^*. We have:

(3) If $\dim V \geq 3$ then the following hold:

    (a) $\Sigma := \Sigma(W^*)$ is a non-degenerate class of abstract root subgroups of $T(V, W^*)$. Further $\mathcal{F}(\Sigma)$ is connected.

    (b) $T(V, W^*)$ is perfect.

Proof. By (1) and (2) Σ is a set of abstract root subgroups of $T(V, W^*)$. Let $T_{P,\Phi}$ and $T_{Q,\Psi} \in \Sigma$. Then, since $\operatorname{Ann}_V(W^*) = 0$ and since W^* is not the union of two proper subspaces, there exists a point $\Lambda \subseteq W^*$ with $P \not\subseteq \operatorname{Kern} \Lambda$ and $Q \not\subseteq \operatorname{Kern} \Lambda$. By the same reason there exists a point $R \subseteq \operatorname{Kern} \Lambda$ with $R \not\subseteq \operatorname{Kern} \Phi$ and $R \not\subseteq \operatorname{Kern} \Psi$. Hence by (2) $T_{R,\Lambda}$ is a neighbor of $T_{P,\Phi}$ and $T_{Q,\Psi}$ in $\mathcal{F}(\Sigma)$. Since vertices of $\mathcal{F}(\Sigma)$ on a common edge are by (2)(c) conjugate in the group they generate, this proves (a).
(b) is a consequence of (2)(b). □

We will see later on that the groups $T(V, W^*)$ are quasi-simple.

(1.4)   Example

Isotropic transvection subgroups [CS99]

(1) *Notation.* Let K be a division ring with antiautomorphism σ satisfying $\sigma^2 = \operatorname{id}$. We set

$$\Lambda := \Lambda_{\min} = \{c + c^\sigma \mid c \in K\} \text{ and } \Lambda_{\max} = \{c \in K \mid c^\sigma = c\}.$$

Let V be a (left) vector space over K, endowed with one of the following forms:

    (i) A pseudo-quadratic form $q : V \to K/\Lambda$ with associated (trace valued) $(\sigma, -1)$ hermitian form $f : V \times V \to K$ or

    (ii) a $(\sigma, -1)$ hermitian form f, where $\Lambda = \Lambda_{\max}$.

(We use the definition of pseudo-quadratic and (σ, ϵ)-hermitian forms of [Tit74, §8].) Assume that V is generated by isotropic* vectors. For a subspace U of V

---

  *In this example we write uniformly isotropic, also when this means "singular" with respect to the given pseudo-quadratic form.

we set

$$U^\perp := \{v \in V \mid f(v, u) = 0 \text{ for all } u \in U\}$$
$$\mathrm{Rad}(U, f) := \{u \in U \mid f(w, u) = 0 \text{ for all } w \in U\}.$$

In case (i) let $R := \{r \in \mathrm{Rad}(V, f) \mid q(r) = 0 + \Lambda\}$ and call R the isotropic radical of (V, q). Fix a complement V_0 to R in V. Then $q \mid_{V_0}$ is non-degenerate. Set $L := \Lambda_{\max}$ in case (ii) and

$$L := \{c \in K \mid \text{ there exists a } v_c \in \mathrm{Rad}(V_0, f) \text{ with } q(v_c) = c + \Lambda\}$$

in case (i). Then by I(1.6)(1) and (3) L satisfies as a subset of K conditions (1)–(3) of I(1.5).

(2) Isotropic transvection subgroups.
Let P be an isotropic point of V not contained in $\mathrm{Rad}(V, f)$ and choose a vector $0 \neq v_p \in P$.

  (i) Assume we are in case (1)(i). Then for each $c \in L$ and $v_c \in \mathrm{Rad}(V_0, f)$ with $q(v_c) = c + \Lambda$ the map

$$t_c : v \to v + f(v, v_p)(cv_p + v_c) \text{ for } v \in V$$

  is a transvection preserving the form q with axis $P^\perp$ and center $\langle cv_p + v_c \rangle$. (By example I(1.6)(3) v_c is uniquely determined in $\mathrm{Rad}(V_0, f)$ with $q(v_c) = c + \Lambda$!). For $c, d \in L$ we have $t_c t_d = t_{c+d}$ so that

$$T_P := \{t_c \mid c \in L\} \simeq (L, +).$$

  We call T_P the *isotropic transvection subgroup* associated to the point P on $V = V_0 \oplus R$. Since for $d \in K$, $c \in L$ we have $dv_c \in \mathrm{Rad}(V_0, f)$ and

$$q(dv_c) = dcd^\sigma + \Lambda,$$

  it follows that $dcd^\sigma \in L$. This shows that the description of T_P is independent of the chosen vector $v_P \in P$.
  (ii) Assume we are in case (1)(ii). Then for $c \in L$ let

$$t_c : v \to v + f(v, v_p)cv_p \text{ for } v \in V$$

  and $T_P := \{t_c \mid c \in L\}$ and call T_P the *isotropic transvection group* associated to the point P. Again $T_P \simeq (L, +)$ and T_P is independent of the choice of $0 \neq v_p \in P$.

In both cases let

$$\Sigma := \{T_P \mid P \text{ an isotropic point of } V \text{ not contained in } \mathrm{Rad}(V, f).\}$$

(3) Suppose the notation is as in (1) and (2). Then we have:

Proposition. Σ is a class of abstract transvection subgroups of the subgroup $\langle \Sigma \rangle$ of the group of isometries of (V, q) resp. (V, f).

(Notice that, if Σ is a set of abstract transvection groups, then $\mathcal{F}(\Sigma)$ connected and Σ a conjugacy class of G are obviously equivalent. In the non-degenerate case this is not so obvious. See (2.13).)

Proof. Pick isotropic points P and Q of V not contained in $\mathrm{Rad}(V, f)$. If $Q \in P^{\perp}$ then $P^{\perp\perp} \subseteq Q^{\perp}$ so that

$$[V, T_P, T_Q] \subseteq [P^{\perp\perp}, T_Q] = 0$$

and also

$$[V, T_Q, T_P] = 0,$$

so that by the 3-subgroup lemma $[T_P, T_Q] = 1$.

Next assume $Q \not\subseteq P^{\perp}$. Let $H = P + Q$. Then H is non-degenerate and $V = H \perp H^{\perp}$. Let $X = \langle T_P, T_Q \rangle$. Then in case (ii) $X|_{H^{\perp}} = \{\mathrm{id}\}$ so that

$$X = X|_H \simeq SL_2(L) \text{ in the notation of I (1.5) .}$$

Hence by I(1.5) X is a special rank one group with AUS.

Next assume we are in case (i). Let $\overline{V} = V/\mathrm{Rad}(V, f)$. If $x \in X$ with $x|_{\overline{H}} = \mathrm{id}$ then $[P, x] \subseteq \mathrm{Rad}(V, f)$. Since P is isotropic, $[P, x]$ is isotropic. Hence

$$[P, x] \subseteq [V, X] \cap R \subseteq (P + Q + \mathrm{Rad}(V_0, f)) \cap R = 0.$$

Similarly $[Q, x] = 0$ so that $x|_H = \mathrm{id} = x$. This implies as before

$$X = X|_{\overline{H}} \simeq SL_2(L) \text{ in the notation of I(1.5) .}$$

It remains to show that $\mathcal{F}(\Sigma)$ is connected. Let Δ be a connectivity component of $\mathcal{F}(\Sigma)$ and $\Omega = \Sigma - \Delta$. Let $U_1 = \langle P \mid T_P \in \Delta \rangle$ and $U_2 = \langle Q \mid T_Q \in \Omega \rangle$. Then, since f is trace-valued and $\Sigma \neq \emptyset$ we have $V = U_1 \perp U_2 \perp \mathrm{Rad}(V, f)$. But if $U_1 \neq 0 \neq U_2$ there exist clearly isotropic points of $U_1 \perp U_2 - \mathrm{Rad}(V, f)$, which are neither contained in U_1 nor in U_2, a contradiction. Thus $U_2 = 0$ and $\Omega = \emptyset$. $\qquad\qquad\qquad\qquad\qquad\qquad\qquad\qquad$ $\square$

Together with I(1.5) we obtain:

Corollary. Suppose $|L| > 3$. Then $\langle \Sigma \rangle$ is perfect. Moreover for isotropic points P, Q of V not contained in $\mathrm{Rad}(V, f)$ with $Q \not\subseteq P^{\perp}$ it follows that $\langle T_P, T_Q \rangle$ is a quasi-simple special rank one group with AUS.

The second part of the Corollary follows from I(1.5) and (1.10). Now $\langle \Sigma \rangle$ is perfect since $\mathcal{F}(\Sigma)$ is connected.

We will later on see that $\langle \Sigma \rangle$ is quasi-simple if $R = 0$ in (i) or $\mathrm{Rad}(V, f) = 0$ in (ii) and $|L| > 3$.

(1.5)   Example

Siegel transvections in orthogonal groups. [Tim91, §9]
Let in this example V be a vector space over the (commutative) field k endowed with a quadratic form $q : V \to k$ with associated bilinear form $(,)$. Let for a subspace U of V:

$$U^\perp := \{u \in U \mid (u, w) = 0 \text{ for all } w \in U\},$$
$$\operatorname{Rad} U := U^\perp \cap q^{-1}(0).$$

A subspace U is (totally) singular if $q|_U = 0$. Then it is well known, see [Tit74, 8.2.7], that:

(1) Suppose that not all totally singular subspaces of U are in $U^\perp$. Then U is generated by totally singular subspaces.

We assume now for the rest of this example that:

(∗)   There exists a singular line ℓ (i. e. 2-dimensional singular subspace) with $\ell \cap V^\perp = 0$.

From (∗) we obtain with (1) that the following holds:

(2) (a) Each hyperplane of V is generated by singular subspaces.

    (b) V is generated by singular lines.

    (c) If ℓ is a singular line of V with $\ell \cap V^\perp = 0$, then there exists a singular line s with $\ell^\perp \cap s = 0$.

Proof. (a) is clear, since $\ell \cap H \not\subseteq H^\perp$ for any hyperplane H of V. (b) is a consequence of (a). To prove (c) pick a singular vector $u \in V - \ell^\perp$. Then $u^\perp \ne \ell^\perp \oplus \langle u \rangle$. Hence there exists by (a) a singular vector

$$w \in u^\perp - (\ell^\perp \oplus \langle u \rangle)$$

and $s = \langle u, w \rangle$ is the desired line.   □

Fix now a singular line ℓ with $\ell \not\subseteq V^\perp$ and let v_1, v_2 be a basis of ℓ with $V = v_1^\perp + v_2^\perp$ (such a basis exists!) For $c \in k$ let t_c be the map:

$$t_c : w \to w + (w, cv_2)v_1 - (w, cv_1)v_2, \qquad w \in V.$$

Then t_c is an isometry of V (i.e. element of $O(V, q)$) and the following holds:

(3) (a) $t_c t_d = t_{c+d}$ and $T_\ell = T := \{t_c \mid c \in k\} \simeq (k, +)$.

    (b) T_ℓ is independent of chosen basis of ℓ.

    (c) $T_\ell = \{\sigma \in O(V, q) \mid (\sigma - \mathrm{id})(v_1^\perp) \subseteq kv_1 \text{ and } (\sigma - \mathrm{id})(v_2^\perp) \subseteq kv_2\}$.

    (d) If $V^\perp = 0$ then $T_\ell = \{\sigma \in O(V, q) \mid \sigma|_{\ell^\perp} = \mathrm{id}\}$.

Proof. (a) is an easy computation using the definition of t_c. To prove (b) notice first that T_ℓ is not changed if one replaces v_1, v_2 by the basis $c_1 v_1, c_2 v_2, c_i \in k^*$, since the map t_c equals $t_{(c_1 c_2)^{-1} c}$ defined with respect to the new basis. Hence, if w_1, w_2 is another basis of ℓ we may without loss of generality assume $w_1 = c_1 v_1 + v_2, w_2 = c_2 v_1 + v_2$. Now direct computation yields that the map t_d defined with respect to w_1, w_2 equals to $t_{d(c_1 - c_2)}$ defined with respect to v_1, v_2.

To prove (c) assume $\sigma \in O(V, q)$ satisfies the hypothesis of (c). Then

For $w \in v_1^\perp$ we have $\sigma(w) = w + \lambda(w)v_1, \lambda \in \mathrm{Hom}(v_1^\perp, k)$ with Kern $\lambda \supseteq v_1^\perp \cap v_2^\perp$.

For $v \in v_2^\perp$ we have $\sigma(v) = v + \mu(v)v_2, \mu \in \mathrm{Hom}(v_2^\perp, k)$ with Kern $\mu \supseteq v_1^\perp \cap v_2^\perp$.

Since

$$\{\lambda \in \mathrm{Hom}(v_1^\perp, k) \mid \mathrm{Kern}\ \lambda \supseteq v_1^\perp \cap v_2^\perp\} \simeq \{\lambda \in \mathrm{Hom}(V, k) \mid \mathrm{Kern}\ \lambda \supseteq v_2^\perp\}$$

and similarly for μ, we may assume $\lambda \in \mathrm{Hom}(V, k)$ with Kern $\lambda \supseteq v_2^\perp$ and $\mu \in \mathrm{Hom}(V, k)$ with Kern $\mu \supseteq v_1^\perp$. We obtain

$$\sigma(v + w) = \sigma(v) + \sigma(w) = (v + w) + \mu(v + w)v_2 + \lambda(v + w)v_1.$$

Since $V = v_1^\perp + v_2^\perp$ this implies

$$\sigma(u) = u + \lambda(u)v_1 + \mu(u)v_2 \text{ for } u \in V.$$

Now there exist $c, d \in k$ with $\lambda(u) = c(u, v_2), \mu(u) = d(u, v_1)$ for arbitrary $u \in V$. Since $q(u) = q(\sigma(u))$ we have

$$\lambda(u)(u, v_1) + \mu(u)(u, v_2) = 0.$$

Hence, picking $u \notin v_1^\perp \cup v_2^\perp$, this implies $d = -c$ and thus $\sigma = t_c \in T_c$.

Finally, to prove (d), we must show that $\sigma \in O(V, q)$ with $\sigma|_{\ell^\perp} = \mathrm{id}$ satisfies $(\sigma - \mathrm{id})(v_1^\perp) \subseteq kv_1$ and $(\sigma - \mathrm{id})(v_2^\perp) \subseteq kv_2$ if $V^\perp = 0$. Now for $v \in \ell^\perp$ and $w \in V$ we have $(w, v) = (\sigma(w), \sigma(v)) = (\sigma(w), v)$. Hence

$$((\sigma - \mathrm{id})(w), v) = 0 \text{ for all } w \in V, v \in \ell^\perp$$

and thus $(\sigma - \mathrm{id})(w) \in \ell^{\perp\perp} = \ell$, since $V^\perp = 0$. This implies that there exist $\lambda, \mu \in \mathrm{Hom}(V, k)$ with

$$\sigma : w \to w + \lambda(w)v_1 + \mu(w)v_2.$$

As $q(w) = q(\sigma(w))$ we obtain as before

$$\lambda(w)(w, v_1) + \mu(w)(w, v_2) = 0$$

which again implies Kern $\lambda \supseteq v_2^\perp$ and Kern $\mu \supseteq v_1^\perp$. But then $\sigma \in T_\ell$ by (c). $\qquad\square$

The group $T_\ell \simeq (k, +)$ will be called the *Siegel transvection group* corresponding to the singular line ℓ and its elements $\neq 1$ *Siegel transvections*. (The usual name in the literature is Siegel transformations. But to be able to speak of (Siegel) transvections, when we mean either ordinary transvections or the Siegel transvections of (1.5) I have chosen the latter name!) Let $\Omega(V, q) := \langle T_\ell \mid \ell \text{ a singular line of } V \rangle$.

Next we show:

(4) Let ℓ and s be singular lines of V, which are both not contained in $V^\perp$. Then the following hold:

   (a) $[T_\ell, T_s] = 1$ if $s \subseteq \ell^\perp$.

   (b) $[T_\ell, T_s] = 1$ if $\ell \cap s \neq 0$.

   (c) $X = \langle T_\ell, T_s \rangle \simeq SL_2(k)$ with T_ℓ, T_s unipotent subgroups of X, if $s \cap \ell^\perp = 0$.

   (d) Suppose $\ell \cap s = 0$ and $\dim(s \cap \ell^\perp) = 1$. Then $g = (s \cap \ell^\perp) \oplus (\ell \cap s^\perp)$ is a singular line and one of the following holds:

   (i) $g \subseteq V^\perp$ and $[T_\ell, T_s] = 1$.

   (ii) $g \not\subseteq V^\perp$ and

$$[T_\ell, T_s] = [t, T_s] = [T_\ell, \tau] = T_g \text{ for all } t \in T_\ell^\#, \tau \in T_s^\#.$$

Proof. (a) is an immediate consequence of the definition of T_ℓ. Let in (b) $A = [T_\ell, T_s]$ and $P = \ell \cap s$. Then the 3-subgroup lemma implies that A centralizes $P^\perp$ and $[V, A] \subseteq P$. Hence we may assume $P \not\subseteq V^\perp$. Pick now a singular vector $v \in V - P^\perp$ and $1 \neq a \in A$. Then $v^a = p + v$ for some $0 \neq p \in P$. But

$$0 \neq (p, v) = q(p + v) = q(v) = 0,$$

a contradiction. Thus $A = 1$, which proves (b).

Next assume $s \cap \ell^\perp = 0$, as in the hypothesis of (c). Then $W = \ell + s$ is a non-degenerate 4-dimensional orthogonal space of Witt index 2 and $V = W \perp W^\perp$. Hence

$$X = X|_W \leq SO^+(4, k) \simeq SL_2(k) * SL_2(k).$$

Pick now points P and Q in ℓ and R, S in s with $\ell = P + Q$, $s = R + S$ and $R \subseteq P^\perp, S \subseteq Q^\perp$. Then

$$W = (P \oplus R) \oplus (Q \oplus S)$$

and, by definition of T_ℓ and T_s, X induces the $SL_2(k)$ on $P \oplus R$ and on $Q \oplus S$. Let now X_1 be the kernel of the action of X on $P \oplus R$ and X_2 the kernel of the action of X on $Q \oplus S$. Then $X_1 \cap X_2 = \mathrm{id}$. Moreover, if $X_1 \not\leq X_2$ then

$$X_1 = X_1/X_1 \cap X_2 \simeq X_1 X_2/X_2 = X/X_2 \simeq SL_2(k),$$

since $X_1 X_2 \trianglelefteq X$. But then X_1 would contain transvections with center Q and axis $P \oplus R \oplus Q$ on W, which is obviously not the case. With symmetry we obtain $X_1 = X_2 = \mathrm{id}$. Hence $X \simeq SL_2(k)$ and (c) holds.

Finally in (d) $\ell \cap s = 0$ and $\dim(s \cap \ell^\perp) = 1$ implies $\dim(\ell \cap s^\perp) = 1$ and g is a singular line. If $g \subseteq V^\perp$, then $\ell \cap V^\perp \neq 0 \neq s \cap V^\perp$ and $[V, T_\ell] \leq \ell \cap V^\perp$, $[V, T_s] \leq s \cap V^\perp$. Hence the 3-subgroup lemma implies $[T_\ell, T_s] = 1$. So we may assume $g \not\subseteq V^\perp$. Let $P = s \cap \ell^\perp$ and $Q = \ell \cap s^\perp$ and $A = [T_\ell, T_s]$. Then $P^\perp = s^\perp + \ell, Q^\perp = \ell^\perp + s$ and we have

$(*)$
$$[P^\perp, T_\ell, T_s] \leq [\ell, T_s] \leq P$$
$$[P^\perp, T_s, T_s] \leq [P, T_\ell] = 0.$$

Hence $[P^\perp, A] \leq P$ and similarly $[Q^\perp, A] \leq Q$. If now $P^\perp + Q^\perp = V$, then (3)(c) implies $A \leq T_g$. So assume $P^\perp = Q^\perp = g^\perp$. Then the equations $(*)$ show $[g^\perp, A] = 0$. Moreover

$$[V, T_\ell, T_s] \leq [\ell, T_s] \leq P$$
$$[V, T_s, T_\ell] \leq [s, T_\ell] \leq Q$$

so that $[V, A] \leq g$. On the other hand, since $g = (g \cap V^\perp) + P$, we have for a singular vector $w \in V - g^\perp$ and $a \in A^\#$ that $w^a = w + p + v$ for $p \in P$ and $v \in g \cap V^\perp$. Now

$$0 = q(w^a) = q(w) + q(p + v) + (w, p + v) = (w, p)$$

so that $p = 0$ and $[w, a] = v \in g \cap V^\perp$. This shows $[V, A] \leq g \cap V^\perp$ and again by (3)(c) $A \leq T_g$.

It remains to show that, if $g \not\subseteq V^\perp$, then

$$[\sigma, T_s] = [T_\ell, \tau] = T_g \text{ for each } \sigma \in T_\ell^\#, \tau \in T_s^\#.$$

For this assume $P \not\subseteq V^\perp$ and pick a vector $v \in V - P^\perp$. Then there exist points $R \subseteq v^\perp \cap g, S \subseteq v^\perp \cap s$.

By definition of T_g the map

$$\alpha \to [v, \alpha], \alpha \in T_g$$

is an isomorphism of T_g onto R. In particular, if $A_0 = [\sigma, T_s] \leq T_g$, then $R_0 = [v, A_0] \leq R$. On the other hand $[v, T_s] = S$. Moreover $P = P^\sigma \subseteq s^\sigma$ so that $[v, T_s^\sigma] = S_1 \neq S, S_1$ a point on s^σ. We obtain

$$[v, T_s T_s^\sigma] = S \oplus S_1 \text{ is a 2-space .}$$

On the other hand $T_s T_s^\sigma = T_s A_0$ and $[v, T_s A_0] = S \times R_0$, so that R_0 must be a 1-space, whence $R_0 = R$ and $A_0 = T_g$, which was to be shown. With symmetry this proves (d).   □

From (4) we obtain as a Corollary

(5) Let $\Sigma = \{T_\ell \mid \ell$ a singular line of V with $\ell \not\subseteq V^\perp\}$. Then Σ is a set of k-root subgroups of the normal subgroup $G = \langle \Sigma \rangle$ of $O(V, q)$. Moreover, if $\Delta = \{T_\ell \in \Sigma \mid \ell \cap V^\perp \neq 0\}$, then $\langle \Delta \rangle$ is an abelian normal subgroup of G.

Since $(T_\ell)^\sigma = T_{\ell^\sigma}$, $\sigma \in O(V, q)$, Σ is closed under conjugation. Now we exhausted in (4) all possibilities for pairs T_ℓ, T_s. Hence Σ is a set of k-root subgroups of G.

Clearly also $\Delta^g \subseteq \Delta$ for $g \in G$, since $O(V, q)$ leaves $V^\perp$ invariant. Hence $\langle \Delta \rangle$ is by (4)(b) and (d)(i) an abelian normal subgroup of G.

(6) Suppose $\mathrm{Rad}V = V^\perp \cap q^{-1}(0) = 0$ and $\dim V > 4$. Then $\mathcal{F}(\Sigma)$ is connected.

Proof. We show first:

(∗)   If ℓ and g are singular lines with $\ell \cap g \neq 0$, then there exists a singular line s with $\ell^\perp \cap s = 0 = g^\perp \cap s$.

Indeed let $Q = \ell \cap g$ and P a singular point not in $Q^\perp$. Then $W = Q^\perp \cap P^\perp$ is a non-degenerate orthogonal space of Witt index ≥ 1 and $\dim W > 2$. Hence if $\ell = Q + S$, $g = Q + R$ with $S + R \subseteq W$, then there exists a singular point F in W with $F \not\subseteq S^\perp$ and $F \not\subseteq R^\perp$. Hence the line $s = P + F$ satisfies $s \cap \ell^\perp = 0 = s \cap g^\perp$.

Now given any two singular lines ℓ, g, if $\ell \cap g^\perp = 0$ then ℓ and g are adjacent in $\mathcal{F}(\Sigma)$ by (4)(c). Suppose $\ell \cap g^\perp \neq 0$. Pick a point $P \subseteq \ell \cap g^\perp$ and a point Q in g, with $Q \neq P$ if $P \subseteq g$. Let $m := P + Q$. Then, by (∗) applied to the pairs of lines (ℓ, m) and (m, g), we get that m has distance at most 2 in $\mathcal{F}(\Sigma)$ from each of ℓ and g. Hence ℓ and g have distance at most 4 in $\mathcal{F}(\Sigma)$. Thus, $\mathcal{F}(\Sigma)$ is connected (with diameter $d \leq 4$).   □

Finally a proposition which allows us to determine an element of the projective orthogonal group as an image of a Siegel transvection.

(7) **Proposition.** Suppose V is a non-degenerate orthogonal space of Witt index ≥ 3 with quadratic form $q : V \to k$. (Nondegenerate means $V^{\perp} \cap q^{-1}(0) = 0$!) Denote by $\pi : O(V, q) \to PO(V, q)$ the natural homomorphism of the orthogonal group onto the projective orthogonal group and let $\sigma \in PO(V, q)$ satisfying:

(i) There exists a singular line ℓ such that σ fixes each singular point in $\ell^{\perp}$.

(ii) If P is a point on ℓ, then σ fixes each singular line through P.

Then there exists a $\varphi \in T_{\ell}$ with $\sigma = \pi \varphi$.

Proof. Let E be a singular plane through P with $E \not\subseteq \ell^{\perp}$. Then $s = \ell^{\perp} \cap E$ is a line, which is fixed pointwise by σ. Moreover each line g in E through P is fixed by σ. Hence σ induces a transvection with axis s and center P on E.

Let $\varphi \in O(V, q)$ be a coimage of σ. Then there exists a basis of E such that $\varphi \mid_E$ has matrix

$$\begin{pmatrix} c & & a \\ & c & \\ & & c \end{pmatrix} ; c \in k^*, a \in k$$

with respect to this basis. Now the diagonal entry c of this matrix is already determined by the action of φ on P. Since $P^{\perp}$ is generated by such planes, this implies

$$\varphi \mid_{P^{\perp}} = c \cdot \mathrm{id}_{P^{\perp}} + \alpha, \qquad \alpha \in \mathrm{Hom}(P^{\perp}, P).$$

Since $\varphi \mid_{\ell^{\perp}} \in O(\ell^{\perp})$ we have $c = \pm 1$. On the other hand, if $\ell = P + Q$ then $V = P^{\perp} + Q^{\perp}$, so that multiplying with $-\mathrm{id}_V \in O(V)$ if necessary, we may assume

$$\varphi \mid_{P^{\perp}} = \mathrm{id}_{P^{\perp}} + \alpha; \varphi \mid_{Q^{\perp}} = \mathrm{id}_{Q^{\perp}} + \beta; \qquad \beta \in \mathrm{Hom}(Q^{\perp}, Q).$$

Now (3)(c) implies $\varphi \in T_{\ell}$.   □

(1.6) Exercises

(1) [Tim99, 11.5] Let $G = T(V, W^*)$ and $\Sigma = \Sigma(W^*)$, where W^* is a subspace of V^* with $\mathrm{Ann}_V(W^*) = 0$ in the notation of (1.3). Fix a point P of V and a point Φ of W^* with $P \subseteq \mathrm{Kern}\ \Phi$ and let

$$\Lambda \quad := \quad \{T_{P,\Psi} \mid \Psi \subseteq W^* \text{ with } P \subseteq \mathrm{Kern}\ \Psi\}$$
$$\Lambda^* \quad := \quad \{T_{Q,\Phi} \mid Q \subseteq \mathrm{Kern}\ \Phi\}.$$

Show:

(a) Λ and Λ^* are sets of imprimitivity under the conjugation action of G on Σ.

(b) $T = \bigcup D \cup \{1\}, D \in \Lambda$ and $T^* = \bigcup D \cup \{1\}, D \in \Lambda^*$ are subgroups of G.

(c) $T \cap T^* = T_{P,\Phi}$.

(d) Let $g \in G$ with $T^g \neq T$. Let $Y = \langle T, T^g \rangle$, $N = N_T(T^g)N_{T^g}(T)$ and pick $A \in \Lambda - N_\Lambda(T^g)$, $B \in \Lambda^g - N_{\Lambda^g}(T)$ and set $X = \langle A, B \rangle$. Then the following holds:

(i) $N = N_T(T^g) \times N_{T^g}(T) \lhd Y$.

(ii) $Y = N \cdot X, X \simeq SL_2(K), K$ the division ring over which G is defined.

(iii) Suppose K is a field. Pick $C \in N_\Lambda(T^g)$. Use I(3.4) to show that $\langle C^X \rangle$ is a natural $\mathbb{Z}X$-module. (The last statement is true in general. But to obtain it in case K is non-commutative one needs to apply I(3.2), which is more difficult.)

(2) Use the notation of Example (1.4) and let $\Sigma = \{T_P \mid P$ an isotropic point of V not in $\mathrm{Rad}(V, f)\}$ and $G = \langle \Sigma \rangle$. For $A = T_P \in \Sigma$ set

$$\underline{A} = \{B \in \Sigma \mid C_\Sigma(A) = C_\Sigma(B)\}.$$

(a) Show $\underline{A} = \{T_Q \mid \mathrm{Rad}(V, f) + Q = \mathrm{Rad}(V, f) + P\}$. In particular if $\mathrm{Rad}(V, f) \cap q^{-1}(0)$ resp. $\mathrm{Rad}(V, f) = 0$ in case f is $(\sigma, -1)$ hermitian, then $\underline{A} = \{A\}$. ($q : V \to K/\Lambda$ is the pseudoquadratic form!)

(b) Let $\Sigma_{\underline{A}} = C_\Sigma(A) - \underline{A}$. Show that either $\Sigma_{\underline{A}} = \emptyset$ or $\Sigma_{\underline{A}}$ is a class of abstract transvection groups of $\langle \Sigma_{\underline{A}} \rangle$.
(The elements of $\Sigma_{\underline{A}}$ act as isotropic transvection groups on $P^\perp$!)

(c) For $B \in \Sigma_{\underline{A}}$ let $\ell_{\underline{A},\underline{B}} = \{\underline{C} \mid C \in \Sigma$ and $C \le Z(\langle C_\Sigma(A) \cap C_\Sigma(B) \rangle)$.

Show

(i) $\ell_{\underline{A},\underline{B}} = \ell_{\underline{B},\underline{A}}$.

(ii) $\ell_{\underline{A},\underline{B}} = \ell_{\underline{A},\underline{C}}$ for each $\underline{C} \in \ell_{\underline{A},\underline{B}}$ with $\underline{A} \neq \underline{C}$.

(iii) $C_\Sigma(A) \cap C_\Sigma(B) = C_\Sigma(A) \cap C_\Sigma(C) = C_\Sigma(B) \cap C_\Sigma(C)$ for each $\underline{C} \in \ell_{\underline{A},\underline{B}}$ with $\underline{A} \neq \underline{C} \neq \underline{B}$.

(3) Let V be an orthogonal space with non-degenerate quadratic form $q : V \to k$ of Witt index ≥ 3. Let E be a maximal totally singular subspace of V and

$$\Lambda = \{T_\ell \mid \ell \text{ a singular line in } E\},$$

T_ℓ the Siegel transvection group corresponding to ℓ. Let $\Sigma = \{T_\ell \mid \ell$ a singular line of $V\}$ and $G = \langle \Sigma \rangle$. Show

(a) $\Lambda \cup \{1\} = \langle \Lambda \rangle$ if and only if the Witt index of q is 3.

(b) Let $D(\Sigma) = \{d \in D^{\#} \mid D \in \Sigma\}$. Then, if the Witt index of q is 3,

$$\Lambda \cap \Lambda^d = \begin{cases} \Lambda \\ \emptyset \end{cases} \text{ for each } d \in D(\Sigma),$$

but Λ is not a set of imprimitivity under the conjugation action of G.

(c) If the Witt index of q is ≥ 4, then there exists a $d \in D(\Sigma)$ with

$$\Lambda \neq \Lambda \cap \Lambda^d \neq \emptyset.$$

(4) Suppose L is a field of Char 2 , W an L vector space and $f : W \times W \to L$ a non-degenerate symplectic form on W. Let $(w_i; i \in I)$ be an ordered basis of W and define a form g on W by

$$g(w_i, w_j) = \begin{cases} f(w_i, w_j) & i < j \\ 0 & i \geq j \end{cases}$$

and bilinear extension. Let $\widehat{K}$ be the algebraic closure of L and $K = \{c \in \widehat{K} \mid c^2 \in L\}$. Consider the L vector space $V = K \oplus W$ and define a quadratic form $q : V \to L$ by

$$q(c + w) = c^2 + g(w, w) \text{ for } c \in K, w \in W.$$

Show

(a) q is a non-degenerate quadratic form on V with associated bilinear form $\widehat{f}$ where:

$$\widehat{f}(c_1 + w_1, c_2 + w_2) = g(w_1, w_2) + g(w_2, w_1).$$

(b) $V^{\perp} = K$ (with resp. to $\widehat{f}$!) Moreover for $w_1, w_2 \in W$ we have

$$\widehat{f}(w_1, w_2) = 0 \text{ if and only if } f(w_1, w_2) = 0.$$

(5) Let the hypothesis and notation be as in (4). Show

(a) f and $\widehat{f}$ define the same polarity. (If one identifies V/K with W).

(b) $PSp(W, f) \simeq PO(V, q) \simeq PSp(V/K, \widehat{f})$. (Here PSp resp. PO is always the group induced by $Sp(W, f)$ resp. $O(V, q)$ on the projective space of W resp. V.)

(c) If dim $W \geq 4$, then there exists a class of abstract root subgroups of $PSp(W, f)$ the elements of which correspond to the isotropic lines of W.

§ 2 Basic properties of groups generated by abstract root subgroups

We assume in this section that Σ is a set of abstract root subgroups of the group $G = \langle \Sigma \rangle$. We start to develop the more elementary theory of such groups. For this we will need properties of rank one groups provided in chapter I. The results in this section are from [Tim99].

(2.1) Lemma

Let $A, C \in \Sigma$, $B \in \Omega_A$ and $D \in \Omega_C$. Suppose $X = \langle A, B \rangle \leq Y = \langle C, D \rangle$. Then $X = Y$ and

$$\{A, B, C, D\} \subseteq A^X.$$

Proof. Let $\Delta = C^Y$. If A commutes with each element of Δ, then $A \leq X \cap Z(Y) \leq Z(X)$, which is impossible since $A \in \Omega_B$. If $A \in \Psi_E$ for some $E \in \Delta$, then, as $[E, a] = [E, A] \in C_\Sigma(A)$, (1.1) ($\gamma$) implies

$$E \neq E^a \in \Delta \cap \Sigma_E \text{ for some } a \in A^\#,$$

a contradiction to $Y = \langle E, E^a \rangle$ by I (1.4). Hence there exists an $E \in \Delta \cap \Omega_A$. But then E and A are conjugate in $\langle E, A \rangle \leq Y$, whence $A \in \Delta$. Similarly $B \in \Delta$. Now, I(1.4) again implies $X = Y$ and (2.1) holds. □

(2.2) Lemma

Let $A \in \Sigma$, $B \in \Omega_A$, $X = \langle A, B \rangle$ and $1 \neq z \in C(X)$. Then $D(\Sigma) \cap z A^\# = \emptyset$.

Proof. Suppose false and let $d \in z A^\# \cap D(\Sigma)$ and $D \in \Sigma$ with $d \in D^\#$. Then $d = za, a \in A^\#$ and $B^d = B^a \in \Omega_B$. Hence $D \in \Omega_B$ and

$$Y = \langle B, D \rangle = \langle B, d \rangle = \langle B, B^d \rangle = X$$

by I(1.4) (or (2.1)!). Hence $D \in A^X$, which is again by I(1.4) a contradiction to $D \neq A$ and $[A, D] = 1$. □

As a direct consequence of (2.2) we obtain:

(2.3) Corollary

There exists no triple $A, B, C \in \Sigma$ with

$$\underset{\substack{A \quad\quad B \\ \circ\!\!-\!\!\!-\!\!\!-\!\!\circ}}{} \sim \underset{\substack{C \\ \circ}}{}$$

(2.4)   Lemma

Suppose Σ is degenerate. Let $a, b \in D(\Sigma)$ with $ab = ba \neq 1$. Then $C_\Sigma(ab) = C_\Sigma(a) \cap C_\Sigma(b)$.

Proof. If $C \in \Omega_A \cap C_\Sigma(ab)$, where $a \in A^\#$ and $b \in B^\#$, then

$$b \in D(\Sigma) \cap (ab)A^\#$$

and $1 \neq ab \in C(X), X = \langle A, C \rangle$, a contradiction to (2.2).   □

(2.5)   Lemma

Suppose Σ is degenerate. Let $A, C \in \Sigma$ with $C_\Sigma(A) \neq C_\Sigma(C)$ and $[A, C] = 1$. Then $D(\Sigma) \cap AC = A^\# \cup C^\#$.

Proof. Pick $D \in C_\Sigma(A) - C_\Sigma(C)$ resp. $\in C_\Sigma(C) - C_\Sigma(A)$. Then by (2.2)

$$\text{either}\quad D(\Sigma) \cap aC^\# = \emptyset \text{ for each } a \in A^\#,$$
$$\text{or}\quad D(\Sigma) \cap cA^\# = \emptyset \text{ for each } c \in C^\#.$$ □

(2.6)   Lemma

Suppose $\mathcal{F}(\Sigma)$ has no isolated vertices. Then $Z(G) = Z_2(G)$.

Proof. Suppose $x \in Z_2(G)$ and $A \in \Sigma$ with $[A, x] \neq 1$. Then there exists an $a \in A^\#$ with $1 \neq a^{-1}a^x \in Z(G)$. By hypothesis there exists a $B \in \Omega_A$, a contradiction to

$$a^x \in D(\Sigma) \cap (a^{-1}a^x)A^\#$$

by (2.2). Thus $Z_2(G)$ centralizes each element of Σ and so $Z_2(G) \leq Z(G)$. □

(2.7)   Lemma

Suppose $A \in \Sigma$ is not isolated in $\mathcal{F}(\Sigma)$. Then

$$A \cap B = 1 \text{ or } A = B \text{ for each } B \in \Sigma.$$

Proof. Pick $C \in \Omega_A$ and $B \in \Sigma$ with $A \cap B \neq 1$. Let $1 \neq a \in A \cap B$. Then $\langle A, C \rangle = \langle a, C \rangle \leq \langle B, C \rangle$ by I(1.4). In particular $B \in \Omega_C$ by I(2.1) and thus $\langle A, C \rangle = \langle B, C \rangle$ and $A = B$ by (2.1).   □

(2.8) Lemma

Let $A \in \Sigma$, $B \in \Psi_A$ and $C = [A, B]$. Then the following hold:

(1) If A is not isolated in $\mathcal{F}(\Sigma)$ then $A^B \cup C = \Sigma \cap AC$ is a partition of AC. In particular $A \in \Lambda_C$. Moreover for each $a \in A^{\#}$ the map

$$\chi_a : b \to [a, b], b \in B$$

is an isomorphism of B onto C.

(2) If A and B are not isolated in $\mathcal{F}(\Sigma)$ then $C \in \Lambda_A \cap \Lambda_B$.

Proof. Pick $a \in A^{\#}$. Then we have by definition of abstract root subgroups

$$[a, B] = C, \text{ whence } a^B = aC.$$

This shows that each element of AC is contained in some element of $A^B \cup C$. Now all elements of A^B are not isolated in $\mathcal{F}(\Sigma)$. Hence (2.7) implies that $A^B \cup C$ is a partition of AC.

Pick now $D \in \Sigma \cap AC$. If $D \cap A^b \neq 1$ for some $b \in B$, then $D = A^b$ by (2.7). Hence we may assume $D \leq C$ and then show $D = C$. Suppose $D < C$ and pick $E \in \Omega_A$. Then by (2.7) applied to C and D, $E \notin \Omega_C$ and by (2.2) $E \notin \Sigma_C$. Hence $E \in \Psi_C$. But then we obtain for $d \in D^{\#} \subset C^{\#}$ and $e \in E^{\#}$:

$$[D, E] = [D, e] < [C, e] = [C, E] = [d, E] = [D, E]$$

by (1.1) II(γ), since by (2.7) $C_C(e) \leq N(E)$ and thus $C_C(e) = 1$. This contradiction shows $D = C$ and $\Sigma \cap AC = A^B \cup C$. In particular $A \in \Lambda_C$.

Now the map χ_a is a surjective homomorphism of B onto C. If $1 \neq b \in$ Kern χ_a, then $a = a^b$ and thus $A = A^b$, a contradiction to $C = [A, b] \not\leq A$. This proves (1). (2) is an immediate consequence of (1) applied to AC and BC. □

(2.9) Lemma

Let $A, B, C \in \Sigma$ with

$$\overset{A}{\circ}\!\!\!\!\rule[0.5ex]{1.2cm}{0.4pt}\!\!\!\!\overset{B}{\circ}\!\!\!\!\rule[0.5ex]{1.2cm}{0.4pt}\!\!\!\!\overset{C}{\circ} \; .$$

Pick $b \in B^{\#}$ and let $a = a(b), c = c(b)$ in the notation of I(1.2). Then $x = ba^{-1}cb^{-1} \in C(B)$ and $A^x = C$.

Proof. We have $A^b = B^a$ and $C^b = B^c$ by I(1.2). Thus $B^{a^{-1}c} = B^{ca^{-1}} \in C_\Sigma(B)$ and $x \in C(B)$. Further

$$A^x = A^{ba^{-1}cb^{-1}} = B^{cb^{-1}} = C^{bb^{-1}} = C.$$ □

(2.10) Lemma

Let $A, B, C \in \Sigma$ with

$$\begin{array}{ccc} A & B & C \\ \circ\!\!=\!\!\!=\!\!\!=\!\!\circ\!\!=\!\!\!=\!\!\!=\!\!\circ \end{array} .$$

Set $R = \langle A, B, C \rangle$, $X = \langle A, C^b \rangle$ and $Y = \langle A^b, C \rangle$ for $b \in B^{\#}$. Then the following hold:

1. X and Y are both abelian or both nilpotent of class 2.
2. $[X, Y] = 1$, $XY \lhd R$ and $R = (XY)B$. Further R is nilpotent.
3. Suppose $\mathcal{F}(\Sigma)$ has no isolated vertex and $ac \in D(\Sigma)$ for some $a \in A^{\#}, c \in C^{\#}$. Then

$$[A, C^b] = X' = Y' = [A^b, C].$$

Proof. By (2.3) and (2.8) $C^b \notin \Omega_A$ and $A^b \notin \Omega_C$. Now $AA^b = A[A, B]$. Thus if C centralizes A^b for some $b \in B^{\#}$, then C centralizes $A[A, B]$ and so X and Y are both abelian. With symmetry this proves (1).

Clearly $[X, Y] = 1$. Since by the above $A^{\bar{b}} \leq XY$ and $C^{\bar{b}} \leq XY$ for each $\bar{b} \in B$ we have $XY \lhd R$. Hence $R = (XY)B$. Let $D = X'$ and $E = Y'$. Then by (1) either $D = 1 = E$ or $D, E \in \Sigma$ and $DE = Z(XY)$. In any case DE is B-invariant and thus $D \notin \Omega_B$ and $E \notin \Omega_B$. Hence it is easy to see that

$$1 \leq [DE, B] \leq DE \leq [XY, B] \leq XY \leq R$$

is a central series of R. This proves (2).

Finally assume $ac \in D(\Sigma)$ for some $a \in A^{\#}, c \in C^{\#}$. Then, since by (2) $\langle ac, B \rangle$ is nilpotent, it is not a rank one group. Hence $[ac, (ac)^b] = 1$ for each $b \in B^{\#}$. Hence

$$\begin{aligned} 1 &= [ac, a^b c^b] = [ac, c^b][ac, a^b]^{c^b} \\ &= [a, c^b]^c [c, c^b][a, a^b]^{cc^b} [c, a^b]^{c^b} \\ &= [a, c^b][c, a^b]. \end{aligned}$$

Thus $[a, c^b] = [c, a^b]^{-1}$. Now (2.7) implies $[A, C^b] = [C, A^b]$, whence (3) holds. $\qquad\square$

(2.11) Lemma

Let $A, B, C \in \Sigma$ with

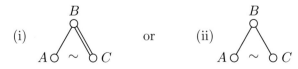

(i) B over $A \circ \sim \circ C$ or   (ii) B over $A \circ \sim \circ C$

Then the following hold:

(1) $|AC \cap \Psi_B| = 1$

(2) For each $D \in AC \cap \Sigma$ with $A \neq D \neq C$ we have $AC = AD = CD$.

Proof. Suppose first we are in case (i). If $C' \in AC \cap \Psi_B$, then by (2.10) $CC' \cap A = 1$ since $\langle A, CC' \rangle$ is nilpotent. Hence $C' \leq C$. Pick $b \in B^{\#}$ and let $E = [B, C]$. Then, as shown in (2.8)(1), the map

$$c \to [b, c]$$

is an isomorphism from C onto E. As $E = [b, C']$ by (1.1)II(γ) this implies $C = C'$. This shows that C is the only element of $AC \cap \Psi_B$. Hence, if $D \in AC \cap \Sigma$ with $A \neq D \neq C$ we have

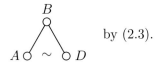

by (2.3).

Let $M = \{a(b)^{-1}d(b) \mid b \in B^{\#}\} \cup \{1\}$. Then by (2.9) $[B, B^m] = 1$ for each $m \in M$ and, since by I(1.2)

$$A^{\#} = \{a(b) \mid b \in B^{\#}\}, D^{\#} = \{d(b) \mid b \in B^{\#}\}$$

we have $AM = AD = DM$. In particular

$$M \leq \{x \in AD \mid [B, B^x] = 1\} \leq C.$$

(Otherwise $[B, B^x] = 1$ for some $x \in D' \in AC \cap \Omega_B$!) Now, also by (2.9), we have for $m = a(b)^{-1}d(b) \in C^{\#}$ and $x = b^m b^{-1} \in C(B) \cap C(C)$ that $A^x = D$, whence $AC = DC$. Moreover, as $m \in C$, we have

$$x \in [B, m] = [B, C] = E \in \Sigma.$$

By (2.3) applied to $\langle A, B \rangle$ and E (resp. $\langle A, E \rangle$ and C) we obtain $E \in \Psi_A$ and $[A, E] = [A, x] \leq AD \leq AC$. But since C is the only element of $AC \cap \Sigma$ centralizing E, we obtain

$$C = [A, E] = [A, x] \leq AD, \text{ whence } AC \leq AD.$$

Since $D \leq AM \leq AC$ this proves (2.11) in case (i).

In case (ii) let $M = \{a(b)^{-1}c(b) \mid b \in B^{\#}\} \cup \{1\}$. Then as in (i) $AM = CM = AC$. Let $b \in B^{\#}$ be fixed and $m = a(b)^{-1}c(b) \in D \in \Sigma$. ($D$ exists, since AC

is partitioned by $AC \cap \Sigma$.) As $[B^m, B] = 1$ we obtain $D \in \Psi_B$. Thus again by (2.9) we have for $x = b^m b^{-1} \in C(B) \cap C(D)$ that $A^x = C$. Hence $AD = CD$ and thus $AC \leq ADC \leq CD$. Since $m \in AC$ and AC is partitioned by $\Sigma \cap AC$, we also have $D \leq AC$ and thus

$$AC = CD = AD.$$

Now the case (i) treated already shows that (2.11) also holds in case (ii). □

(2.12) Lemma

Suppose Δ_1 and Δ_2 are different connectivity components of $\mathcal{F}(\Sigma)$ of cardinality greater than one. Then $B \notin \Lambda_A$ for each $A \in \Delta_1$ and $B \in \Delta_2$.

Proof. Suppose $A \in \Delta_1, B \in \Delta_2$ and $B \in \Lambda_A$. Pick $C \in \Delta_1 \cap \Omega_A, D \in \Delta_2 \cap \Omega_B$. Then by (2.3) $[C, B] \neq 1$ and $[A, D] \neq 1$. By (2.11)

$$(AB \cap \Sigma) - B \subseteq \Omega_C$$

and

$$(AB \cap \Sigma) - A \subseteq \Omega_D.$$

Now $|AB \cap \Sigma| \geq 3$, since $AB \cap \Sigma$ is a partition of AB. Hence $\Delta_1 \cap AB \cap \Delta_2 \neq \emptyset$, a contradiction to $\Delta_1 \cap \Delta_2 = \emptyset$. □

(2.13) Lemma

Suppose $|\Sigma| > 1$. Then the following are equivalent:

(1) $\mathcal{F}(\Sigma)$ is connected.
(2) Σ is a conjugacy class in G and there exists an $A \in \Sigma$, which is not isolated in $\mathcal{F}(\Sigma)$.

Proof. It is obvious that (1) implies (2), since if $X = \langle A, B \rangle$ is a rank one group, then A and B are conjugate in X.

Suppose now (2) holds. If $\mathcal{F}(\Sigma)$ contains isolated vertices, then all vertices of $\mathcal{F}(\Sigma)$ are isolated, contradicting (2). Let $\Delta_1 \neq \Delta_2$ be connectivity components of $\mathcal{F}(\Sigma)$. If $A \in \Delta_1$ and $B \in \Delta_2 \cap \Psi_A$, then by (2.8) $C = [A, B] \in \Lambda_A \cap \Lambda_B$. Hence (2.12) implies $C \in \Delta_1 \cap \Delta_2 = \emptyset$, since each connectivity component of $\mathcal{F}(\Sigma)$ has cardinality greater than 1.

This shows that different connectivity components of $\mathcal{F}(\Sigma)$ centralize each other. Since $G = \langle \Sigma \rangle$ and Σ is a conjugacy class in G this implies $\mathcal{F}(\Sigma)$ connected. □

(2.14)   Theorem

Suppose $|\Sigma| > 1$ and the following holds:

(1) $\mathcal{F}(\Sigma)$ is connected.
(2) There exists no pair $A \neq C \in \Sigma$ with $C_\Sigma(A) = C_\Sigma(C)$.
(3) If Σ is degenerate, then $|A| > 3$ for $A \in \Sigma$.
(4) If $A \in \Sigma$, $B \in \Omega_A$ then $a^{b(a)} = b(a)^{-a}$ for each $a \in A^{\#}$.

Then $G = G'A$, where $A \in \Sigma$ with $A \cap G' \neq 1$ and G' is quasi-simple. Moreover, in the following two cases $G = G'$:

(i) If Σ is non-degenerate.
(ii) If X is perfect for some rank one group $X = \langle A, B \rangle$; $A, B \in \Sigma$.

Remark. As will be shown (2.14) gives a uniform (quasi) simplicity criterion for most of the classical and Lie type groups. Moreover, if there exist different commuting elements in Σ, it will be shown in (2.15) and (3.15) with the help of I(2.5), that condition (4) of (2.14) follows from the other.

Proof of (2.14). If Σ is non-degenerate, then obviously $G = G'$ since Σ is by (1) and (2.13) a conjugacy class of G. If Σ is degenerate, then (3) and I(2.10) imply $G = G'A$ and $A \cap G' \neq 1$ for each $A \in \Sigma$. Let $N \lhd G'$. We claim

$$(*) \qquad\qquad\qquad N \leq Z(G).$$

First assume $A \cap N \neq 1$ for some $A \in \Sigma$. Let $B \in \Omega_A$ and pick $1 \neq a \in A \cap N$. Then, since a and $b(a)^{-1}$ are conjugate in G', also $b(a) \in B \cap N$ and thus A and B are conjugate in N. Hence the connectedness of $\mathcal{F}(\Sigma)$ implies that N is transitive on Σ and then $G = NN(A)$ by the Frattini argument. But then $NA \unlhd G$ and thus $G = NA$, since $G = \langle \Sigma \rangle$. In particular $G' \leq N$, contradicting the choice of N.

Hence $A \cap N = 1$ for each $A \in \Sigma$. If N normalizes each element of Σ, then N centralizes $A \cap G'$ for each $A \in \Sigma$. Now, again the connectedness of $\mathcal{F}(\Sigma)$ implies as above that $G' = \langle A \cap G' \mid A \in \Sigma \rangle$. Hence $N \leq Z(G')$. But, as $A \cap G' \neq 1$ and thus $Z(G') \leq N(A)$ by (2.7), we have $[A, Z(G')] \leq A \cap Z(G') = 1$, since $\mathcal{F}(\Sigma)$ is connected. Hence $N \leq Z(G') \leq Z(G)$, which is to show. Thus we may assume that there exists an $n \in N$ such that $C = A^n \neq A$ for some $A \in \Sigma$.

For $B \in \Sigma$ let $B_0 = B \cap G'$. Then clearly $C \notin \Psi_A$, since otherwise $1 \neq [C_0, A_0] \leq [C, A] \cap N$. If $C \in \Omega_A$, then $\langle A_0, C_0 \rangle \leq NA_0$ and thus by I (2.10) applied to $\langle A_0, C_0 \rangle$ we obtain $A_0 \cap N \neq 1$ if $|A| > 3$. So we may by (3) assume Σ is non-degenerate and $|A| \leq 3$. Hence there exists by (2.8) a $B \in \Lambda_A$.

By (2.11) and (2.3) we may assume $B \in \Psi_C$. We obtain

$$1 \neq [B_0, C_0] \leq [B_0, NA_0] \leq N,$$

a contradiction to $[B_0, C_0] \leq [B, C] \in \Sigma$.

This shows $[C, A] = 1$. Let $E \in C_\Sigma(A)$ and assume $E \notin C_\Sigma(C)$. If Σ is degenerate, then $E \in \Omega_C$ and we obtain by (3) and I(2.10) applied to $\langle E_0, C_0 \rangle$:

$$1 \neq E_0 \cap \langle E_0, C_0 \rangle' \leq E_0 \cap \langle E_0, NA_0 \rangle' \leq E_0 \cap N,$$

a contradiction. So Σ is non-degenerate. Clearly $E \notin \Psi_C$ since

$$[E_0, C_0] \leq [E_0, NA_0] \leq N.$$

Hence in this case also $E \in \Omega_C$. By (2.8) $\Lambda_C \neq \emptyset$. Hence by (2.3) and (2.11) there exists, as above, an $F \in \Lambda_C \cap \Psi_E$. This implies

$$1 \neq [F_0, E_0] \leq [F_0, NC_0] \leq N,$$

since $NA_0 = NC_0$ and since E_0 and C_0 are conjugate in $\langle E_0, C_0 \rangle \leq C(A_0)$ and thus $NC_0 = NE_0$. But this is a contradiction to $L = [F, E] \in \Sigma$ and $L \cap N = 1$.

We obtain $C_\Sigma(A) \subseteq C_\Sigma(C)$ and with symmetry $C_\Sigma(A) = C_\Sigma(C)$, a contradiction to (2). This proves $(*)$.

Now $(*)$ shows that each normal subgroup of G' lies in $Z(G)$.

If now $G'' < G'$ then $G'' < Z(G)$ and G is solvable, a contradiction to I(2.10) and (3) and (4). Thus in any case G' is quasi-simple. $\qquad \square$

The next proposition shows that in the non-degenerate case condition (4) of (2.14) is superfluous.

(2.15) Proposition

The following hold:

(1) Suppose $\mathcal{F}(\Sigma)$ has no isolated vertices. Let $A \in \Sigma$, $B \in \Omega_A$ and $C \in \Lambda_A \cap \Psi_B$. Then, if $X = \langle A, B \rangle$, $Y = \langle X, C \rangle$ and $N = \langle C^X \rangle$, we have $N \triangleleft Y$ and N is abelian.

(2) Suppose Σ is a non-degenerate class of abstract root subgroups of G, $A \in \Sigma$ and $B \in \Omega_A$. Then there exists a $C \in \Lambda_A \cap \Psi_B$. Moreover

$$a^{b(a)} = b(a)^{-a} \text{ for each } a \in A^\#.$$

Proof. Suppose Σ is a non-degenerate class of abstract root subgroups of G. Then there exists by (2.8) a $\overline{C} \in \Lambda_A$. By (2.3) $C \notin \Sigma_B$. Hence by (2.11) there exists a $C \in A\overline{C} \cap \Lambda_A \cap \Psi_B$.

Now, suppose that (1) holds. Then $C^X \subseteq \Sigma_A \cup \Psi_A$, whence $[C^x, A] \leq C_N(A)$ for each $x \in X$. This implies $[N, A] \leq C_N(A)$. Since clearly $[N, X, X] \neq 0$, (2) is now a consequence of I(2.5).

So (1) remains to be proved. Define recursively

$$\begin{aligned} C_1(B) &:= [C, B], \; C_1(A) := [C_1(B), A] \text{ and} \\ C_i(B) &:= [C_{i-1}(A), B] \text{ and } C_i(A) := [C_i(B), A]. \end{aligned}$$

Then we show first, by induction on i:

(a) $C_i(B) \in \Psi_A \cap \Lambda_B \cap \Lambda_{C_i(A)}$,
   $C_i(A) \in \Psi_B \cap \Lambda_A \cap \Lambda_{C_{i-1}(A)}$.

Moreover, for $i > 1$, $C_i(B) \in \Lambda_{C_{i-1}(A)}$.

Clearly by (2.8) $C_1(B) \in \Lambda_B \cap \Lambda_C$. Hence by (2.3) $C_1(B) \in \Psi_A$ and thus again by (2.8) $C_1(B) \in \Lambda_{C_1(A)}$. This also shows $C_1(A) \in \Lambda_A \cap \Lambda_{C_1(B)}$ and thus, again by (2.3) $C_1(A) \in \Psi_B$. This shows that (a) holds for $i = 1$. The proof that (a) holds for i, if it holds for $i - 1$, is the same. (The statement $C_i(B) \in \Lambda_{C_{i-1}(A)}$ follows directly from the definition of $C_i(B)$!)

Now pick elements $a_1, \ldots, a_n \in A^{\#}$; $b_1, \ldots, b_n \in B^{\#}$ and define for $i \leq n$ recursively:

$$\begin{aligned} M_1 &= \langle CC^{b_1} \rangle, N_1 = \langle M_1, M_1^{a_1} \rangle \text{ and } M_i = \langle N_{i-1}, N_{i-1}^{b_i} \rangle, \\ N_i &= \langle M_i, M_i^{a_i} \rangle. \end{aligned}$$

Then we show by induction on i that the following hold:

(b) $M_i = N_{i-1}C_i(B)$ (for $i > 1$) and $N_i = M_i C_i(A)$.
(c) M_i and N_i are abelian.
(d) M_i is B-invariant and N_i is A-invariant.

We have $M_1 = C[C, b_1] = CC_1(B) = N_0 C_1(B)$ for $N_0 = C$. In particular M_1 is abelian and B-invariant. Now

$$\begin{aligned} N_1 &= CC_1(B)C_1(B)^{a_1} = CC_1(B)[C_1(B), a_1] = CC_1(B)C_1(A) \\ &= M_1 C_1(A) \end{aligned}$$

by (a). Moreover, since $C_1(B)^{a_1}$ commutes with C, N_1 is by (a) abelian. Further, as $C_1(A) = [C_1(B), A] = [M_1, A]$, N_1 is A-invariant. Hence (b)–(d) hold for $i = 1$.

Now suppose $i > 1$ and (b)–(d) hold for $i - 1$. Then by (a), (b) and (d):

$$
\begin{aligned}
M_i &= N_{i-1}N_{i-1}^{b_i} = M_{i-1}C_{i-1}(A)C_{i-1}(A)^{b_i} \\
&= M_{i-1}C_{i-1}(A)[C_{i-1}(A), b_i] = M_{i-1}C_{i-1}(A)C_i(B) \\
&= N_{i-1}C_i(B),
\end{aligned}
$$

since by definition $C_i(B) = [C_{i-1}(A), B] = [C_{i-1}(A), b_i]$. As M_{i-1} is B-invariant, this equation also shows that M_i is B-invariant. Finally, since N_{i-1} is abelian by induction assumption and since $C_{i-1}(A)^{b_i}$ centralizes M_{i-1} by (d) and $C_{i-1}(A)$ by (a), this equation also implies that M_i is abelian. Hence (b)–(d) hold for M_i.

Similarly we obtain:

$$
\begin{aligned}
N_i &= M_iM_i^{a_i} = N_{i-1}C_i(B)C_i(B)^{a_i} = N_{i-1}C_i(B)[C_i(B), a_i] \\
&= N_{i-1}C_i(B)C_i(A) = M_iC_i(A)
\end{aligned}
$$

and N_i is A-invariant and abelian, since

$$
C_i(A) = [C_i(B), A] \text{ and } N_{i-1} \text{ is } A\text{-invariant}
$$

and since $C_i(B)^{a_i}$ centralizes $C_i(B)$ by (a). This shows that (b)–(d) hold.

Let now $x \in X$. Then x is a product of a finite number of $a's \in A$ and $b's \in B$. Hence (c) implies $[C, C^x] = 1$. This in turn shows that $\langle C^X \rangle$ is abelian. But then $N = \langle C^X \rangle = \langle C^Y \rangle$ and (1) holds. $\qquad\square$

As a corollary to (2.14) we obtain a (quasi)-simplicity proof for most of the classical groups.

(2.16) Corollary

The following groups are quasi-simple:

(a) The groups $T(V, W^*)$ of (1.3), when $\dim V \geq 3$.

(b) The groups G of (1.4) generated by isotropic transvections, in the case when the pseudoquadratic form $q : V \to K/\Lambda$ resp. the $(\sigma, -1)$-hermitian form $f : V \times V \to K$ is non-degenerate and $|L| > 3$. ($L = \{c \in K \mid c^\sigma = c \}$ in the second case and $L = \{c \in K \mid \text{There exists an } r \in \text{Rad}(V, f) \text{ with } q(r) = c + \Lambda\}$ in the first.

(c) The groups $\Omega(V, q)$ of (1.5), where V is a vector space over some field k, $q : V \to k$ a non-degenerate quadratic form of Witt index ≥ 3 and $\Omega(V, q)$ is the (normal) subgroup of $O(V, q)$ generated by all Siegel transvections.

(d) The groups $\Omega(V, q)$ of (c) also in case when the Witt index of q is equal to 2, but $|k| > 3$ and $\dim V > 4$.

Proof. Let G be one of the groups of (a)–(d) and $\Sigma = \Sigma(W^*)$ in case (a), Σ the set of isotropic transvection groups of G in (b) and Σ the set of Siegel transvection groups of $O(V, q)$ in (c) and (d). Then by (1.3), (1.4) and (1.5) Σ is in any case a set of abstract root subgroups of G with $\mathcal{F}(\Sigma)$ connected. Moreover, in case (a) and (c) Σ is non-degenerate, while in case (b) and (d) Σ is degenerate. Since $|L| > 3$ resp. $|k| > 3$, Σ satisfies (2.14)(3). Moreover in any case Σ satisfies (2.14)(4), since the rank one groups are isomorphic to $SL_2(K)$ resp. to $SL_2(L)$ as defined in I(1.5). Since $SL_2(L)$ is by I(1.5) perfect, G is also perfect. Thus it remains by (2.14) to show that $C_\Sigma(A) = C_\Sigma(B)$ for $A, B \in \Sigma$ implies $A = B$.

Now suppose $C_\Sigma(A) = C_\Sigma(B)$. Then in case (a) A and B are transvection groups corresponding to the same points P of V and Φ of W^*, whence $A = T_{P,\Phi} = B$. In case (b) we have $[V, A]^\perp = [V, B]^\perp$. Hence, since q resp. f is non-degenerate, we have $[V, B] + V^\perp = [V, A] + V^\perp$. But when A and B are defined with respect to the pseudoquadratic form q, then $[V, A]$ is the only singular point in $[V, A] + V^\perp$, since $q(v) \neq 0$ for $0 \neq v \in V^\perp$. Hence $[V, A] = [V, B]$ and $A = B$.

Finally in case (c) and (d) we also have

$$[V, A] + V^\perp = [V, B] + V^\perp$$

and $[V, A]$ is the only singular line in $[V, A] + V^\perp$. Hence $A = B$. Now (2.14) implies (2.16). $\square$

(2.17) Lemma

Suppose Σ is a set of abstract root subgroups of G. Then the following hold:

(1) If U is a Σ-subgroup, then $U \cap \Sigma$ is a set of abstract root subgroups of U.

(2) Suppose $\mathcal{F}(\Sigma)$ is connected and $- : G \to \overline{G}$ is a homomorphism of G. Then $\overline{\Sigma} = \{\overline{A} \in \Sigma\}$ is a set of abstract root subgroups of $\overline{G}$ with $\mathcal{F}(\overline{\Sigma})$ connected or $\overline{G} = \overline{A}$ for $A \in \Sigma$.

Proof. (1) is completely obvious. To prove (2) let $N = \mathrm{Kern}\ -$. Then either $\Sigma \subseteq N$ in which case $\overline{G} = 1$ or $\Sigma \cap N = \emptyset$, since Σ is a conjugacy class in G. We may assume that the second case holds. If $A \cap N \neq 1$ for some $A \in \Sigma$ pick $1 \neq a \in A \cap N$ and $B \in \Omega_A$. Then $\langle A, B \rangle = \langle a, B \rangle \leq NB = NA$ by I (1.4). Hence $G = NA$, which was to be shown.

So we may assume $A \cap N = 1$ for each $A \in \Sigma$. If now $B \in \Psi_A$, and $C = [A, B]$ we have for $a \in A^\#$:

$$[\overline{a}, \overline{B}] = \overline{[a, B]} = \overline{[A, B]} = \overline{C},$$

which shows that $\overline{\Sigma}$ is a set of abstract root subgroups of $\overline{G}$. $\square$

(2.18) Lemma

Let $A \in \Sigma$, $B \in \Psi_A$, $C = [A, B]$ and $D \in \Psi_C$ with $[D, C] = A$. Set $X = \langle B, D \rangle$ and $N = AC$. Then the following hold:

(1) X is a rank one group. Further $\overline{X} = X/C_X(N) \simeq SL_2(K)$, K a division ring or a Cayley division algebra and N is the natural $\overline{X}$-module. (See I(1.8)!)

(2) $A^X = C^X = A^B \cup C = C^D \cup A$ is a partition of N.

(3) $C_X(N) \leq Z(X)$.

Proof. We verify the hypothesis of I(3.2) for the action of $\overline{X}$ on N. Clearly N is abelian and X-invariant. Suppose $B \in \Psi_D$ and set $E = [B, D]$. Then E normalizes $C = C_N(B)$ and $A = C_N(D)$. Since $E \in \Sigma$ this implies $[N, E] = 1$. But then B normalizes $C_N(D) = A$, since $[B, D] \equiv 1 \bmod C_X(N)$. But this is impossible as $B \in \Psi_A$.

The same argument shows $B \notin \Sigma_D$, so that $B \in \Omega_D$ and X is a rank one group. Since each element $n \in N - C$ is of the form $n = ac, a \in A^\#, c \in C$ we have

$$[n, B] = [a, B] = [A, B] = C = C_N(B).$$

As $\overline{X}$ is also a rank one group with unipotent subgroups $\overline{B}$ and $\overline{D}$, this shows that the hypothesis of I(3.2) is satisfied. Hence (1) holds, (3) is now a consequence of I(1.10). Finally, as $[a, B] = C$ for $a \in A^\#$ we have

$$a^B = aC \text{ and thus } A^B \cup C \text{ is a partition of } N. \qquad \square$$

As final results of this section we determine the structure of Σ-subgroups of the form

○———○══○

and derive some consequences. Notice that, to be able to do this, we must have (2.15) first, since the proof depends on the fact that a rank one group $\langle A, B \rangle$, $A, B \in \Sigma$, is special.

(2.19) Proposition

Suppose $\mathcal{F}(\Sigma)$ contains no isolated vertices. Let $A, B, C \in \Sigma$ with

Set $X = \langle A, B \rangle$, $N = \langle C^X \rangle$ and $Y = \langle X, C \rangle$. Then the following hold:

(1) $Y = NX, N \lhd Y$ and $N \cap X = 1$.
(2) $\overline{X} = X/C_X(N) \simeq SL_2(K)$, K a division ring or a Cayley division algebra. Further $C_X(N) \leq Z(X)$.
(3) $N' = 1$, N is the natural $\mathbb{Z}\overline{X}$-module and $C^X = \Sigma \cap N$ is a partition of N.
(4) $\Sigma \cap AC = A^N \cup C$ is a partition of AC.

Proof. Pick $c \in C^{\#}$, $a \in A^{\#}$ and let $b = b(a) \in B^{\#}$ given by I(1.2). Then $a^b = b^{-a}$ by (2.15). Set

$$n = c^{-1}c^{b^{-1}}c^{-b^{-1}a}.$$

Then $n \in C_Y(A)$, since $n = c^{-1}n_0$, $n_0 = [(c^{b^{-1}})^{-1}, a] \in [C^{b^{-1}}, A]$ and since $C^{b^{-1}} \in \Psi_A$ by (2.3) applied to $C^{b^{-1}}$ and $\langle A, B^{b^{-1}} \rangle$. Moreover

$$(c^{-1})^{b^{-1}ab} = (c^{-1})^{a^b} = (c^{-1})^{b^{-a}} = (c^{-1})^{a^{-1}b^{-1}a} = (c^{-1})^{b^{-1}a}.$$

Hence $n \in C_Y(b)$. Since $X = \langle A, b \rangle$ by I(1.4) we obtain $n \in C_Y(X)$. Now

$$na^{-1} = c^{-1}(a^{-1})^{(c^{-1})^{b^{-1}}} = (ca)^{-(c^{-1})^{b^{-1}}} \in D(\Sigma) \cap nA^{\#}.$$

Hence $n = 1$ by (2.2) and thus

$$c^{b^{-1}a} = c^{-1}c^{b^{-1}} \quad \text{for all } c \in C^{\#}, a \in A^{\#} \text{ and } b = b(a).$$

Since $C[C, B] = CC^{b^{-1}}$ for arbitrary $b \in B^{\#}$ we obtain $C^{b^{-1}a} \leq C[C, B]$. This shows that $C[C, B]$ is A-invariant. Since it is also B-invariant, we obtain $N = C[C, B]$. As $[C, B] \in \Psi_A$ by (2.3) and as $C_N(A) = C$ we have $[C, B, A] = C$. This shows that for the action of X on N the hypothesis of (2.18) is satisfied. Now (1), (2) and (3) are consequences of (2.18). ($N \cap X = 1$ since $N \cap X$ is X-invariant, but X is irreducible on N!) But (4) follows immediately from (2.8). □

(2.20) Theorem

Suppose $\mathcal{F}(\Sigma)$ has no isolated vertices. Let $A, B, C \in \Sigma$ with

$$\underset{\circ}{A} \overset{}{\underset{\circ}{\quad}} \underset{\circ}{B} \overset{}{\underset{\circ}{\quad}} \underset{\circ}{C}.$$

Set $X = \langle A, B \rangle$, $Y = \langle X, C \rangle$, $M = \langle C^X \rangle$, $D = [C, B]$ and $N = \langle D^X \rangle$. Then the following hold:

(1) $M \lhd Y$, $Y = MX$ and $M = N \times C_M(Y)$.

(2) $\overline{X} = X/C_X(M) \simeq SL_2(K)$, K a division ring or a Cayley division algebra. Further $C_X(M) \le Z(X)$.

(3) $N \lhd Y$ and N is the natural $\mathbb{Z}\overline{X}$-module.

(4) $(\Sigma \cap M) \cup C_M(Y)$ is a partition of M. In particular $C \in \Lambda_A$ if and only if $C_M(Y) = 1$.

Proof. By (2.8) we have $D \in \Lambda_B \cap \Lambda_C$. Now by (2.3) applied to A, D, C resp. A, B, D we obtain

Further NC is X invariant by (2.19). Since $[N, C] = 1$ we have $M = NC$ and X centralizes M/N. As $C_X(N) \le Z(X)$ by (2.19) the 3-subgroup lemma implies

$$[M, C_X(N), X] = 1.$$

Hence $[M, C_X(N)] \le C_N(X) = 1$ and thus $C_X(N) = C_X(M)$. Now (2), (3) and the first part of (1) are direct consequences of (2.19). If now $C \in \Lambda_A$, then $N = M$ by (2.19), and so all parts of (2.20) hold. So we may assume $C \notin \Lambda_A$. Let $E = [D, A]$. Then $E \cap C = 1$ by (2.7). Since $E = C_N(A)$ this implies $N \cap C = 1$. Pick $b \in B^\#$. Then $[C, b] = D = [E, b]$. Hence $CE = E \times C_{CE}(b)$. As $X = \langle A, b \rangle$ we have $C_{CE}(b) \le C_M(X)$ and

$$NC_M(X) \ge NC_{CE}(b) = NEC_{CE}(b) = NC = M,$$

which shows $M = N \times C_M(X)$.

To prove (4) let $H = N_X(A) \cap N_X(B)$. Then it follows from the description of the natural $\mathbb{Z}\overline{X}$-module in I(1.3) that there exist elements $h(t) \in H, t \in K^\#$, which act on D by right-multiplication with t and on E by right-multiplication with t^{-1}, if one identifies N with the natural $\overline{X}$-module. Hence H acts transitively on $E^\#$ and $D^\#$. Now

$$C_M(A) = E \times C_M(Y) = E \times C$$

by the above. Let $em \in C_M(A), e \in E^\#, m \in C_M(Y)^\#$. Then $e'em \in C$ for some $e' \in E$. Since H centralizes m and acts transitively on $E^\#$ this implies

$$(E^\#)m \subseteq D(C^H) \subseteq D(\Sigma).$$

Since this holds for arbitrary $m \in C_M(Y)^\#$ we obtain

$$(E^\#)C_M(Y) \subseteq D(C^H) \cup D(E).$$

Now (3) and (2.7) imply (4). $\qquad\qquad\qquad\qquad\qquad\qquad\qquad\square$

(2.21) Lemma

Let $\Delta \subseteq \Sigma$ satisfying:

(a) $\Delta^g \subseteq \Delta$ for each $g \in R = \langle \Delta \rangle$.

(b) $[A, B] \in \Delta$ for all $A \in \Delta, B \in \Psi_A \cap \Delta$.

Let Ω be the set of isolated vertices of $\mathcal{F}(\Delta)$ and $\Delta_1 \neq \Delta_2$ two connectivity components of $\mathcal{F}(\Delta)$ of cardinality > 1. Then the following hold:

(1) If $A \in \Omega, B \in \Delta \cap \Psi_A$, then $[A, B] \in \Omega$.

(2) If $A \in \Delta_1, B \in \Delta_2$ then $[A, B] = 1$ or $[A, B] \in \Omega$.

(3) Let $N = \langle \Omega \rangle$ and $\overline{R} = R/N$. Then $\overline{\Delta}_1$ is a class of abstract root subgroups of $\langle \overline{\Delta}_1 \rangle$. (With $\mathcal{F}(\overline{\Delta}_1)$ connected !)

(4) Δ_1 is a normal set in R.

Proof. (2). Let $1 \neq C = [A, B]$. Then by (2.8) $C \in \Lambda_A \cap \Lambda_B \cap \Delta$. Now (2) is a consequence of (2.12) applied to Δ. To prove (1) assume $C = [A, B] \notin \Omega$. Then there exists a $D \in \Omega_C \cap \Delta$. As $A^b \in \Omega$ for $b \in B^\#$, (2.10) implies that $\langle D, A, A^b \rangle$ is nilpotent, a contradiction to $C = [A, b] \leq AA^b$ and I(2.1).

To prove (3), notice that if $A \in \Delta_1, B \in \Delta_1 \cap \Psi_A$ then by (2.12) and (2.8) either $[A, B] \in \Omega$ or $[A, B] \in \Delta_1$. Now (3) is immediate. Finally to prove (4) let $A \in \Delta_1$ and $B \in \Psi_A \cap (\Delta - \Delta_1)$. Then by (2.8) $A^b \in \Lambda_A$ for $b \in B^\#$ and thence by (2.12) $A^b \in \Delta_1 \cup \Omega$. If now $A^b \in \Omega$, then $A \leq A^b[A, B]$, which is by (2) a contradiction to (2.10)(2) and I(2.1). Thus $A^b \in \Delta_1$, which proves (4).$\square$

(2.22) Lemma

Suppose Σ is a non-degenerate class of abstract root subgroups and let $A \neq B \in \Sigma$. Then the following hold:

(1) If $C_\Sigma(A) \subseteq C_\Sigma(B)$, then $C_\Sigma(A) = C_\Sigma(B)$.

(2) If $\Omega_A \subseteq \Omega_B$, then $[A, B] = 1$ and $\Omega_A = \Omega_B$.

(3) Let $\underline{A} = Z(\langle C_\Sigma(A) \rangle) \cap \Sigma$. Then

$$
\begin{aligned}
\underline{A} &= \{ D \in \Sigma \mid C_\Sigma(A) = C_\Sigma(D) \} = \{ D \in \Sigma \mid \Omega_A = \Omega_D \} \\
&= \{ D \in \Sigma \mid \Psi_A = \Psi_D \}.
\end{aligned}
$$

Proof. (1) Pick $E \in \Omega_B$. Then by (2.20) $E \in \Omega_A$. Pick $e \in E^\#$ and let $a = a(e), b = b(e)$ in the notation of I(1.2). Then

$$E^b = B^e \text{ and } E^a = A^e$$

whence

$$(*) \qquad\qquad C_\Sigma(E^a) = C_\Sigma(A^e) \subseteq C_\Sigma(B^e) = C_\Sigma(E^b).$$

Now by I(2.2) and (2.15)(3) $b^{-1} = b(e^{-1})$ and $a^{-1} = a(e^{-1})$ so that $E^{b^{-1}} = B^{e^{-1}}$ and $E^{a^{-1}} = A^{e^{-1}}$. Hence we obtain

$$(**) \qquad\qquad C_\Sigma(E^{a^{-1}}) = C_\Sigma(A^{e^{-1}}) \subseteq C_\Sigma(B^{e^{-1}}) = C_\Sigma(E^{b^{-1}}).$$

Now $(*)$ and $(**)$ together imply

$$C_\Sigma(E^{ab^{-1}}) \subseteq C_\Sigma(E) \subseteq C_\Sigma(E^{b^{-1}a}).$$

Since a and b^{-1} commute, this shows $C_\Sigma(E^a) = C_\Sigma(E^b)$, whence $C_\Sigma(A^e) = C_\Sigma(B^e)$ and thus (1) holds.

To prove (2) it suffices to show $[A, B] = 1$, since from this point on, picking $E \in \Omega_A$, the proof is the same as in (1). Clearly $B \notin \Omega_A$. So assume $B \in \Psi_A$. Then $C = [A, B] \in \Lambda_A \cap \Lambda_B$ by (2.8). Pick $D \in \Omega_C$. Then by (2.3) and (2.11) $D \in \Omega_{A^b}$ for some $b \in B$. Hence, replacing D by $D^{b^{-1}}$ if necessary, we may assume $D \in \Omega_C \cap \Omega_A$. By hypothesis also $D \in \Omega_B$. Since

$$\Omega_A = (\Omega_A)^a \subseteq (\Omega_B)^a = \Omega_{B^a} \text{ for each } a \in A,$$

and $BC \cap \Sigma = B^A \cup C$ by (2.8) this shows $D \in \Omega_E$ for each $E \in BC \cap \Sigma$, a contradiction to (2.11). This shows $B \in \Sigma_A$ and thus (2) holds.

Now the equation $\underline{A} = \{D \in \Sigma \mid C_\Sigma(A) = C_\Sigma(D)\}$ is a consequence of (1). Let $\Omega(A) = \{D \in \Sigma \mid \Omega_A = \Omega_D\}$. Then by (2) the elements of $\Omega(A)$ commute pairwise. Further, if $D \in \Omega(A)$ and $E \in C_\Sigma(A) \cap \Psi_D$, then $D^E \subseteq \Omega(A)$. If now $C = [D, E]$ and $F \in \Omega_C$ then there exist by (2.3) and (2.11) $e_1, e_2 \in E$ such that

$$F \in \Omega_{D^{e_1}} \cap \Psi_{D^{e_2}},$$

since by (2.8)(1) $C \cup D^E = \Sigma \cap CD$. But this is a contradiction to $\Omega_{D^{e_1}} = \Omega_A = \Omega_{D^{e_2}}$.

This shows $C_\Sigma(A) \subseteq C_\Sigma(D)$ for each $D \in \Omega(A)$ and thus $\Omega(A) \subseteq \underline{A}$ by (1). That $\underline{A} \subseteq \Omega(A)$ is a direct consequence of (2.20). This shows the second equation of (3). The third is a consequence of the other two. $\square$

(2.23) Lemma

Let $A \in \Sigma$ and $B \in \Lambda_A$ and suppose that $\mathcal{F}(\Sigma)$ has no isolated vertices. Then $R = \langle N_\Sigma(AB) \rangle$ induces the $SL_2(K)$ on AB, K a division ring or a Cayley division algebra, such that AB is the natural $SL_2(K)$-module. In particular R acts doubly transitively on $AB \cap \Sigma$.

Proof. We verify the hypothesis of (2.18). Pick $C \in \Omega_A$. Then there exist by (2.3) and (2.11) a $B_0 \in AB \cap \Psi_C$. Hence if $D = [B_0, C]$, then (2.19) implies $[AB, D] = B_0$ since $AB = AB_0$. Now, picking some $E \in \Omega_{B_0}$ we obtain as before that for $A_0 \in AB \cap \Psi_E$, $[AB, F] = A_0$ for $F = [A_0, E]$. Since $A_0 B_0 = AB$ (2.18) now implies that $X = \langle D, F \rangle$ induces the $SL_2(K)$ on AB. Especially X is doubly transitive on $\Sigma \cap AB$.

Now let $L \in N_\Sigma(AB) - C_\Sigma(AB)$. Then, since there exists a $B' \in AB \cap \Psi_L$ and since $[B', L] \in AB \cap \Sigma_L$, it follows that $[AB, L] \in \Sigma$. So we may without loss of generality assume $[AB, L] = A_0$ and then show that

$$(*) \qquad\qquad L \equiv F \mod C(AB).$$

(If $(*)$ is shown, we obtain $\langle N_\Sigma(AB) \rangle \le XC(AB)$, which remains to be shown!)

Now for $e \in L$ there exists by (2.18) an $f \in F$ such that $B_0^e = B_0^f$. Hence $[B_0, ef^{-1}] \le B_0 \cap A_0 = 1$ and $ef^{-1} \in C(AB)$. This proves $(*)$ and thus (2.23). $\qquad\square$

(2.24) Exercises

Suppose for all the exercises that $\mathcal{F}(\Sigma)$ has no isolated vertices.

(1) Let Σ be a non-degenerate set of abstract root subgroups of G. Show

    (a) If $A \in \Sigma$ and $B \in \Lambda_A$, then $\Omega_A \cap \Omega_B \ne \emptyset$ and $\Omega_A \cap \Psi_B \ne \emptyset$.

    Hint. Use (2.3), (2.11) and (2.23).

    (b) Show that the connectivity components of $\mathcal{E}(\Sigma)$ are contained in connectivity components of $\mathcal{F}(\Sigma)$.

    (c) Show: If Δ is a connectivity component of $\mathcal{E}(\Sigma)$ with $|\Delta| \ge 2$, then Δ is also a connectivity component of $\mathcal{F}(\Sigma)$.

    Hint. Use again (2.3) and (2.11).

    Finally conclude, that if $\mathcal{F}(\Sigma)$ is connected, then also $\mathcal{E}(\Sigma)$ is connected.

(2) Let $A, B, C, D \in \Sigma$ with

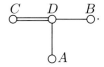

    and $\{A, B\} \subseteq \Lambda_C$ and $B \in \Lambda_A$.

    Let $N = C[C, D]$, $X = \langle A, D \rangle$ and $R = \langle A, B, C, D \rangle$. Show:

    (a) N is R-invariant. Moreover, there exists an $E \in AB \cap C_\Sigma(N)$.

Hint. Use (2.19) and (2.23).

(b) $E \in \Psi_D$. If $M = [E, D]E$, then $[N, M] = 1$.

(c) $N \cdot M \lhd R$ and $R = (N \cdot M)X$.

(d) N and M are natural modules for X.

(3) Let $A, B, C, D \in \Sigma$ with

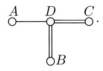

and $\{A, B\} \subseteq \Lambda_C, B \in \Lambda_A$ and $|AB \cap \Lambda_C| > 2$. Let $N = C[C, D], M = B[B, D], X = \langle A, D \rangle$ and $R = \langle A, B, C, D \rangle$. Then the following hold:

(a) $N \lhd R$ and $M \lhd R$. Further $[N, M] = 1$.

(b) $R = (NM)X$.

(c) N and M are natural modules for X.

Hint. Pick $B_1 \in AB \cap \Lambda_C$ with $A \neq B_1 \neq B$. Then we have by (2.19)

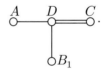

Now apply exercise (2).

(4) Let $A \in \Sigma$ with $\Lambda_A \neq \emptyset$ and pick $B \in \Omega_A$. Set $R = \langle \Lambda_A, B \rangle$ and $\Delta = B^R$. Then the following hold:

(a) $A \cup \Lambda_A \subseteq \Delta$, whence $\Delta = A^R$.

Hint. By (2.23) there exist $D, C \in \Lambda_A$ with $A = [D, C]$. Now apply (2.11) and (2.19).

(b) Δ is a connectivity component of $\mathcal{E}(\Sigma)$.

(c) If Σ is a conjugacy class in G, then $R = G$.

Hint: Use exercise (1) and (2.13)

(5) Let R be a group such that each element of R has order 2 or 3. Show that one of the following holds:

(a) R is an elementary abelian 2-group.

(b) Each element of R has order 3.

(c) $R = R'\langle t\rangle$, R' an elementary abelian 3-group and t an involution inverting R'.

(d) $R = R'\langle t\rangle$, R' an elementary abelian 2-group and t an element of order 3 acting fixed-point-freely on R'.

(6) Let G be a group generated by a class D of 3-transpositions. (i.e. $D = D(\Sigma)$, Σ a class of $GF(2)$-transvections!) Let $a \in D$ and U a subgroup of G with $[U, U^a] = 1 = U \cap U^a$. Then all elements of U have order 2 or 3.

(7) Let G be a group generated by the class D of 3-transpositions such that G is primitive on D. Show that G' is quasisimple.

Hint. One may assume $Z(G') = 1$ and $G = G'\langle d\rangle$, $d \in D$. If $N \trianglelefteq G'$ with $N \trianglelefteq G$, then $G = N\langle d\rangle$ and $G' = N$ by assumption. Hence one may apply (6) for each $M \triangleleft G'$.

(8)* Formulate and prove the same statement as in (7) for G generated by the class Σ of $GF(3)$-transvections.

Hint. Generalize (6) to subgroups U with $U \cap U^a = U^a \cap U^{a^2} = U^{a^2} \cap U = 1$ and $[U, U^a] = [U^a, U^{a^2}] = [U^{a^2}, U] = 1$ for $a \in D(\Sigma)$ and Σ a class of $GF(3)$-transvections.

§ 3 Triangle groups

We assume in this section that Σ is a set of abstract root subgroups of the group G. A *triangle* is a triple (A, B, C) of elements of Σ with

$$\overset{\displaystyle A}{\circ}\!\!-\!\!-\!\!-\!\!\overset{\displaystyle B}{\circ}\!\!-\!\!-\!\!-\!\!\overset{\displaystyle C}{\circ}$$

(and $A \neq C!$). The group $Y = \langle A, B, C\rangle$ is the corresponding *triangle group*. This notation, which is somewhat misleading since it would be more natural to call a triple with

a triangle, goes back to M. Aschbacher's first paper on odd transpositions [Asc72].

If Σ is degenerate, i.e. a set of abstract transvection groups, we say that Σ satisfies hypothesis (H), if for each pair $A \in \Sigma$, $B \in \Omega_A$ there exist $A_0 \leq$

$A, B_0 \leq B$ such that $X_0 = \langle A_0, B_0 \rangle \simeq (P)SL_2(k)$, k a commutative field with $|k| \geq 4$ and A_0, B_0 are full unipotent subgroups of X_0. Notice that if Σ is non-degenerate and $\mathcal{F}(\Sigma)$ connected $A \in \Sigma$ and $B \in \Omega_A$, then $X/Z(X) \simeq (P)SL_2(K)$ for $X = \langle A, B \rangle$ by (2.19) and (2.15)(1). Hence if $|A| \geq 4$ condition (H) holds anyway by I(2.12).

Together with (2.20) and (2.10) the determination of the structure of triangle groups yields a knowledge of all Σ-subgroups of G generated by 3-elements of Σ, which contain a commuting pair of elements of Σ.

(3.1) Lemma

Let (A, B, C) be a triangle. Then the following hold:

1. There exists a $t \in AC$ with $B \neq D = B^t \in C_\Sigma(B)$.
2. There exists a $\tau \in BD$ with $A^\tau = C$.
3. t and τ are unique in AC resp. BD with $B^t = D$ resp. $A^\tau = C$.

Proof. Pick $b \in B^\#$ and let $a = a(b) \in B^\#, c = c(b) \in C^\#$ with $A^b = B^a$ and $C^b = B^c$. Then $A^{ba^{-1}} = C^{bc^{-1}}$. Let $y = ba^{-1}cb^{-1} = (a^{-1})^{b^{-1}}c^{b^{-1}}$. Then $A^y = C$. By I(1.3) there exists an $e \in A$ with $A^{b^{-1}e} = B$. Set $D = C^{b^{-1}e}$. Then $[B, D] = 1$ since $[A, C] = 1$. Moreover $\tau = y^e \in BD$ and

$$A^\tau = A^{e^{-1}ye} = A^{ye} = C^e = C.$$

This implies (2). Further, if $c' = c(b^{-1})$, then $C^{b^{-1}} = B^{c'}$ and $D = B^{c'e}$. Hence for $t = ec' \in AC$ (1) holds.

If now $t' \in AC$ with $B^{t'} = D$, then $t't^{-1} \in N(B)$. Since $t't^{-1} \in C(A)$ and $t't^{-1}$ centralizes with each $a \in A^\#$ also the element $b(a) \in B^\#$, we obtain $\overline{t}'t^{-1} \in AC \cap C(Y)$. Hence (2.2) implies $t't^{-1} = 1$ and $t' = t$. The proof of the uniqueness of τ is the same. $\qquad\square$

(3.2) Lemma

Let (A, B, C) be a triangle and t, τ and D as in (3.1). Then the following hold:

1. $F := [A, \tau] = \{a^{-1}a^\tau \mid a \in A\} = \{a(b)^{-1}c(b) \mid b \in B^\#\} \cup \{1\}$ is the set of all elements $\sigma \in AC$ with $[B, B^\sigma] = 1$.
2. $L := [B, t] = \{b^{-1}b^t \mid b \in B\} = \{b(a)^{-1}d(a) \mid a \in A^\#\} \cup \{1\}$ is the set of all elements $\varphi \in BD$ with $[A, A^\varphi] = 1$.

Proof. Because of $a^{-1}a^\tau \overline{a}^{-1}\overline{a}^\tau = (a\overline{a})^{-1}(a\overline{a})^\tau$ it follows that $\{a^{-1}a^\tau \mid a \in A\}$ is a subgroup of AC. Hence $F = \{a^{-1}a^\tau \mid a \in A\}$. Since $\tau \in C(B)$, we have $a(b)^\tau = c(b), b \in B^\#$. Hence the equation in (1) holds.

Now $B^{a(b)} = A^b$ and $B^{c(b)} = C^b$ so that B and $B^{a(b)^{-1}c(b)}$ commute for each $b \in B^\#$. Hence $[B, B^\sigma] = 1$ for each $\sigma \in F$. Moreover $[B^\alpha, B^{\alpha\sigma}] = 1$ for $\alpha \in A$ since $\alpha\sigma = \sigma\alpha$. Suppose there exists a $\rho \in AC - F$ with $[B, B^\rho] = 1$. Then, as $AC = AF$, there exists a $1 \neq \alpha \in A$, $\sigma \in F$ with $\rho = \alpha\sigma$. Hence B^ρ commutes with

$$X = \langle B^\alpha, B \rangle = \langle A, B \rangle \ (\text{by I}(1.4)!) \ ,$$

which is impossible as $A^\rho = A$.
This shows (1). The proof of (2) is the same.   □

(3.3) Corollary

Assume the same hypothesis and notation as in (3.2). Then the following hold:

1. $t \in F, \tau \in L$.
2. $[B^\sigma, D] = 1 = [A^\varphi, C]$ for each $\sigma \in F, \varphi \in L$.

Proof. (1) is a direct consequence of (3.2). Now as $D = B^t, t \in F$ and F is abelian we obtain $[B^\sigma, D] = 1$ for each $\sigma \in F$.   □

(3.4) Lemma

Let (A, B, C) be a triangle, $Y = \langle A, B, C \rangle$ and t, τ, D and F, L as in (3.1), (3.2). Then $[F, L] \leq Z(Y)$.

Proof. Let $L_1 = [F, B]$ and $F_1 = [L, A]$. Then, since $t \in F$ and $L = [B, t] = [t, B] \leq L_1$ (as $L = L^{-1} = \{\ell^{-1} \mid \ell \in L\}$) we obtain $L \leq L_1$ and similarly $F \leq F_1$. Pick now $f \in F^\#$. Then by (3.2) $[B, B^f] = 1$ and, since (B, A, B^f) is a triangle, (3.2) implies that

$$\{b^{-1}b^f \mid b \in B\} = [B, f] = [f, B]$$

is the set of all elements $\sigma \in BB^f$ with $[A, A^\sigma] = 1$. Hence we obtain

(1) $\langle A^{[f,B]} \rangle$ is abelian for each $f \in F$.

Next we show

(2) $[A^{[f,b]}, C] = 1$ for each $f \in F, b \in B$.

Namely, as $C = A^\tau$ (2) holds if and only if

$$[A^{f^b\tau^{-1}}, A] = 1.$$

Now $f = a^{-1}a^\tau$ for some $a \in A$. Hence

$$f^b\tau^{-1} = a^{-b}a^{\tau b}\tau^{-1} = a^{-b}a^{b\tau}\tau^{-1} = a^{-b}\tau^{-1}a^b$$

and (2) holds if and only if

$$[A^{a^{-b}\tau}, A^{a^{-b}}] = 1.$$

Now $A^{a^{-b}} \in A^B \cup B$ by I(1.3) and, as $[A, A^\tau] = 1$ and $b\tau = \tau b$ also $[A^b, A^{b\tau}] = 1$ for each $b \in B$. This proves (2). Now by (3.2) $\langle B^F \rangle$ is abelian and F-invariant. Since $L_1 \leq \langle B^F \rangle$, L_1 is also abelian and F-invariant.

Next we show

(3) $[L_1, F] \leq Z(Y)$.

Since $L \leq L_1$ this proves (3.4). To prove (3) it suffices to show

$$[f_1, b, f_2] \in Z(Y) \text{ for all } f_1, f_2 \in F, b \in B.$$

Let $x = [f_1, b]$. Then by (1), (2) and since (2) also holds if one exchanges the roles of A and C, $T = \langle A, A^x, C, C^x \rangle$ is abelian. Now

$$[x, f_2] = f_2^{-x} f_2 \in F^x F \leq T.$$

Hence $[x, f_2] \in C(A) \cap C(C)$. On the other hand $x \in L_1$ and thus $[x, f_2] \in L_1 \leq C(B)$ by the above. This shows $[x, f_2] \in C(Y)$, which proves (3) and so (3.4). □

In fact (3) actually shows:

(3.5) Corollary

Let (A, B, C) be a triangle and choose notation as in (3.4). Then

$$[B, F, F] \leq Z(Y).$$

(3.6) Corollary

Let (A, B, C) be a triangle, $Y = \langle A, B, C \rangle$ and choose the notation as in (3.4). Then for each $f \in F^\#$ there exists a $b \in B^\#$ such that $[b, f] \in LZ(Y)$. Moreover, $B^\# = \{b \in B^\# \mid [b, f] \in LZ(Y) \text{ for some } f \in F^\#\}$.

Proof. By (3.2) $f = a(b)^{-1}c(b)$ for some $b \in B^\#$. Set $a = a(b), c = c(b)$. Then $A^b = B^a$ and $C^b = B^c$. Hence $A^{ba^{-1}cb^{-1}} = C$, and so we obtain for $\tau' = f^{-1}f^{b^{-1}}$ that $A^{\tau'} = C$ and $\tau' \in L_1 = [F, B]$. Now $L_1 \leq \langle B^F \rangle$ and the latter is by (3.2) abelian. Hence $L_1 \leq C(B)$. Thus, if $\varphi \in L_1$ with $A^\varphi = C$, then $\tau'\varphi^{-1} \in Z(Y)$. In particular $\tau'\tau^{-1} \in Z(Y)$ with τ as in (3.2). Hence

$$\tau' \in \tau Z(Y) \leq LZ(Y)$$

which proves the first part of (3.6). The second follows from

$$B^\# = \{b \in B^\# \mid a(b)^{-1}c(b) \in F^\#\}.$$ □

(3.7)   Lemma

Let (A, B, C) be a triangle, $Y = \langle A, B, C \rangle$ and assume $Z(Y) = 1$. Choose notation as in (3.2) and let $S = C_Y(AC) \cap C_Y(L), T = C_Y(BD) \cap C_Y(F)$. Then the following hold:

(1) $F \leq S, L \leq T$ and $S \cap T = 1$.
(2) If $\langle B^S \rangle \leq C(B)$ and $\langle A^T \rangle \leq C(A)$, then $[S, B] \leq T$ and $[A, T] \leq S$.

Proof. Since $Z(Y) = 1$, (1) holds by definition of S, T and (3.4). To prove (2) notice that $D \leq BL$, so that $\langle B^S \rangle$ is abelian and $\langle B^S \rangle \leq C(BD)$. Hence, as $F \leq S$,

$$(*) \qquad\qquad [S, B, F] \leq [S, B] \leq C(BD).$$

Pick $f \in F^\#$. Then there exists by (3.6) a $b \in B^\#$ such that $[f, b] \in L$. Hence $[f, b, S] = 1 = [f, S, b]$ and thus $[b, S] \leq C(f)$. Let $s \in S^\#$. Then (B, A, B^s) is a triangle. Since $S \leq C(A)$, s maps an element $b = b(a), a \in A^\#$ on the corresponding element of B^s. Hence, as $[b, s] = b^{-1} b^s$, (3.2) implies that $[A, A^{[b,s]}] = 1$. Now $A^{[b,s]} \leq C(f)$ by the above and $f = a^{-1}c$ for some $a \in A^\#$ and $c \in C^\#$ by (3.2). Hence $A^{[b,s]} \leq C(c)$ and thus $[C, A^{[b,s]}] = 1$. Now with symmetry also $[A, C^{[b,s]}] = 1$ and $[C, C^{[b,s]}] = 1$. Together we obtain:

$$[b, s, F] \leq \langle F^{[b,s]}, F \rangle \leq \langle (AC)^{[b,s]}, AC \rangle \leq C(A).$$

Hence by $(*)$ $[b, s, F] \leq Z(Y) = 1$. Now $s \in S^\#$ is arbitrary and b runs by (3.6) over $B^\#$ when f runs over $F^\#$. Hence $[b, s, F] = 1$ for each $b \in B^\#$ and $s \in S^\#$, which with $(*)$ implies that $[S, B] \leq T$. The proof of the second inequality in (2) is the same. $\qquad\square$

(3.8)   Remark

Suppose $Z(Y) = 1$ and B^Y is degenerate. (I.e. a set of abstract transvection groups of Y.) Then as $\langle B^S \rangle \leq C(L)$, (2.4) and (3.2) imply that $\langle B^S \rangle \leq C(B)$, since the elements of $L^\#$ are products of elements of $B^\#$ and $D^\#$. Hence the additional hypothesis of (3.7) (2) is satisfied. This will be used to determine the structure of a triangle group Y in the case when $\Sigma \cap Y$ is degenerate. But there are also other cases when one can verify this additional hypothesis.

(3.9)   Theorem

Let (A, B, C) be a triangle, $X = \langle A, B \rangle$, $Y = \langle X, C \rangle$ and suppose that A^Y is a set of abstract transvection groups of Y. Then there exists a normal subgroup N of Y satisfying:

1. $Y = NX$ and $N \cap X \leq Z(X)$.
2. $[N, A] \leq C_N(A)$ and $[N, A]$ is abelian.
3. Let $Z = [N, A] \cap [N, B]$ and $\overline{N} = N/Z$. Then $Z \leq Z(Y), \overline{N}' = 1$ and $\overline{N} = [\overline{N}, A] \oplus [\overline{N}, B]$ with $[\overline{N}, A] = [\overline{N}, a] = C_{\overline{N}}(a)$ for each $a \in A^{\#}$.

Proof. Let $\widetilde{Y} = Y/Z(Y)$. Then by (2.6) $Z(\widetilde{Y}) = 1$ and $(\widetilde{A}, \widetilde{B}, \widetilde{C})$ is a triangle. Let $\widetilde{S} = C_{\widetilde{Y}}(\widetilde{A}\widetilde{C}) \cap C_{\widetilde{Y}}(\widetilde{L})$, $\widetilde{T} = C_{\widetilde{Y}}(\widetilde{B}\widetilde{D}) \cap C_{\widetilde{Y}}(\widetilde{F})$, where F and L are as in (3.2) and $\widetilde{N}_1 = [\widetilde{T}, \widetilde{A}], \widetilde{N}_2 = [\widetilde{S}, \widetilde{B}]$. Then by (3.7) and (3.8) $\widetilde{N}_1 \cap \widetilde{N}_2 = 1$ and $\widetilde{N}_1 \leq \widetilde{S}, \widetilde{N}_2 \leq \widetilde{T}$. Further, as shown in the proof of (3.7) $\widetilde{N}_1$ and $\widetilde{N}_2$ are abelian. (As $\langle \widetilde{B}^{\widetilde{S}} \rangle$ is abelian!) Moreover, since

$$[\widetilde{N}_1, \widetilde{B}] \leq [\widetilde{S}, \widetilde{B}] = \widetilde{N}_2,$$

$\widetilde{N} = \widetilde{N}_1 \widetilde{N}_2$ is $\widetilde{X}$-invariant. By (3.2)(1) and (3.7)(1)

$$\widetilde{F} = [\widetilde{A}, \widetilde{\tau}] \leq [\widetilde{A}, \widetilde{L}] \leq [\widetilde{A}, \widetilde{T}] = \widetilde{N}_1$$

so that $\widetilde{Y} = \widetilde{N}\widetilde{X}$ and $\widetilde{N} \triangleleft \widetilde{Y}$.

Let N be the coimage of $[\widetilde{N}, \widetilde{X}]$. Then by (3.2) $FL \leq N$ and thus $NX = Y$. Further, as $\widetilde{N}_2 \leq C_{\widetilde{N}}(\widetilde{B}), \langle \widetilde{B}^{\widetilde{N}} \rangle \leq \langle \widetilde{B}^{\widetilde{S}} \rangle$ is abelian and thus by hypothesis $\langle B^N \rangle$ and $\langle A^N \rangle$ are abelian. Hence obviously $N \cap X \leq Z(X)$ and (1) and (2) hold.

Clearly $Z \leq Z(Y)$. Since $\overline{N} = [\overline{N}, X]$ by definition of N we have $\overline{N} = [\overline{N}, A][\overline{N}, B]$. As $[\overline{N}, A] = [\overline{N}, A]$ also $[\overline{N}, A] \cap [\overline{N}, B] = 1$ and $\overline{N} = [\overline{N}, A] \oplus [\overline{N}, B]$. Let $R = [\overline{N}, A] \cap C_{\overline{N}}(b)$ for some $b \in B^{\#}$. Then, as $X = \langle A, b \rangle, R \leq C_{\overline{N}}(X)$. Hence $R \leq [\overline{N}, A] \cap [\overline{N}, B] = 1$, since A and B are conjugate in X. This implies

$$[\overline{N}, B] = C_{\overline{N}}(b) \text{ for each } b \in B^{\#}.$$

If now $[\overline{N}, b] < [\overline{N}, B]$ for some $b \in B^{\#}$, then $\overline{N}_0 = [\overline{N}, A][\overline{N}, b] < \overline{N}$ and $\overline{N}_0$ is $X = \langle A, b \rangle$-invariant. But this is impossible as

$$[\overline{N}, B] = [\overline{N}, A, B] \leq N_0.$$

Since also $[\overline{N}, A] = [\overline{N}, a] = C_{\overline{N}}(a)$ for $a \in A^{\#}$ this proves (3). $\qquad\square$

(3.10) Corollary

Let (A, B, C) be a triangle, $X = \langle A, B \rangle$, $Y = \langle X, C \rangle$ and suppose that A^Y is degenerate. Then the following hold:

(1) X is a special rank one group.
(2) All elements of $A^{\#}, B^{\#}$ and $C^{\#}$ have the same order. Hence A, B and C are either torsion free or elementary abelian p-groups.

Proof. (1) is a consequence of I(2.5) and (3.9). To prove (2) we may assume $Z(Y) = 1$. Then by (3.1) there exists a unique $\tau \in L$ with $A^\tau = C$. Moreover by the proof of (3.1) τ is a conjugate of $a(b)^{-1}c(b)$ for an arbitrary $b \in B^\#$. Now by (1) b and $a(b)^{-1}$ and b and $c(b)^{-1}$ are conjugate in Y, so that

$$o(b) = o(a(b)^{-1}) = o(c(b)).$$

Since $AC = A \times C$ we obtain

$$o(\tau) = o(a(b)^{-1}c(b)) = o(b) \text{ for each } b \in B^\#. \qquad \square$$

(3.11) Theorem

Suppose Σ is a class of abstract root subgroups of G and $A \neq C \in \Sigma$. Then the following hold:

(1) If $C_\Sigma(A) \subseteq C_\Sigma(C)$, then $C_\Sigma(A) = C_\Sigma(C)$.

(2) If $\Omega_A \subseteq \Omega_C$, then $[A, C] = 1$ and $\Omega_A = \Omega_C$.

(3) Let $\underline{A} = Z(\langle C_\Sigma(A)\rangle) \cap \Sigma$. Then

$$\underline{A} = \{D \in \Sigma \mid C_\Sigma(A) = C_\Sigma(D)\}.$$

Proof. If Σ is non-degenerate then (3.11) is (2.22). So assume that Σ is degenerate. Then by exercise I(2.13)(7) there exists a $B \in \Sigma$ such that (A, B, C) is a triangle. Pick $b \in B^\#$ and let $a = a(b), c = c(b)$. Then $B^a = A^b, B^c = C^b$ whence

$$(*) \qquad C_\Sigma(B^a) = C_\Sigma(A^b) \subseteq C_\Sigma(C^b) = C_\Sigma(B^c).$$

Now by (3.10)(1) and I(2.2) $a^{-1} = a(b^{-1})$ and $c^{-1} = c(b^{-1})$ so that $A^{b^{-1}} = B^{a^{-1}}$ and $C^{b^{-1}} = B^{c^{-1}}$. Hence

$$(**) \qquad C_\Sigma(B^{a^{-1}}) = C_\Sigma(A^{b^{-1}}) \subseteq C_\Sigma(C^{b^{-1}}) = C_\Sigma(B^{c^{-1}}).$$

Now $(*)$ and $(**)$ together imply

$$C_\Sigma(B^{ac^{-1}}) \subseteq C_\Sigma(B) \subseteq C_\Sigma(B^{c^{-1}a}).$$

Since c^{-1} and a commute, this shows that we have equality in $(*)$ and $(**)$, whence (1) holds. (2) is equivalent to (1) when Σ is degenerate. Finally (3) is, as in (2.22), a consequence of (1).

(3.12) Lemma

Let (A, B, C) be a triangle and assume that Σ is degenerate. Suppose that $o(a) \geq 5$ for some $a \in A^{\#}$ and pick τ as in (3.1) with $A^{\tau} = C$. Then $A \neq C^{\tau} \neq C$ and

$$C_{\Sigma}(A) \cap C_{\Sigma}(C) = C_{\Sigma}(A) \cap C_{\Sigma}(C^{\tau}) = C_{\Sigma}(C) \cap C_{\Sigma}(C^{\tau}).$$

Proof. Choose the notation as in (3.9). Then by (3.9) $\tau \in L = BD \cap N$. Since by (3.9)(1) $D = B^{n}, n \in N$ and since $B \cap N = 1$, we have $\tau \in [B, n] \leq [B, N]$. By (3.10)(2) $o(\tau) = o(a) \geq 5$. Hence $\tau^{n} \notin Z$ for $n \leq 4$ and so by (3.1)(3) $A \neq C^{\tau} \neq C$ and $A^{\tau^{n}} \neq A$ for $n \leq 4$. By (3.4) $[F, \tau] \leq Z$ where Z is as in (3.9)(3) and $F = [A, \tau]$ by (3.2)(1). Hence

$$(+) \qquad\qquad R = AA^{\tau}A^{\tau^{2}} = AF[F, \tau] \text{ is } \tau\text{-invariant .}$$

Assume that there exists a $D \in (C_{\Sigma}(A) \cap C_{\Sigma}(C)) - C_{\Sigma}(C^{\tau})$. If $A^{\tau^{3}} \leq AC$ then clearly $A^{\tau^{4}} \leq CA^{\tau^{2}} = CC^{\tau}$. Hence $\langle D, C^{\tau} \rangle$ is a rank one group, $[D, C] = 1$ and

$$(A^{\tau^{4}})^{\#} \subseteq D(\Sigma) \cap C \cdot (C^{\tau})^{\#},$$

a contradiction to (2.2). Thus $A^{\tau^{3}} \not\leq AC$ but $A^{\tau^{3}} \leq R$. But, since $R = (AC) \times C^{\tau}$, $C^{\tau} \in \Omega_{D}$ and $AC \leq C(D)$, this is again a contradiction to (2.2). We have shown:

$$(*) \qquad\qquad C_{\Sigma}(A) \cap C_{\Sigma}(C) \subseteq C_{\Sigma}(R).$$

Now $(*)$ and $(+)$ imply that

$$C_{\Sigma}(C) \cap C_{\Sigma}(C^{\tau}) = C_{\Sigma}(R^{\tau}) = C_{\Sigma}(R)$$

which proves (3.12). $\qquad\qquad\square$

Remark. If Σ is degenerate and $o(a) \geq 5$ for some $a \in A^{\#}$, then (3.12) will show that C^{τ} is a "third point" on the line through A and C. This is an important step in the construction of a polar space.

If now $o(a) \leq 3$ for all $a \in A$ we need an additional hypothesis to prove the existence of this third point on the line through A and C.

(3.13) Notation

Suppose Σ is a set of abstract transvection groups of G. Then we say that condition (H) holds, if for $A \in \Sigma$ with $o(a) \leq 3$ for each $a \in A$ and each $B \in \Omega_{A}$, there exists an $A_{0} \leq A$, $B_{0} \leq B$ such that:

$$\langle A_{0}, B_{0} \rangle \simeq (P)SL_{2}(k), \ k \text{ a commutative field with } |k| \geq 4$$

and A_{0}, B_{0} are full unipotent subgroups of $\langle A_{0}, B_{0} \rangle$.

If $o(a) \leq 3$ for each $a \in A$, $A \in \Sigma$ we will prove in the next two lemmata a statement similar to (3.12) if condition (H) holds. The proof depends on I(3.4).

(3.14) Lemma

Let (A, B, C) be a triangle. Assume that $o(a) \leq 3$ for each $a \in A$ and that condition (H) holds. Choose notation as in (3.9) and let $A_0 \leq A$, $B_0 \leq B$ such that $\langle A_0, B_0 \rangle \simeq (P)SL_2(k)$. Let $\tau \in BD$ as in (3.1) with $C = A^\tau$ and set $C_0 = A_0^\tau$, $X_0 = \langle A_0, B_0 \rangle$ and $Y_0 = \langle X_0, C_0 \rangle$. Then the following hold:

(1) $Y_0 = N_0 X_0$, $N_0 = N \cap Y_0$.
(2) $N_0/N_0 \cap Z$ is a natural $\mathbb{Z}X_0$-module and $X_0 \simeq SL_2(k)$.

Proof. Let $\overline{N} = N/Z$ as in (3.9). Then by (3.9)(3)

$$\overline{N} = [\overline{N}, A_0] \oplus [\overline{N}, B_0] \text{ and } [\overline{N}, A_0] = C_{\overline{N}}(a), a \in A_0^\#.$$

Hence it is easy to see that the hypothesis of I(3.4) holds for the action of X_0 on $\overline{N}$. Hence by I(3.4) $\overline{N}_0 = \langle \overline{\tau}^{X_0} \rangle$ is a natural kX_0-module for some k action and $X_0 \simeq SL_2(k)$. Now, if we set $\overline{Y} = Y/Z$, then $\overline{Y}_0 \leq \overline{N}_0 \overline{X}_0$. Since clearly $\overline{N}_0 \overline{X}_0 = \langle \overline{C}_0, \overline{X}_0 \rangle$ this implies $\overline{Y}_0 = \overline{N}_0 \overline{X}_0$. Hence $Y_0 = N_0 X_0$ for $N_0 = \langle \tau^{X_0} \rangle$ and $N_0 = N \cap Y_0$. Since $N_0/N_0 \cap Z \simeq \overline{N}_0$ as X_0-module, this proves (3.14).□

(3.15) Lemma

Assume the same hypothesis and notation as in (3.14) and set $Z_0 = Z \cap N_0$. Then for each $n \in N_0$ with $A_0 \neq A_0^n \neq C_0$ one has:

(1) $\langle A_0^{N_0} \rangle = A_0 C_0 Z_0 = A_0 C_0 A_0^n$.
(2) $C_\Sigma(A) \cap C_\Sigma(C) = C_\Sigma(A) \cap C_\Sigma(A^n) = C_\Sigma(C) \cap C_\Sigma(A^n)$.

Proof. Since N_0/Z_0 is by (3.14) the natural X_0-module it is obvious that:

$$A_0 C_0 A_0^n \leq \langle A_0^{N_0} \rangle \leq A_0 C_0 Z_0.$$

Now let $N_1 = [N_0, A_0][N_0, B_0]$. Then $N_1 \triangleleft Y_0$ and

$$\widetilde{N}_0 = N_0/N_1 \leq Z(\widetilde{Y}_0), \widetilde{Y}_0 = Y_0/N_1.$$

Since $Z(X_0)$ inverts N_0/Z_0 we have $N_0 \cap X_0 = 1$ and thus

$$\widetilde{Y}_0 = \widetilde{N}_0 \times \widetilde{X}_0 = \langle \widetilde{C}_0, \widetilde{X}_0 \rangle \text{ with } \widetilde{C}_0 \leq \widetilde{N}_0 \times \widetilde{A}_0.$$

Now, as $\widetilde{A}_0^{\widetilde{Y}_0}$ is a set of abstract root subgroups of $\widetilde{Y}_0$, (2.2) implies $\widetilde{N}_0 = 1$ and $\widetilde{C}_0 = \widetilde{A}_0$. Hence $N_0 = N_1$ and thus $Z_0 \leq [N_0, A_0] \leq \langle A_0^{N_0} \rangle$. This implies $\langle A_0^{N_0} \rangle = A_0 C_0 Z_0$.

Let $H = N_{X_0}(A_0) \cap N_{X_0}(B_0)$. Then, since N_0/Z_0 is a natural X_0-module, H acts transitively on $(N_0/C_{N_0}(A_0))^\#$. Hence

$$(*) \qquad A_0^{N_0} = A_0 \cup \{A_0^{n^h} \mid h \in H\} = A_0 \cup (A_0^n)^H = A_0 \cup C_0^H.$$

Let $L = \langle A_0^{N_0} \rangle$. Then, since L is abelian and since each $h \in H - Z(X_0)$ acts fix-point-freely on L/Z_0, as $L = A_0(L \cap N_0)$, we have $L = [L, h] \cdot Z_0$. Now H is abelian (since k is commutative!) and thus normalizes $[L, h]$. This implies $[L, h] = [L, H]$ for each $h \in H - Z(X_0)$. Hence by $(*)$ above

$$L = A_0 \langle C_0^H \rangle = A_0 C_0 [C_0, H] = A_0 C_0 [L, h] = A_0 C_0 [C_0, h]$$

for each $h \in H - Z(X_0)$. We need to show that this equation also holds if h is the central involution of X_0. In this case h centralizes A_0 and Z_0 and inverts $L \cap N_0/Z_0$. Hence

$$[C_0, h] = [C_0 A_0 \cap N_0, h] = [L \cap N_0, h]$$

is inverted by h. Hence $[C_0, h]$ is H-invariant and $L \cap N_0 = [C_0, h] Z_0$. This implies that

$$C_0 A_0 [C_0, h] = A_0 [C_0, h](A_0 C_0 [C_0, h] \cap Z_0)$$

and thus is H-invariant. Hence $\langle C_0^H \rangle \leq A_0 C_0 [C_0, h]$ and by $(*)$ $L = A_0 C_0 [C_0, h]$ also in case $1 \neq h \in Z(X_0)$. Again by $(*)$ above there exists an $h \in H^\#$ with $C_0^h = A_0^n$. Hence $C_0 A_0^n = C_0 C_0^h = C_0 [C_0, h]$ and thus

$$A_0 C_0 A_0^n = A_0 C_0 [C_0, h] = L$$

which proves (1).

Now, as $|k| \geq 4, |A_0^{N_0}| \geq 4$. So let $m \in N_0$ with $A_0^m \notin \{A_0, C_0, A_0^n\}$. Assume by the way of contradiction $D \in (C_\Sigma(A) \cap C_\Sigma(C)) - C_\Sigma(A^n)$. Since also $L = A_0 C_0 A_0^m$ we have $A^m \not\leq AC$ but $A^m \cap ACA^n \neq 1$. Hence there exists an $\alpha \in (A^m)^\#$ with $\alpha = z \cdot a, z \in (AC)^\#$ and $a \in (A^n)^\#$. (Since $A^m \cap A^n = 1$ by (2.7)!) But, as $A^n \in \Omega_D$, this is a contradiction to (2.2). We obtain $C_\Sigma(A) \cap C_\Sigma(C) \subseteq C_\Sigma(A^n)$. Exchanging the roles of A^n and C this implies (2).
$\qquad\qquad\qquad\qquad\qquad\qquad\qquad\qquad\qquad\qquad\qquad\qquad\qquad\qquad$ □

Finally, combining (3.12), (3.15) and (2.15) we obtain

(3.16) Theorem

Suppose Σ is a class of abstract root subgroups of G, which contains different commuting elements. Suppose further that, if Σ is degenerate and $o(a) \le 3$ for all $a \in A$, $A \in \Sigma$, condition (H) holds. Then for each $A \in \Sigma$ and $B \in \Omega_A$ the group $\langle A, B \rangle$ is a special rank one group.

Proof. If Σ is non-degenerate this is (2.15). So assume Σ is degenerate. If there exists a $C \in \Sigma$ such that (A, B, C) is a triangle (3.16) is a consequence of (3.9) and I(2.5). Hence we may assume

$$C_\Sigma(A) - A \subseteq C_\Sigma(B) - B.$$

By hypothesis there exists a $C \in C_\Sigma(A) - A$. By exercise I(2.13)(7) there exists a $D \in \Omega_A \cap \Omega_C$. Hence (A, D, C) is a triangle. Now by (3.12) and (3.15) there exists an $E \in \Sigma$ with $A \ne E \ne C$ and

$$C_\Sigma(A) \cap C_\Sigma(C) = C_\Sigma(A) \cap C_\Sigma(E) = C_\Sigma(C) \cap C_\Sigma(E).$$

In particular $E \in C_\Sigma(A) - A \subseteq C_\Sigma(B)$ and thus $B \in C_\Sigma(E) \cap C_\Sigma(C) \subseteq C_\Sigma(A)$, a contradiction to $B \in \Omega_A$. $\qquad\square$

(3.17) Exercises

(1) Let Σ be a class of abstract root subgroups of G, $A \in \Sigma$, $N \lhd G$ and $g \in G$ with $A^g N = AN$. Then there exists an $n \in N$ with $A^g = A^n$.

Hint: Pick $B \in \Omega_A$. Then $B \in \Omega_{A^g}$. Set $X = \langle A, B \rangle$ and $Y = NX$. Then, as $N_X(A)$ is maximal in X, $N_X(A)N = N_Y(AN)$. Now there exists an $h_1 \in X$ with $A^{h_1} = B$ and $h_2 \in \langle B, A^g \rangle$ with $B^{h_2} = A^g$. Then $h_1 h_2 \in N_X(A)N$ and $n = x h_1 h_2 \in N$ for some $x \in N_X(A)$.

In the following two exercises let (A, B, C) be a triangle, Y the corresponding triangle group and suppose that A^Y is a class of k-transvection groups of Y, k a commutative field. Show:

(2) If N, Z, F and L are as in (3.2) and (3.9) then $N = ZLF$.
   Hint: Show, as in the proof of (3.14), that N/Z is a natural $\mathbb{Z}X$-module, $X = \langle A, B \rangle$.

(3) $A^N \subseteq Z(\langle C_\Sigma(A) \cap C_\Sigma(C) \rangle) \cap \Sigma$. In particular, if $n \in N$ with $A \ne A^n \ne C$, then

$$C_\Sigma(A) \cap C_\Sigma(C) = C_\Sigma(A) \cap C_\Sigma(A^n) = C_\Sigma(C) \cap C_\Sigma(A^n).$$

   Hint: Argue as in (3.15).

(4) Let (A, B, C) be a triangle and S, T as in (3.7). Suppose $A^t \notin C_\Sigma(A)$ for some $t \in T$. Then $A^t \in \Psi_A \cap \Psi_C$ and if $E = [A, A^t]$, then $E = [A^t, C] \in \Psi_B$. Further
$$M = E[E, B] \lhd Y = \langle A, B, C \rangle.$$

Hint: $A^t \leq C(F)$, so $A^t \notin \Omega_A$ by (2.2). If $E \in \Omega_B$, then there exists an $\overline{E} \in AE \cap \Psi_B$ by (2.11). But then $Q = \overline{E}[\overline{E}, B][\overline{E}, B, A]$ is by (2.20) Y-invariant and $EQ = AQ$, a contradiction to $E = [A, A^t]$. Thus $E \in \Lambda_A \cap \Lambda_C \cap \Psi_B$ and $M \lhd Y$ by (2.19).

For the next three exercises assume that Σ is a class of k-root subgroups of G, k a commutative field, (A, B, C) is a triangle of Σ, $X = \langle A, B \rangle$, $Y = \langle A, B, C \rangle$. Also use the notation of (3.2) and (3.7); that is we use the letters F, L, T and S.

(5) For each $t_i \in T$ with $E_i = [A, A^{t_i}] \neq 1$ let by (4) $M_i = E_i[E_i, B]$ and let $M = \oplus M_i$. Show:

   (i) $M \lhd Y$ and M is a direct sum of natural $\mathbb{Z}X$-modules.

   (ii) Let $\widehat{Y} = Y/M$. Then $\langle \widehat{A}^{\widehat{T}} \rangle \leq C(\widehat{A})$ and $\langle \widehat{B}^{\widehat{S}} \rangle \leq C(\widehat{B})$.

   (iii) $\widehat{Y} = \widehat{N}\widehat{X}, \widehat{N} \cap \widehat{X} \leq Z(\widehat{X})$ and $\widehat{N}$ satisfies the properties of (3.9) in $\widehat{Y}$. ($\widehat{N}$ replacing N!)

   (iv) $\widehat{A}^{\widehat{Y}}$ is a class of k-transvections of $\widehat{Y}$.

Hint for the proof of (iii). For the proof of (3.9) we just used the fact that A^Y is degenerate, to conclude from (3.8) that the hypothesis of (3.7)(2) holds. But here this hypothesis holds by (ii).

(6) Let N be the coimage of $\widehat{N}$ in Y and M_1 the coimage of $Z(\widehat{Y})$ in Y. Then
$$N = M_1 F L.$$

Hint. Use (3.4) and exercise (2) applied to $\widehat{Y}$.

(7) A^Y is a class of k-transvection groups of Y. In particular
$$N = Z(Y)F L.$$

Hint. We have $M = C_M(A) \oplus C_M(B)$ and $\langle A^{M_1} \rangle = \langle A^M \rangle \leq AC_M(A)$. Hence by exercise (6) and (3.6) (with the roles of A and B exchanged)
$$\langle A^N \rangle = \langle A^{ML} \rangle \leq \langle A^L \rangle C_M(A) \leq AC_N(A).$$

Now apply exercise (1) and (2).

Remark. The minimal parabolic subgroup of $G_2(k)$, which is not the normalizer of a long root group, has a similar structure as in (6). But as (7) shows, this group is not a triangle group.

(It is just generated by 3-root subgroups satisfying

 .)

§ 4 The radical $R(G)$

Suppose Σ is a class of abstract root subgroups of the group G. Then it is the purpose of this section to prove the existence of a normal subgroup $R(G)$ satisfying:

(i) $R(G)/R(G) \cap Z(G)$ is abelian.
(ii) $(G/R(G))'$ is quasi-simple.

Although the definition and existence proof for $R(G)$ is the same in the degenerate and non-degenerate case, we need certain auxiliary generational results, the proof of which is different in the separate cases.

For $A \in \Sigma$ let $\underline{A} = Z(\langle C_\Sigma(A) \rangle) \cap \Sigma$. Then by (3.11)

$$\begin{aligned} \underline{A} &= \{D \in \Sigma \mid C_\Sigma(A) = C_\Sigma(D)\} \\ &= \{D \in \Sigma \mid \Omega_A = \Omega_D\}. \end{aligned}$$

Let $\Sigma_{\underline{A}} = C_\Sigma(A) - \underline{A}$. Then we have

(4.1) Lemma

Suppose Σ is degenerate, $\Sigma_{\underline{A}} \neq \emptyset$ and condition (H) (of section 3) holds. Then $\mathcal{F}(\Sigma_{\underline{A}})$ is connected.

Proof. Suppose (4.1) is false and pick C, D in different connectivity components of $\mathcal{F}(\Sigma_{\underline{A}})$. Let Δ_1, Δ_2 be the connectivity components of $\mathcal{F}(\Sigma_{\underline{A}})$ containing C (resp. D). By exercise I(2.13)(8) there exist a $B \in \Sigma$ such that (C, B, D) is a triangle. Now (3.12) (resp. (3.15) in case $o(a) \leq 3$ for each $a \in A$) imply that there exists an $E \in \Sigma$ with $C \neq E \neq D$ and

$(*)$ $C_\Sigma(C) \cap C_\Sigma(D) = C_\Sigma(C) \cap C_\Sigma(E) = C_\Sigma(D) \cap C_\Sigma(E).$

Hence $E \in C_\Sigma(A)$ and as $[\Delta_1, \Delta_2] = 1, E \notin \underline{A}$. Thus $E \in \Sigma_{\underline{A}}$. Let Δ_3 be the connectivity component of $\mathcal{F}(\Sigma_{\underline{A}})$ containing E. If $\Delta_1 \neq \Delta_3$, then $\Delta_1 \subseteq C_\Sigma(D) \cap C_\Sigma(E)$, contradicting $(*)$. Thus $\Delta_1 = \Delta_3$ and, by the same argument $\Delta_2 = \Delta_3$, which proves (4.1). $\qquad \square$

(4.2)   Lemma

Assume the same hypothesis as in (4.1). Then

$$\Delta_A = \{E \in \Sigma \mid \Sigma_{\underline{A}} \subseteq \Sigma_{\underline{E}}\} = \underline{A}.$$

Proof. Assume $\underline{A} \neq \underline{E}$ and $\Sigma_{\underline{A}} \subseteq \Sigma_{\underline{E}}$. Then $[A, E] \neq 1$ by (4.1). Since $\langle A, E \rangle$ is a rank one group, there exists a $g \in G$ interchanging $\underline{A}$ and $\underline{E}$. This implies $\Sigma_{\underline{A}} = \Sigma_{\underline{E}}$. Hence we obtain:

(1) $\Delta_A = \{E \in \Sigma \mid \Sigma_{\underline{A}} = \Sigma_{\underline{E}}\} = \Delta_E$ for each $E \in \Delta_A$.

In particular Δ_A is a *TI-subset* of Σ. (Trivial intersection subset, i.e. $\Delta_A \cap \Delta_A^g = \Delta_A$ or $\emptyset$ for each $g \in G$!) Suppose now $D \in \Sigma - \Delta_A$ commutes with some $E \in \Delta_A$. Then, as $\underline{A} \subseteq \Delta_A$ and Δ_A is a TI-set, $\underline{E} \subseteq \Delta_A$. Hence $D \in \Sigma_{\underline{E}}$. Now by (1) $D \in \Sigma_{\underline{F}}$ for each $F \in \Delta_A$ and $[\Delta_A, D] = 1$. This implies:

(2) $[\Delta_A, \Delta_C] = 1$ if an element of Δ_A and an element of Δ_C commute .

Pick now $C \in \Sigma_{\underline{A}}$. Then there exists a $B \in \Sigma$ such that (A, B, C) is a triangle. Hence again (3.12) and (3.15) imply that there exists a $D \in \Sigma$ with $A \neq D \neq C$ and

(*) $C_\Sigma(A) \cap C_\Sigma(C) = C_\Sigma(A) \cap C_\Sigma(D) = C_\Sigma(C) \cap C_\Sigma(D).$

Clearly $D \notin \underline{A}$ since otherwise by (*) $C \in \underline{A}$, contradicting the choice of C. By the same argument $D \notin \underline{C}$. Thus by (1) and (2)

$$\Delta_A \cap \Delta_C = \emptyset = \Delta_A \cap \Delta_D = \Delta_C \cap \Delta_D$$

and

$$[\Delta_A, \Delta_C] = [\Delta_A, \Delta_D] = [\Delta_C, \Delta_D] = 1.$$

Hence if $E \in \Delta_A - \underline{A}$, then $\{D, C\} \subseteq C_\Sigma(E)$, whence $E \in C_\Sigma(A)$ by (*). But then (1) implies $C_\Sigma(A) \subseteq C_\Sigma(E)$, contradicting $E \notin \underline{A}$. This shows $\Delta_A = \underline{A}$.
$\square$

(4.3)   Lemma

Assume the same hypothesis as in (4.1). Then $\mathcal{D}(\Sigma)$ is connected.

Proof. We will prove (4.3) in several steps. Suppose for the whole proof that Δ is a connectivity component of $\mathcal{D}(\Sigma)$. We show first:

(1) $\mathcal{F}(\Delta)$ is connected.

Clearly $\Delta \neq C_{\Sigma}(A)$ for $A \in \Delta$. Indeed, if this would not be the case, pick $C \in \Sigma_A$. Then $C_{\Sigma}(C) \subseteq \Delta = C_{\Sigma}(A)$, contradicting $C \in \underline{A}$. Now pick $A \in \Delta$ and $E \in \Delta - C_{\Sigma}(A)$. Then there exists by (4.2) an $F \in \Sigma_A$ such that (E, F) is an edge in $\mathcal{F}(\Delta)$. Since (A, E) is also an edge (4.1) implies (1). Now let $\Lambda \neq \Delta$ be another connectivity component of $\mathcal{D}(\Sigma)$. Then

(2) $A \cap N(\Lambda) = 1$ for each $A \in \Delta$.

Indeed suppose $1 \neq a \in N_A(\Lambda)$ and pick $B \in \Lambda$. Then, as $[A, B] = 1, X = \langle A, B \rangle = \langle a, B \rangle \leq N(\Lambda)$. But since $A \sim B$ in X, this implies $A \in \Lambda \cap \Delta = \emptyset$, a contradiction.

(3) Let $R = \langle \Delta \rangle$ and $U = N_R(\Lambda)$. Then $R = UE$ for each $E \in \Delta$.

Pick $C \in \Lambda$ and $B \in \Delta$ such that $[E, B] \neq 1$. For each $c \in C^{\#}$ let $e(c) \in E^{\#}$ and $b(c) \in B^{\#}$ be the elements satisfying:

$$C^{e(c)} = E^c, C^{b(c)} = B^c. \qquad \text{(See I(1.2)!)}$$

Then, since Λ and Δ are TI-subsets, we obtain

$$\Lambda^{e(c)} = \Delta^c = \Lambda^{b(c)}$$

and $e(c)b(c)^{-1} \in U$. Let $X = \langle E, B \rangle$. Then, since

$$E^{\#} = \{e(c) \mid c \in C^{\#}\}, B^{\#} = \{b(c) \mid c \in C^{\#}\}$$

there exists for each $e \in E$ a $b \in B$ such that

$$(U \cap X)e = (U \cap X)b.$$

Hence $(U \cap X)E = (U \cap X)B$ and thus $X = (U \cap X)E$, since $X = \langle E, B \rangle$. Now the connectivity of $\mathcal{F}(\Delta)$ implies (3).

We now prove (4.3). Pick $A \in \Delta$ and let $Y = \langle C_{\Sigma}(A) \rangle$. Then (3) and the Dedekind law imply $Y = (Y \cap U)A$. Thus by (2) $Y \cap U \lhd Y$ and $A \cap (Y \cap U) = 1$. Hence $Y = (Y \cap U) \times A$ and $Y' \leq Y \cap U$. On the other hand there exist $E, F \in \Sigma_A$ with $[E, F] \neq 1$. Let $X = \langle E, F \rangle$. Then hypothesis (H) and I(2.10) imply

$$1 \neq E \cap X' \leq E \cap Y' \leq E \cap (Y \cap U),$$

a contradiction to (2).   □

We are now able to prove the desired generational result:

(4.4) Lemma

Suppose Σ is degenerate, $\Sigma_A \neq \emptyset$ and condition (H) holds. Then $G = \langle C_\Sigma(A), B \rangle$ for each $B \in \Omega_A$.

Proof. Let $R = \langle C_\Sigma(A), B \rangle$ and $\Delta = A^R$. By (4.2) there exists an $E \in \Sigma_A \cap \Omega_B$. Hence (4.1) implies $C_\Sigma(A) \cup \{B\} \subseteq \Delta$ and $\mathcal{F}(\Delta)$ is connected. As $C_\Sigma(A) \subseteq \Delta$ this implies

$$C_\Sigma(C) \subseteq \Delta \text{ for each } C \in \Delta.$$

But then the connectivity of $\mathcal{D}(\Sigma)$ by (4.3) implies $\Delta = \Sigma$ and $R = G$. $\square$

(4.5) Notation

A subset $\Delta \subseteq \Sigma$ satisfying

   (i) $|\Delta| > 1$ and $\Delta \neq \Sigma$,
   (ii) $\Delta \cap \Delta^d = \emptyset$ or Δ for all $d \in D(\Sigma)$

is called a *weak TI-subset* of Σ. Clearly TI-subsets are weak TI-subsets. Moreover, if Σ is degenerate satisfying condition (H) and Δ is a weak TI-subset of Σ, then

$$\Delta \subseteq \underline{A} \text{ for } A \in \Delta.$$

Indeed if there exists a $B \in \Delta \cap \Omega_A, A \in \Delta$, then by (4.4) $G = \langle C_\Sigma(A), B \rangle \leq N(\Delta)$, a contradiction to $\Delta \neq \Sigma$. Hence Δ is abelian and $\Delta \subseteq \underline{A}$.

But in the non-degenerate case the notion of weak TI-subsets is important. We show:

(4.6) Lemma

Suppose Σ is non-degenerate and Δ is a weak TI-subset of Σ. Then Δ is abelian.

Proof. We prove (4.6) in several steps. First

   (1) If $A \in \Delta$ and $B \in \Lambda_A \cap \Delta$, then $AB \cap \Sigma \subseteq \Delta$ and $\langle N_\Sigma(\Delta) \rangle$ acts transitively on $AB \cap \Sigma$.

(1) is a direct consequence of (2.23) and the description of the natural $SL_2(K)$-module in I(1.8). Indeed if $F \in AB \cap \Sigma$ with $A \neq F \neq B$, then there exists an $x \in D(\Sigma) \cap C(A)$ with $B^x = F$. Let Ω be the set of isolated vertices of $\mathcal{F}(\Delta)$. Then:

   (2) Ω is abelian.

Pick $A \in \Omega$ and $B \in \Psi_A \cap \Omega$ and set $C = [A, B]$. Then there exists by (2.23) a $D \in \Omega_B$ with $[D, C] = A$. On the other hand:

$$D \in C_\Sigma(A) \subseteq N_\Sigma(\Delta) \subseteq N_\Sigma(\Omega),$$

so that $D \in \Omega$ since $B \in \Omega$. But this is impossible by definition of Ω. This shows $\Omega \subseteq C_\Sigma(A)$ for each $A \in \Omega$, which implies (2).

(3) $\Delta = \Omega$.

Indeed if (3) is false, there exists a connectivity component Δ_1 of $\mathcal{F}(\Delta)$ with $|\Delta_1| > 1$. Since Δ_1 is again a weak TI-subset of Σ, we may assume $\Delta = \Delta_1$. Now pick $A \in \Delta$, $B \in \Lambda_A - \Delta$ and $C \in \Delta \cap \Omega_A$. Then by (2.3) $C \in \Psi_B$ since $B \notin \Delta$. As $B \in N_\Sigma(\Delta)$ we obtain $C^B \subseteq \Delta$. Hence by (1) $[C, B] \in \Delta$. But then by (2.19) there exists an $x \in X = \langle A, C \rangle$ with $B = [C, B]^x$, a contradiction to $X \leq N(\Delta)$ and $B \notin \Delta$. This shows $\Lambda_A \subseteq \Delta$ for each $A \in \Delta$ and thus $\Lambda_A \cup \Psi_A \subseteq \Delta$ for each $A \in \Delta$ by (2.8). Now (2.11) implies $\Omega_A \subseteq \Delta$ for $A \in \Delta$, which shows $\Delta = \Sigma$, a contradiction.
Now (2) and (3) together prove (4.6).   □

As an immediate consequence of (4.6) we obtain:

(4.7) Corollary

Suppose Σ is non-degenerate. Then $\mathcal{E}(\Sigma)$ and $\mathcal{D}(\Sigma)$ are connected.

This is an immediate consequence of (2.11) and (4.6), since a proper connectivity component of $\mathcal{E}(\Sigma)$ is a TI-subset of Σ.

(4.8) Lemma

Suppose Σ is non-degenerate, $A \in \Sigma$ and $B \in \Omega_A$. Then $G = \langle \Lambda_A, B \rangle$.

Proof. Let $Y = \langle \Lambda_A, B \rangle$ and $\Delta = B^Y$. By (2.3) and (2.11) $|AC \cap \Omega_B| \geq 2$ for $C \in \Lambda_A$. Hence (2.11) and (2.19) imply $A \in \Delta$. (If $C \neq D \in AC \cap \Psi_B$ and $X = \langle A, B, C \rangle = \langle A, B, D \rangle$, then $A \in D^X$ and $X \leq Y$!) Moreover by the same argument $A \cup \Lambda_A \subseteq \Delta$. Thus

$$C \cup \Lambda_C \subseteq \Delta \text{ for each } C \in \Delta$$

since Y is transitive on Δ. Now the connectivity of $\mathcal{E}(\Sigma)$ implies $\Delta = \Sigma$ and $Y = G$.   □

For the rest of this section we assume that Σ satisfies hypothesis (H), if Σ is degenerate. We start to define the radical $R(G)$.

(4.9) Lemma

Let $A \in \Sigma$, $B \in \Omega_A$ and $N_A = N_{\langle A \rangle}(\underline{B})$ and similarly N_B. Then the following hold:

(1) $[N_A, \langle \underline{B} \rangle] \leq N_B$, $[N_B, \langle \underline{A} \rangle] \leq N_A$ and

$$N = N_A N_B \lhd R = \langle \underline{A}, \underline{B} \rangle.$$

(2) N is nilpotent of class at most 2 with

$$[N_A, N_B] \leq N_A \cap B_B \leq Z(R).$$

(3) $R = NX, X = \langle A, B \rangle$ for each $A \in \underline{A}$ and $B \in \underline{B}$. Further $\langle \underline{A} \rangle = A N_A$ and $A \cap N_A = 1$.

Proof. (1) Let $n \in N_A$ and $b \in B^{\#}$, $B \in \underline{B}$. Then there exists an $a = a(b) \in A^{\#}$ with $A^b = B^a$. Let $\bar{b} = b^n \in B^n \in \underline{B}$. Then:

$$A^{\bar{b}} = A^{b^n} = A^{bn} = B^{an} = (B^n)^a \in \underline{B}^a$$

since $\langle \underline{A} \rangle$ is abelian. We obtain:

$$\underline{A}^b = \underline{B}^a = \underline{A}^{\bar{b}} \text{ and } [b, n] = b^{-1}\bar{b} \in N_B.$$

This shows $[N_A, \langle \underline{B} \rangle] \leq N_B$. With symmetry this implies (1).

(2) is an immediate consequence of (1). To prove (3) it suffices to show that, if $A \neq A_1 \in \underline{A}$, then $A N_A = A_1 N_A$. For this pick $a \in A^{\#}, b = b(a) \in B^{\#}$ and $a_1 = a(b) \in A_1^{\#}$ with $B^{a_1} = A_1^b$. Then as above

$$\underline{B}^{a_1} = \underline{A}^b = \underline{B}^a$$

and thus $aa_1^{-1} \in N_A$. Hence $a N_A = a_1 N_A$. Since for each $a \in A^{\#}$ there exists such an $a_1 \in A_1^{\#}$, this implies $A N_A \subseteq A_1 N_A$, which with symmetry proves (3).   □

(4.10) Lemma

Assume the same hypothesis and notation as in (4.9). Then one of the following holds:

(1) $\underline{A} = \{A\}, N = 1$ and $R = \langle A, B \rangle$.
(2) $\underline{A} \neq \{A\}, N \not\leq Z(R)$ and if $Z = N_A \cap N_B, \overline{R} = R/Z$ then the following hold:

(a) $\overline{N}' = 1, N = [N, R]$ and $C_{\overline{N}}(R) = 1$.

(b) $[\overline{N}, A, A] = 1$.

(c) $\overline{N}_A = [\overline{N}_B, a]$ for each $a \in \langle \underline{A} \rangle - N_A$.

(d) $R = \langle \underline{B}, B^x \rangle$ for each $x \in \langle \underline{A} \rangle - N_A$.

(e) $D(\underline{A}) \cap xZ \neq \emptyset$ for each $x \in \langle \underline{A} \rangle - N_A$.

Proof. If $\underline{A} = \{A\}$ then $\underline{B} = \{B\}$ and (1) holds. So assume $\{A, A_1\} \subseteq \underline{A}$ with $A \neq A_1$. Then, as shown in (4.9), there exists for each $a \in A^{\#}$ an $a_1 \in A_1^{\#}$ with $aN_A = a_1 N_A$. By (2.2) $aa_1^{-1} \notin Z(R)$, whence $N \not\leq Z(R)$.

Since $[N_A, B_B] \leq Z, \overline{N}$ is abelian. Let $M = [N, R], \Delta = \Sigma \cap R$ and $\widetilde{R} = R/M$. Then $[\widetilde{N}, \widetilde{R}] = 1$. Hence by (4.9)(3) $\widetilde{R}$ is a central extension of the rank one group $\widetilde{X}$. Since $\widetilde{\Delta}$ is a class of abstract root subgroups of $\widetilde{R}$, (2.2) again implies $\widetilde{\underline{A}} = \{\widetilde{A}\}$. Hence $\widetilde{R} = \widetilde{X}, \widetilde{N} = 1$ and $N = M$.

Now $\overline{N} = \overline{N}_A \times \overline{N}_B$ with $\overline{N}_A \sim \overline{N}_B$ in $\overline{R}$, since $A \sim B$ in X. As

$$\overline{N}_A C_{\overline{N}}(R) \cap \overline{N}_B \leq C_{\overline{N}_B}(\overline{R}) = 1,$$

this shows $C_{\overline{N}}(R) = 1$ and (a) holds. (b) is obvious using (4.9).

To prove (2)(c) we may by (4.9)(3) assume $a \in A^{\#}$. Hence $X = \langle B, B^a \rangle$ and $\overline{N}_0 = [\overline{N}_B, a] \overline{N}_B$ is X-invariant and thus $\overline{N}_0 \triangleleft \overline{R}$. But since $\overline{N}_B \leq \overline{N}_0$ and $\overline{N}_A \sim \overline{N}_B$ in $\overline{R}$ we obtain $\overline{N} = \overline{N}_0$ and thus $\overline{N}_A = [\overline{N}_B, a]$ since $\overline{N} = \overline{N}_A \times \overline{N}_B$.

Let x be as in (2)(d). Then by (4.9)(3) $R = N \langle B, B^x \rangle$, since X is a rank one group. Hence

$$\underline{A} \subseteq \underline{B}^{\langle B, B^x \rangle}$$

and (2)(d) holds.

Let $x \in \langle \underline{A} \rangle - N_A$. Then $x = a \cdot n, a \in A^{\#}$ and $n \in N_A$ by (4.9)(3). Now by (2) (c)

$$a^{-1} a^m = [a, m] = n \cdot z \text{ for some } m \in N_B, z \in Z.$$

This implies $xz = anz = a^m \in D(\underline{A})$, which proves (e). $\qquad\square$

(4.11) Lemma

Assume the same hypothesis and notation as in (4.9). Then the following hold:

(1) $Z \leq Z(G)$.

(2) If $Z(G) = 1$, then $\langle \underline{A} \rangle - N_A = D(\underline{A})$ and

$$N_A = \{x \in \langle \underline{A} \rangle \mid x \notin D(\underline{A})\}.$$

(3) If $Z(G) = 1$ and C is another element of Ω_A, then $N_{\langle A \rangle}(\underline{B}) = N_{\langle A \rangle}(\underline{C})$.

Proof. Since $C_\Sigma(A) = C_\Sigma(\langle \underline{A} \rangle)$, (1) is a consequence of (4.4), (4.8) and the definition of Z. Thus if $Z(G) = 1$, then $Z = 1$ and $\langle \underline{A} \rangle - N_A = D(\underline{A})$ by (4.10) (2)(e). Since $\langle \underline{A} \rangle = N_A \dot{\cup} (\langle \underline{A} \rangle) - N_A)$ this implies (2). But (2) shows that N_A is independent of the chosen $B \in \Omega_A$, whence (3) holds.   □

(4.12)   Notation

Suppose $Z(G) = 1$. Then (4.11) (3) shows that N_A is independent of $B \in \Omega_A$. Set

$$R(G) := \langle N_A \mid A \in \Sigma \rangle.$$

Then $R(G)$ is well defined and $R(G) \trianglelefteq G$. Finally, if $Z(G) \neq 1$ set $\overline{G} = G/Z(G)$. Then $Z(\overline{G}) = 1$ by an easy application of (2.2). Now let $R(G)$ be the maximal coimage of $R(\overline{G})$.

(4.13)   Lemma

Suppose $Z(G) = 1$. Then $R(G)$ is an abelian normal subgroup of G and one of the following holds:

(1) $G/R(G)$ is quasi-simple.

(2) Σ is degenerate, $o(a) \leq 3$ for $a \in A \in \Sigma$ and $(G/R(G))'$ is quasi-simple.

Proof. It suffices to show that $R(G)$ is an abelian normal subgroup of G. Indeed if this is the case and $\overline{G} = G/R(G)$, then by definition of $R(G)$ we have $\overline{A} = \{\overline{A}\}$. Hence the hypotheses (1)–(4) of (2.14) are satisfied for $\overline{G}$ and $\overline{\Sigma}$. (Hypothesis (3) since we assume that condition (H) holds if Σ is degenerate.) Since by I(5.6) and (2.14) $\overline{G} = \overline{G}'$, except when $\overline{\Sigma}$ is degenerate and $o(a) \leq 3$ for each $a \in A \in \Sigma$, this shows that (4.13) holds.

Now when Σ is degenerate, $R(G)$ is abelian since $N_A \cap N_B = 1$ for $A \in \Sigma, B \in \Omega_A$ by (4.9) and (4.4). So assume Σ is non-degenerate. For $A \in \Sigma, B \in \Omega_A$ let

$$L_A = N_A \langle N_C \mid C \in \Lambda_A \rangle, L_{A,B} = N_A \langle N_C \mid C \in \Lambda_A \cap \Omega_B \rangle.$$

Then $L_{A,B} \leq L_A$. We first show that (4.13) holds if

$(*)$ $L_A = L_{A,B}.$

Indeed let $M = \langle L_A, N_B \rangle$. Then for $C \in \Lambda_A \cap \Omega_B$ we have $N_C \cap N_B = 1$ by (4.9) and (4.8). Hence M is abelian. Again by (4.9) $[L_{A,B}, B] \leq N_B$ and $[C, N_B] \leq N_C \leq L_A$ for each $C \in \Lambda_A \cap \Omega_B$. Since $M_A = \langle \Lambda_A \cap \Omega_B \rangle$ for example by (2.11), this shows $M \triangleleft G$. The transitivity of G on Σ then implies $M = R(G)$.

Thus $(*)$ remains to be shown. For this pick $C \in \Lambda_A \cap \Psi_B$. Then it suffices to show

$(**)$ $\qquad\qquad\qquad\qquad x \in L_{A,B}$ for each $x \in N_C$.

Suppose this is false for some $x \in N_C$. By $(4.10)(2)(c)$ $x = cc_1$ for some $c \in C^\#$ and $c_1 \in C_1 \in \underline{C}$. By (2.23) there exist $E \in \Lambda_A, F \in \Lambda_C$ such that $X = \langle E, F \rangle$ induces an $SL_2(K)$, K a division ring or Cayley division algebra on $N = AC$. Further, since X is a rank one group, $C_X(N) \leq Z(X)$.

Consider $L = \langle C_1^X \rangle$. By (2.20) $L = L_1 \times C_L(X)$, where $L_1 = [L, X]$ is a natural $\mathbb{Z}X$-module. Since $L \leq \langle C_\Sigma(C_1) \rangle = \langle C_\Sigma(C) \rangle$ we obtain $[L, N] = 1$ and $LN = L_1 \times N \times C_L(X)$ with N (resp. L_1) natural modules for $\overline{X} = X/C_X(N)$ (resp. $\widetilde{X} = X/C_X(L_1)$). Further $C_X(N)C_X(L_1) \leq Z(X)$ by I(1.10).

Suppose now $|A| > 2$ for $A \in \Sigma$ and let $K = N_X(E) \cap N_X(F)$. Then as shown in I(1.8)(11) K acts transitively on $C^\#$ and $[L_1, F]^\#$. Moreover $C_1^K \subseteq \underline{C}$. Now by (2.20):

$$C_1 C_L(X) = C_L(F) = [L_1, F]C_L(X) = C_1[L_1, F].$$

Hence $[C_L(F), K] = [L_1, F]$ and $\langle C_1^K \rangle = C_L(F)$. This implies $D(\Sigma) \cap C_L(F) \subseteq D(\underline{C})$. Hence by $(2.22)(3)$ $D(\Sigma) \cap C_L(X) = \emptyset$ and so by $(4.10)(e)$

$(+)$ $\qquad\qquad\qquad\qquad C_L(X) \leq N_A \cap N_C \leq L_{A,B}$

(as $Z = 1$ in the notation of (4.10) !) Let $C_0 = [L_1, F]$, $A_0 = [L_1, E]$. Then as $C_L(F) \cap \Sigma \subseteq \underline{C}$ by the above, we have $C_0 \in \underline{C}$ and $A_0 \in \underline{A}$. Let $M = L_1 N$. Then, as $x \in N_C - C_L(X)$, $(+)$ implies $R_C = N_C \cap M \neq 1$. Now since R_C is K-invariant and since K acts transitively on $C^\#$ and $C_0^\#$ we obtain

$$M \cap \langle \underline{C} \rangle = CC_0 = CR_C = C_0 R_C.$$

Hence K acts transitively on $R_C^\#$ and $R_A^\#$, where $R_A = M \cap N_A$, since R_C is equivalent to C as a K-module.

We claim that $R_A \times R_C$ is X-invariant. As above $(4.10)(e)$ implies

$$R_C = N_C \cap M = \{x \in CC_0 \mid x \notin D(\Sigma)\}$$

and similarly

$$R_A = \{y \in AA_0 \mid y \notin D(\Sigma)\}.$$

Fix $f \in F^\#$ and let $Q(A) = \{y \in AA_0 \mid [y, f] \in R_C\}$. Then $Q(A) \leq R_A$ and the commutator map

$$y \rightarrow [y, f], \quad y \in Q(A)$$

is an isomorphism from $Q(A)$ onto R_C. Since

$$a \to [a, f], \quad a \in A$$

is an isomorphism from A into C, this implies that the commutator map $ay \to [ay, f]$ is an isomorphism from $AQ(A)$ onto $CR_C = CC_0$. On the other hand $C_{AA_0}(f) = 1$ and $[AA_0, f] = CC_0$. Hence $AQ(A) = AA_0$ and thus $Q(A) = R_A$. This shows $[R_A, f] = R_C$ for each $f \in F^\#$ and so with symmetry that $R_A \times R_C$ is X-invariant.

Now by (2.11), $N = AD, D \in \Lambda_A \cap \Omega_B \cap N$ and, since the same statement holds for L_1, we obtain

$$(LN \cap N_A)(LN \cap N_C) = R_A R_C C_L(X) = R_A R_D C_L(X) \le L_{A,B}$$

for some $D \in \Lambda_A \cap \Omega_B \cap N$ and $R_D = N_D \cap M$. Since $x \in LN \cap N_C$ this proves $(**)$ and whence $(*)$ in case $|A| > 2$.

So the case $|A| = 2$ remains to be treated. Let in this case $\langle a \rangle = A, \langle a_1 \rangle \in \underline{A}$ with $a \ne a_1$ and $d = ac$ for $\langle c \rangle = C$. Then we have for $\langle b \rangle = B$:

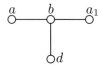

since $\langle d \rangle \in \Omega_B$ as

$$
\begin{array}{ccc}
A & B & C \\
\circ\!\!-\!\!\!-\!\!\!-\!\!\circ\!\!-\!\!\!-\!\!\!-\!\!\circ
\end{array}
$$

and $C \in \Lambda_A$.

Hence if $R = \langle a, b, a_1, d \rangle$, then R is a factor group of $W(D_4)$. By (4.10)(e) $n = aa_1 \in N_A$, whence $m = bb^n \in N_B$ and thus $d_1 = d^m \in D(\langle \underline{d} \rangle)$. Assume $aa_1 = dd_1$ for each choice of $\langle a_1 \rangle \in \underline{A}$ with $a \ne a_1$. (I.e. R is isomorphic to $W^*(D_4)$, the center factor group of $W(D_4)$.) Then, since $N_A = \langle aa_1 \mid \langle a_1 \rangle \in \underline{A} \rangle$ by (4.10), we obtain $N_A \le N_D$ for $D = \langle d \rangle$. Since A and D are conjugate by an involution, this implies $N_A = N_D$. But D and C are conjugate in M_A, whence we obtain $N_C = N_A = N_D \le L_{A,B}$ and thus $(**)$ and $(*)$ hold.

So we may assume $aa_1 \ne dd_1$ for some $\langle a_1 \rangle \in \underline{A}$. Hence $z = aa_1 dd_1$ is the central involution of R. (I.e. $R \simeq W(D_4)$!) Let $c_1 = a_1 d_1$ and $C_1 = \langle c_1 \rangle$. Then $C_1 \sim C$ in R. If now $E \in \Omega_C$, then $E \in \Omega_A \cap \Psi_D$ or $E \in \Psi_A \cap \Omega_D$ by (2.3). Hence $E \in \Omega_{A_1} \cap \Psi_{D_1}$ or $E \in \Psi_{A_1} \cap \Omega_{D_1}$ by (2.22)(3), $D_1 = \langle d_1 \rangle$ and $A_1 = \langle a_1 \rangle$. Hence in any case $E \in \Omega_{C_1}$. We have shown $\Omega_C \subseteq \Omega_{C_1}$. Now (2.22) implies $\Omega_C = \Omega_{C_1}$ and $C_1 \in \underline{C}$. Thus $z = cc_1 \in N_C$ by (4.10). We have shown:

$$(++) \qquad
\begin{array}{l}
\text{For each } \alpha \in N_A - N_D \text{ there exists} \\
\text{a } \delta \in N_D - N_A \text{ such that } \alpha\delta \in N_C.
\end{array}
$$

Since $D \sim C$ in $C(\langle \underline{A} \rangle)$ as above, we have $N_A \cap N_D \leq N_C$. Now $(++)$ implies $N_A N_D \leq N_C N_D$, whence with symmetry $N_A N_D = N_C N_D$ and thus $N_C \leq N_A N_D \leq L_{A,B}$, which finally proves (4.13). □

Collecting the results obtained so far in this section we now can prove the following theorem for the radical $R(G)$.

(4.14) Theorem

The following hold:

(1) $\widetilde{R} = R(G)/R(G) \cap Z(G)$ is abelian.
(2) Either $\overline{G} = G/R(G)$ is quasi-simple or Σ is degenerate, $o(a) \leq 3$ for each $a \in A \in \Sigma$ and $\overline{G}'$ is quasi-simple.
(3) $C_{\widetilde{R}}(\overline{G}) = 1$ and $\widetilde{R} = [\widetilde{R}, \overline{G}]$.
(4) $[\widetilde{R}, \overline{A}, \overline{C}] = 1$ for all commuting $\overline{A}, \overline{C} \in \Sigma$.
(5) $\underline{\overline{A}} = \{\overline{A}\}$ for each $\overline{A} \in \Sigma$.

Proof. (1) and (2) are direct consequences of (4.13) and the fact that $Z(G/Z(G)) = 1$. By (4.10)(c) we have $[\overline{A}, \widetilde{N}_B] = \widetilde{N}_A$ for $B \in \Omega_A$. Since $\widetilde{R} = \Pi \widetilde{N}_A, A \in \Sigma$ this implies (3). ($C_{\widetilde{R}}(\overline{G}) = 1$ is another way to express $Z(G/Z(G)) = 1$!) Since Σ is a conjugacy class, $[\overline{A}, \overline{C}] = 1$ if and only if $[A, C] = 1$. Hence (4) is a consequence of $[\widetilde{R}, \overline{A}] \leq \widetilde{N}$. (5) follows from the definition of $R(G)$. □

Condition (3) and (4) of (4.14) make it possible to determine the structure of $\widetilde{R}$ as $\mathbb{Z}\overline{G}$-module in certain cases, see [Tim90b]. We will finally show that actually in many cases $R(G) = Z(G)$.

(4.15) Lemma

Suppose the group G is generated by a non-degenerate class Σ of abstract root subgroups, satisfying $\langle C_\Sigma(A) \rangle$ is Σ-maximal for $A \in \Sigma$. (I.e. maximal among subgroups of G generated by elements of Σ.) Then there exists no non-trivial $\mathbb{Z}G$-module V satisfying

$(*)$ $\qquad\qquad [V, E, F] = 0$ for all commuting $E, F \in \Sigma$.

Proof. Suppose V is a non-trivial $\mathbb{Z}G$-module satisfying $(*)$. Then we may, since G is perfect, assume $C_V(G) = 0$ and $V = [V, G]$. (Passing to $V/C_V(G)$ and $[V, G]$!) Pick $B \in \Psi_A, A \in \Sigma$. Then $G = \langle C_\Sigma(A), B \rangle$ by assumption. Hence

$$[V, A] \cap [V, B] \subseteq C_V(G) = 0 \text{ by } (*).$$

Now let $W = [V, A] \oplus [V, B]$ and $W_0 = [W, A] \oplus [W, B]$. Then, since $D = [A, B]$ centralizes W by $(*)$, the group $\overline{L}$ induced by $L = \langle A, B \rangle$ on W is abelian. Hence

$$[W, A, B] = [W, \overline{A}, \overline{B}] \subseteq [W, \overline{A}] \cap [W, \overline{B}] = 0.$$

This shows that L centralizes W_0, whence $[W, A] \subseteq C_V(G) = 0$, since $G = \langle C_\Sigma(A), B \rangle$. But then $[V, B, A] \subseteq [W, A] = 0$ and thus $[V, B] \subseteq C_V(G) = 0$ since $G = \langle C_\Sigma(B), A \rangle$. This shows that V is a trivial $\mathbb{Z}G$-module, contradicting the choice of V. $\square$

(4.16)   Remark

(4.14) and (4.15) applied to $\overline{G}$ show that $\tilde{R} = 1$, if Σ is non-degenerate and $\langle C_{\overline{\Sigma}}(\overline{A}) \rangle$ is $\overline{\Sigma}$-maximal for $\overline{A} \in \overline{\Sigma}$, where $\overline{G} = G/R(G)$. But as the classification of simple groups generated by a non-degenerate class of abstract root subgroups shows, (see the next chapter) this is always the case except when:

  (1) $\overline{G}/Z(\overline{G})$ is a (generalized) linear group.
  (2) $\overline{G}/Z(\overline{G})$ is of type G_2.
  (3) $\overline{G}/Z(\overline{G})$ is an orthogonal group of Witt index 3 over a commutative field.

Hence in all other cases $G/Z(G)$ is simple.

(4.17)   Corollary

Suppose $A \neq C \in \underline{A}$. Then there exists a $g \in R(G)$ with $A^g = C$.

Proof. Pick $B \in \Omega_A$. Then as $C \equiv A \mod R(G)$ by (4.14) also $B \in \Omega_C$ and $Y = \langle A, B, C \rangle \leq R(G)X, X = \langle A, B \rangle$. Since $A \sim B$ in X and $B \sim C$ in $\langle B, C \rangle$, there exists $y \in Y$ with $A^y = C$. Since $Y \leq R(G)X$ we have $y \in R(G)N_X(A)$, which proves (4.17). $\square$

(4.18)   Exercises

  (1) Let $R = \mathrm{Sp}(4, k), |k| > 2$, and Σ be the class of transvection groups of R. Show that there exist $\{A_1, A_2, A_3, A_4\} \subseteq \Sigma$ satisfying:

  (a)
  $$\overset{A_1}{\circ}\!\!-\!\!-\!\!\overset{A_2}{\circ}\!\!-\!\!-\!\!\overset{A_3}{\circ}\!\!-\!\!-\!\!\overset{A_4}{\circ}$$
  (b) $Y = \langle A_1, A_2, A_3 \rangle = \langle C_\Sigma(D) \rangle$ for some $D \in \Sigma$.
  (c) $\langle A_1, \ldots, A_4 \rangle = R$.

  Hint. $\langle C_\Sigma(D) \rangle$ is Σ-maximal. Also, if U is the unipotent radical of $\langle C_\Sigma(D) \rangle$ and L a Levi complement, then $U = [U, L] = [u, L]$ for $u \in U - Z(U)$.

  (2) Show that (1) is false if $|k| = 2$.

(3) Let V be a vector space over the field k and $q : V \to k$ a non-degenerate quadratic form of Witt index ≥ 4. Let $G = \Omega(V, q)$ and Σ be the class of Siegel transvection groups of G in the notation of (1.5). Let P be the stabilizer of a singular point of V in G and $\Delta = \Sigma \cap P$. Show:

    (a) Δ is a set of abstract root subgroups of $P' = \langle \Delta \rangle$.

    (b) $\Delta = \Lambda \dot{\cup} \Omega$ with $\Lambda = \Delta \cap U_P$ (U_P the unipotent radical of P) and $\Omega = \Delta - \Lambda$ a conjugacy class in P'. (I.e. $\mathcal{F}(\Omega)$ is connected).

    (c) Show that Ω is not a class of abstract root subgroups of P' although $P' = \langle \Omega \rangle$.

    Hint. There exist $A, B \in \Omega$ with $1 \neq [A, B] \in \Lambda$.

(4) Let $G = D_3(k) \simeq A_3(k)$, k a field.
Show that $k^4 \cdot G$ is generated by a class of k-root subgroups.

(5) Show that the split extension of $G = G_2(k)$ with its 7-dimensional kG-module is also generated by a class of abstract root subgroups.

(6) Let R be a special symplectic or unitary group of Witt index ≥ 2 (i.e. the subgroup generated by the isotropic transvection groups as in (1.4)) and let V be the natural R-module and $G = V \cdot R$. Let $A \leq R$ be an isotropic transvection group as defined in (1.4) and let $\Sigma = A^G$. Show:

    (1) Σ is a class of abstract transvection groups of G.

    (2) $R(G) = V$.

    (3) $N_A = [V, A] = N_{\langle A \rangle}(B)$ for each $B \in \Sigma$ with $[A, B] \neq 1$.

§ 5   Abstract root subgroups and Lie type groups*

In this section we will use the notion of buildings, roots and root subgroups introduced in §4 of Chapter I. So let in this section $\mathcal{B}$ be a thick, irreducible, spherical Moufang building over $I = \{1, \dots, \ell\}, \ell \geq 2$, different from an octagon, $\mathcal{A}$ an apartment of $\mathcal{B}$ and $c \in \mathcal{A}$ a fixed chamber. We fix the following notation:

$$\Delta_J(c) \text{ is the residue of type } J \text{ containing } c \text{ for } J \subseteq I.$$

Then, if $|J| \geq 2, \Delta_J(c)$ is a building over J with apartment $\mathcal{A} \cap \Delta_J(c)$. Let
Φ – the set of roots of $\mathcal{A}$ (half apartments)
Φ^+ – the set of roots of $\mathcal{A}$ containing c.

For $r \neq \pm s; r, s \in \Phi$ let $[r, s] := \{\alpha \in \Phi \mid r \cap s \subseteq \alpha\}$ and $(r, s) = [r, s] - \{r, s\}$. $-r$ is the opposite of r. r_i is the unique root of Φ^+ with $r_i \cap \Delta_i(c) = \{c\}$

for $i \in I$. $\Pi = \{r_1, \ldots, r_\ell\}$. Using the geometric realization of the Weyl group of $\mathcal{A}$, one can identify Φ with a root system in the original sense (i.e. set of vectors in $\mathbb{R}^\ell$) such that Π is a basis of $\mathbb{R}^\ell$ and each vector in Φ^+ is a linear combination of the roots in Π with non-negative integral coefficients.

We often use the description of the irreducible reduced root systems (of type $A_\ell, B_\ell, C_\ell, D_\ell, E_\ell, F_4, G_2$) given in the appendix of [Bou81]. For the convenience of the reader we describe shortly the irreducible non-reduced root system of type BC_ℓ, since it will be used later on.

Let $(e_i; i = 1, \ldots, \ell)$ be an orthonormal basis of $\mathbb{R}^\ell$. Then the roots of a root system $\widetilde{\Phi}$ of type BC_ℓ are

$$\pm e_i, \ \pm 2e_i, \ \pm e_i \pm e_j \text{ with } i < j; 1 \leq i, j \leq \ell.$$

Fundamental roots are:

$$\alpha_1 = e_1 - e_2, \ldots, \alpha_{\ell-1} = e_{\ell-1} - e_\ell, \alpha_\ell = e_\ell.$$

Positive roots are:

$$e_i = \Sigma_{i \leq k \leq \ell} \alpha_k, \ 2e_i$$

$$e_i - e_j = \Sigma_{i \leq k < j} \alpha_k, \ e_i + e_j = \Sigma_{i \leq k < j} \alpha_k + 2\Sigma_{j \leq k \leq \ell} \alpha_k.$$

The highest root in $\widetilde{\Phi}^+$ is

$$h = 2e_1 = \Sigma_{k=1}^\ell 2\alpha_k.$$

The extended Dynkin diagram is

$$
\begin{array}{ccccc}
-h & \alpha_1 & \alpha_2 & \alpha_{\ell-1} & \alpha_\ell \\
\circ\!\!=\!\!\!=\!\!\circ & \!\!-\!\!\circ & \cdots & \circ\!\!=\!\!\!=\!\!\circ &
\end{array} .
$$

For each $r \in \Phi$ let

$$
\begin{aligned}
A_r &:= \text{the root subgroup of } \mathrm{Aut}(\mathcal{B}) \text{ corresponding to } r \\
X_r &:= \langle A_r, A_{-r} \rangle \\
H_r &:= N_{X_r}(A_r) \cap N_{X_r}(A_{-r}) \\
G &:= \langle A_r \mid r \in \Phi \rangle \\
U &:= \langle A_r \mid r \in \Phi^+ \rangle \\
H &:= \langle H_r \mid r \in \Phi \rangle.
\end{aligned}
$$

Then by I(4.12) X_r is a rank one group with unipotent subgroups A_r and A_{-r} and diagonal subgroup H_r. We call the subgroup G of $\mathrm{Aut}(\mathcal{B})$ the group of *Lie type* $\mathcal{B}$. It will be seen that $G \trianglelefteq \mathrm{Aut}(\mathcal{B})$ and that the definition of G is independent of the special apartment $\mathcal{A}$. Our notion of a Lie type group generalizes the usual one as for example given in [Car72], since it includes the exceptional algebraic group (for example $E_{6,2}, E_{7,3}$ and $E_{8,4}$) and the "mixed groups" of [Tit74].

Notice that we also include Moufang polygons (with the exception of octagons). The main purpose of this section is to show that a normal subgroup G_0 of G is generated by a class of abstract root subgroups and, using (2.14) and exercise (2.24)(7) and (8), that G_0' is simple. To be able to do this we need to develop the theory of such Lie type groups. This theory depends, similarly as the theory of Chevalley groups depends on the Chevalley commutator relations, on the commutator relations of root groups due to [Tit94]. The results in this section are from [Tim00b, §3].

We will assume a certain familiarity with the theory of (finite) Coxeter groups and root systems, see [Car72], Chapter 2 or [Hum90], in this section.

(5.1) Lemma

The following hold:

  (1) G acts chamber transitively on $\mathcal{B}$.

  (2) c is the unique chamber of $\mathcal{B}$ fixed by U.

Proof. (1) We show with induction on $d(c, d)$ that for each $d \in \mathcal{B}$ there exists a $g \in G$ with $d^g = c$. If $d(c, d) = 1$, then there exists an $r \in \Phi^+$ with $d \notin r$. Suppose $d \overset{i}{\sim} c$. Then I(4.12) implies that X_r is transitive on $\Delta_i(c)$, whence $d^g = c$ for some $g \in X_r$.

Suppose now $d(c, d) = n > 1$, $c = c_0 \sim c_1 \sim \cdots \sim c_n = d$ is a minimal gallery from c to d and the statement holds for chambers d' with $d(c, d') < n$. Then there exists a $g \in G$ with $c_1^g = c$. Hence $d(c, d^g) \leq n - 1$ and $d^{gh} = c$ for some $h \in G$. This proves (1).

To prove (2) let $c \neq d \in \mathcal{A}$. Then there exists a root $r \in \Phi^+$ with $d \notin r$, but d adjacent to some chamber $d' \in r$. Hence by I(4.12)(1) $d^\alpha \notin \mathcal{A}$ for each $1 \neq \alpha \in A_r$. Since $A_r \leq U$, U does not fix d.

Next suppose the chamber d is not in $\mathcal{A}$ but i-adjacent to some $d \in \mathcal{A}$. Then $\Delta_i(d) \cap \mathcal{A} = \{d_1, d_2\}$ with $d_2 \neq d_1$. Choose the enumeration so that $d(c, d_1) < d(c, d_2)$, where $d(\ ,\ )$ is the distance in the chamber graph. (Clearly $d(c, d_1) \neq d(c, d_2)$ since $d_1 \overset{i}{\sim} d_2$.) Then c and d_1 are not opposite and there exists a root $s \in \Phi^+$ with $d_1 \in s$, but $d_2 \notin s$. (By the proof of I(4.11)(d) each pair of chambers of distance $d-1$ of $\mathcal{A}$ is contained in a unique root of $\mathcal{A}$. This implies the existence of s!) Now by I(4.12)(1) there exists an $\alpha \in A_s$ with $d^\alpha = d_2$, which again implies that U does not fix d.

Finally let $d \in \mathcal{B} - \mathcal{A}$ be arbitrary and $k = \mathrm{Min}\{d(d,e) \mid e \in \mathcal{A}\}$. Let $e \in \mathcal{A}$ with $d(d,e) = k$ and $d \sim d_1 \sim \cdots \sim d_{k-1} \overset{i}{\sim} e$ a gallery from d to e of length k. Then $d_{k-1} \notin \mathcal{A}$ and, as above, there exists an i-adjacent $e' \neq e$ in $\mathcal{A}$. Assume as before $d(c,e') < d(c,e)$ and choose a root $r \in \Phi^+$ with $e' \in r$ but $e \notin r$. Then there exists an $\alpha \in A_r$ with $e = d_{k-1}^\alpha$ by I(4.12)(1). Hence

$$d(d^\alpha, e) = d(d^\alpha, d_{k-1}^\alpha) = d(d, d_{k-1}) = k - 1,$$

so that by induction we may assume that U does not fix d^α. But then U does not fix d. □

(5.2) Lemma

Suppose $A_r' \neq 1$ for $r \in \Phi$. Then $\mathcal{B}$ is of type C_ℓ :

$$\begin{array}{cccc} 1 & 2 & \ell-1 & \ell \\ \circ\!\!-\!\!-\!\!-\!\!\circ & \cdots & \circ\!\!=\!\!=\!\!\circ \end{array}$$

and r is conjugate to r_ℓ under the Weyl group of $\mathcal{A}$.

Proof. By [Car72, 2.1.8] r is conjugate to some fundamental root r_i under W. Suppose there exists a $j \in I$ such that $\Delta_{ij}(c)$ is (the chamber system of) a projective plane. Then by I(4.11) $\Delta_{ij}(c)$ is a Moufang plane and A_{r_i} acts faithfully on $\Delta_{ij}(c)$ and induces a root group on this projective plane. (See comment after the proof of I(4.11).) Hence it is well known, see for example [VM98, section 5.4], that A_{r_i} and whence A_r are abelian.

So we may assume that there exists no such j. But then by the list of Dynkin diagrams, either $\mathcal{B}$ is of type C_ℓ and $i = \ell$ or $\mathcal{B}$ is of type G_2. But in the latter case it follows from [Tit94], see also [VM98, section 5.4], that A_r is abelian too. □

(5.3) Notation

Let $\mathcal{B}$ be a building of type C_ℓ. Then $\mathcal{B}$ can be considered by [Tit74, 7.4] as a polar space. If now P is a point (resp. ℓ a line) of this polar space, let

$$\begin{aligned} T_P \quad :=& \quad \text{the set of all \textit{central elations} corresponding to } P \\ :=& \quad \{\sigma \in \mathrm{Aut}(\mathcal{B}) \mid \sigma \text{ fixes all points on all lines through } P\} \end{aligned}$$

and similarly

$$T_\ell \quad := \quad \text{the set of all } \textit{axial elations} \text{ corresponding to } \ell.$$

Let now $\mathcal{B}$ be of type C_ℓ and r a root of $\mathcal{A}$ conjugate to r_ℓ under W. Then r corresponds to a point P of $\mathcal{B}$, considering $\mathcal{B}$ as a polar space (I.e. if $r = r_\ell$ then $\Delta_{\ell-1,\ell}(c)$ is a generalized quadrangle and $r \cap \Delta_{\ell-1,\ell}(c)$ is a root with a point in the "middle".) or $\ell = 2$ and r corresponds to a line. But in the latter case we may, passing to the dual quadrangle, also assume that r corresponds to a point P. There are three cases to distinguish:

(a) A_r contains no central elations corresponding to P. In this case we choose Φ to be of type B_ℓ. (So far Φ is a set of half-apartments of $\mathcal{A}$. Identifying Φ with a root system in the original sense, i.e. set of vectors of an ℓ-dimensional euclidian space, we have still the freedom to choose the length of the roots appropriately.)

(b) A_r consists of central elations corresponding to P. In this case choose Φ to be of type C_ℓ.

(c) Let A_{2r} be the set of central elations corresponding to P contained in A_r and suppose

$$1 \neq A_{2r} < A_r.$$

(Clearly A_{2r} is a subgroup of A_r!). In this case we choose Φ to be of type B_ℓ and set

(∗) $$\widetilde{\Phi} = \Phi \cup \{2r \mid r \in \Phi \text{ conjugate to } r_\ell \text{ under } W\}.$$

Then $\widetilde{\Phi}$ is of type BC_ℓ. We will show that in this case $A'_r \leq A_{2r} \leq Z(A_r)$.

To have a uniform notation we set $\Phi = \widetilde{\Phi}$ in all other cases. (I.e. $\mathcal{B}$ not of type C_ℓ or in case (a) and (b).) Define $\widetilde{\Phi}^+$ in the usual way. (I.e. $\widetilde{\Phi}^+$ is the set of all roots in $\widetilde{\Phi}$, which are linear combinations of roots in Π with non-negative coefficients.) Define a *height function* $h(s), s \in \widetilde{\Phi}^+$ by setting:

$h(s)$ is the sum of the coefficients in the expression of s as a linear combinations of fundamental roots.

For the convenience of the reader we list, without proof, the (global) commutator relations for the root groups of a Moufang quadrangle from [Tit94, 4.7] or [TW, chapter21], since they will be used several times in this section. For this purpose we enumerate the roots of the apartment $\mathcal{A}$ in the following way:

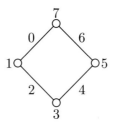

Let U_i, $0 \le i \le 7$, be the corresponding root subgroups and read the indices mod 8 (i.e. $U_8 = U_0$ and so on.)

If there exist non-abelian root subgroups we assume without loss of generality that the U_j, j odd, are non-abelian. For j odd let

$$Y_j = Z(\langle U_{j-2}, U_{j-1}, U_j, U_{j+1}, U_{j+2}\rangle) \cap U_j.$$

Then we have

(1) $U_i' = 1$ for i even.
(2) $U_j' \le Y_j \le Z(U_j)$ for j odd.
(3) $[U_i, U_{i+1}] = 1$.
(4) $[U_{2i}, U_{2i+2}] \le Y_{2i+1}$.
(5) $[U_{2i-1}, U_{2i+1}] \le U_{2i}$.
(6) $[U_i, U_{i+3}] \le U_{i+1} U_{i+2}$.
(7) $[U_{2i-2}, Y_{2i+1}] \le Y_{2i-1} U_{2i}$.

(5.4) Lemma

Suppose Φ is of type BC_ℓ and $r \in \Phi$ is conjugate to r_ℓ under W. Then the following hold:

(1) $A_{2r} = T_P$ for some point P of $\mathcal{B}$. (Considering $\mathcal{B}$ as polar space.)
(2) If $\ell = 2$ then, without loss of generality, $A_r = U_1$ and $A_{2r} = Y_1$ in the notation of (5.3).
(3) In any case $A_r' \le A_{2r} \le Z(A_r)$.

Proof. Suppose first $\ell \ge 3$. Then $\mathcal{B}$ is a classical polar space, since by [Tit74, (9.1),(9.3)] the root system corresponding to a non-embeddable polar space of rank 3 is of type C_3. Let V be the vector space underlying $\mathcal{B}$. With slightly incorrect notation, we identify G and its subgroups with subgroups of $GL(V)$. Then, if P is the isotropic 1-space to which r corresponds, and

$$U_P = \{\sigma \in G \mid [V, \sigma] \subseteq P^\perp, [P^\perp, \sigma] \subseteq P, [P, \sigma] = 0\},$$

where $\perp$ is defined with respect to the underlying (σ, ϵ)- hermitian form, the definition of root subgroup implies $A_r \leq U_P$. (One has to identify $\mathcal{A}$ with a polar frame of V!) Hence by definition:

$$A_{2r} = T_P = \text{the set of all isotropic transvections associated}$$
$$\text{to } P \text{ in the notation of } (1.4).$$

Applying the 3-subgroup lemma to the action of U_P on V one obtains

$$[P^\perp, U_P'] = 0, [V, U_P'] \subseteq P.$$

Hence

$$A_r' \leq U_P' \leq T_P \text{ and } T_P \leq Z(U_P) \cap A_r.$$

This proves (1) and (3) in case $\ell \geq 3$.

Next assume $\ell = 2$. We use the following self-explanatory notation for the apartment $\mathcal{A}$:

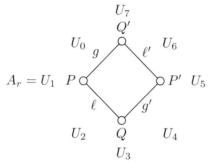

where P, P', Q, Q' are points; g, g', ℓ, ℓ' lines and $U_0, \ldots, U_7$ root subgroups with $A_r = U_1$. (Passing to the dual quadrangle if necessary, we may choose the notation accordingly!) Then we have to show:

$$(*) \qquad\qquad A_{2r} = T_P = Y_1.$$

By the Moufang property we have:

U_7 fixes P and acts transitively on the lines through P different from g.

U_0 fixes P and ℓ and acts transitively on the points of ℓ different from P.

U_2 fixes P and g and acts transitively on the points of g different from P.

U_3 fixes P and ℓ and acts transitively on the lines through P different from ℓ.

Let $R = \langle U_{-1}, \ldots, U_3 \rangle$. Since $Y_1 \leq Z(R)$ this immediately implies $Y_1 \leq T_P$. Let K be the kernel of the action of R on:

$$\mathcal{W} := \text{The set of points, which lie on some line through } P.$$

Then $T_P = K \leq U_1$, whence $T_P \triangleleft R$. Now $[U_7, K] \leq K \cap U_0 = 1$ by I(4.9). Hence $K \leq C(\langle U_7, U_3 \rangle)$. By (5.3) we have $[U_0, U_1] = 1 = [U_1, U_2]$. So $T_P \leq Z(U_1)$ remains to be shown. As $T_P = K$ we have $T_P \trianglelefteq U_1$. We will show:

$$(**) \qquad\qquad U_1 \leq Y_1 U_2 Y_3 [U_0, U_3]$$

using (5.3). By the above $(**)$ shows $T_P \leq Z(R)$, which proves $(*)$ and (5.4) in case $\ell = 2$.

For the proof of $(**)$ let $M = U_1 U_2 U_3$ and $M_0 = Y_1 U_2 Y_3$. Then by (5.3) $M_0 \trianglelefteq M$ and both groups are $F = \langle U_0, U_4 \rangle$-invariant. Now we have

$$U_1 \cap Y_1 U_2 U_3 = Y_1 (U_1 \cap U_2 U_3) = Y_1,$$

since by I(4.9) $U_1 \cap U_2 U_3 = 1$. Thus, if $\widetilde{M} = M/M_0$ then $\widetilde{M} = \widetilde{U}_1 \oplus \widetilde{U}_3$ is a $\mathbb{Z}F$-module. (By (5.3) $U_1' \leq Y_1$!) Since F is by I(4.12) a rank one group we have $F = \langle u, U_4 \rangle$ for each $1 \neq u \in U_0$. Hence $\widetilde{U}_3 [\widetilde{U}_3, u]$ is F-invariant. Since there exist a $w \in F$ with $U_4^w = U_0, U_3^w = U_1$ we have $\widetilde{U}_1 \subseteq \widetilde{U}_3 [\widetilde{U}_3, u]$ and thus $\widetilde{U}_1 = [\widetilde{U}_3, u] \subseteq [\widetilde{U}_3, U_0]$ which proves $(**)$. $\qquad\square$

We need an additional lemma on buildings:

(5.5) Lemma

Suppose $\ell \geq 3$ and $\alpha \neq \pm\beta$ are roots in Φ with walls $\partial\alpha$ resp. $\partial\beta$. Then there exists a rank 2 residue Δ_0 of $\mathcal{A}$ in $\partial\alpha \cap \partial\beta$ and for each such Δ_0 the following hold:

(1) If Δ is the rank 2 residue of $\mathcal{B}$ containing Δ_0, then $\Delta_0 = \Delta \cap \mathcal{A}$ is an apartment of Δ.

(2) Let $\gamma \in [\alpha, \beta]$. Then $\gamma \cap \Delta = \gamma \cap \Delta_0$ is a root of the apartment Δ_0 and

$$\gamma \cap \Delta_0 \in [\alpha \cap \Delta_0, \beta \cap \Delta_0].$$

(3) $V = \langle A_\gamma \mid \gamma \in [\alpha, \beta] \rangle$ acts faithfully on Δ.

Proof. To prove the existence of Δ_0 identify $\mathcal{A}$ with the set of elements of the Coxeter group $W = \langle w_i \mid i \in I \rangle$. Then it is well known and easy to show that

$$\langle w_\alpha, w_\beta \rangle^g \leq \langle w_i, w_j \rangle \text{ for some } g \in W; i, j \in I.$$

Hence $\Delta_{i,j}(1)^g$ is a residue of $\mathcal{A}$ of type i, j invariant under w_α and w_β, whence contained in $\partial\alpha \cap \partial\beta$.

Now (1) is obvious and (2) is an immediate consequence of (1). The proof of (3) uses I(4.8):

$$\text{Let } \Delta_0' = \{c' \in \mathcal{A} \mid c' \text{ opposite to some } c \in \Delta_0\}.$$

Then I(4.7)(b) shows that Δ_0' is a rank 2 residue of $\mathcal{A}$ contained in $\partial\alpha \cap \partial\beta$. Let Δ' be the rank 2 residue of $\mathcal{B}$ containing Δ_0'. Then by I(4.12) A_γ fixes Δ and Δ' for each $\gamma \in [\alpha,\beta]$. In particular V acts on Δ. Let now $g \in V$ be in the kernel of the action of V on Δ and $c \in \Delta_0 \cap \alpha \cap \beta$ and $e \in \Delta'$ with $d(c,e)$ minimal. Then e is uniquely determined and $e \in \Delta_0'$. (This is the so called gatedness property, see [Sch95, (5.1.7)]). Hence $e^g = e$. Now, as $e \in \Delta_0'$ the opposite e' of e lies in Δ_0, whence $(e')^g = e'$ and g fixes $\mathcal{A}$ by I(4.6)(5), hence is the identity on $\mathcal{A}$.

In particular g fixes the opposite $c' \in \Delta_0'$ of c and, since $c \in \alpha \cap \beta$, V and whence g fixes all chambers adjacent to c. Now I(4.8) implies $g = \mathrm{id}_{\mathcal{B}}$. $\square$

As a Corollary of (5.5) one obtains:

(5.6) Corollary

Assume the same hypothesis and notation as in (5.5). Then the commutator $[A_\alpha, A_\beta]$ can be computed in $V\mid_\Delta$.

This is an immediate consequence of (5.5), since the map $x \to x\mid_\Delta, x \in V$ is an isomorphism with $A_\gamma \to A_{\gamma\cap\Delta_0}$.

With (5.4)–(5.6) we obtain:

(5.7) Theorem

The following hold:

(1) Let $r,s \in \Phi$ with $r \neq \pm s$. Then
$$[A_r, A_s] \le \langle A_\alpha \mid \alpha \in (r,s)\rangle.$$

(2) Let $r,s \in \widetilde{\Phi}$ be linearly independent. Then
$$[A_r, A_s] \le \langle A_\alpha \mid \alpha \in \widetilde{\Phi} \text{ and } \alpha = \lambda r + \mu s; \lambda,\mu \in \mathbb{N}\rangle.$$

(3) Let $U_i := \langle A_s \mid s \in \widetilde{\Phi}^+, h(s) \ge i\rangle$ and h the greatest height of a root in $\widetilde{\Phi}^+$. Then
$$U = U_1 \ge U_2 \ge \cdots \ge U_h > 1$$
is a descending central series of U. In particular U is nilpotent of class at most h and $U_h \le Z(U)$.

Proof. (1) By (5.6) the commutator $[A_r, A_s]$ can be computed in the action of $V = \langle A_r, A_s\rangle$ on some rank 2 residue Δ of $\mathcal{B}$, which is by I(4.11) a Moufang polygon, since for $\alpha \in (r,s)$ we have $A_\alpha = A_{\alpha \cap \Delta}$ by the comment after the proof of I(4.11). Hence (1) is [VM98, (5.2.3)].

To prove (2) let $r', s' \in \Phi$ with $r' = r$ or $r = 2r'$ (in case $2r'$ exists!) and $s' = s$ resp. $s = 2s'$. Then again by (5.6) $[A_{r'}, A_{s'}]$ can be computed in the action of $V = \langle A_{r'}, A_{s'} \rangle$ on some rank 2 residue Δ, which is a Moufang polygon. (I.e., if the commutator formula of (2) holds for $A_r|_\Delta$ and $A_s|_\Delta$ it also holds for A_r and A_s, since by the comment after the proof of I(4.11) we have $A_\alpha = A_{\alpha \cap \Delta}$ for all roots α occurring in commutator formula (2).)

Now by the assumption of this section and a theorem of Tits and Weiss, see [VM98, (5.3.4)] either Δ is not irreducible or it is an n-gon with $n = 3, 4$ or 6. In the first case $[A_r, A_s] \leq [A_{r'}, A_{s'}] = 1$ by (1). In the second case we may by the above assume $\Delta = \mathcal{B}$. Now (2) follows from a theorem of [Tit94], see also [VM98, (5.4.6)], which says that the $\{A_r \mid r \in \widetilde{\Phi}\}$ is a "root datum" of A_2, B_2, BC_2, C_2 or G_2. (For the most difficult case, when Δ is a Moufang quadrangle this follows from the commutator relations listed under (5.3) and (5.4)(2), after making the proper identifications of the root groups. I.e., if $\Pi = \{\alpha, \beta\}$, then $U_0 = A_\alpha, U_3 = A_\beta, Y_3 = A_{2\beta}$ and so on!)

Finally (2) implies, that if $r \in \widetilde{\Phi}^+$ and $s \in \widetilde{\Phi}^+$ with $h(s) \geq i$, then $[A_r, A_s] \leq U_{i+1}$. This in turn shows that $[U, A_s] \leq U_{i+1}$ for each $s \in \widetilde{\Phi}$ with $h(s) \geq i$ and then that $[U, U_i] \leq U_{i+1}$. This proves (3).   □

(5.8) Lemma

For $r \in \Phi$ the following hold:

    (1) H_r fixes all chambers in $\mathcal{A}$.

    (2) H_r fixes all root $s \in \Phi$.

    (3) H_r normalizes all $A_s, s \in \widetilde{\Phi}$.

Proof. By I(4.12)(2) H_r fixes the roots r and $-r$ globally, whence also $\mathcal{A} = r \cup (-r)$. On the other hand H_r fixes by I(4.12)(1) some chamber in $\mathcal{A}$ and thus all chambers in $\mathcal{A}$. (Since by I(4.3) the full automorphism group of $\mathcal{A}$ is the corresponding Coxeter group and thence acts regularly on $\mathcal{A}$.)
This proves (1). (2) is an immediate consequence of (1). Clearly (2) implies that H_r normalizes all $A_s, s \in \Phi$. If now $s \in \Phi$ with $2s \in \widetilde{\Phi}$, then the definition of A_{2s} implies that H_r normalizes with A_s also A_{2s}. This proves (3).   □

(5.9) Corollary

For $r, s \in \Phi$ we have $[H_r, H_s] \leq H_r \cap H_s$.

Proof. By (5.8) H_r normalizes A_s and A_{-s} and thus also $H_s = N_{X_s}(A_s) \cap N_{X_s}(A_{-s})$. $\qquad\square$

(5.10) Notation

For each $r \in \Phi$ choose by I(4.12) a $w_r \in X_r$ with $A_r^{w_r} = A_{-r}$ and $w_r^2 \in H_r$. Then w_r interchanges A_r and A_{-r} and obviously

$$H_r w_r = \{n_r \in X_r \mid n_r \text{ interchanges } A_r \text{ and } A_{-r}\}.$$

Hence for $s \in \Phi$, H_s normalizes $H_r w_r$ and thus also $\langle H_r w_r \rangle = H_r \langle w_r \rangle$.

Let $N := \langle H_r, w_r \mid r \in \Phi \rangle$. By (5.8) (3) each H_r normalizes U, whence H normalizes U. Set $B := U \cdot H$. Then we obtain:

(5.11) Lemma

The following hold:

(1) $[H_r, w_s] \le H_s$ for $r, s \in \Phi$.
(2) $H \trianglelefteq N$ and $W := \overline{N} = N/H$ is generated by the involutions $\overline{w}_r \in \Phi$.
(3) H fixes all chambers of $\mathcal{A}$.
(4) w_r fixes $\mathcal{A}$ and interchanges r and $-r$.
(5) W acts as $W(\Phi)$ on $\mathcal{A}$.

Proof. Since H_s normalizes $H_r \langle w_r \rangle$ and $|H_r \langle w_r \rangle : H_r| = 2$ we obtain (1). (2) is a consequence of (1). (3) is a consequence of (5.8)(1). (4) follows from I(4.12)(2), since $\mathcal{A} = r \cup (-r)$. (3) and (4) imply that W acts on $\mathcal{A}$ as the group generated by the reflections on the walls $\partial r, r \in \Phi$. Hence W acts as $W(\Phi)$ on $\mathcal{A}$. (I.e. I(4.6).) $\qquad\square$

Notice that we have not shown yet, that H is the kernel of the action of N on $\mathcal{A}$. This will be shown in (5.17), when we know that $B = G_c$.

(5.12) Lemma

The following hold:

(1) $N = H \langle w_r \mid r \in \Pi \rangle$.
(2) $H = \Pi H_r, r \in \Pi$.
(3) $G = \langle X_r \mid r \in \Pi \rangle$.
(4) If $x \in U$ normalizes A_r and A_{-r} for all $r \in \Pi$, then $x = 1$. In particular $B \cap N = H$ and $U \cap H = 1$.

Proof. (1) It follows from (5.11)(5) and the proof of [Car72, (2.1.8)], that each root $r \in \Phi$ is an image of some fundamental root under some element of $N_0 = \langle w_r \mid r \in \Pi \rangle$. This shows $w_r \in HN_0$ for each $r \in \Phi$, which proves (1).

Now by (5.11) and the definition of the A_r, X_r and $H_r, r \in \Phi$ we have for each $w \in N$:

$$(*) \qquad\qquad A_r^w = A_{rw}, X_{rw} = X_r^w, H_r^w = H_{rw}.$$

Further for fundamental roots r_i, r_j we have by (5.11)(1)

$$(H_{r_i})^{w_{r_j}} \leq H_{r_i} H_{r_j}.$$

This implies that $\Pi H_r, r \in \Pi$ is N_0 invariant. Now (1) together with (5.9) imply (2).

Now by (1) and (2) $N \leq \langle X_r \mid r \in \Pi \rangle$. Hence $(*)$ implies (3). To prove (4) let for $i \in I$:

$$U_i = \langle A_r \mid r \in \widetilde{\Phi}^+ - \{r_i, 2r_i\} \rangle.$$

Then by (5.7)(2) $[U_i, A_{r_i}] \leq U_i$, since a root of the form $s + \lambda r_i, \lambda \in \mathbb{N}$ is in $\widetilde{\Phi}^+$, since each root α in $\widetilde{\Phi}$ can be uniquely expressed as a linear combination of fundamental roots and $\alpha \in \widetilde{\Phi}^+$ if and only if the coefficient of some fundamental root in this expression is positive. This implies $U_i \lhd U$ and $U = U_i A_{r_i}$. Now $U_i^{w_{r_i}} = U_i$ and $A_{r_i}^{w_{r_i}} = A_{-r_i}$, since w_{r_i} maps each positive root different from r_i onto some positive root. Hence (5.7)(4) shows that $U_i A_{r_i} = U$ and $U_i A_{-r_i}$ are nilpotent. But X_{r_i} is by I(2.1) not nilpotent. Since by I(1.4) $X_{r_i} = \langle y, A_{-r_i} \rangle$ for each $y \in A_{r_i}^{\#}$, this implies $A_{r_i} \cap U_i = 1$.

Suppose now $x = u\alpha$, $u \in U_i$ and $1 \neq \alpha \in A_{r_i}$ satisfies (4). Then, again by I(1.4):

$$U_i X_{r_i} = U_i \langle A_{-r_i}, A_{-r_i}^\alpha \rangle = U_i \langle A_{-r_i}, A_{-r_i}^x \rangle \leq U_i A_{-r_i},$$

since x normalizes A_{-r_i}, a contradiction. Hence $x \in U_i$ for $x \in U$ satisfying (4) and each $i \in I$. This implies $[x, A_{r_i}] \leq U_i \cap A_{r_i} = 1$. But then x centralizes each A_{r_i} and thus also each $X_{r_i}, r_i \in \Pi$. Now (3) implies $x \in Z(G)$. Since G acts faithfully and by (5.1) chamber transitively on $\mathcal{B}$, this shows $x = 1$.

Now clearly $H \leq B \cap N$. If $B \cap N \neq H$, then $U \cap N \neq 1$ by definition of B. (I.e. $B = U \cdot H!$) But an element $1 \neq x \in U \cap N$ fixes $\mathcal{A}$ and c and thus all chambers in $\mathcal{A}$. Thus x fixes all roots and thence normalizes all root subgroups, a contradiction to the above. $\qquad\qquad\qquad\square$

(5.13) Theorem

(B, N) is a BN-pair of G. (For definition of a BN-pair see [Car72, 8.2].)

Proof. We verify the BN-pair axioms:

(1) $G = \langle B, N \rangle$.

This is obvious, since $A_r \le U$ and since $A_{-r} = A_r^{w_r}$ for $r \in \Phi$.

(2) $B \cap N = H$.
  This was already shown in (5.12)(4).

(3) W is generated by $w_{r_1}, \ldots, w_{r_\ell}$. (Identifying w_r with $\overline{w}_r$!)

This is (5.12)(1).

(4) $B \ne B^{w_r}$ for $r \in \Phi$.

B^{w_r} fixes the chamber $c^{w_r} \ne c$. Hence (4) follows from (5.1).

(5) For $n \in N$ and $r \in \Pi$ we have:

$$(Bw_r B)(BnB) \subseteq Bw_r nB \cup BnB.$$

Proof. For $r \in \Pi$ let $U_r = \langle A_s \mid s \in \tilde{\Phi}^+ - \{r, 2r\}\rangle$. ($2r$ only if $\Phi \ne \tilde{\Phi}$!) Then as shown in (5.12)(4) $U_r \lhd U$ and $U = U_r A_r$. This implies:

$$
\begin{aligned}
(Bw_r B)(BnB) &= Bw_r HU_r A_r nB = Bw_r HU_r nA_r^n B\\
&= Bw_r(HU_r)w_r(w_r n)A_{r^n} B = B(HU_r)^{w_r} w_r nA_{r^n} B.
\end{aligned}
$$

Now $s^{w_r} \in \tilde{\Phi}^+ - \{r, 2r\}$ for each $s \in \tilde{\Phi}^+$ with $s \ne \lambda r, \lambda \in \mathbb{N}$. Hence $(HU_r)^{w_r} = HU_r \le B$. If now $r^n \in \Phi^+$, then $A_{r^n} \le U \le B$ and we obtain:

$$(Bw_r B)(BnB) \le Bw_r nB.$$

Suppose $r^n \in \Phi^- = \Phi - \Phi^+$. Set $n' = w_r n$. Then $r^{n'} = (-r)^n = -r^n \in \Phi^+$. Now by I(4.12) we have:

$$X_r = A_r H_r \cup (A_r H_r)w_r(A_r H_r).$$

Hence

$$A_{-r}^{\#} \subseteq (A_r H_r)w_r(A_r H_r) \subseteq Bw_r B,$$

since $A_r H_r = N_{X_r}(A_r)$ and $A_{-r} \cap (N(A_r)) = 1$ by I(1.2). This implies

$$
\begin{aligned}
w_r Bw_r &= w_r^{-1}(A_r U_r H)w_r = A_{-r}U_r H\\
B \cup (Bw_r B)B &\subseteq B \cup Bw_r B,
\end{aligned}
$$

which shows that $B \cup Bw_r B$ is a subgroup (It is clearly closed under inverses!).

Now as $r^{n'} \in \Phi^+$ we have by the above

$$(Bw_r B)(Bn'B) = Bw_r n'B.$$

Thus

$$
\begin{aligned}
(Bw_r B)(BnB) &= (Bw_r B)(Bw_r n'B) \\
&= (Bw_r B)(Bw_r B)(Bn'B) \subseteq (B \cup Bw_r B)(Bn'B) \\
&\subseteq Bn'B \cup (Bw_r B)(Bn'B) = Bn'B \cup Bw_r n'B \\
&= Bw_r nB \cup BnB.
\end{aligned}
$$

This proves (5) and Theorem (5.13).   □

Remark. It is shown in [Ron89, 6.16] that $(\widehat{B}, N)$ is a BN-pair of G, where $\widehat{B} = \langle A_\alpha, \widehat{H} \mid \alpha \in \Phi^+ \rangle$ and $\widehat{H}$ is the subgroup of G fixing all chambers in $\mathcal{A}$. By (5.8)(1) $H \leq \widehat{H}$. Our Theorem (5.13) actually shows, see (5.17), that $H = \widehat{H}$, an information well-known for Chevalley groups. Insofar as (5.13) is stronger than [Ron89, 6.16].

(5.14) Notation

For $J \subseteq I$ let Φ_J resp. $\widetilde{\Phi}_J$ be the set of roots in Φ (resp. $\widetilde{\Phi}$), which are linear combinations of fundamental roots $r_j, j \in J$. Set:

$$
\begin{aligned}
P_J &:= \langle B, X_{r_j} \mid j \in J \rangle \text{ with } P_\emptyset = B \text{ and } P_j = P_{\{j\}} \\
L_J &:= \langle X_r \mid r \in \Phi_J \rangle \text{ and} \\
U_J &:= \langle A_r \mid r \in \widetilde{\Phi}^+ - \widetilde{\Phi}_J \rangle.
\end{aligned}
$$

If now $r \in \widetilde{\Phi}^+ - \widetilde{\Phi}_J$ and $s \in \widetilde{\Phi}^+ \cup \widetilde{\Phi}_J$ then by (5.7)(2)

$$(*) \qquad [A_r, A_s] \leq \langle A_{\lambda r + \mu s} \mid \lambda, \mu \in \mathbb{N} \text{ and } \lambda r + \mu s \in \widetilde{\Phi} \rangle.$$

Since all coefficients in an expression of a root of $\widetilde{\Phi}$ as a linear combination of fundamental roots are either non-negative or non-positive and since $\lambda \in \mathbb{N}$, clearly $\lambda r + \mu s \in \widetilde{\Phi}^+$. If now $\lambda r + \mu s \in \widetilde{\Phi}_J$, then also $r \in \widetilde{\Phi}_J$ by the uniqueness of an expression as a linear combination of fundamental roots. Hence $\lambda r + \mu s \notin \widetilde{\Phi}_J$, which implies:

$$[A_r, A_s] \leq U_J \text{ for all such } r \text{ and } s.$$

Since obviously $P_J \leq \langle A_s \mid s \in \widetilde{\Phi}^+ \cup \widetilde{\Phi}_J \rangle H$ and since $L_J = \langle X_{r_j} \mid j \in J \rangle$ by (5.12) we obtain:

(5.15) Lemma

The following hold for $J \subseteq I$.

(1) $U_J \trianglelefteq P_J$.
(2) $P_J = (U_J L_J) H$.

As a Corollary to (5.13) we obtain:

(5.16) Proposition

The following hold:

(1) $\mathcal{B} = \mathcal{C}(G, B, (P_i)_{i \in I})$ (In the notation of I section 4.)
(2) G acts transitively on the pairs $(c, \mathcal{A})$; where $c \in \mathcal{B}$ is a chamber and $\mathcal{A}$ is an apartment of $\mathcal{B}$ containing c.

Proof. (1) By (5.1) we have $B \leq G_c$ (c the unique chamber fixed by U!). Hence G_c is one of the parabolic subgroups, i.e. $G_c = P_J$ for $\emptyset \subseteq J \subseteq I$. (It is a well known fact, see [Car72, 8.3], that in a group with a BN-pair each subgroup containing B is of the form P_J for $\emptyset \subseteq J \subseteq I$.) Suppose $J \neq \emptyset$ and let $j \in J$. Then $w_{r_j} \in P_J$, a contradiction since w_{r_j} does not fix c since it interchanges r_j and $-r_j$. Hence $G_c = B$.

Next we show $G_{\Delta_i(c)} = P_i = B \cup B w_{r_i} B$. For this let M_{r_i} be the "wall" of $\mathcal{A}$ corresponding to the reflection w_{r_i}. Then $M_{r_i} \cap \Delta_i(c) = \{c, d\}$ and $c^{w_{r_i}} = d$. Hence $w_{r_i} \in G_{\Delta_i(c)}$ and thus $P_i = B \cup B w_{r_i} B \leq G_{\Delta_i(c)}$.

Suppose $P_i \neq G_{\Delta_i(c)}$. Then, as above, there exists a $j \neq i$ such that $w_{r_j} \in G_{\Delta_i(c)}$. But $c \neq c^{w_{r_j}} \overset{j}{\sim} c$. Hence $c^{w_{r_j}} \in \Delta_j(c)$ and thus $c^{w_{r_j}} \notin \Delta_i(c)$, since $\Delta_i(c) \cap \Delta_j(c) = \{c\}$, a contradiction.

Now (5.1)(1) and I(4.2) show that (1) holds. To prove (2) we show by descending induction on

$$b(\mathcal{A}, \mathcal{A}') = \mathrm{Max}\{d(c, e) \mid e \in \mathcal{A} \cap \mathcal{A}'\}$$

that

(∗) U is transitive on the set of apartments $\mathcal{A}'$ of $\mathcal{B}$ containing c.

Indeed, if $b(\mathcal{A}, \mathcal{A}') = d$, then $\mathcal{A} \cap \mathcal{A}'$ contains a pair of opposite chambers, whence $\mathcal{A} = \mathcal{A}'$ by I(4.6)(5). Suppose now $b = b(\mathcal{A}, \mathcal{A}') < d$ and, if $\overline{\mathcal{A}}$ is an apartment containing c with $b(\mathcal{A}, \overline{\mathcal{A}}) > b$, then there exists an $u \in U$ with

$\overline{\mathcal{A}}^u = \mathcal{A}$. Let $e \in \mathcal{A} \cap \mathcal{A}'$ with $d(c, e) = b$. Then there exists an $\overline{e} \in \mathcal{A}$ with $d(c, \overline{e}) = b + 1$ and $e \overset{i}{\sim} \overline{e}$ for some $i \in I$. Hence $\overline{e} \notin \mathcal{A}'$. As

$$\{e\} = \Delta_i(e) \cap \mathcal{A} \cap \mathcal{A}',$$

there exists an $e' \in \mathcal{A}' \cap \Delta_i(e)$ with $e' \notin \mathcal{A}$. Since $\mathcal{A}$ is convex (see I(4.6)(5)) we have $d(c, e') = b + 1$. By I(4.6)(4) there exists a root r of $\mathcal{A}$ containing $\{c, e\}$ but not $\overline{e}$. Hence by I(4.12)(1) there exists an $x \in A_r \le U$ with $(e')^x = \overline{e}$. Thus

$$b(\mathcal{A}, (\mathcal{A}')^x) \ge b + 1 > b,$$

since $\{c, \overline{e}\} \subseteq \mathcal{A} \cap \mathcal{A}'^x$. Hence by induction assumption there exists a $u \in U$ with $\mathcal{A}'^{xu} = \mathcal{A}$, which proves $(*)$.

Now $(*)$ and $(5.1)(1)$ obviously imply (2). □

(5.17) Corollary

The following hold:

(1) $B = G_c$.
(2) $H = \{\sigma \in G \mid d^\sigma = d \text{ for each } d \in \mathcal{A}\}$.
(3) N is the stabilizer of $\mathcal{A}$ in G.

Proof. (1) was shown in (5.16). Suppose σ fixes all chambers of $\mathcal{A}$. Then $\sigma \in G_c = B = U \cdot H$. Let $\sigma = uh, u \in U$ and $h \in H$. Then, since h fixes by (5.11)(3) all chambers of $\mathcal{A}$, also u fixes all chambers of $\mathcal{A}$. Hence u fixes all roots and thus normalizes all root subgroups. Hence $u = 1$ by (5.12)(4). This proves (2). Finally, if $x \in G_\mathcal{A}$, then there exists a $y \in N$ such that $xy \in G_c = B$, since N is transitive on $\mathcal{A}$. Hence xy fixes all chambers of $\mathcal{A}$ and thus $xy \in H$ by (2). Thus $x \in Hy^{-1} \le N$, which proves (3). □

As a Corollary we obtain:

(5.18) Corollary

Let $\mathrm{Aut}(\mathcal{B})$ be the group of all type-preserving automorphisms of $\mathcal{B}$. Then

$$G \trianglelefteq \mathrm{Aut}(\mathcal{B}) = G\widehat{H},$$

where $\widehat{H} = \{\sigma \in \mathrm{Aut}(\mathcal{B}) \mid \sigma \text{ fixes each chamber of } \mathcal{A}\}$.

Proof. By (5.16) G is transitive on the pairs $(c, \mathcal{A})$, where c is a chamber and $\mathcal{A}$ an apartment containing c. Hence the Frattini argument implies

$$\mathrm{Aut}(\mathcal{B}) = G \cdot \widehat{H}, \widehat{H} \text{ the stabilizer of } (c, \mathcal{A}).$$

Since $\mathcal{A}$ is a Coxeter complex, $\widehat{H}$ fixes all chambers of $\mathcal{A}$. But then $\widehat{H}$ fixes all roots of $\mathcal{A}$, whence normalizes all $X_r, r \in \Phi$, and thus also normalizes $G = \langle X_r \mid r \in \Phi \rangle$. $\square$

(5.18) shows that the definition of the Lie type group G is independent of the apartment $\mathcal{A}$ we started with.

Finally, before we can prove the main result of this section, a lemma which substitutes for I(4.12) in case $\Phi \neq \widetilde{\Phi}$.

(5.19)   Lemma

Suppose $\widetilde{\Phi}$ is of type C_ℓ or BC_ℓ, $\Pi = \{r_1, \ldots, r_\ell\}$ and $r = r_\ell$ if $\widetilde{\Phi}$ is of type C_ℓ (resp. $r = 2r_\ell$ if $\widetilde{\Phi}$ is of type BC_ℓ). Then $X_r = \langle A_r, A_{-r} \rangle$ is a special rank one group with AUS.

Remark. Even in case $\widetilde{\Phi}$ of type C_ℓ (5.19) gives more information than I(4.12). Namely it shows that X_r is special. Of course (5.19) also holds for all conjugates s of r_ℓ under W. This allows us to pick the element w_s interchanging A_s, A_{-s} and A_{2s}, A_{-2s} in $X_{2s} = \langle A_{2s}, A_{-2s} \rangle$ (if $\widetilde{\Phi}$ is of type BC_ℓ). Hence w_s normalizes H_s and $H_{2s} = N_{X_{2s}}(A_{2s}) \cap N_{X_{2s}}(A_{-2s})$, a fact we will use.

Proof of (5.19). By I(4.12) X_{r_ℓ} acts doubly transitively on $\Delta_\ell(c)$. Hence X_{r_ℓ} acts on the rank 2 residue $\Delta = \Delta_{\ell-1,\ell}(c)$, which is by I(4.11) a Moufang quadrangle, such that A_{r_ℓ} and A_{-r_ℓ} act as root groups corresponding to $r_\ell \cap \Delta$ (resp. $-r_\ell \cap \Delta$) on Δ. (Comment after the proof of I(4.11)!). Moreover, since X_{r_ℓ} is a rank one group, the kernel K of the action of X_{r_ℓ} on Δ lies in $Z(X_{r_\ell})$. (I(1.10)!) We will verify the hypothesis of I(2.4) for the action of $X_{r_\ell}, X_r, A_{r_\ell}$ and A_r on some abelian quotient of a nilpotent subgroup of G. Clearly by (5.4) and definition of A_{2r_ℓ}, A_r is H_{r_ℓ}-invariant.

To enumerate the roots of $\mathcal{A} \cap \Delta$ we use the notation of (5.3) and without loss of generality $A_{r_\ell} = U_1$ and $A_r = Y_1$. (It may be the case that $A_r = A_{r_\ell}$, whence $U_1 = Y_1$. This situation is included in our proof with obvious notation!) Then $A_{-r_\ell} = U_5, A_{-r} = Y_5$. Let $N = U_2 U_3 U_4 / Y_3$. Then N is abelian by equations (1)–(4) of (5.3). Moreover

$$[N, U_1] \leq U_2 U_3 / Y_3, \quad [N, U_5] \leq U_3 U_4 / Y_3$$

by equation (6) of (5.3), so that N is X_{r_ℓ}-invariant. Now by equation (7) of (5.3) $[N, Y_1] \leq U_2 Y_3 / Y_3$ so that

$$[N, Y_1, U_1] = 0.$$

It remains to show that $[N, X_{r_\ell}, X_r] \neq 0$. Now w_{r_ℓ} interchanges U_2 and U_4. Since $U_2 \cap Y_3 U_4 \leq Y_3$ (for example since $\langle U_2, U_6 \rangle$ is a rank one group!) we

obtain
$$U_2U_4Y_3/Y_3 = [U_2, w_{r_\ell}]U_2Y_3/Y_3 = [U_4, w_{r_\ell}]U_4Y_3/Y_3.$$

If now Y_1 centralizes $[N, X_{r_\ell}]$, then it centralizes N, since it centralizes by definition U_2U_3. But this would by I(1.10) imply $X_{r_\ell} = C_{X_{r_\ell}}(N)A_{r_\ell}$, which is, by what we have shown, not the case.

Now I(2.4) shows that the group induced by X_r on Δ is a special rank one group with AUS. Since by I(1.4) A_{r_ℓ} is strongly closed in $A_{r_\ell}K \le A_{r_\ell}Z(X_{r_\ell})$ (i.e. each element $a^x, 1 \ne a \in A_{r_\ell}$ and $x \in X_{r_\ell}$, which lies in $A_{r_\ell}K$ lies already in A_{r_ℓ}!) this implies that X_r is a special rank one group. (I.e. if the equation $a^b = b^{-a}, a \in A_r^\#, b \in A_{-r}^\#$ holds mod$K$, it holds already in X_r!)   □

Now the main result of this section:

(5.20) Theorem

Let h be a highest root in $\widetilde{\Phi}^+$ and $A = A_h$. Then $\Sigma = A^G$ is a class of abstract root subgroups of $G_0 = \langle \Sigma \rangle$.

Remark. If G is a Chevalley-group, then the proof of (5.20) is much easier. Namely let $g = b_1nb_2 \in BNB = G$. Then, since $B \le N(A_h)$, we have

$$\langle A_h, A_h^g \rangle = \langle A_h, A_h^n \rangle^{b_2} = \langle A_h, A_r \rangle^{b_2}$$

with $r = h^n \in \Phi$. Hence (5.20) follows from the Chevalley-commutator-relations and the fact that $\langle A_h, A_{-h} \rangle \simeq SL_2(k)$.

Our proof of (5.20) is complicated by the fact that we do not have exact commutator-relations, but just an inequality in (5.7)(2).

Proof. By (5.7)(3) $A \le Z(U)$, whence $A \lhd B$. This shows that $P = N(A)$ is a parabolic subgroup of G and thus $P = P_J$ for some $J \subseteq I$. If $h \in \Phi$, w_h is well defined. If $h = 2s, s \in \Phi$ set $w_h = w_s \in X_{2s}$. In both cases let

$$W_0 := \langle w_r \mid r \in \Pi \text{ and } w_rw_h = w_hw_r \rangle.$$

Then W_0 is a parabolic subgroup of W and the structure of W_0 can be read of the extended Dynkin diagram of W. (The diagram of $\Pi \cup \{-h\}$ is the corresponding affine diagram, see [Bou81]). If now $r \in \Pi$ with $w_rw_h = w_hw_r$, then w_r fixes h, since r is the only positive root mapped by w_r onto a negative root. Hence w_r normalizes A_h. Since $[A_r, A_h] = 1$ we obtain:

$$(*) \qquad\qquad X_r = \langle w_r, A_r \rangle = \langle A_r, A_r^{w_r} \rangle \le C(A_h).$$

$(*)$ implies $BW_0B \le P$ and $\langle U, W_0 \rangle \le C(A_h)$. On the other hand the list of affine diagrams shows that W_0 is, except Φ which is of type A_ℓ, already a maximal parabolic subgroup of W. Hence we obtain in any case $P_J = P = BW_0B$ and $U_JL_J \leq C(A)$. (In case Φ of type A_ℓ the extended Dynkin diagram is

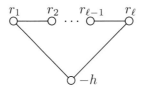

Hence $W_0 = \langle w_{r_2}, \ldots, w_{r_{\ell-1}} \rangle$. Since it is easy to see that neither w_{r_1} nor W_{r_ℓ} fix h, we obtain $P = BW_0B$ also in this case.)

Now it is easy to show, see [Asc86, Exercise 6 for chapter 14] that the map

$$W_0wW_0 \to PwP, w \in W$$

is a bijection of the double cosets of W_0 in W onto the double cosets of P in G. Hence the permutation rank of G on Σ is the same as the permutation rank of W on $\{W_0w \mid w \in W\}$.

Now assume for sake of simplicity that $\widetilde{\Phi}$ is not of type $A_\ell, \ell \geq 2, B_\ell, \ell \geq 3$ or $D_\ell, \ell \geq 4$. Indeed if $\widetilde{\Phi}$ is of type $A_\ell, \ell \geq 3$, then G is a linear group and Σ the class of transvection groups. Moreover, if $\widetilde{\Phi}$ is of type $B_\ell, \ell \geq 3$ or $D_\ell, \ell \geq 4$, then G is an orthogonal group of Witt index ℓ and Σ is the class of Siegel transvection groups. Hence it follows from (1.3) resp. (1.5) that Σ is a class of abstract root subgroups of G. (The case Φ of type A_2 will be treated in the exercises, see also [Tim94a]. It is possible, and in the exercises partially will be done, to treat also the cases A_ℓ, B_ℓ, D_ℓ in our general concept. But since the permutation rank of G on Σ is higher, this is more complicated.) So we are left with the cases:

(i) Φ of type $C_\ell, \ell \geq 2$ or $\widetilde{\Phi}$ of type BC_ℓ.
(ii) Φ of type G_2.
(iii) Φ of type F_4, E_6, E_7 or E_8.

If Φ is of type C_ℓ or $\widetilde{\Phi}$ of type BC_ℓ, then the extended diagram is

$$\underset{h}{\circ}\!=\!=\!\underset{r_1}{\circ}\!-\!-\!\underset{r_2}{\circ}\cdots\underset{r_{\ell-1}}{\circ}\!-\!-\!\underset{r_\ell}{\circ}$$

and $W_0 = \langle w_{r_2}, \ldots, w_{r_\ell} \rangle \simeq W(C_{\ell-1})$. It is now easy to see that the permutation rank of W on $\{W_0w \mid w \in W\}$ is equal to the permutation rank of W on w_h^W is 3. (There are three suborbits $\{w_h\}, w_h^W \cap W_0, w_h^W - W_0$!). Thus the permutation rank of G on Σ is 3. Since W has just two orbits on Φ, there exists a conjugate $A_h \neq A_h^w \leq U \leq C(A_h)$. On the other hand $\langle A_h, A_{-h} \rangle$ is by (5.19) a rank one group. This implies that Σ is a set of abstract transvection groups of G_0. Now P normalizes the connectivity component of $\mathcal{F}(\Sigma)$ containing A. Since P is maximal in G and A is not isolated in $\mathcal{F}(\Sigma)$, this shows that G normalizes this connectivity component, whence $\mathcal{F}(\Sigma)$ is connected and Σ is a conjugacy class of abstract transvection groups of G_0.

Since the other cases (ii) and (iii) are similar we will just sketch the proof. In case (ii) it follows from the description of the root system of type G_2 in [Car72, p.46] that $W_0 = \langle w_r \rangle$, r a short fundamental root. Hence the permutation rank of W on $\{W_0 w \mid w \in W\}$ is 4. On the other hand, there are four natural suborbits of the action of P on Σ with representatives:

(a) A_h, $h = 3r + 2s$ where $\Pi = \{r, s\}$, s long.
(b) A_s with $[A_h, A_s] = 1$.
(c) A_{-h} with $\langle A_h, A_{-h} \rangle$ a rank one group by I(4.12).
(d) A_{-s} with $[A_h, A_{-s}] \leq A_{3r+s}$.

(For example since by [Tit94], see also [VM98, (5.4.6)] $\{A_r \mid r \in \Phi\}$ forms a root datum of type G_2.) Thus, to show that Σ is a set of abstract root subgroups of G_0, it remains to prove in case (d) that also the elementwise commutator relations (1.1) II (γ) hold. For this purpose we use (1.2). We have $\Psi = \{r \in \Phi \mid r \text{ long } \}$ is a root system of type A_2. Hence by (1.2)(2) either

$$[\alpha, A_{-s}] = [A_h, \beta] = A_{3r+s} \text{ for all } \alpha \in A_h^{\#}, \beta \in A_s^{\#}$$

or

$$[A_h, A_{-s}] = 1 = [A_h, A_s].$$

But in the second case $w_s \in C(A_h)$ and thence $A_h \lhd G$, which is obviously impossible.

This shows that Σ is a set of abstract root subgroups of G_0. Since, again by the maximality of P, $\mathcal{F}(\Sigma)$ is connected, this proves (5.20) in case (ii).

In the remaining case (iii) we argue that the permutation rank r of W on $\{W_0 w \mid w \in W\}$ is smaller than or equal to 5. Now in case Φ of type E_6, E_7 or E_8 it is well known that $D = w_h^W$ is a class of 3-transpositions of W, whence the permutation rank of W on D is 3. Since $C_W(w_h) = \langle w_h \rangle W_0$, this shows $r \leq 5$. ($w_r \in W_0$, if $r \in \Phi$ with $(r, h) = 0$. Hence W_0 is transitive on $\{r \in \Phi \mid (r, h) = 0\}$. Since $\langle w_h \rangle W_0$ has just two orbits on the reflections different from w_h, this shows that W_0 has permutation rank at most 5 on Φ). In case Φ is of type F_4 we have $W_0 \simeq W(B_3) \simeq \Sigma_4 \times \mathbb{Z}_2$. Now the proof that the permutation rank of W on $\{W_0 w \mid w \in W\}$ is 5 is the same as before.

Now there are five natural suborbits of the action of P on Σ with representatives:

(a) A_h.

(b) A_r with $\ell(h) = \ell(r)$ and there exists an $s \in \Phi$ with $\ell(h) = \ell(s)$ and $r + s = h$. In this case

$$[A_r, A_h] = 1 \text{ and } A_h \geq [A_r, A_s].$$

(c) $A_s, s \in \Phi$ with $(h, s) = 0$ and $\ell(h) = \ell(s)$. In this case $[A_h, A_s] = 1$ and there exists no $C \in \Sigma$ with $[A_s, C] = A_h$.

(d) A_{-h}.

(e) $A_\alpha, \alpha \in \Phi$ with $\ell(\alpha) = \ell(h)$ and $\alpha + h \in \Phi$. In this case $[A_h, A_\alpha] \leq A_{h+\alpha}$.

Now in case (b) r, s and in case (e) h, α span a subsystem of type A_2. Hence by $(1.2)(2)$ we have:

$$[a, A_\alpha] = [A_h, b] = A_{h+\alpha} \text{ for each } a \in A_h^{\#}, b \in A_\alpha^{\#},$$

since otherwise $[A_h, A_\alpha] = 1$ and then $A_h \lhd G$. The same argument shows $[A_r, A_s] = A_h$ in case (b), whence the cases (b) and (c) are disjoint. This implies that Σ is a set of abstract root subgroups of G_0 and then, as in case (ii) that $\mathcal{F}(\Sigma)$ is connected. This proves (5.20). □

(5.21) Corollary

Let G_0 and h be as in (5.20). Then the following hold:

(1) G_0' is simple.

(2) If $h \in \Phi$, then $G = G_0$ if all roots have the same length, resp. $G = G_0 A_s$ if $s \in \Phi$ with $\ell(s) \neq \ell(h)$.

(3) If $h = 2r, r \in \Phi$ and $s \in \Phi$ with $\ell(s) \neq \ell(r)$ then $G = (G_0' A_r) A_s$.

Proof. $G_0' B$ is a parabolic subgroup of G. Hence either $G_0' B = G$ or $G_0' B = P_J \leq P_{I_i}$ where $I_i = I - \{i\}$ for some $i \in I$. Now by (5.15) $U_{I_i} \lhd P_{I_i}$ and $A_h \leq U_{I_i}$ by definition of $U_J, J \subseteq I$, since all fundamental roots occur in the expression of h as a linear combination of fundamental roots. Thus in the second case we obtain

$$G_0 = \langle \Sigma \rangle = \langle A_h^{G_0} \rangle = \langle A_h^{G_0'} \rangle \leq U_{I_i}$$

since $G_0 = G_0' A_h$. Since this is impossible we obtain $G = G_0' B$. (At this point we used that $\mathcal{B}$ is irreducible, whence Φ is indecomposable.) As $U \lhd B$ we obtain $G_0' U \lhd G$. Since A_r and A_{-r} are conjugate in G for $r \in \Phi$ this implies $G = G_0' U$. If now all roots have the same length, then as $G_0 = G_0' A_h \lhd G$ and W acts transitively on Φ, we obtain $G = G_0$.

Next assume that there are roots of different length in $\widetilde{\Phi}$. Pick $s \in \widetilde{\Phi}^+$ with $\ell(s) \neq \ell(h)$ and $h(s)$ as big as possible under this constraint. Then by (5.7)(2)

$$(*) \qquad\qquad [U, A_s] \leq \langle A_r \mid r \in \widetilde{\Phi}^+, h(r) > h(s) \rangle.$$

Now by choice of s all these r satisfy $\ell(r) = \ell(h)$, whence they are all conjugate to h under the action of W. Since $G_0 \trianglelefteq G$ this implies $G_0 A_s \trianglelefteq G_0 U = G$.

If now $h \in \Phi$, then W has two orbits on Φ, whence $G = G_0 A_s = (G_0' A_h) A_s$. So suppose $h = 2r, r \in \Phi$. Then it follows from (5.6) and equation (5) of (5.3) that $[U, A_r] \leq A_h A_s$. Hence $(G_0 A_s) A_r \trianglelefteq G$, which then implies $G = (G_0 A_s) A_r = (G_0' A_r) A_s$ since all roots of Φ are either conjugate to r or to s by W.

This proves (2) and (3). Next we show $Z(G_0) = 1$. Clearly $Z(G_0) B$ is a parabolic subgroup. Suppose $Z(G_0) B = P_J$ for $\emptyset \neq J \subseteq I$. If $U \cap Z(G_0) \neq 1$, then also $Z(U) \cap Z(G_0) \neq 1$, since $[U, U \cap Z(G_0)] \leq U \cap Z(G_0)$. But by (2) and (3) $Z(U) \cap Z(G_0) \leq Z(G) \cap U = 1$. Hence $[Z(G_0), U_J] = 1$ and thus by (2) and (3) $Z(G_0) \leq Z(G)$, since each root $r \in \Phi$ is conjugate to some root in $\widetilde{\Phi}^+ - \Phi_J$. But clearly $Z(G) B \neq P_J$ for $J \neq \emptyset$ by (5.15)(2)

To show that G_0' is quasi-simple if Σ is non-degenerate or $|A_h| > 3$ we use (2.14) in combination with (2.15) and (5.19). Since by (2.16)(a) and (c) we may assume that the permutation rank of G on Σ is ≤ 5, it is obvious that condition (2) of (2.14) holds. This proves (1) in that case.

So it finally remains to treat the case $|A_h| \leq 3$ and $\widetilde{\Phi}$ of type C_ℓ or BC_ℓ. Then B is a thick polar space. Now it follows as in (4.1) that $\mathcal{F}(\Sigma_A)$ is connected, for $A = A_h$. Suppose $G_0 > R > P \cap G_0$. Then $\Lambda = A^R$ is a set imprimitivity of the action of G_0 on Σ, which contains no commuting elements. (As $\mathcal{F}(\Sigma_A)$ is connected!) Hence for each pair $A, B \in \Lambda$ we have $\Sigma_A = \Sigma_B$. But then $\Delta_A = \{B \in \Sigma \mid \Sigma_A = \Sigma_B\}$ is a non-trivial set of imprimitivity of the action of G on Σ, which contradicts the maximality of P in G.

So we obtain that $P \cap G_0$ is maximal in G_0 and the simplicity of G_0' follows from exercise (2.24) (7) and (8).   □

(5.22) Corollary

Let G_0 be as in (5.20). Then one of the following holds:

(1) If all roots of Φ have the same length, then $G = G_0'$ is simple.

(2) If Φ is of type F_4 or $B_\ell, \ell \geq 3$, then $G = G_0'$ is simple.

(3) If Φ is of type G_2, then $G_0 = G_0'$ is simple and $G = G_0 A_s$, s as in (5.21)(2).

(4) If $\widetilde{\Phi}$ is of type C_ℓ or BC_ℓ, $\ell \geq 2$, then $G = G_0' A_r$, where either $h = r \in \Phi$ or $h = 2r, r \in \Phi$.

Proof. (1) follows from (5.21) and (2.14), since in that case Σ is non-degenerate. Next, if Φ is of type F_4, then each root $r \in \Phi$ can be embedded in a subsystem of type A_2. Hence by (1.2)(2) $A_r \leq G'$. This implies $G = G' \leq G_0$, whence G is simple.

Next assume Φ is of type $B_\ell, \ell \geq 3$. Let s be as in (5.21)(2). Then we can embed $\{h, s\}$ in a subsystem of type $B_2 = C_2$ such that:

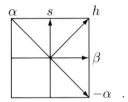

Then the group X_α acts on $N = A_s A_h A_\beta$ and $\widetilde{N} = N/A_h$. Since X_α is a rank one group and since w_α interchanges s and β, it follows that $\widetilde{A}_s \leq [\widetilde{A}_\beta, A_\alpha]$ since $\widetilde{A}_\beta[\widetilde{A}_\beta, A_\alpha]$ is X_α-invariant. Hence $A_s \leq [A_\beta, A_\alpha]A_h$. Now (5.21)(2) shows $G \leq G'A_h \leq G'$ and thus $G = G'$, since Σ is non-degenerate.

In case (3) we have $A_h \leq G_0'$, since Σ is non-degenerate. Now (3) is a consequence of (5.21)(2).

Finally in (4) let s be as in (5.21)(2) or (3). Then the same argument as in case (2) (see also the proof of (5.19)) shows that $A_s \leq G'A_r$. Since A_s is abelian and $G_0'A_r \trianglelefteq G$ as shown in (5.21), we also have $G' \leq G_0'A_r$. Hence

$$G \leq G_0'A_sA_r \leq G'A_r \leq G_0'A_r$$

which proves (4).   □

(5.23) Remark

With more detailed arguments of this type, using the commutator relations of [Tit94], it can be shown that $G = G_0'$, when either

(i) $|A_h| > 2$ and Σ is non-degenerate

   or

(ii) $|A_h| > 3$.

(See §4 of [Tim00b]). Using the classification of groups generated by 3-transposition, $\{3, 4\}^+$-transpositions or $GF(3)$-transvections, this implies that either G is simple or $G \simeq \mathrm{Sp}(4, 2)$ or $G_2(2)$.

(5.24) Exercises

(1) Show, using (1.2), that if $r, s \in \Phi$ span a subsystem of type A_2 (i.e. the set of roots of Φ, which are linear combinations of r and s, form a root system of type A_2) then

$$A_{r+s} = [A_r, A_s] = [a, A_s] = [A_r, b] \text{ for all } a \in A_r^\#, b \in A_s^\#.$$

(2) Let $r \neq \pm s \in \Phi$. Show $[X_r, X_s] = 1 \iff w_r w_s = w_s w_r$.

(3) Let $J \subseteq I, |J| \geq 2, U^J = \langle A_r \mid r \in \tilde\Phi_J^+ \rangle, H^J = \langle H_r \mid r \in \Phi_J \rangle$, $B^J = U^J H^J$ and $N^J = H^J \langle w_r \mid r \in \Phi_J \rangle$. Show

  (a) (B^J, N^J) is a BN-pair of L_J.

  (b) The building $\mathcal{B}^J = \mathcal{C}(L_J, B^J, (R_j)_{j \in J})$; where $R_j = \langle B^J, w_j \rangle$ is isomorphic to the residue $\Delta_J(c)$.

(4) Assume $\tilde\Phi$ is of type BC_ℓ. Let $\Pi^\circ = \{r_1, \ldots, r_{\ell-1}, 2r_\ell\}$,

$$\Phi^\circ = \{r \in \tilde\Phi \mid r \text{ is a linear combination of } \Pi^\circ\}.$$

$(\Phi^\circ)^+ = \Phi^\circ \cap \tilde\Phi^+, U^\circ = \langle A_r \mid r \in (\Phi^\circ)^+ \rangle$ $H^\circ = \langle H_r \mid r \in \Phi^\circ \rangle, B^\circ = U^\circ H^\circ$ and $N^\circ = H^\circ \langle w_r \mid r \in \Phi^\circ \rangle$. (We have the convention $w_{r_\ell} = w_{2r_\ell} \in X_{2r_\ell}$!). Show:

  (a) (B°, N°) is a BN-pair of $G^\circ = \langle X_r \mid r \in \Pi^\circ \rangle$.

  (b) $\mathcal{B}^\circ = \mathcal{C}(G^\circ, B^\circ, (R_i)_{i \in I})$ is a building of type C_ℓ where $R_i = \langle B^\circ, w_{r_i} \rangle$ for $r_i \in \Pi^\circ$.

  (c) The groups $A_r, r \in \Phi^\circ$ act as root groups on $\mathcal{B}^\circ$.
    (In (c) the only problem is to show that A_{2r_ℓ} acts as a root group on $\mathcal{B}^\circ$!)

Remark. (4) describes the embedding of a unitary group of dimension 2ℓ and Witt index ℓ in an arbitrary unitary group of Witt index ℓ.

(5) Suppose $\mathcal{B}$ is a Moufang plane and let $\Phi = \{\pm r, \pm s, \pm(r + s)\}$. Show:

  (a) $X_s \simeq SL_2(K)$, K a division ring or a Cayley division algebra, and X_s acts naturally on $A_r A_{r+s}$ and $A_{-r} A_{-r-s}$.
    Hint: Use I(3.2) and show $C_{X_s}(A_r A_{r+s}) = C_{X_s}(A_{-r} A_{-r-s}) = 1$.

  (b) Let $N = A_r A_{r+s}$. Show that G acts doubly transitively on the conjugates of N.
    Hint. Use the fact that $N(N) = N X_s H =: P$ is a maximal parabolic subgroup and that G acts by the argument of (5.20) doubly transitively on $\{Pg \mid g \in G\}$.

(c) $M = A_{-r}A_s$ is conjugate to N and $\langle M, N \rangle = (A_s A_{r+s})X_r, X_r \simeq SL_2(K)$ acting naturally on $A_s A_{r+s}$.

(d) Conclude from (b) and (c) that A_r^G is a class of abstract root subgroups of G.

If K is a Cayley division algebra the group G is called E_6^K.

(6) Suppose Φ is of type $B_\ell, \ell \geq 4$ or $D_\ell, \ell \geq 5$. Let h be the highest root of Φ^+ and $W_0 = \langle w_r \mid r \in \Pi \text{ and } w_r w_h = w_h w \rangle$. Show:

(a) W_0 has two orbits on

$$\Lambda = \{r \in \Phi \mid r \text{ long and } (r, h) = 0\}.$$

(b) W_0 has six orbits on $\Psi = \{r \in \Phi \mid r \text{ long }\}$, namely $\{h\}, \{-h\}$, two orbits on Λ and two more orbits.

(7) Let Φ be as in (6) and $G = \langle A_r \mid r \in \Phi \rangle, P = N(A_h)$ and $\Sigma = A_h^G, G_0 = \langle \Sigma \rangle$. Show:

(a) G has permutation rank 6 on Σ.

(b) P has four orbits on $\Delta = \{C \in \Sigma \mid [A_h, C] = 1\}$.

(c) Σ is a class of abstract root subgroups of G_0.

Chapter III
Classification Theory

In this chapter we will describe the classification of groups G generated by a class of abstract root subgroups. This classification which is relatively long and difficult, works in principal as follows: First one constructs from the group-theoretic conditions a point-line geometry, which is either a polar space, a projective space, a generalized n-gon or a parapolar space. Then one uses certain geometric classifications of these geometries, i.e. theorems showing that under some additional conditions the flag complex of these geometries is a spherical building. Finally, to determine the group, one shows that the abstract root subgroups act as (centers) of long root subgroups on this building, whence G is the uniquely determined normal subgroup of the automorphism group of the building generated by the class of (centers) of the long root subgroups.

Although the construction is, except in case $D_4(k)$ where we construct a chamber system of type D_4, in principal the same in every case, the arguments are quite different in the separate cases. This is reasonable, since from a geometric point of view, one just starts with information on the points of the geometry (resp. singular lines in case of orthogonal groups) and then has to reconstruct the whole geometry from this information.

In §9, at the end of this chapter, we will give for the convenience of the reader, a short outline of the case subdivision for the classification proof.

If not mentioned otherwise the results of this chapter are from [Tim90a], [Tim91] or [Tim99], except in case of buildings of type F_4, E_6, E_7 or E_8, where we give an easier argument as in the original papers. Since Tits and Weiss are writing down the classification of Moufang n-gons at the time of this writing, we will formulate the classification in the rank 2 case (i.e. case corresponding to Moufang n-gons) so that the classification of G will be a consequence of the results of Tits and Weiss. We first start with the case of abstract transvection groups, which is the easiest, since in this case one can construct a polar space in a uniform way.

§1 Abstract transvection groups

In this section we assume that Σ is a class of abstract transvection groups of the group G, satisfying condition (H) of II §3, containing different commuting points $\underline{A}$ and $\underline{B}$. (I.e. $A, B \in \Sigma$ and $\underline{A} = Z(\langle C_\Sigma(A)\rangle) \cap \Sigma$, $\underline{B} = Z(\langle C_\Sigma(B)\rangle) \cap \Sigma$, $\underline{A} \neq \underline{B}$ and $[\underline{A}, \underline{B}] = 1$, see II§4.) Then the line $\ell_{A,B}$ through $\underline{A}$ and $\underline{B}$ is defined by:

$$\ell_{A,B} := \{\underline{C} \mid C \in \Sigma \text{ and } \underline{C} \subseteq Z(\langle C_\Sigma(A) \cap C_\Sigma(B)\rangle) \cap \Sigma\}.$$

Let $\Sigma_{\underline{A}} := C_\Sigma(A) - \underline{A}$. Then by II(4.1), $\Sigma_{\underline{A}}$ is a class of abstract transvection groups of $C_A = \langle \Sigma_{\underline{A}}\rangle$ which also satisfies (H). Hence the radical $R(C_A)$ is by II(4.12) well defined. We set

$$M_A := R(C_A)\langle \underline{A}\rangle \text{ for } A \in \Sigma.$$

Then by II(4.14), $M'_A \leq Z(\langle C_\Sigma(A)\rangle)$. We first show:

(1.1) Lemma

Let $\underline{A}$, $\underline{B}$ be two different commuting points. Then the following holds:

(1) $\ell_{A,B} = \underline{A} \cup \underline{B}^{M_A} = \underline{B} \cup \underline{A}^{M_B}$.
(2) $\ell_{A,B} = \ell_{C,D}$ for all pairs of points $\underline{C} \neq \underline{D} \in \ell_{A,B}$.
(3) $|\ell_{A,B}| \geq 3$.

Proof. Suppose $\underline{A} \neq \underline{D} \in \ell_{A,B}$. Then by definition of $\ell_{A,B}$:

$$C_{\Sigma_{\underline{A}}}(B) = C_{\Sigma_{\underline{A}}}(\underline{B}) \subseteq C_{\Sigma_{\underline{A}}}(\underline{D}) = C_{\Sigma_{\underline{A}}}(D).$$

Hence by II(3.11) applied to C_A and $\Sigma_{\underline{A}}$ we obtain $C_{\Sigma_{\underline{A}}}(B) = C_{\Sigma_{\underline{A}}}(D)$. Thus by II(4.17) applied to C_A and $\Sigma_{\underline{A}}$ we obtain $D \in B^{M_A}$. Hence $\underline{D} \in \underline{B}^{M_A}$ and (1) holds.

To prove (2) let $\underline{A} \neq \underline{D} \neq \underline{B}$, $\underline{D} \in \ell_{A,B}$. Then by (1)

$$\ell_{A,B} = \underline{A} \cup \underline{B}^{M_A} = \underline{A} \cup \underline{D}^{M_A} = \ell_{A,D}$$

and similarly $\ell_{A,B} = \ell_{B,D}$. This proves (2). (3) is a consequence of II(3.12) and II(3.15). □

(1.2) Lemma

Let $\ell_{A,C}, \ell_{A,D}$ be lines through $\underline{A}$ with $[C, D] \neq 1$ and let $B \in \Omega_A$. Suppose B commutes with some point of $\ell_{A,C}$. Then B commutes with some point of $\ell_{A,D}$.

Proof. For the proof of (1.2) we may assume that $R(G) = 1$. Indeed if the statement holds for $\overline{G} = G/R(G)$, then $\overline{B}$ commutes with an element of $\ell_{\overline{A},\overline{D}}$, whence B commutes with some point of $\ell_{A,D}$.

By (1.1)(2) we may assume $[B, C] = 1$. Now assume that (1.2) is false. Then (A, B, D) is a triangle. Let $Y = \langle A, B, D \rangle$ be the corresponding triangle-group, $N = R(Y)$ and $Z = Z(Y) \cap \langle \ell_{A,D} \rangle$. Then there exists by II(3.4), II(3.12) and II(3.15) an $n \in N$ with $A \neq D^n \neq D$, $C_\Sigma(A) \cap C_\Sigma(D) = C_\Sigma(D) \cap C_\Sigma(D^n)$ and $ADD^n \leq ADZ$. In particular, since $D^n \not\leq AD$ by II(2.5), $Z \neq 1$. Further $D^n \in \ell_{A,D}$. (Since $\underline{D} = \{D\}$, as $R(G) = 1$, we identify the elements of Σ with the points!)

Claim

$$(*) \qquad\qquad Z \leq C(C).$$

If $Z \leq N(C)$ then $(*)$ holds, since $\langle C, D \rangle$ is a rank one group. So, to prove $(*)$, we may assume that there exists a $1 \neq z \in Z$ with $C^z \neq C$. As $[C^z, B] = 1$, (1.1) implies $C^z \notin \ell_{A,C}$. Now $\langle \ell_{A,D} \rangle = D(\langle \ell_{A,D} \rangle \cap M_A)$ by definition of $\ell_{A,D}$ and M_A. Hence $z = m \cdot d$ with $m \in M_A$ and $1 \neq d \in D$. Since $[C, D] \neq 1$ this implies $[C, C^z] \neq 1$.

Let $\widetilde{C}_A = C_A M_A / M_A$. Then by II(2.1) applied to $\widetilde{\Sigma}_A$ we obtain $\langle C, C^z \rangle M_A = \langle C, D \rangle M_A$. Moreover there exists an $E \in \Sigma \cap \langle C, C^z \rangle$ with $EM_A = DM_A$. By definition of $\ell_{A,D}$ this implies $E \in \ell_{A,D}$, a contradiction to $E \leq \langle C, C^z \rangle \leq C(B)$.

This shows that $(*)$ holds. Let $R = \langle Y, C \rangle$ and $\Delta = A^R$. Then $\{A, B, C, D, D^n\} \subseteq \Delta$ and $D^n \leq ADZ$. But this is a contradiction to II(2.5) applied to $R^* = R/Z$ and Δ^*. □

(1.3) Definition

A point-line space $(\mathcal{P}, \mathcal{L}, \in)$ with point set $\mathcal{P}$ and line set $\mathcal{L}$ is called a polar space if for each point $P \in \mathcal{P}$ and line $\ell \in \mathcal{L}$, P is adjacent either to exactly one point of ℓ or to all points of $\mathcal{L}$, where two points are called adjacent if there exists a line through this pair of points. Such a polar space is called non-degenerate, if there exists no point adjacent to all others. The lines are "thick", if there are at least three points on the line.

For the definition of subspaces, rank and the classification of polar spaces the reader should look into [Coh95].

(1.4)   Theorem

Let $\mathcal{P}(\Sigma)$ be the point-line space with points $\underline{A}$, $A \in \Sigma$ and lines $\ell_{A,B}$. Then $\mathcal{P}(\Sigma)$ is a non-degenerate polar space with thick lines of not necessary finite rank $r \geq 2$. Moreover each subspace of $\mathcal{P}(\Sigma)$ which is contained in two maximal subspaces is contained in at least three.

Proof. By $(1.1)(1)$ it is clear that, if a point commutes with two different points of a line, it commutes with all. Thus, to verify the polar space property it suffices to show that each point $\underline{B}$ commutes with some point on each line $\ell_{A,D}$. To prove this we may assume $[A, B] \neq 1$. Since $\mathcal{D}(\Sigma)$ is by II(4.3) connected, B commutes with some $C \in \Sigma_A$. Now, as $\mathcal{F}(\Sigma_A)$ is connected by II(4.1), there exist $C_i \in \Sigma_A$, $i = 0, \dots, n$ with $C_0 = C, C_n = D$ and $[C_i, C_{i+1}] \neq 1$ for $0 \leq i \leq n - 1$. Hence by (1.2) B commutes with some point on each line $\ell_{A,C_i}, i = 0, \dots, n$ and thus with some point on $\ell_{A,D}$.

Since $\mathcal{F}(\Sigma)$ is connected, $\mathcal{P}(\Sigma)$ is non-degenerate. By II(3.12) and II(3.15) the lines of $\mathcal{P}(\Sigma)$ are thick. If now the subspace Λ of $\mathcal{P}(\Sigma)$ is contained in the maximal subspaces $\Delta_1 \neq \Delta_2$, then by maximality of Δ_i there exists an $A \in \Delta_1$ and $B \in \Delta_2$ with $[A, B] \neq 1$. But then all $\Delta_1^b, b \in B$ and all $\Delta_2^a, a \in A$ contain Λ, which proves (1.4). □

(1.5)   Corollary

The following hold:

  (1) $R(G)$ is in the kernel of the action of G on $\mathcal{P}(\Sigma)$.
  (2) Let $\overline{G} = G/R(G)$. Then $\mathcal{P}(\Sigma) \simeq \mathcal{P}(\overline{\Sigma})$.
  (3) The points of $\mathcal{P}(\overline{\Sigma})$ consist of elements of $\overline{\Sigma}$.

Proof. The definition of $R(G)$ implies that $R(G)$ stabilizes all points of $\mathcal{P}(\Sigma)$, whence (1) holds. The definition of points and lines of $\mathcal{P}(\Sigma)$ and $\mathcal{P}(\overline{\Sigma})$ shows that the natural homomorphism: $G \to \overline{G}$ induces an isomorphism from $\mathcal{P}(\Sigma)$ onto $\mathcal{P}(\overline{\Sigma})$. (3) is obvious as $R(\overline{G}) = 1$ by II(4.14). □

(1.6)   Corollary

Let $A \in \Sigma$ and $B \in \Omega_A$. Then $\langle C_\Sigma(A) \rangle = M_A \langle C_\Sigma(A) \cap C_\Sigma(B) \rangle$. Further M_A acts transitively on Ω_A.

Proof. Let $C \in \Sigma_A$. Then, since $\mathcal{P}(\Sigma)$ is a polar space, there exists a point $\underline{D}$ on $\ell_{A,C}$ with $[D, B] = 1$. Since $M_A C = M_A D$ by definition of M_A this proves the first part of (1.6).

By the first part of (1.6) it suffices to show that $\langle C_\Sigma(A) \rangle$ is transitive on Ω_A. To do this pick $C \in \Omega_A$. If $[B, C] = 1$ then there exists by II(2.9) an

$x \in \langle C_\Sigma(A) \rangle$ with $B^x = C$. Next, if $[B,C] \neq 1$ but $C_\Sigma(B) \cap C_\Sigma(C) \not\subseteq C_\Sigma(A)$, we may apply the same argument twice. Thus to prove (1.6) we may assume $C_\Sigma(B) \cap C_\Sigma(C) \subseteq C_\Sigma(A)$. Since $\mathcal{P}(\Sigma)$ is a polar space, there exist $D \neq E$ in $C_\Sigma(B) \cap C_\Sigma(C)$ with $[D,E] \neq 1$. Since the lines of $\mathcal{P}(\Sigma)$ are thick, there exists a point $\underline{D}' \in \ell_{D,C} - \{\underline{D}, \underline{C}\}$. Clearly $D' \in \Omega_A$ since $D \in C_\Sigma(A)$. Let $\underline{E}'$ be the point on $\ell_{B,E}$ commuting with $\underline{D}'$. Then $\underline{E}' \neq E$, since otherwise $\underline{E}$ commutes with two points on $\ell_{D,C}$ and thus also with $\underline{D}$, contradicting $[E,D] \neq 1$. Hence $E' \in \Omega_A$ and thus $\{B, E', D', C\} \subseteq \Omega_A$ with $[B,E'] = [E',D'] = [D',C] = 1$. Now the above argument proves (1.6). □

The proof of (1.6) is from [Stb98a, (3.1)]. In (3.2), (3.3) of the same paper a proof of the Moufang condition when $\mathrm{rk}(\mathcal{P}(\Sigma)) = 2$ has been given. We will use the argument of [Tim90a, (7.5)], where the Moufang condition has been proved in the more special situation of k-transvections.

(1.7) Theorem

Suppose $\mathrm{rk}(\mathcal{P}(\Sigma)) = 2$. Then $\mathcal{P}(\Sigma)$ satisfies the Moufang condition. ($\mathcal{P}(\Sigma)$ is a generalized quadrangle and whence a building, see I(4.5).)

Proof. We will show that G induces a group on $\mathcal{P}(\Sigma)$ satisfying the Moufang condition. Since $R(G)$ is in the kernel of the action of G on $\mathcal{P}(\Sigma)$ and since $\mathcal{P}(\Sigma) \simeq \mathcal{P}(\overline{\Sigma})$, where $\overline{G} = G/R(G)$ we may assume that $R(G) = 1$. Hence points are just elements of Σ.

(1) "Moufang condition for points".

Let $A \in \Sigma$, ℓ, s different lines through A and $A \neq B$ be a point on ℓ and $A \neq C$ a point on s. Then the set of chambers

$$r = \{(B,\ell), (A,\ell), (A,s), (C,s)\}$$

is a half-apartment in the sense of I§4. Let $T = M_A \cap N(B) \cap N(C)$. Then, since $T \leq M_A$, T stabilizes all lines through A. Since each point on $\ell - \{A\}$ is by (1.1) conjugate to B in M_A and since $M'_A \leq Z(\langle C_\Sigma(A) \rangle)$, T stabilizes all points on ℓ and s. Hence T is contained in the root group A_r corresponding to r. Thus it suffices to show that T acts transitively on the lines through B different from ℓ. (See I(4.12)).

Let m, n be lines through B different from ℓ and X, Y points on m resp. n adjacent to C. Then $X, Y \in \Omega_A$. Hence there exists by (1.6) a $g \in M_A$ with $X^g = Y$. Since B is the only point on ℓ adjacent to X and Y clearly $B^g = B$. By the same reason $C^g = C$ and $g \in T$ with $m^g = n$.

(2) "Moufang condition for lines".

Let ℓ be a line, $A \neq B$ points on ℓ and s a line through A different from ℓ and g a line through B different from ℓ. Then the set of chambers

$$r = \{(s, A), (\ell, A), (\ell, B), (g, B)\}$$

is a half-apartment. Let $T = \langle \ell \rangle \cap M_A \cap M_B$. Then T stabilizes all points on ℓ and all lines through A and B. Hence $T \leq A_r$ – the root group corresponding to r. Thus it suffices to show that T acts transitively on the points of s different from A. Let $D \neq C$ be two such points. Then (D, B, C) is a triangle. Thus by II(3.1) there exists a $t \in DC$ with $[B^t, B] = 1$ and a $\tau \in BB^t$ with $C^\tau = D$. Since $\mathrm{rk}(\mathcal{P}(\Sigma)) = 2$ we have $B^t \in \ell$ and $\tau \in \langle \ell \rangle$. Now by (1.1)

$$\langle \ell \rangle = (\langle \ell \rangle \cap M_A \cap M_B) AB = TAB.$$

Hence there exist $a \in A, b \in B$ with $x = \tau a b \in T$. As $x \in T$ we have $C^x = D^b \in s$. Since clearly D is the only element in $\langle D, B \rangle \cap s$, this implies $C^x = D^b = D$, which was to be shown. $\qquad \square$

The next Proposition, [Tim99, (7.1)], is crucial for the determination of $\overline{G}/Z(\overline{G})$, where $\overline{G} = G/R(G)$, as a subgroup of the automorphism group of $\mathcal{P}(\Sigma)$.

(1.8) Proposition

Let $X = \langle A, B \rangle$ be a special rank one group and $Y = \langle \Omega \rangle \leq X$ satisfying:

(1) Ω is a set of abstract root subgroups of Y.
(2) For each $C \in \Delta = A^X$ there exists a $C_1 \in \Omega$ with $C_1 \leq C$.

Then $X' \leq Y$ and $X = Y \cdot A$.

Proof. If $X \simeq SL_2(2)$ or $(P)SL_2(3)$, (1.7) is obvious. So we may by I(2.10) assume that X' is quasi-simple and $|X' \cap A| > 3$. Pick $A_1, B_1 \in \Omega$ with $A_1 \leq A$ and $B_1 \leq B$. Then, as $A_1 \neq 1 \neq B_1$, also $\langle A_1, B_1 \rangle$ is a rank one group. Thus if for $a \in A^\#$, $b(a) \in B^\#$ is the element with $A^{b(a)} = B^a$ we obtain by I(1.2) (See also exercise I(2.13) (2)!)

$$(*) \qquad\qquad B_1^\# = \{b(a) \mid a \in A_1^\#\}.$$

Let $a \in A - A_1$ and $B_1^a \leq B^a = C \in \Delta$. Then by $(*)$, $C_1^\# = \{c(a) \mid a \in A_1^\#\}$ where $C_1 \in \Omega$ with $C_1 \leq C$. On the other hand we have for $a_1 \in A_1^\#$ and $b_1 = b(a_1)$:

$$a_1^{b_1^a} = (b_1^{-a_1})^a = (b_1^{-1})^{a a_1} = (b_1^a)^{-a_1}.$$

Hence the uniqueness of $c(a_1)$ implies:

$$C_1^\# = (B_1^a)^\# \text{ and } C_1 = B_1^a.$$

In particular $B_1^a \in \Omega$ for each $a \in A$ and each $A_1 \neq B_1 \in \Omega$. Hence A and similarly B normalize Ω and thus $Y \trianglelefteq X$. Thus by I(1.10) $X = YA$ and $X' \leq Y$, since A is abelian as X is special. $\qquad\square$

(1.9) Remark

Let $\mathcal{S} = (\mathcal{P}, \mathcal{L}, \in)$ be a non-degenerate polar space with thick lines of rank ≥ 2 and $P \in \mathcal{P}$ a point. Then, according to II(5.3) a central elation corresponding to P is an automorphism of $\mathcal{S}$, which fixes all lines through P pointwise. Let T_P be the set of all such central elations corresponding to P.

Let now P and Q be non-adjacent points and assume $T_P \neq 1 \neq T_Q$. Then it follows from II(5.4) and II(5.19) that $\langle T_P, T_Q \rangle$ is a special rank one group with unipotent subgroups T_P and T_Q if $\mathcal{S}$ satisfies the Moufang condition. Indeed if A_r is the root subgroup of $\mathrm{Aut}(\mathcal{S})$ corresponding to P then $T_P \leq A_r$ and $T_P = A_{2r}$ in the notation of II(5.4) if $T_P \neq A_r$. Thus in any case either $T_P = A_{2r}$ or $T_P = A_r$. Since $T_Q \leq A_s$ for some root s, which is opposite to r in some apartment, the statement follows from II(5.19).

(1.10) Theorem

Let $\widetilde{G} = \overline{G}/Z(\overline{G})$, where $\overline{G} = G/R(G)$. Then the following hold:

(1) $\widetilde{A} \leq T_{\underline{A}}$ for each $A \in \Sigma$, where $T_{\underline{A}}$ is the group of central elations of $\mathcal{P}(\Sigma)$ corresponding to the point $\underline{A}$.

(2) $R' \leq \widetilde{G} \leq R$ and $R = \widetilde{G} T_{\underline{A}}$ for $A \in \Sigma$, where R is the normal subgroup of $\mathrm{Aut}(\mathcal{P}(\Sigma))$ generated by all central elations.

Proof. It follows immediately from the definition of lines of $\mathcal{P}(\Sigma)$ that $\widetilde{A}$ stabilizes all points on all lines through $\underline{A}$, whence (1) holds.

Let A, B be non-commuting elements of Σ. Then by (1.9) $X = \langle T_{\underline{A}}, T_{\underline{B}} \rangle$ is a special rank one group with unipotent subgroups $T_{\underline{A}}, T_{\underline{B}}$. Let

$$\Omega = \{ \widetilde{E} \in \widetilde{\Sigma} \mid \widetilde{E} \leq T_{\underline{A}}^x \text{ for some } x \in X \} \text{ and } Y = \langle \Omega \rangle.$$

Then, since for each point $\underline{A}^x, x \in X$ of $\mathcal{P}(\Sigma)$ there exists by (1) an $\widetilde{E} \in \widetilde{\Sigma}$ with $\widetilde{E} \leq T_{\underline{A}}^x$, Ω and Y satisfy (1.8)(2). Let $y \in Y$ and $\widetilde{E} \in \Omega$ with $\widetilde{E} \leq T_{\underline{A}}^x$. Then $\widetilde{E}^y \leq T_{\underline{A}}^{xy}$ and $xy \in X$. Since $\widetilde{\Sigma}^y = \widetilde{\Sigma}$ this implies $\widetilde{E}^y \in \Omega$, whence $\Omega^y \subseteq \Omega$ for each $y \in Y$. This shows that Ω is a set of abstract transvection groups of Y and thus $X' \leq Y$ and $Y T_{\underline{A}} = X$ by (1.8).

From this we obtain $T_{\underline{B}} \leq \widetilde{G} T_{\underline{A}}$ and $\widetilde{G} T_{\underline{A}} = \widetilde{G} T_{\underline{B}}$. Since $\mathcal{F}(\Sigma)$ is connected this implies $R = \widetilde{G} T_{\underline{A}}$ for each $A \in \Sigma$. Since $T_{\underline{A}}$ is abelian as $\mathcal{P}(\Sigma)$ satisfies by (1.7) the Moufang condition, it remains to show that $\tilde{G} \trianglelefteq R$. For this pick $h \in T_{\underline{A}}$ and $C \in \Sigma$. Then, as $R = \tilde{G} T_{\underline{C}}$, $h = cg$ with $g \in \tilde{G}$ and $c \in T_{\underline{C}}$. Hence

$$\tilde{C}^h = \tilde{C}^{cg} = \tilde{C}^g \in \tilde{\Sigma},$$

which shows that h normalizes $\tilde{G} = \langle \tilde{\Sigma} \rangle$. $\square$

(1.11) Remark

Notice that (1.10) determines $\tilde{G}$ as a subgroup of $\mathrm{Aut}(\mathcal{P}(\Sigma))$ without any contribution to the classification of polar spaces. This is important for the rank 2 case, since as noted earlier Tits and Weiss are writing down the classification of thick generalized quadrangles satisfying the Moufang condition.

If $\mathrm{rk}(\mathcal{P}(\Sigma)) \geq 3$ then it follows from the classification of polar spaces by Buekenhout-Shult [BS74], Tits-Veldkamp [Tit74] and Cuypers, Johnson and Pasini [CJP92] in the infinite rank case that either $\mathcal{P}(\Sigma)$ is classical, i.e. isomorphic to the polar space obtained from isotropic points and lines of some vector space with some form or is isomorphic to the uniquely determined polar space of rank 3, the planes of which are Moufang planes over some Cayley division algebra K. For details see [Coh95]. In the latter case it follows from [Tit74, 9.1] that the group $X = \langle T_{\underline{A}}, T_{\underline{B}} \rangle \simeq SL_2(k)$, when $\underline{A}$ and $\underline{B}$ are non-adjacent points, where k is the field over which K is defined. Hence X is quasi-simple and thus $X = Y$ in (1.8). From this it follows that $\tilde{G} = R = E_7^K$ in (1.10), where E_7^K denotes the subgroup of $\mathrm{Aut}(\mathcal{P}(\Sigma))$ generated by all central elations.

In the case of a classical polar space we will show that the central elations groups are induced by isotropic transvection groups in the sense of II(1.4). This shows by I(1.5) and I(1.6), that the group X of (1.8) is perfect, when $|L| > 3$, L defined as in I(1.6) (1)–(3). Hence again $\tilde{G} = R$ and the latter is a certain projective special classical group. (Special stands for the normal subgroup generated by isotropic transvections!) For the possible groups R see II(1.4).

(1.12) Lemma

Let $\mathcal{S}$ be a thick non-degenerate classical polar space of rank ≥ 2 obtained from the vector space V over some division ring K together with some (σ, ϵ)-hermitian form f or pseudoquadratic form q. (For possible forms see II(1.4).) Let $P : GL(V) \to PGL(V)$ be the natural homomorphism and Q a point of $\mathcal{S}$. Then, if σ is an automorphism of $\mathcal{S}$ fixing all points adjacent to Q, then there exists an isotropic transvection φ of V with $\varphi \mid_{Q^\perp} = \mathrm{id}$ such that $\sigma = P\varphi$.

Proof. We use the crucial fact from [Tit74, (8.6)II] that $\mathrm{Aut}(\mathcal{S}) = P\Gamma O(V, q)$ (resp. $P\Gamma U(V, f)$ in the notation of Tits) where $\Gamma O(V, q)$ (resp. $\Gamma U(V, f)$) is the group of all semilinear mappings of V, which map q (resp. f) onto some proportional form. (See [Tit74, 8.2.8]). Let now $\varphi \in \Gamma O(V, q)$ (resp. $\Gamma U(V, f)$) with $P\varphi = \sigma$. Then, since $Q^\perp$ is generated by isotropic points, $\varphi \mid_{Q^\perp}$ is a scalar map $c \cdot \mathrm{id}$. Let α be the companion automorphism of φ and $d \in K$ such that

$$f(\varphi(v), \varphi(w)) = (df(v, w))^\alpha \text{ for all } v, w \in V$$

(resp. the same property holds for the pseudoquadratic form q). Applying this to a hyperbolic pair in $Q^\perp$ we see that $d^\alpha = c \cdot c^{-\sigma}$, where σ is the anti-automorphism of K in the definition of f (resp. q). Now it follows that the map $\psi : v \to c^{-1}\varphi(v)$ is an isometry of V with $P\Psi = P\varphi$ and $\psi \mid_{Q^\perp} = \mathrm{id}$. Hence we may assume that $\varphi \mid_{Q^\perp} = \mathrm{id}$ and φ is an isometry.

We now use the notation of II(1.4) for (V, q) (resp. (V, f)). Then, since $\mathcal{S}$ is non-degenerate, in the first case $R = \{r \in \mathrm{Rad}(V, f) \mid q(r) = 0 + \Lambda\} = 0$ and in the second case $\mathrm{Rad}(V, f) = 0$. Since φ is an isometry it follows that $(\varphi - \mathrm{id})(V) \subseteq Q^{\perp\perp} = Q \oplus \mathrm{Rad}(V, f)$. Let $Q = \langle q \rangle$ with $q \in V$ and choose a vector $x \in V$ with $f(x, q) = 1$. Then

$$\varphi(x) = x + a \cdot q + r_a \text{ with } a \in K \text{ and } r_a \in \mathrm{Rad}(V, f).$$

Since codim $Q^\perp = 1$ now an easy computation shows that $\varphi = t_a$ in the notation of II(1.4).   □

(1.13)   Remark

(1.10)–(1.12) show that the group $\widetilde{G}$ (i.e. $\widetilde{G} = \overline{G}/Z(\overline{G}), \overline{G} = G/R(G)$) generated by the class $\widetilde{\Sigma}$ of abstract transvection groups is (essentially) equal to the normal subgroup R of $\mathrm{Aut}(\mathcal{P}(\Sigma))$ generated by all central elations and that the elements of $\widetilde{\Sigma}$ are contained in central elation groups. Moreover these central elation groups are images of isotropic transvection groups under the homomorphism $P : GL(V) \to PGL(V)$. But it does not show that the elements of $\widetilde{\Sigma}$ are full central elation groups.

Indeed, as the following exercises will show, this is the case when the characteristic of the underlying division ring is 0 or odd. But in characteristic 2 this is not necessarily the case.

(1.14)   Exercises

(1) Let K be a field of characteristic 2, $k = K^2$, and $k \subseteq L \subseteq K$ a k vector space. Let $X = \langle A, B \rangle = SL_2(K)$, where $A = \{a(c) \mid c \in K\}$ and $B = \{b(c) \mid c \in K\}$ in the notation of I(1.5). Let $A_0 = \{a(c) \mid c \in L\}$,

$\Omega = A_0^X$ and $Y = \langle \Omega \rangle$. Show that the pair X, Y satisfies the hypothesis of (1.8).

Hint. Because of $a(e)^{h(c)} = a(c^2 e)$, $H = N_X(A) \cap N_X(B)$ normalizes A_0. Let $w \in SL_2(L)$ interchanging A_0 and $B_0 = \{b(c) \mid c \in L\}$. Then, as $X = AH \cup (AH)wA$, we have $\Omega = \{A_0\} \cup B_0^A$.

(2) Let K and L be as in (1), V a K-vector space together with some non-degenerate symplectic form $(,)$. For each point P of V let $T_P = \{t_c \mid c \in K\}$ be the symplectic transvection group corresponding to P, see II(1.4), $\Sigma = \{T_P \mid P \text{ point of } V\}$ and $G = \langle \Sigma \rangle$. Let $A_0 = \{t_c \mid c \in L\} \leq T_P$ and $\Omega = A_0^G$. Show that Ω is a class of abstract transvection groups of G.

(3) Suppose that X and Y satisfy the hypothesis of (1.8). Show that for each $C \in \Delta = A^X$ the element $C_1 \in \Omega$ with $C_1 \leq C$ is uniquely determined.

Hint. Suppose $A_1, A_2 \in \Omega$ with $A_1 \leq A \geq A_2$. Let $B_1 \in \Omega$ with $B_1 \leq B$. Then $\langle A_1, B_1 \rangle$ and $\langle A_2, B_1 \rangle$ are rank one groups. Since for each $b \in B_1^{\#}$ the element $a(b) \in A^{\#}$ is uniquely determined, this implies $A_1 = A_2$.

(4) Let R and L be as in I(1.5) with $1 + 1 \neq 0$ (i.e. Char $R \neq 2$). Let $L_0 \subseteq L$ satisfying (1)–(3) of I(1.5) and in addition:

$$tct \in L_0 \text{ for each } c \in L_0 \text{ and } t \in L.$$

Show $L = L_0$.

Hint. For $t \in L$ we have $1 + 2t + t^2 = (1+t)^2 \in L_0$. Hence $2t \in L_0$. Thus $\frac{t^{-1}}{2} = (2t)^{-1} \in L_0$ and so $t^{-1} \in L_0$.

(5) Let K be a division ring with Char $K = 0$ or odd and let L be as in I(1.6) (1) or (3). Let $X = \langle A, B \rangle$ with $A = \{a(c) \mid c \in L\}$ and $B = \{b(c) \mid c \in L\}$ in the notation of I(1.5). Let $Y = \langle \Omega \rangle \leq X$, Ω a set of abstract transvection groups of Y, such that the pair X, Y satisfies the hypothesis of (1.8). Show that if $C_1 \in \Omega$ with $C_1 \leq C \in A^X$, then $C = C_1$.

Hint. Let $A_1 \in \Omega$ with $A_1 \leq A$. Then it suffices to show that $A = A_1$. For this let

$$L_0 = \{c \in L \mid a(c) \in A_1\}.$$

Show, using exercise (3), that L_0 satisfies the hypothesis of exercise (4).

(6) Let V be a vector space over the division ring K, Char $K = 0$ or odd, together with some non-degenerate $(\sigma, -1)$ hermitian form f or pseudoquadratic form q. Let Σ be the set of isotropic transvection groups as defined in II(1.4) and $G = \langle \Sigma \rangle$ (As a subgroup of the group of isometries of (V, f) (resp. (V, q))!). Let $\overline{G}$ and $\overline{\Sigma}$ be the images under

$P : GL(V) \to PGL(V)$ and suppose there exists a set Ω of abstract transvection groups of $\overline{G}$ satisfying:

(∗)   For each $\overline{A} \in \overline{\Sigma}$ there exists an $A_1 \in \Omega$ with $A_1 \leq \overline{A}$.

Then $\Omega = \overline{\Sigma}$.

Hint. Apply exercise (5) for $\overline{X} = \langle \overline{A}, \overline{B} \rangle, \overline{A} \neq \overline{B}$ non-commuting elements of $\overline{\Sigma}$.

(7) Suppose $R(G) = Z(G) = 1$ and $\mathcal{P}(\Sigma)$ is a classical polar space over the division ring K with Char $\neq 2$. Show $A = Z(\langle C_\Sigma(A) \rangle)$ for $A \in \Sigma$.
   Hint. $A^0 = Z(\langle C_\Sigma(A) \rangle)$ consists of central elations corresponding to the point A on $\mathcal{P}(\Sigma)$. Now $A = A^0$ is a consequence of exercise (6) and (1.12).

(8) Suppose Σ is a class of k-transvections of G and (A, B, C) is a triangle of Σ with triangle group Y. Let $\underline{E}$ be the point on $\ell_{A,C}$ commuting with B. Show $N = R(Y) \leq M_E$.
   Hint. By exercise II(3.17)(3) we have $\underline{A}^N \subseteq \ell_{A,C} = \ell_{A,E}$. Let $A \neq A^n \neq C$, $n \in N$ and $N_A = AA^nC \cap N$. Then, as in II(3.15) one has $N_A = \langle A^N \rangle \cap N \geq Z$ and by II(3.9) $N = N_A N_A^b$, $b \in B^\#$. Hence

$$N_A \leq \langle \ell_{A,E} \rangle \cap N(\ell_{B,E}) \leq M_E$$

by definition of M_E.

(9) Assume the same hypothesis and notation as in (8) and, in addition, Char $k \neq 2$. Show

$$Z(Y) \leq M_E' \leq Z(\langle C_\Sigma(E) \rangle).$$

   Hint. Use exercise II(3.17)(8).

(10) Suppose Σ is a class of k-transvections of G, $R(G) = Z(G) = 1$, Char $k \neq 2$ and $\mathcal{P}(\Sigma)$ is a classical polar space. Let (A, B, C) be a triangle of Σ with triangle group Y and let E be the point on $\ell_{A,C}$ commuting with B. Show $Z(Y) \leq E$. (Clearly $Z(Y) \neq 1$!)
   Hint. By (9) $Z(Y) \leq Z(\langle C_\Sigma(E) \rangle) = E^0$ and by (7) $E^0 = E$.

§2   The action of G on Σ

Let in this section Σ be a non-degenerate set of abstract root subgroups of G and assume that $\mathcal{F}(\Sigma)$ is connected. The purpose of this section is to develop a more general theory, which will be needed for classification purposes.

We will use freely the notation introduced in chapter II, in particular the notion of a weak TI-set of II(4.5).

(2.1)   Proposition

Let $A, B, C \in \Sigma$ with $C \in \Lambda_B$ and $BC \cap \Sigma \subseteq \Lambda_A$. Set $N = ABC$ and $\Delta = \Sigma \cap N$. Then the following hold.

1. N is a 3-dimensional vector space over the division ring $K = \{\sigma \in \text{End } K \mid D^\sigma \subseteq D$ for all $D \in \Delta\}$. Further, Δ is the set of all 1-spaces of N.
2. $\langle N_\Sigma(N) \rangle$ induces the $SL_3(K)$ on N.

Proof. Let $n = abc \in N$ with $a \in A^\#, b \in B^\#$ and $c \in C^\#$. Then, as $C \in \Lambda_B$, $bc \in B' \in \Sigma \cap BC$. Since $B' \in \Lambda_A$ by hypothesis, we obtain $n \in D \in \Lambda$. Together with II(2.7) this implies that Δ is a partition of N. Let $\mathcal{W} = 1 \cup N \cup \Delta \cup \{EF \mid E, F \in \Delta$ and $E \neq F\}$. Then it is easy to see that $\mathcal{W}$ satisfies:

(1) $1, N \in \mathcal{W}$.
(2) $\Pi_{X \in \mathcal{M}} X \in \mathcal{W}, \bigcap_{X \in \mathcal{M}} X \in \mathcal{W}$, if $\mathcal{M} \subseteq \mathcal{W}$.
(3) If $X \in \mathcal{W}$ with $X \neq 1$ and $X \neq N$, then there exists a $Y \in \mathcal{W}$ with either $X \subset Y \subset N$ or $Y \subset X \subset N$.
(4) If $1 \neq x, y \in N$ and $x \in X \in \mathcal{W}, y \notin X$, then there exists a $Y \in \mathcal{W}$ with $y \in Y$ but $x \notin Y$.

Now the theorem of [Pic55, p. 315] shows that K is a division ring, N is a 3-dimensional vector space over K and $\mathcal{W}$ is the set of all K-subspaces of N. This implies (1).

For the proof of (2) let $R = \langle N_\Sigma(N) \rangle$ and $\overline{R} = R/C_R(N)$. Let $D \in N_\Sigma(N) - C_\Sigma(N)$ and $E \in \Delta - C_\Sigma(D)$. Then $F = [D, E] \in \Delta \cap \Delta_D$ by (1). Let $T \in \Delta$ with $T \not\leq EF$. Then, since D normalizes the 2-subspace EF of N we have $[T, D] \leq EF \cap C_N(D) = F$. This shows $[N, D] = F$ and $C_N(D)$ is a 2-subspace of N. In particular D induces the set of all transvections corresponding to the point F and hyperplane $C_N(D)$ on N, whence $\overline{D}$ is a transvection group on N.

Let now $A \neq B \in \Delta$. Then there exist by II(2.3), II(2.11) and II(2.23) an $E_1 \in \Psi_A \cap \Omega_B$. Let $L = [A, E_1]$. Then $B^L = (AB \cap \Sigma) - \{A\}$ by II(2.19)(4). Let $C \in \Delta - \{B\}$. Then there exists by II(2.3) and II(2.11) a unique $C_1 \in$

$\Delta \cap BC \cap \Psi_{E_1}$. We now can apply exercise II(2.24)(3) to $\langle A, B, C_1, E_1 \rangle$. Part (a) of this exercise implies $[C_1, L] = 1$. Thus if $H = AC_1$, then $N = H \cdot B$ and $[H, L] = 1$. Hence by the above L induces the set of all transvections corresponding to the point A and hyperplane H on N. Repeating the above argument with $B_1 \in (\Delta \cap H) - \{A\}$, $E_2 \in \Psi_A \cap \Omega_{B_1}$ and $L_1 = [A, E_2]$ we see that L_1 induces the set of all transvections corresponding to the point A and hyperplane $H_1 = \langle C \mid C \in \Delta \cap \Psi_{E_2} \rangle$ on N, whence $L_1 L_2$ induces the set of all transvections corresponding to A on N. Since A is an arbitrary point of N this implies that $\overline{R}$ induces the $SL_3(K)$ on N. $\qquad\square$

(2.2) Lemma

Let $A \in \Sigma$, $B \in \Sigma_A$ and assume $aB \subseteq D(\Sigma)$ for some $a \in A^{\#}$. Then $B \in \Lambda_A$.

Proof. Pick $C \in \Omega_A$. Then by II(2.2) $C \notin \Sigma_B$. Suppose first $C \in \Psi_B$ and let $X = \langle A, C \rangle$, $Y = \langle X, B \rangle$, $N = [B, C][B, C, A]$ and $M = \langle B^X \rangle$. Then by II(2.20) $M, N \lhd Y$ and $M = N \times C_M(Y) = NB$ and $B \le N$ if and only if $B \in \Lambda_A$. Thus if $B \notin \Lambda_A$, then $N \cap B = 1$ since $\Sigma \cap N$ is a partition of N. Hence

$$\overline{Y} = Y/N = \overline{X} * \overline{B}.$$

Let $d = ab$, $b \in B^{\#}$ and $d \in D \in \Sigma$. Then, since $\langle C, D \rangle$ is not nilpotent, it is a rank one group. Hence

$$R = \langle C, D \rangle = \langle C, d \rangle \le Y.$$

But then

$$\overline{R} = \langle \overline{C}, \overline{C^d} \rangle = \langle \overline{C}, \overline{C^a} \rangle = \overline{X},$$

and $\overline{D} \in \overline{C}^{\overline{X}}$ since $\overline{X}$ is a rank one group. This implies $\overline{D} = \overline{A}$ as $[\overline{D}, \overline{A}] = 1$, and thus $b \in AN \cap M = N$, a contradiction to $B \cap N = 1$.

So we may assume $C \in \Omega_B$. Let $c = c(a)$ and $b = b(c)$ in the notation of I(1.2). Then $d = ab^{-1} \in D(\Sigma)$ and $[C, C^d] = 1$ since $[A^c, B^c] = 1$. Hence by II(2.2) $D \in \Psi_C$ for $d \in D \in \Sigma$. If now $D \in \Lambda_A$, then as $b^{-1} \in AD$, also $B \in \Lambda_A$. So we may assume $D \notin \Lambda_A$. But then, as $dB = aB \subseteq D(\Sigma)$, we get a contradiction as before. (With $Y = \langle B, C, D \rangle$!) $\qquad\square$

(2.3) Lemma

$C_\Sigma(M_A) = \underline{A}$.

Proof. Clearly $\underline{A} \subseteq C_\Sigma(M_A)$. So, assuming that (2.3) is false, pick $B \in C_\Sigma(M_A) - \underline{A}$. If $C \in \Omega_B$ then $C \notin \Psi_A$ by II(2.20). If $\Omega_B \subseteq \Omega_A$, then $B \in \underline{A}$ by II(2.22), contradicting the choice of B. Thus there exists a $C \in \Omega_B \cap \Sigma_A$. Hence also $C \in C_\Sigma(M_A)$, since $B \sim C$ in $\langle B, C \rangle$.

Pick $E \in \Omega_A$. Then by II(4.8) $E \notin \Sigma_B$. Suppose $E \in \Psi_B$ and let $F = [B, E], L = [F, A]$. Then, if $|A| \neq 2$, we obtain by II(2.20) (4) applied to $\langle A, E, B \rangle = Y$:

$$D(\Sigma) \cap BL \nsubseteq B \cup L.$$

But as $L \in \Lambda_A$ and whence $L \in \Sigma_C$, this is a contradiction to II(2.2). So we may assume $|A| = 2$. Then $Y \simeq \Sigma_4 \times \mathbb{Z}_2$. Let $\langle b \rangle = B, F = \langle f \rangle$ and so on. Then $b\ell \in Z(Y)$ by the structure of Y. Since $L \in \Lambda_A$ we have $[B, L] = 1 = [C, L]$. Thus by II(2.3) applied to $\langle B, C, L \rangle$ we have $b\ell \notin D(\Sigma)$. On the other hand by II(2.20) $b\ell f \in D(\Sigma)$. Again by II(2.3) $[C, F] \neq 1$. If $C \in \Omega_F$, then as $L \in \Lambda_F$, this also contradicts II(2.3). So we obtain $C \in \Psi_F$. Now II(2.19) applied to $\langle B, C, F \rangle \leq C(L)$ implies that $b \sim bf$ in $C(L)$. Hence $b\ell \sim b\ell f$, a contradiction to $b\ell \notin D(\Sigma)$.

This finally shows $E \in \Omega_B$ and thus $\Omega_A \subseteq \Omega_B$. Now again II(2.22) implies $B \in \underline{A}$, a final contradiction.   □

(2.4) Notation

We assume for the rest of this section that $R(G) = 1$. For $A \in \Sigma$ let

$$C_A := \langle C_\Sigma(A) \rangle$$
$$\Delta_A := \text{the set of isolated vertices of } \mathcal{F}(C_\Sigma(A)).$$

Then by II(2.3) $\Lambda_A \subseteq \Delta_A$ and thus by II(2.21)(1) applied to $\Delta = C_\Sigma(A)$ also $\Sigma \cap M_A \subseteq \Delta_A$. A Σ-subgroup R is Σ-*maximal* , if it is not properly contained in a proper Σ-subgroup.

Next we show:

(2.5) Lemma

$M'_A = A$ for $A \in \Sigma$. (We assume $R(G) = 1$!)

Proof. Suppose (2.5) is false. Then there exist $B, C \in \Lambda_A$ with $1 \neq E = [B, C] \neq A$. As $E \leq M_A$ we have $E \in \Delta_A$. Thus, if $F \in \Omega_E$, then $F \notin \Sigma_A$. If $\Omega_E \subseteq \Omega_A$, then by II(2.22) $E \in \underline{A}$ and thus $E = A$ as $R(G) = 1$, a contradiction. Hence there exists an $F \in \Omega_E \cap \Psi_A$. By II(2.11) $F \in \Omega_{B^c}$ for some $c \in C$ and so we may, conjugating with c^{-1}, assume $F \in \Omega_B$. We obtain:

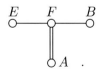

If now $|A| > 2$, then as $(EB \cap \Sigma) - E = B^C \subseteq \Lambda_A$, the hypothesis of exercise II(2.24)(2) is satisfied for $R = \langle E, F, B, A \rangle$. Hence E and B are conjugate in $C_R(A)$. This implies $E \in \Lambda_A$ and thus, if $N = ABE$, then N is partitioned by $\Sigma \cap N$. But then by (2.1) and the action of $SL_3(K)$ on its natural module we find an $L \in N_\Sigma(N) \cap \Sigma_A \cap \Omega_C$, a contradiction to II(2.3).

So we may assume $|A| = 2$. Let $A = \langle a \rangle$, $B = \langle b \rangle$ and so on. Then we have:

and thus either $R \simeq W(D_4)$ or to the center-factor group $W^*(D_4)$ of $W(D_4)$. In the second case also $E \sim B$ in $C_R(A)$, a contradiction as above. Now the structure of $W(D_4)$ implies $z = [a, f, eb] \in Z(R)$, $N = \langle a, [a, f], z \rangle$ has order 8 and $N^\# \subseteq D(\Sigma)$. Further

$$B^c = \langle eb \rangle \in \Lambda_A \cap \Lambda_{\langle z \rangle} \text{ and } B^c \notin C_\Sigma(N).$$

Hence applying (2.1) again we get a contradiction as before.   □

Now we can show:

(2.6) Theorem

Let $A \in \Sigma$. Then the following hold:

1. $A = M'_A \le Z(M_A)$
2. $\Delta_A = \Lambda_A \cup \{A\}$
3. $C_\Sigma(M_A) = \{A\}$.

(We assume $R(G) = 1$!)

Proof. As $\{A\} = \underline{A}$, since $R(G) = 1$, (3) is (2.3). By (2.5) $M'_A = A \le Z(M_A)$. Thus (2) remains to be proved.

Suppose (2) is false and pick $C \in \Delta_A - (\Lambda_A \cup \{A\})$. Then there exists by (2.3) a $B \in \Lambda_A \cap \Psi_C$. By (1) and (3) $D = [B, C] \in \Lambda_A$, since $D \in (\Sigma \cap M_A) - \{A\}$. Let $N = ABD$. Then the hypothesis of (2.1) is satisfied for N. Hence there exists an

$$F \in N_\Sigma(N) \cap \Sigma_A \cap \Omega_C,$$

a contradiction to $C \in \Delta_A$.   □

(2.7) Corollary

Let $A, B, C \in \Sigma$ with

$$\overset{A}{\circ}\!\!-\!\!-\!\!-\!\!\overset{B}{\circ}\!\!=\!\!=\!\!\overset{C}{\circ} \; ,$$

$X = \langle A, B \rangle$ and $N = \langle C^X \rangle$. Then $\Sigma \cap N$ is a partition of N. In particular either $C_N(X) \in \Sigma$ or $C_N(X) = 1$ and N is the natural module for $\overline{X} = X/C_X(N) \simeq SL_2(K)$. (See II(2.20)!)

Proof. By II(2.20) $N = C_N(X) \times M$, $M = [N, X]$ the natural module for $\overline{X}$. Let $D = [B, C]$. Then $C \in \Lambda_D$ and $M \leq M_D$ since $\Sigma \cap M$ is a partition of M. As $N = CM$ we obtain $N \leq M_D$.

Let $E = [D, A]$. Then $E = C_M(A)$ and $N \leq M_E$ since X is transitive on $\Sigma \cap M$. Hence $\Sigma \cap N \subseteq \Lambda_E \cup \{E\}$ by (2.6). Now $C_N(A) = EC = EC_N(X)$ and $(\Sigma \cap C_N(A)) \cup C_N(X)$ is by II(2.20) a partition of $C_N(A)$. Since $C \in \Lambda_E$ also $\Sigma \cap N$ is a partition of $C_N(A)$. Hence either $C_N(X) \in \Sigma$ or $C_N(X) = 1$, $E = C$ and $M = N$.   □

(2.8) Proposition

Let $A \in \Sigma$. Then the following are equivalent:

(1) C_A is not Σ-maximal.

(2) $\Lambda = A^R$ is a weak TI-subset of Σ for each Σ-subgroup R with $G > R > C_A$.

Proof. It is obvious that we just have to show that (1) implies (2). For this assume that (1) holds and let R be a proper Σ-subgroup containing C_A properly. Set $\Lambda = A^R$, $\Delta = \Sigma \cap R$ and Ω the set of isolated vertices of $\mathcal{F}(\Delta)$. Claim $\Lambda \subseteq \Omega$.

If the claim is false, then since Ω is a normal subset of R, there exists a $D \in \Delta \cap \Omega_A$. But then by II(4.8) $G = \langle \Lambda_A, D \rangle \leq R$, a contradiction.

Now $C_E \leq R$ for each $E \in \Lambda$. Hence if $E \in \Lambda$ and $F \in \Lambda \cap \Lambda_E$, then by II(2.23) $EF \cap \Sigma \subseteq \Lambda$. Let now $c \in D(\Sigma)$ with $E^c \in \Lambda$ for some $E \in \Lambda$. If $E = E^c$, then $c \in C_E \leq R$ and thus $\Lambda = \Lambda^c$. So suppose $E \neq E^c$ and let $c \in C \in \Sigma$. Then $C \notin \Omega_E$, since $\{E, E^c\} \subseteq \Lambda \subseteq \Omega$. Hence $C \in \Psi_E$ and $E^c \in \Lambda_E$. But then $[E, c] = [E, C] \in \Lambda$, $c \in C_{[E,C]} \leq R$ and thus $\Lambda = \Lambda^c$, which was to be shown.   □

(2.9) Corollary

Suppose Λ is a weak TI-subset of Σ. Then the following hold:

1. $\langle \Lambda \rangle$ is abelian.
2. $\mathcal{E}(\Lambda)$ is connected and $\langle N_\Sigma(\Lambda) \rangle$ is transitive on Λ.
3. If $A \in \Lambda, B \in \Lambda \cap \Lambda_A$, then $AB \cap \Sigma \subseteq \Lambda$.

Proof. (1) is II(4.6). (3) was shown in the proof of (2.8). To prove (2) let $A \neq B \in \Lambda$. Then there exists by (2.6) a $C \in \Lambda_A \cap \Psi_B$. Again by (2.6) either $D = [C, B] = A$ or $C \in \Lambda_A \cap \Lambda_B$. In any case A and B are connected in $\mathcal{E}(\Lambda)$. Further, as shown in (2.8), A is conjugate to B in $\langle N_\Sigma(\Lambda) \rangle$. $\qquad\square$

Next a special lemma

(2.10) Lemma

Suppose Λ is a weak TI-subset of Σ and $A, B, C \in \Lambda$ and $E \in \Sigma$ with

Let $X = \langle A, E \rangle, R = \langle X, B, C \rangle$ and $N = \langle (BC)^R \rangle$. Then N is abelian and $R = NX$.

Proof. Let $N_1 = \langle B^X \rangle$ and $N_2 = \langle C^X \rangle$. If $[N_1, N_2] = 1$, then $R = (N_1 N_2)X$ by II(2.20) and (2.10) holds. So we may assume $[N_1, N_2] \neq 1$. By II(2.20) $N_1 = B[B, E][B, E, A]$ and $[B, E, A] \in \Lambda$. Thus, if $[B, E] \in C_\Sigma(C)$, then $N_1 \leq C(C)$ and $[N_1, N_2] = 1$. Thus by II(2.10)

$$Z_1 = [B, E, C] \neq 1 \neq Z_2 = [C, E, B]$$

and $Z_1, Z_2 \in \Lambda$. Now for $e \in E^\#$ we have $Z_1 = [B^e, C], Z_2 = [C^{e^{-1}}, B]$ since $CC^e = CC^{e^{-1}} = C[C, E]$. Hence $Z_2^e = Z_1$. Thus $Z_2 \in \Psi_E, \Sigma \cap Z_1 Z_2 \subseteq \Lambda$ by (2.9) (3) and $E \in N_\Sigma(\Lambda)$, a contradiction to $E \in \Omega_A$. $\qquad\square$

(2.11) Proposition

Suppose Λ is a weak TI-subset of Σ. Then Λ is a partition of $\langle \Lambda \rangle$.

Proof. Define a relation $\sim$ on Λ by $A \sim B$ if and only if $B \in \Lambda_A \cup \{A\}$. Then, to prove (2.11), it suffices by (2.9)(3) to show that $\sim$ is transitive.

Let therefore $A, B, C \in \Lambda$ with $A \sim B \sim C$ and $A \neq B \neq C$. By II(2.3) and (2.22) there exists an $E \in \Omega_B \cap \Psi_A$. Hence, replacing C by some $C' \in BC \cap \Sigma$ we obtain by II(2.11):

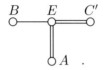

Hence by (2.10) B is conjugate to some $B \neq \overline{C} \in BC \cap \Sigma$ in $C(A)$ and thus $BC \leq M_A$. Now (2.6)(2) implies $BC \cap \Sigma \subseteq \Lambda_A \cup \{A\}$ and thus $A \sim C$, which was to be shown.   □

(2.12)   Theorem

Suppose Λ is a weak TI-subset of Σ. Then the following hold:

1. $\langle N_\Sigma(\Lambda) \rangle$ is Σ-maximal.
2. $C_\Sigma(\Lambda) = \Lambda$ and $N_\Sigma(\Lambda) = \Lambda \dot\cup \Delta$ with Δ a conjugacy class in $\langle N_\Sigma(\Lambda) \rangle$ and $\mathcal{F}(\Delta)$ connected.

Proof. (1) Suppose $G > R = \langle R \cap \Sigma \rangle > \langle N_\Sigma(\Lambda) \rangle$. Then by (2.8) and (2.9) (2) we have

$$\Lambda \subset \Lambda' = D^R \text{ for each } D \in \Lambda.$$

Pick $A \neq B \in \Lambda$ and $C \in \Lambda' - \Lambda$. Then by (2.11) applied to Λ' we have $\Lambda' \cap N$ is a partition of $N = ABC$. Thus by (2.1) $\langle N_\Sigma(N) \rangle \leq R$ and induces an $SL_3(K)$ on N. In particular there exists an $h \in D(N_\Sigma(N))$ with $B^h = C$ and $h \in C(A)$, a contradiction to $h \in N(\Lambda)$ but $C \notin \Lambda$.

To prove (2) let Ω be the set of isolated vertices of $N_\Sigma(\Lambda)$ and claim $\Omega = \Lambda$. Clearly $\Lambda \subseteq \Omega$ and by (2.1), (2.11) $\Omega \subseteq C_\Sigma(\Lambda)$. Hence by (2.6) $\Omega \subseteq \Lambda_A \cup \{A\}$ for each $A \in \Lambda$. Now picking $A \neq B \in \Lambda$ and $C \in \Omega - \Lambda$ the same argument as in the proof of (1) provides a contradiction.

Thus $\Omega = \Lambda$. Let $\Delta = N_\Sigma(\Lambda) - C_\Sigma(\Lambda)$ and pick $E, F \in \Delta$. Then by (2.6) and (2.11) $E \in \Lambda_A$, $F \in \Lambda_B$ for $A, B \in \Lambda$ and there exists a $C \in \Lambda$ with $[C, E] = A, [C, F] = B$. Now, applying (2.1) to $N = ABC$ we see that E and F are connected in $\mathcal{F}(\Delta \cap N_\Sigma(N))$ and whence in Δ.

This shows that Δ is a connectivity component of $\mathcal{F}(N_\Sigma(\Lambda))$. Suppose $\Theta \subseteq C_\Sigma(\Lambda)$ is another connectivity component of $\mathcal{F}(N_\Sigma(\Lambda))$ of cardinality > 1. Then $[\Delta, \Theta] \leq \langle \Lambda \rangle$ by II(2.21) (2). On the other hand $\Theta \cap \Lambda_A = \emptyset$ for each $A \in \Lambda$ by II(2.3), so that by II(2.21) $[\Delta, \Theta] = 1$ and by the same reason $N_\Sigma(\Lambda) - \Theta \subseteq C_\Sigma(\Theta)$ since $\Lambda = \Omega$. By II(2.3) this shows $N_\Sigma(\Lambda) \cap \Lambda_C \subseteq \Theta$ for each $C \in \Theta$.

Let now Λ^g be a conjugate of Λ containing $C \in \Theta$. If $[A, F] \neq 1$ for some $A \in \Lambda, F \in \Lambda^g$ then

$$[A, F] \in \Lambda^g \cap \Lambda_A \subseteq N_\Sigma(\Lambda) \cap \Lambda_C \subseteq \Theta$$

by (2.11) applied to Λ^g, a contradiction to $\Theta \cap \Lambda_A = \emptyset$. Hence $[\Lambda, \Lambda^g] = 1$ and $\Lambda^g \subseteq \Theta$. Let now $E \in \Lambda_C - \Lambda^g$. Then by what we have shown applied to $N_\Sigma(\Lambda^g)$ we obtain $E \in N_\Sigma(\Lambda^g) - C_\Sigma(\Lambda^g)$. Moreover $\Lambda \subseteq C_\Sigma(\Lambda^g) - \Lambda^g$, so that by the above $[E, \Lambda] = 1$. This shows $\Lambda_C \subseteq N_\Sigma(\Lambda)$ and thus $\Lambda_C \subseteq \Theta$ for each $C \in \Theta$. But this is a contradiction to II(4.8).

Hence no such Θ exists and $C_\Sigma(\Lambda) = \Omega = \Lambda$.   □

(2.13) Notation

For the rest of this section we fix $A \in \Sigma, A_0 \in \Omega_A, X = \langle A, A_0 \rangle, C_A = \langle C_\Sigma(A) \rangle,$ $\Delta_0 = C_\Sigma(X)$ and $V = \langle \Delta_0 \rangle$. Let $\Lambda_A^* = \Lambda_A \cap \Psi_{A_0}$. Then II(2.11) implies

$$\Lambda_A \subseteq \bigcup_{C \in \Lambda_A^*} (AC \cap \Sigma).$$

Hence by (2.6) $M_A = \langle \Lambda_A^* \rangle$.

We say that G satisfies hypothesis (M) if either C_A is Σ-maximal or $\Sigma_A \not\subseteq M_A$. Notice that, if G satisfies (M), then $AB \cap \Sigma$ is not a weak TI-subset of Σ for each $B \in \Lambda_A$. Indeed if $\Lambda = AB \cap \Sigma$ is a weak TI-subset of Σ, then by (2.12) $\Sigma_A \subseteq \Lambda \cup (N_\Sigma(\Lambda) - C_\Sigma(\Lambda))$. Hence $[C, \Lambda] = A$ for each $C \in \Sigma_A - \Lambda$ and thus $\Sigma_A \subseteq M_A$.

From hypothesis (M) we get:

(2.14) Lemma

Suppose G satisfies hypothesis (M) and let $E \in \Sigma, F \in \Lambda_E$. Then there exists an abelian subgroup N containing EF satisfying:

1. $\Sigma \cap N$ is a partition of N.
2. N is a 3-dimensional vector space over the division ring $K = \{\sigma \in \operatorname{End} V \mid D^\sigma \subseteq D \text{ for each } D \in \Sigma \cap N\}$.

Proof. By (2.13) $\Lambda = EF \cap \Sigma$ is not a weak TI-subset. Hence $\emptyset \neq \Lambda \cap \Lambda^d \neq \Lambda$ for some $d \in D \in \Sigma$. But then $L^d \in \Lambda$ for some $L \in \Lambda$ and either $L = L^d$ or $LL^d = EF$. In the second case $[L, d] = [L, D] \in \Lambda \cap \Lambda_L$, whence $[\Lambda, D] = [L, D]$, a contradiction to $\Lambda \neq \Lambda^d$. Thus $D \leq C(L)$ and, since $\Lambda \subseteq \Lambda_L$, we have $E \in \Psi_D$ resp. $F \in \Psi_D$. Hence $N = LEE^d$ resp. LFF^d satisfies (2.14) by (2.1).   □

(2.15)   Theorem

Suppose G satisfies (M). Then $\mathcal{E}(\Omega_A)$ is connected and M_A acts transitively on Ω_A.

Proof. We will prove the theorem in several steps. Let for the whole proof Δ be a connectivity component of $\mathcal{E}(\Omega_A)$ containing A_0. We first show:

(1) Δ is an orbit of M_A on Ω_A.

Indeed if $B \in \Delta$ and $C \in \Delta \cap \Lambda_B$ then there exists by II(2.11) and (2.19) an $m \in M_A$ with $B^m = C$. Hence $\Delta \subseteq \Delta'$, Δ' an orbit of M_A on Ω_A. On the other hand, also by II(2.19), $\Lambda_A^* \subseteq N(\Delta)$ so that (1) holds.

(2) $C_A \leq N(\Delta)$. Further Δ is a connectivity component of $\mathcal{D}(\Omega_A)$.

Suppose $C \in \Sigma_A - N_\Sigma(\Delta)$. Then clearly $\Delta \subseteq \Omega_C$, whence $\Delta \subseteq \Delta'$, Δ' a connectivity component of $\mathcal{E}(\Omega_C)$. As $C \not\leq N(\Delta)$ we have $\Delta \neq \Delta'$. Hence there exist $B \in \Delta$ and $E \in \Delta' \cap \Lambda_B$ with $E \notin \Delta$. We obtain

$$(\alpha) \qquad \overset{A}{\underset{B \ \sim \ E}{\bigwedge}} \qquad \text{and} \qquad (\beta) \qquad \overset{C}{\underset{B \ \sim \ E}{\bigwedge}}$$

by II(2.3) and (1). Now II(2.11) implies that there exists a unique $D \in BE \cap \Psi_C$. Since $D \neq E$ also by II(2.11) $D \in \Omega_A$. Hence $D \in \Delta \subseteq \Delta' \subseteq \Omega_C$, a contradiction.

This proves the first part of (2). To prove the second let $B \in \Delta, C \in \Omega_A \cap \Sigma_B$. Then $B^x = C$ for some $x \in C_A$ by II(2.11) and (2.20). Hence $C \in \Delta$ by the first part of (2), which proves the second. The proof of (2) also shows

(3) $\Delta \not\subseteq \Omega_C$ for each $C \in \Omega_A$.

Indeed if $\Delta \subseteq \Omega_C$, then $\Delta \subseteq \Delta'$, Δ' a connectivity component of $\mathcal{E}(\Omega_C)$. By II(4.8) and (1), $\Delta \neq \Delta'$. Now the same argument as in the proof of (2) yields a contradiction.

Now pick $B \in \Delta$ and $C \in (\Omega_A \cap \Psi_B) - \Delta$ (Exist by (2)!) Let $D = [B, C]$. Then $D \in \Psi_A$, since otherwise $D \in \Delta$ and thus $C \in \Delta$. Let by (2.14) N be an abelian subgroup of G containing BD partitioned by $\Sigma \cap N$, which is a 3-dimensional K-vector space. Then $C \in N_\Sigma(N) - C_\Sigma(N)$ by (2.6). By (2.1) $R = \langle N_\Sigma(N) \rangle$ induces the $SL_3(K)$ on N. Now $[C, N] = D$ by (2.16), whence C is a transvection group corresponding to the point D and some plane on N. Now the structure of $SL_3(K)$ implies that there exists an

$$E \in (N_\Sigma(N) \cap C_\Sigma(D) - C_\Sigma(N)) \cap \Psi_C.$$

Thus by II(1.3) CC' induces the set of all transvections corresponding to D on N, where $C' = [C, E]$. Hence

$$(*) \qquad\qquad C_N(C) \neq C_N(\overline{C}) \text{ for each } C \neq \overline{C} \in CC' \cap \Sigma.$$

By II(2.3) $C' \in \Psi_A \cup \Omega_A$. Hence by II(2.11) there exists a $C \neq \overline{C} \in CC' \cap \Omega_A$. Clearly $C, C' \notin \Delta$. Pick $F \in C_{\Sigma \cap N}(C)$ or $\in C_{\Sigma \cap N}(\overline{C})$. If $F \in \Omega_A$, then $F \in \Delta$ and thus also $C \in \Delta$ resp. $\overline{C} \in \Delta$ by (2), a contradiction. By II(2.3) $F \notin \Sigma_A$. This shows

$$C_{\Sigma \cap N}(C) \cup C_{\Sigma \cap N}(\overline{C}) \subseteq \Psi_A.$$

Now II(2.10) implies $\Sigma \cap N \subseteq \Psi_A \cup \Sigma_A$, contradicting $B \in \Omega_A$. This proves (2.15). $\qquad\square$

Remark. (2.15) is false in $G_2(2)' \simeq U_3(3)$. This is the reason why we need hypothesis (M), which comes into the proof through (2.14).

(2.16) Corollary

Suppose G satisfies (M). Then M_A acts regularly on Ω_A.

Proof. By (2.15) it suffices to show $N_{M_A}(A_0) = 1$. Since X is a rank one group clearly $N_{M_A}(A_0) \leq C(A_0)$. Now for $B \in \Lambda_A^*$ and $m \in N_{M_A}(A_0)$ we have:

$$B^m \in BA \cap \Lambda_A^* = \{B\}.$$

Hence $B^m = B$ and thus $m \in C(B)$, since $[B, m] \leq M_A' = A$. This implies $N_{M_A}(A_0) \leq C_{M_A}(\Lambda_A^*) = Z(M_A)$. Now II(4.8) implies $N_{M_A}(A_0) \leq C(\langle M_A, A_0 \rangle) \leq Z(G) = 1$. (As $Z(G) \leq R(G)$ by definition of $R(G)$!) $\qquad\square$

(2.17) Proposition

Suppose G satisfies (M). Then the following hold:

1. $N(A) = M_A N(X)$, $M_A \cap N(X) = 1$ and $C(A) = M_A C(X)$.
2. $C_A = M_A \cdot V$.
3. $\Delta_0 = \bigcup_{i \in I} \Delta_i$ with

   (a) $|\Delta_i| > 1$, Δ_i a conjugacy class in $V_i = \langle \Delta_i \rangle$ and $\mathcal{F}(\Delta_i)$ is connected.

   (b) $[\Delta_i, \Delta_j] = 1$ for $i \neq j$.

Proof. (1) follows from (2.16), the Frattini argument and the fact that X is a rank one group. To prove (2) it suffices to show:

$$(*) \qquad\qquad \text{If } B \in \Sigma_A - M_A, \text{ then } B^m \in \Delta_0 \text{ for some } m \in M_A.$$

Now $A \in \Sigma_B - \Lambda_B$. Hence by (2.6) there exists a $C \in \Sigma_B \cap \Omega_A$. By (2.16) $C^m = A_0$ for some $m \in M_A$ and thus $B^m \in \Sigma_A \cap \Sigma_{A_0} = \Delta$.

Finally let $\Delta_i, i \in I$ be the connectivity components of $\mathcal{F}(\Delta_0)$. Then, since Λ_A is by (2.6) the set of isolated vertices of $\mathcal{F}(\Sigma_A)$, there are no isolated vertices in $\mathcal{F}(\Delta_0)$. Hence $|\Delta_i| > 1$ for $i \in I$ and by II(2.21) $[\Delta_i, \Delta_j] = 1$ for $i \neq j$. $\quad\square$

(2.18)   Lemma

Suppose G satisfies (M). Then the following hold:

1. If $B \in \Sigma_A - \Lambda_A$, then $B^* = M_A B \cap V \in \Delta_0$.
2. Let $T \leq C_A$ such that $T' = 1, \Sigma \cap T$ is a partition of T and $T \cap M_A = 1$. Let $T* = M_A T \cap V$. Then $\Delta_0 \cap T*$ is a partition of T^*.

Proof. (1) is essentially (2.17)(∗). Indeed let $m \in M_A$ such that $B^m \in \Delta_0$. Then
$$B^* = M_A B \cap V = M_A B^m \cap V = B^m (M_A \cap V) = B^m.$$

To prove (2) let $1 \neq y \in T^*$. Then there exists a unique $x \in T$ with $M_A x = M_A y$. Let $B \in \Sigma \cap T$ with $x \in B$. Then
$$y \in M_A y \cap V = M_A x \cap V \subseteq M_A B \cap V = B^* \in \Delta_0$$
and $B^* \leq T^*$. $\quad\square$

The next important theorem shows that the absence of weak TI-sets containing A (which of course are normal in C_A) implies transitivity of C_A on Λ_A.

(2.19)   Theorem

Suppose C_A is Σ-maximal. Then C_A is transitive on Λ_A.

Proof. Suppose false and let Λ_1 be an orbit of C_A on Λ_A. Then, if $B \in \Lambda_1$ each element of Λ_1 is of the form $B^{x_1 \cdots x_n}$ with $x_i \in D(\Sigma_A)$. Since clearly $B^{x_1 \cdots x_{i+1}} \in \Lambda_{B^{x_1 \cdots x_i}}$, this implies that $\mathcal{E}(\Lambda_1)$ is connected. On the other hand (2.1) shows that C_A is transitive on each connectivity component of $\mathcal{E}(\Lambda_A)$, whence we obtain:

(1) Λ_1 is a connectivity component of $\mathcal{E}(\Lambda_A)$.

Let $\Lambda_2 = \Lambda_A - \Lambda_1$ and $\Delta_0 = \bigcup \Delta_i$, $i \in I$ as in (2.17). We show next:

(2) Either $\Delta_i \subseteq C(\Lambda_1)$ or $\Lambda_1 \cap \Lambda_B \neq \emptyset$ for each $B \in \Delta_i$.

Indeed suppose $\Delta_i \nsubseteq C(\Lambda_1)$. Then for each $B \in \Delta_i$ there exists a $C \in \Psi_B \cap \Lambda_1$. Hence by (1) $[C, B] \in \Lambda_1 \cap \Lambda_B$.

Now by Zorn's lemma let T be a subgroup of G satisfying:

(i) $T' = 1, A \leq T$ and $\Sigma \cap T$ is a partition of T.

(ii) T is maximal with (i) (with respect to inclusion)

and assume without loss of generality $(\Sigma \cap T) - \{A\} \subseteq \Lambda_1$. Since $T \cap \Sigma$ is no weak TI-set by (2.8), there exists a $C \in \Delta_0$ and $c \in C$ with $T \neq T^c$. $(M_A \leq N(T)$ by (2.6)!). If $T = (T \cap T^c)E$ for some $E \in \Sigma \cap T$, then it follows immediately from (2.6)(2) that $L = (T \cap T^c)EE^c$ satisfies (i), a contradiction to the maximality of T. So there exist $E \neq F \in \Sigma \cap T$ with $EF \cap (T \cap T^c) = 1$. Hence $EF \cap (EF)^c = 1$ and, if $D = [E, C], L = [F, C]$, then $D \neq L$ and $L \in \Lambda_D$ since $C \in \Lambda_D$. Hence by (1) $DL \cap \Sigma \subseteq \Lambda_1 \cap \Lambda_C$. Pick $K \in \Lambda_2$. Then, as $[DL, K] \leq A, DL \cap \Sigma_K \neq \emptyset$. If $K \in \Psi_C$, then $[K, C] \in \Lambda_{D_1}$ for $D_1 \in DL \cap \Sigma_K$, a contradiction to (1). So $[\Lambda_2, C] = 1$ and, if we assume without loss of generality that $C \in \Delta_1$, then $[\Lambda_2, \Delta_1] = 1$. On the other hand $D = [E, C] \in \Lambda_1 \cap C(\Lambda_2)$, so that also $[\Lambda_1, \Lambda_2] = 1$. Together with (2) this implies

(3) $\Sigma_A = C_{\Sigma_A}(\Lambda_1) \dot{\cup} C_{\Sigma_A}(\Lambda_2)$ and both are normal sets in C_A.

Indeed if $S \in \Delta_i$ with $[S, \Lambda_1] \neq 1 \neq [S, \Lambda_2]$, then by (2) $\Lambda_S \cap \Lambda_1 \neq \emptyset$ and thus $[S, T]$ is connected to Λ_1 in $\mathcal{E}(\Lambda_A)$ for $T \in \Lambda_2 \cap \Psi_S$, a contradiction to (1).

Let now Ω be a connectivity component of $\mathcal{F}(\Sigma_A)$ contained in $C_{\Sigma_A}(\Lambda_1)$. (Since $\Lambda_A \neq \Lambda_1$ and Λ_1 is an orbit of C_A, we may choose Λ_1 by (2) accordingly!) Then $\Omega \subseteq C_\Sigma(T)$ and thence is a connectivity component of $\mathcal{F}(N_\Sigma(T))$. Now II(2.21) (4) implies $N_\Sigma(T) \subseteq N(\Omega)$, a contradiction to $G = \langle C_A, N_\Sigma(T) \rangle$ by Σ-maximality of C_A and (2.1) and (2.6). □

(2.20) Corollary

Suppose C_A is Σ-maximal. Then V acts transitively on Λ_A^*.

Proof. Let $C, D \in \Lambda_A^*$. Then there exists a $v \in V, m \in M_A$ with $C^{vm} = D$. Hence $C^v \in DA \cap \Lambda_A^* = \{D\}$. □

(2.21) Lemma

Suppose C_A is not Σ-maximal, but $N(A)$ is maximal. Then $N(A)$ is transitive on Λ_A.

Proof. As $N(A)$ is maximal, G is primitive on Σ. Hence there exists a weak TI-subset Λ which is not a TI-subset of Σ. Hence by (2.1) there exists an $h \in N(A)$ with $\Lambda \neq \Lambda^h$. By (2.6) and (2.12) $\Lambda \cap \Lambda^h = A = [\Lambda, \Lambda^h]$.

Let $C \in \Lambda_A - (\Lambda \cup \Lambda^h)$. Then there exists by (2.12) an $E \in \Lambda$ with $[E, C] = A$. Since Λ^h is also a weak TI-set, there exists by (2.12) and (2.6) a $B \in \Lambda^h$ with $[E, B] = A$. Let $Y = \langle N_\Sigma(EA) \rangle$. Then by II(2.23) $Y = C_Y(EA)X_1, X_1 =$

$\langle F, B \rangle$ a rank one group inducing $SL_2(K)$ on EA. Since $C \equiv B \bmod C_Y(EA)$, there exist a $y \in C_Y(EA)$ with $B^y = C$. As $EA \subseteq \langle \Lambda \rangle \cap \langle \Lambda^y \rangle$ we have $\Lambda = \Lambda^y$ by (2.6) and (2.12). Hence together with (2.1) it follows that

$$(*) \qquad C(A) \cap N(\Lambda) \text{ is already transitive on } \Lambda_A - \Lambda.$$

Applying (2.1) once more this shows that $N(A)$ is transitive on Λ_A and (2.21) holds. □

(2.22) Corollary

Suppose C_A is not Σ-maximal, but $N(A)$ is maximal. Then the following hold:

(1) $N(X)$ is transitive on Λ_A^*.
(2) If there exist more than 2 weak TI-subsets of Σ containing A, then $C(X)$ is already transitive Λ_A^*.

Proof. (1) follows from (2.21) as in (2.20). If there exist at least three weak TI-subsets containing A, then (2.21) $(*)$ shows that $C(A)$ is already transitive on Λ_A. Hence (2) holds. □

From (2.20) and (2.22) we obtain that in the non-degenerate case, contrary to the degenerate case, in most cases Σ is already a class of k-root subgroups of G.

(2.23) Theorem

Suppose that either C_A is Σ-maximal or $N(A)$ is maximal in G, $\Sigma_A \not\subseteq M_A$, and there exist at least three weak TI-subsets of Σ containing A. Then Σ is a class of k-root subgroup, k a fixed commutative field. Moreover $X \simeq SL_2(k)$.

Proof. Let $B \in \Lambda_A^*$ and $C \in \Lambda_A^* \cap \Psi_B$. Then by II(2.19) $N_1 = B[B, A_0]$ and $N_2 = C[C, A_0]$ are natural modules for $X_i = X/C_X(N_i)$ and $X_i \simeq SL_2(K_i)$, K_i a division ring or a Cayley division algebra. By (2.20) resp. (2.22) $C(X)$ is transitive on Λ_A^*. Hence

$$C_X(N_i) \leq C_X(M_A) \cap Z(X) \leq Z(G) \leq R(G) = 1$$

by II(2.19) and since $G = \langle M_A, A_0 \rangle$ by II(4.8). Suppose that K_1 is non-commutative. Then there exists a $1 \neq h \in H_0 = N_X(A) \cap N_X(A_0)$ with $[A, h] \neq 1$ but $[B, h] = 1$, since N_1 can by (2.14) be embedded in a 3-dimensional K-vector space N, K a division ring and X acts by (2.6) as a subgroup of $SL_3(K)$ generated by transvection groups. Hence $K_1 \simeq K$ and we may choose h corresponding to a matrix $\begin{pmatrix} 1 & \\ & c \end{pmatrix}$ with $1 \neq c \in (K^*)'$. But then also $[C, h] = 1$ since B and C are conjugate in $C(X)$, a contradiction to $A = [B, C]$.

This shows together with I(5.7) that $K_1 \simeq K_2 \simeq k$, k a commutative field. Since by (2.15) $C(A)$ is transitive on Ω_A, I(5.7) also shows that $\langle A, D \rangle \simeq X \simeq SL_2(k)$ for each $D \in \Omega_A$. Now the connectivity of $\mathcal{F}(\Sigma)$ implies (2.23). □

Remark. The proof that $X \simeq SL_2(k)$, k a commutative field, does not need the assumption $\Sigma_A \not\subseteq M_A$. It just needs the transitivity of $C(X)$ on Λ_A^*, i.e. (2.20) (resp. (2.22)). But to show that the same commutative fields occur we used (2.15), which relies on hypothesis (M).

We will see in the next two sections that the only groups for which the conclusion of (2.23) does not hold are the groups $PT(V, W^*)$ of II(1.3) and E_6^K of exercise II(5.24)(5). That is in the case $N(A)$ maximal but $\Sigma_A \subseteq M_A$ (which corresponds to groups of type G_2) it will be shown by other means that the fields are the same.

(2.24) Notation

Suppose that the hypothesis of (2.23) is satisfied, i.e. either C_A is Σ-maximal or $N(A)$ is maximal, $\Sigma_A \not\subseteq M_A$ and there exist at least three weak TI-subsets of Σ containing A. Hence by (2.23) $X \simeq SL_2(k)$, k a fixed commutative field, independent of $A_0 \in \Omega_A$. Let $H_0 = N_X(A) \cap N_X(A_0)$. Then we may set:

(1) $H_0 = \{h(\lambda) \mid \lambda \in k^*\}$ with $h(\lambda)h(\mu) = h(\lambda\mu)$.

If now $D \in \Lambda_A^*$, then by II(2.19) $D[D, A_0]$ is a natural X-module. In particular D is H_0-invariant and we may label:

(2) $D = \{d(\lambda) \mid \lambda \in k\}$ with $d(\lambda)d(\mu) = d(\lambda + \mu)$ and $d(\lambda)^{h(\mu)} = d(\mu\lambda)$
    where $\lambda, \mu \in k$ resp. $\lambda \in k, \mu \in k^*$.

In particular D is a 1-dimensional k-vector space with this action, i.e. we have

$$d(\lambda)^{h(\nu + \mu)} = d((\nu + \mu)\lambda) = d(\nu\lambda + \mu\lambda) = d(\lambda)^{h(\nu)}d(\lambda)^{h(\mu)}.$$

(By (2.20), resp. (2.22), all $D \in \Lambda_A^*$ are equivalent H_0-modules!)

Let now $\widetilde{M}_A = M_A/A$. Then $\widetilde{M}_A$ is abelian by (2.6). Further, if $D \in \Lambda_A^*$ then $\widetilde{D}$ is equivalent to D as H_0-module. Hence we have:

(3) $\widetilde{D} = \{\widetilde{d}(\lambda) \mid \lambda \in k\}, \widetilde{d}(\lambda)\widetilde{d}(\mu) = \widetilde{d}(\lambda + \mu)$ and $\widetilde{d}(\lambda)^{h(\mu)} = \widetilde{d}(\mu\lambda)$.
    Moreover $\widetilde{D}$ is with this action a 1-dimensional k-vector space.

Finally for $\widetilde{m} \in \widetilde{M}_A$ and $\lambda \in k^*$ we set

(4) $\lambda\widetilde{m} = \widetilde{m}^{h(\lambda)}$ and $0 \cdot \widetilde{m} = 1$.

Then we get:

(2.25)   Lemma

Suppose that the hypothesis of (2.23) is satisfied. Then $\widetilde{M}_A$ is, with the k-action defined in (2.24), a kV-module.

Proof. Since $[V, H_0] = 1$, it suffices to show that $\widetilde{M}_A$ is with the k-action defined in (2.24) a k-vector space. For this write $\widetilde{M}_A$ additively. Then we have:

(α) $\lambda(\widetilde{m} + \widetilde{n}) = (\widetilde{m} + \widetilde{n})^{h(\lambda)} = \widetilde{m}^{h(\lambda)} + \widetilde{n}^{h(\lambda)} = \lambda\widetilde{m} + \lambda\widetilde{n}$ for $\widetilde{m}, \widetilde{n} \in \widetilde{M}_A$ and $\lambda \in k$.

To show:

(β) $(\lambda + \nu)\widetilde{m} = \lambda\widetilde{m} + \nu\widetilde{m}$

write $\widetilde{m} = \widetilde{d}_1(\lambda_1) + \cdots + \widetilde{d}_k(\lambda_k)$ with $d_i(\lambda_i) \in D_i \in \Lambda_A^*$.
Then we have by (α):

$$
\begin{aligned}
(\lambda + \nu)\widetilde{m} &= (\lambda + \nu)\widetilde{d}_1(\lambda_1) + \cdots + (\lambda + \nu)\widetilde{d}_k(\lambda_k) \\
&= \lambda\widetilde{d}_1(\lambda_1) + \nu\widetilde{d}_1(\lambda_1) + \cdots + \lambda\widetilde{d}_k(\lambda_k) + \nu\widetilde{d}_k(\lambda_k)
\end{aligned}
$$

by (2.24) (2) and (3). Hence by (α)

$$(\lambda + \nu)\widetilde{m} = \lambda\widetilde{m} + \nu\widetilde{m}.$$

Finally

(γ) $(\lambda\nu)\widetilde{m} = \widetilde{m}^{h(\lambda\nu)} = \widetilde{m}^{h(\nu)h(\lambda)} = \lambda(\nu\widetilde{m}).$

$\square$

The next theorem will be needed to show that, contrary to the degenerate case, in the non-degenerate case A is always a full root subgroup on the corresponding building.

(2.26)   Theorem

Suppose that C_A is Σ-maximal. Then $\langle C_\Sigma(V)\rangle = X$.

Proof. Let $B \in C_\Sigma(V) - X$. Then by (2.17) (3) and (2.19) $B \in \Omega_A$. Hence there exists by (2.15) an $m \in M_A$ with $A_0^m = B$. Now for each $v \in V$ we have:

$$B = B^v = A_0^{mv} = A_0^{v\overline{m}} = A_0^{\overline{m}}; \text{ where } \overline{m} = m^v.$$

Hence (2.16) implies $m \in C(V)$. Now (2.14) and (2.6) show that m centralizes some element of Λ_A^* and thus by (2.20):

(*) $m \in M_A \cap Z(C_A) = \overline{A}$

and $m \notin A$ since $B \notin \Sigma \cap X$. Let $\overline{\Sigma} = \{\overline{D} \mid D \in \Sigma\}$. Then $\overline{\Sigma}$ is a conjugacy class of TI-subgroups of G, since $G = \langle C_A, C_B \rangle$ for $A \neq B \in \Sigma$.

Let k be the field of (2.23) and assume first $|k| = 2$. Then $m^2 \in A$ and thus $Y = \langle A, A_0, B \rangle$ is m-invariant. Further, since $Y \leq C(V)$, we have $Y \cap \Sigma_A = \emptyset$. Hence $\Sigma \cap Y \subseteq \Omega_A \cup \{A\}$. If now $m^2 = 1$, then m normalizes $\langle A_0, B \rangle \simeq \Sigma_3$ and thus m fixes some element of Ω_A, a contradiction to (2.16). Thus each element of $\overline{A} - A$ has order 4. Since $A = \Phi(\overline{A})$ by (2.6), this implies $\overline{A} \simeq \mathbb{Z}_4$. On the other hand we have:

and a^Y is a class of 3-transpositions of Y for $A = \langle a \rangle$ by the above. Hence it is easy to see that $|Y| = 3^2 \cdot 2$ or $3^3 \cdot 2$ and $|Y \cap \Omega_A| = 8$. Further, if $A_0 \neq C \in Y \cap \Omega_A$, then $C = A_0^n$ for some $n \in M_A$. Since $C \in C_\Sigma(V)$ the above argument shows $n \in \overline{A}$ and thus $\overline{A}$ is transitive on $\Omega_A \cap Y$, a contradiction to $|\overline{A}| = 4$ and $|\Omega_A \cap Y| = 8$.

Next assume $|k| > 3$. Then (2.25) shows that there exists an $h \in H_0 = N_X(A) \cap N_X(A_0)$ with $C_{M_A}(h) = 1$. Now for $C \in \Lambda_A^*$ we have

$$(+) \qquad\qquad \overline{C} \cap M_A = C.$$

Indeed if $c \in (\overline{C} \cap M_A) - C$, then there exists a $d \in D(\Lambda_A)$ with $1 \neq [c, d] \in A$. Since $[C, d] = A$ this shows that $C_{\overline{C}}(d) \neq 1$, a contradiction to $\langle \Sigma_C, d \rangle = G$. Hence $(+)$ holds.

Now since $\overline{C} \leq M_A C(X)$ by (2.17), we have $[\overline{C}, h] = C$. Together with (2.24) (2) this easily shows:

$$(**) \qquad\qquad \overline{C} = C \times C_1, C_1 = C_{\overline{C}}(h).$$

On the other hand by (2.17)(1):

$$M_A C_1 = M_A \overline{C} = M_A(M_A \overline{C} \cap C(X)),$$

whence $C_1 = M_A \overline{C} \cap C(X) = \overline{C} \cap C(X)$, since $C_{M_A}(h) = 1$. But clearly $C_{\overline{C}}(X) = 1$, since $G = \langle \Sigma_C, X \rangle$ by Σ-maximality of C_A. Hence $\overline{C} = C$ and $\overline{A} = A$, a contradiction to $m \in \overline{A} - A$.

So finally the case $|k| = 3$ remains to be treated. Let in that case h be the central involution of X. Then h inverts $\widetilde{M}_A$ and centralizes A. Now the same argument as before shows that for $C \in \Lambda_A^*$ we have:

$$C_1 A = A C_{\overline{C}A}(X).$$

On the other hand by II(2.19) $N = C[C, A_0]$ is the natural X-module. In particular $\langle C^X \rangle$ is abelian, whence $\langle \overline{C}^X \rangle$ is also abelian. As

$$\overline{C} \le C(C_1 A) \le C A C_{\overline{C} A}(X) \le (NX) \times C_{\overline{C} A}(X),$$

this shows that $\overline{C} \le N C_{\overline{C} A}(X)$. Now h inverts N and centralizes $C_{\overline{C} A}(X)$, whence $C_1 \le C_{\overline{C} A}(X)$, which is a contradiction as before. $\qquad\square$

We close this section with several more specialized lemmata, which will be needed for the classification of the orthogonal groups.

(2.27) Lemma

Suppose G satisfies (M) and $N(A)$ is maximal in G. Let $\Delta_0 = \cup_{i \in I} \Delta_i$ as in (2.17). Then the following hold:

(1) If $|I| \ge 2$, then $\langle F^{V_i} \rangle$ is abelian for each $F \in \Lambda_A^*$ and $i \in I$.

(2) $F^{V_i} \ne F$ for each $i \in I$ and $F \in \Lambda_A^*$.

Proof. Pick $F \in \Lambda_A^*$. If $B \in \Delta_j \cap \Psi_F, j \in I$, then $N = F[F, B]A$ satisfies the hypothesis of (2.1). Hence there exists a $C \in \Sigma_A$ with $[F, B, C] = F$. By (2.1) $C \in \Lambda_F \cap \Omega_B$. Let by (2.18) $m \in M_A$ with $C^m \in \Delta_0$. Then $C^m \in \Omega_B \cap \Delta_0 \subseteq \Delta_j$ and C^m normalize $F[B, F] = \langle N \cap \Psi_{A_0} \rangle$. Since $[F, B, C^m] \le FA$, this implies

(*) If $F \in \Lambda_A^*$ and $B \in \Psi_F \cap \Delta_j$, then there exists a $C \in \Delta_j \cap \Lambda_F$ such that $\langle B, C \rangle$ induces the $SL_2(K)$ on $F[F, B]$.

Now, to prove (2), assume first that C_A is Σ-maximal and $F^{V_i} = F$ for some $i \in I$. Let $T = \Pi_{j \ne i} V_j$. Then, as $V = V_i T$, (2.20) implies

$$\Lambda_A^* = F^V \subseteq C(V_i),$$

a contradiction to $M_A = \langle \Lambda_A^* \rangle$ and (2.6). If now C_A is not Σ-maximal, then $\Sigma \cap FA \subseteq \Lambda$, Λ a weak TI-subset of Σ by (2.21). Now by (2.1) and (2.11) Σ_A is transitive on $\Lambda \cap \Lambda_A$, whence V is transitive on $\Lambda \cap \Lambda_A^*$. Since $\langle \Lambda \rangle = \langle \Lambda \cap \Lambda_A^* \rangle A$ this implies $V_i \subseteq C(\Lambda)$, a contradiction to (2.12)(2).

This proves (2). Now (2) implies $\Delta_j \cap \Psi_F \ne \emptyset$, whence $\Delta_j \cap \Lambda_F \ne \emptyset$ for each $j \in I$. Hence if $i \ne j$ and $C \in \Delta_j \cap \Lambda_F$, then

$$F^{V_i} \subseteq \Lambda_A \cap \Lambda_C$$

and thus $\langle F^{V_i} \rangle$ is abelian by (2.6). $\qquad\square$

(2.28) Lemma

The following hold:

(1) If $E \in \Delta_0$ then $\Lambda_A \cap \Lambda_E \subseteq \Lambda_A^*$.
(2) Suppose that the hypothesis of (2.27) holds and Δ_j is degenerate for $j \in I$. Let $F \in \Delta_j$. Then

$$\Sigma_F \cap \Lambda_A^* = \Lambda_F \cap \Lambda_A^*.$$

Proof. (1) is a consequence of II(2.3). To prove (2) it suffices to show:

$$\Sigma_F \cap \Lambda_A^* \subseteq \Lambda_F \text{ for some } F \in \Delta_j.$$

Suppose this is false and pick $B \in \Sigma_F \cap \Lambda_A^*, B \notin \Lambda_F$. By (2.27)(2) there exists an $E \in \Delta_j \cap \Omega_F \cap \Psi_B$, since $\mathcal{F}(\Delta_j)$ is connected. Let $Y = \langle E, F \rangle$ and $N = \langle B^Y \rangle$. Then (2.7) implies $\Sigma \cap N$ is a partition of N and $1 \neq C_N(Y) \in \Sigma$. Hence N is by (2.1) a 3-dimensional K-vector space. Since $\Sigma \cap N \subseteq \Lambda_A^*$ it follows easily from (2.1) that $\langle N_{\Sigma_A}(N) \rangle$ induces the $SL_3(K)$ on N. But then by (2.18) also $\langle N_\Delta(N) \rangle$ induces the $SL_3(K)$ on N. Since $\{E, F\} \subseteq N_{\Delta_j}(N)$, this shows that $\langle N_{\Delta_j}(N) \rangle$ induces the $SL_3(K)$ on N, a contradiction to Δ_j degenerate. (The class of transvection groups of $SL_3(K)$ is non-degenerate!)□

(2.29) Lemma

Suppose G satisfies (M). Then the following hold:

(1) If $B \in \Lambda_A^*$ and $C \in \Lambda_A^* \cap \Lambda_B$, then there exists a $D \in \Delta_0$ with $[B, D] = C$.
(2) If $B, C \in \Sigma_A - \Lambda_A$ with $M_A B = M_A C$, then $C = B^m$ for some $m \in M_A$.

Proof. Application of (2.1) to $N = ABC$ yields the existence of a $\overline{D} \in \Sigma_A$ with $[B, \overline{D}] = C$. Now set $D = M_A \overline{D} \cap V$. Then, as $[B, D] \leq CA \cap \Lambda_A^*$, (2.18) implies (1). (2) is (2.17)(∗).   □

The last lemma is needed to apply classifications obtained earlier to the groups $V_i, i \in I$.

(2.30) Lemma

Suppose G satisfies hypothesis (M) and $N(A)$ is maximal in G. Then $R(V_i) \leq Z(V)$ for $i \in I$.

Proof. Since $[V_i, V_j] = 1$ for $i \neq j$ it suffices to show $R(V_i) \leq Z(V_i)$. Suppose this is false. Then, by definition of $R(V_i)$ in II §4, there exists a $t \in R(V_i)$ and $E \in \Delta_i$ with $E \neq E^t$ and $C_{\Delta_i}(E) = C_{\Delta_i}(E^t)$. Pick $B \in \Lambda_A^*$. We show:

(∗) $[B, E] = [B, E^t].$

To prove $(*)$ assume first $B \in \Sigma_E$ but $C = [B, E^t] \neq 1$. Then $BC \cap \Sigma \subseteq \Lambda_A^*$ and (2.27) $(*)$ implies that there exists an $F \in \Delta_i$ such that $\langle E^t, F \rangle$ induces an $SL_2(K)$ on BC. Hence $F \in \Omega_E$ by $II(2.22)$ applied to Δ_i, a contradiction to $[BC, E] = 1$ but $[BC, F] \neq 1$.

With symmetry this shows

$$(+) \qquad\qquad\qquad C_{\Lambda_A^*}(E) = C_{\Lambda_A^*}(E^t).$$

Hence, to prove $(*)$, we may assume

$$1 \neq C = [B, E^t] \neq D = [B, E] \neq 1.$$

Application of $(+)$ yields $[C, E] = 1 = [D, E^t]$. Hence $C = [B, E^t] \in \Lambda_D$ by (2.6). Now another application of (2.6) shows that $\Sigma \cap N$ is a partition of N, where $N = BCD$. Now by (2.1) N is a 3-dimensional K-vector space and by $II(2.10)$ $\Sigma \cap N \subseteq \Lambda_A^*$. Now it follows easily from (2.1) and (2.18) that $\langle N_{\Delta_i}(N) \rangle$ induces an $SL_3(K)$ on N. (See exercise $(2.31)(3)$!) But as $[E, t] \leq R(V_i)$, this would imply that $[E, t]$ is in the kernel of the action of $\langle N_{\Delta_i}(N) \rangle$ on N, a contradiction to $E^t \leq E[E, t]$.

This proves $(*)$. Now $(*)$ and $(2.27)(*)$ imply that $[E, t]$ centralizes Λ_A^* and whence M_A, by $II(4.8)$ a contradiction to $[E, t] \leq C(A_0)$. $\qquad\square$

(2.31)   Exercises

(1) Let N be an abelian subgroup of G such that $\Sigma \cap N$ is a partition of N and $N \neq AB$; $A, B \in \Sigma \cap N$. Show that N is a vector space over

$$K = \{\sigma \in \mathrm{End}\, N \mid A^\sigma \subseteq A \text{ for each } A \in \Sigma \cap N\}.$$

Moreover, $\Sigma \cap N$ is the set of all 1-dim. K-subspaces of N.

Hint. Let $\mathcal{W} = \{\Pi_{j \in J} A_j\} \cup \{1\}$; where $\Sigma \cap N = \{A_i \mid i \in I\}$ and $J \subseteq I$. Show that $\mathcal{W}$ satisfies (1)–(4) of (2.1).

(2) Let N be as in exercise (1) and assume that $n = \dim_K N < \infty$. Show that $\langle N_\Sigma(N) \rangle$ induces the $SL_n(K)$ on N.

Hint. Proceed as in (2.1).

(3) Let N be as in (1) with $\dim_K N$ arbitrary and M being an m-dimensional K-subspace of N, $m < \infty$. Show that $\langle N_\Sigma(N) \cap N_\Sigma(M) \rangle$ induces the $SL_m(K)$ on M.

Hint. $M_A \leq \langle N_\Sigma(N) \cap N_\Sigma(M) \rangle$ for $A \in \Sigma \cap M$.

(4) Suppose $\mathcal{B}$ is a Moufang plane over the Cayley division algebra K and $G = E_6^K$ in the notation of exercise II(5.24)(5). Let

$$\mathcal{A}:$$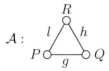

be an apartment of $\mathcal{B}$ and $r = (R, \ell, P, g)$, $r + s = (\ell, P, g, Q)$ and $s = (P, g, Q, h)$ roots of $\mathcal{A}$. Show:

(a) $A_r = \{\sigma \in G \mid \sigma$ fixes all points on ℓ and all lines through $P\}$
$A_{r+s} = \{\sigma \in G \mid \sigma$ fixes all points on g and all lines through $P\}$.

(b) $N = A_r A_{r+s} = \{\varphi \in G \mid \varphi$ fixes all lines through P, but no other line$\}$.

(5) Assume the same hypothesis and notation as in exercise (4) and let $N^* = A_s A_{r+s}$, $\Sigma = A_r^G$ and $\Lambda = \Sigma \cap N$, $\Lambda^* = \Sigma \cap N^*$. Show

(a) Λ and Λ^* are TI-subsets of Σ.

(b) $\Lambda \cup \Lambda^* = \Lambda_A$ where $A = A_r$.

(c) Let $\mathcal{P} = \{N^g \mid g \in G\}$ and $\mathcal{G} = \{\langle N^g, N^h \rangle \mid g, h \in G$ and $N^g \neq N^h\}$. Show that the geometry $(\mathcal{P}, \mathcal{G}, \subseteq)$ is isomorphic to $\mathcal{B}$ ($\mathcal{B}$ considered as point line space.)

(6) Let $\mathcal{B}$ be a Moufang hexagon and G the group of Lie type $\mathcal{B}$ in the notation of II§5. Let

$$\mathcal{A}: Q$$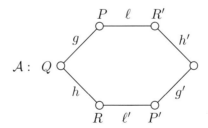

be an apartment of $\mathcal{B}$ and $r = (\ell, P, g, Q, h, R, \ell')$, $s = (g, Q, h, R, \ell', P', g')$ be two roots of $\mathcal{A}$ (with Q resp. R in the "middle") and assume that r and s are long roots in the root system Φ. (Passing to the dual hexagon if necessary we can choose notation so that long roots correspond to points!). Let $\Sigma = A_r^G$, $N = A_r A_s$ and $\Lambda = \Sigma \cap N$. Show:

(a) Λ is a weak TI-subset, but not a TI-subset, of Σ.

(b) Let $\mathcal{P} = \Sigma, \mathcal{B} = \{\Lambda^g \mid g \in G\}$. Then the geometry $(\mathcal{P}, \mathcal{B}, \in)$ is isomorphic to $\mathcal{B}$. ($\mathcal{B}$ considered as point line space.)

Hint. If $\alpha = (h, Q, g, P, \ell, R', h')$, then by the proof of II(5.20) we have $[A_\alpha, A_r] = 1$ and $[A_\alpha, A_s] = A_r$, whence $[A_\alpha, N] = A_r$. Now $\Lambda = \{A_r^x \mid x \in G$ and $Q^x \in h\}$. Hence, if $D \in \Sigma$ centralizes $A_r^x \in \Lambda$, then $[N, D] \leq A_r^x$. Thus Λ is a weak TI-set. That Λ is not a TI-subset of Σ follows from the fact that, if γ is the root of $\mathcal{A}$ corresponding to ℓ, then $X_\gamma = \langle A_\gamma, A_{-\gamma} \rangle$ acts doubly transitively on the lines through Q. (By I(4.12)!)

§3   The linear groups and E_6^K

We assume in this section that Σ is a non-degenerate class of abstract root subgroups of G, $R(G) = 1$ and C_A is not Σ-maximal for $A \in \Sigma$. We use the results and the notation of §2. In particular by (2.8) there exists a weak TI-subset Λ of Σ containing A. By (2.11) Λ is a partition of $T = \langle \Lambda \rangle$ and by (2.12) $C_\Sigma(\Lambda) = \Lambda$ and $N_\Sigma(\Lambda) = \Lambda \dot\cup \Delta$ with $\mathcal{F}(\Delta)$ connected.

By exercise (2.31)(1) either $T = A \cdot B$; $A \neq B \in \Lambda$, in which case we set $m = 2$, or T is a vectorspace over $K = \{\sigma \in \mathrm{End}\,T \mid B^\sigma \subseteq B$ for all $B \in \Lambda\}$ and $m = \dim_K T \geq 3$.

We have

(3.1) Lemma

One of the following holds:

(1) $m \leq 3$.
(2) There exists at most one $\Lambda^g \neq \Lambda$ with $A \in \Lambda^g$.

Proof. Suppose (3.1) is false and let $\Lambda_i, i = 1, 2, 3$ be different conjugates of Λ containing A. Then by (2.6) and the above

$$(*) \qquad \Lambda_i \cap \Lambda_j = A = [\Lambda_i, \Lambda_j] = [\Lambda_i, D]$$

for $i \neq j$ and $D \in \Lambda_j - \{A\}$. In particular, if without loss of generality $\Lambda = \Lambda_1$, there exist $B \in \Lambda_2 - \{A\}$, $C \in C_{\Lambda_3}(B) - \{A\}$ with $C_T(B) \neq C_T(C)$, since $m \geq 4$. (The elements of Λ_2 and Λ_3 induce transvection groups corresponding to A on T!). Hence by Exercise (2.31)(3) there exists an $E \in N_\Sigma(\Lambda) \cap \Omega_B \cap \Psi_C$. Let $Y = \langle B, E \rangle$ and $M = \langle C^Y \rangle$. Then by (2.7) either $B \in \Lambda_C$ or $F = C_M(Y) \in \Sigma$. In the first case $[B, \Lambda_3] \leq C \cap A = 1$ by (2.6), a contradiction to $(*)$. Now we have $T = [T, Y] \times C_T(Y)$ with $A \leq [T, Y]$. Further, since $[C, E, B] \leq M_A$, we have $F \in \Lambda_A$ in the second case. But then $[T, F] = 1$ by the 3-subgroup lemma, whence $F \in C_\Sigma(\Lambda) = \Lambda$ and thus $F = A$, since $C \in \Lambda_A \cap \Lambda_F$. But this contradicts II(2.3). $\square$

(3.2) Lemma

Suppose that there exist at most two conjugates of Λ containing $A \in \Lambda$. Then for each $E \in \Sigma$ there exists a unique element of $\Lambda^{D(\Sigma)}$ containing E.

Proof. We first show the existence of a $\Lambda^d, d \in D(\Sigma)$ containing E. If $E \in \Lambda$ this is clear. If $E \in N_\Sigma(\Lambda) - \Lambda$, then as $[\Lambda, E] \neq 1$, there exists an $F \in \Lambda \cap \Lambda_E$. Now the existence of Λ^d follows from II(2.23). So assume $E \in \Sigma - N_\Sigma(\Lambda)$. If $\Lambda \subseteq \Psi_E \cup \Sigma_E$, then since $\Lambda \cap \Lambda^e = \emptyset$ for $e \in E^{\#}$, we have $[\Lambda, \Lambda^e] \leq \langle \Lambda \rangle \cap \langle \Lambda^e \rangle = 1$, a contradiction to $C_\Sigma(\Lambda) = \Lambda$. Hence $\Lambda \cap \Omega_E \neq \emptyset$, in which case the existence of a Λ^d containing E is also clear.

So the uniqueness remains to be shown. If Λ is a TI-subset of Σ this is also clear. So we may assume

$(*)$ For each $B \in \Lambda$ there exists a unique $\Lambda^g \neq \Lambda$ with $B \in \Lambda^g$.

Let $\mathcal{M} = \Lambda^{D(\Sigma)}, E \in N_\Sigma(\Lambda)$ and $B \in \Lambda \cap \Lambda_E$. If $E \in \Lambda$ the uniqueness is also clear by the weak TI-subset property. So we may assume $E \notin \Lambda$. Since $N(A)$ is maximal as Λ is not a TI-subset, there exists by (2.21) and $(*)$ a unique Λ^g containing B and E. Suppose there exist $d_1, d_2 \in D(\Sigma)$ with $\Lambda^{d_1} \neq \Lambda^{d_2}$ and $E \in \Lambda^{d_1} \cap \Lambda^{d_2}$. Then by $(*)$ $\Lambda^{d_i} = \Lambda^g$ for $i = 1$ or 2, a contradiction to $\Lambda \cap \Lambda^{d_i} = \emptyset$. This shows the uniqueness in case $E \in N_\Sigma(\Lambda)$.

Finally assume $E \in \Sigma - N_\Sigma(\Lambda)$. Then, as above, $\Lambda \cap \Omega_E \neq \emptyset$ and we may without loss of generality assume $A \in \Lambda_E$. Suppose again that there exist $d_1, d_2 \in D(\Sigma)$ with $E \in \Lambda^{d_i}$ and $\Lambda^{d_1} \neq \Lambda^{d_2}$. Then we may without loss of generality assume $d_1 \in D(\Sigma \cap Y), Y = \langle A, E \rangle$ and $A^{d_1} = E$. If $A^{d_2} = E$, then $\langle d_2, A \rangle \geq \langle A, E \rangle = Y$ and thus $\langle d_2, A \rangle = Y$ since $\langle d_2, A \rangle$ is a rank one group. If now $B \in \Lambda \cap \Psi_E$ and $M = \langle B^Y \rangle$, then M is by II(2.19) a natural Y-module. In particular $B^{d_1 d_2^{-1}} \in \Lambda_B \cap \Lambda_A$ and thus $B^{d_1 d_2^{-1}} \in \Lambda$ by (2.6). But then $\Lambda = \Lambda^{d_1 d_2^{-1}}$, a contradiction.

This shows $A^{d_2} \neq E$ and we obtain by II(2.3) one of the following possibilities:

(i) $\overset{E}{\circ}\!\!-\!\!-\!\!\overset{A}{\circ}\!\!-\!\!-\!\!\overset{A^{d_2}}{\circ}$ or (ii) $\overset{E}{\circ}\!\!-\!\!-\!\!\overset{A}{\circ}\!\!-\!\!\!\!=\!\!\!\!-\!\!\overset{A^{d_2}}{\circ}$

Let in any case $R = \langle E, A, A^{d_2} \rangle$. By II(2.11) there exists a unique $E' \in EA^{d_2} \cap \Psi_A$. By II(2.19) $M = \langle (E')^R \rangle$ is a natural module for Y. Moreover by the structure of R described in II(2.19)

$$\Sigma \cap EA^{d_2} = \Sigma \cap EE' = \Lambda^{d_2} \cap R$$

is a TI-subset of R. Now case (ii) is obviously impossible, since then $D_2 \in \Psi_A$ for $d_2 \in D_2 \in \Sigma$. In case (i) $d_2 \in \langle A, A^{d_2} \rangle$ as above. Since $E^{d_1^{-1}d_2} = A^{d_2}$, this shows that $d_1^{-1}d_2 \in N_R(\Lambda^{d_2} \cap R)$. Hence also $d_2^{-1}d_1 = (d_1^{-1}d_2)^{-1} \in N_R(\Lambda^{d_2})$, as $|\Lambda^{d_2} \cap R| \geq 2$, and thus $\Lambda^{d_1} = \Lambda^{d_2}$.

This final contradiction shows the uniqueness of Λ^d, $d \in D(\Sigma)$, containing E. $\qquad \square$

(3.3) Lemma

Suppose that there exist at most two conjugates of Λ containing A. Then Λ is a TI-subset of Σ.

Proof. By (3.2) it suffices to show $\Lambda^{D(\Sigma)} = \Lambda^G$. For this it suffices to show:

$$(*) \qquad \qquad \Lambda^{ef} \in \Lambda^{D(\Sigma)} \text{ for all } e, f \in D(\Sigma).$$

Now suppose that $(*)$ is false for some pair $e, f \in D(\Sigma)$. Set $\Delta = \Lambda^e$. Then $\Delta^f \notin \Lambda^{D(\Sigma)}$. Moreover by (3.2) Δ and Λ are no TI-subsets of Σ. If $D^f \in C_\Sigma(D)$ for each $D \in \Delta$, then since Δ is also a weak TI-subset of Σ, we obtain $[\Delta, \Delta^f] \leq \langle \Delta \rangle \cap \langle \Delta^f \rangle = 1$, a contradiction to (2.12)(2) applied to Δ. Thus there exists a $D \in \Delta$ with $D^f \in \Omega_D$. Pick by II(2.3), (2.11) and (2.19) a $B \in \Delta$ with $D^f \in \Psi_B$ and let $Y = \langle D, D^f \rangle$ and $R = \langle Y, B \rangle$. Then by I(1.3) there exists an $x \in D(\Sigma \cap Y)$ with $D^x = D^f$. By (3.2) applied to Δ and D^x we obtain $\Delta^x = \Delta^f$. Let $E = [B, D^x]$. Then by II(2.19) $D^x E = D^x B^x \leq \langle \Delta^x \rangle$. Claim

$$(+) \qquad \qquad \text{There exists a } g \in G \text{ with } BE \leq \langle \Lambda^g \rangle.$$

Suppose that $(+)$ is satisfied. Then $\Lambda^g \in \Lambda^{D(\Sigma)}$, since $\Delta^f \notin \Lambda^{D(\Sigma)}$ and since by the hypothesis of (3.3) there exist at most two conjugates of Λ containing E, one of which must lie in $\Lambda^{D(\Sigma)}$ by (3.2). But this is by (3.2) impossible as $\Delta \in \Lambda^{D(\Sigma)}$ and $B \in \Delta$.

Thus $(+)$ remains to be proved. Since we may assume that Λ is no TI-subset, we may assume that there exists an $h \in G$ with $B \in \Delta \cap \Delta^h$. By (2.12) there exists an $L \in \Delta^h$ with $[\Delta, L] = B$. Since by II(2.19) $[D, E] = B$ also $[\Delta, E] = B$. By (2.12) there exists a $y \in \langle N_\Sigma(\Delta) \rangle$ with $L^y = E$, whence $B^y = B$. Hence $BE \leq \langle \Delta^{hy} \rangle$ and setting $g = ehy$ $(+)$ holds. $\qquad \square$

(3.4) Theorem

One of the following holds:

(1) $T = AB$ for $A \neq B \in \Lambda$.

(2) T is a 3-dimensional vector space over the division ring $K = \{\sigma \in \mathrm{End}\,T \mid B^\sigma \leq B$ for all $B \in \Lambda\}$.

(3) Λ is a TI-subset of Σ.

Proof. By (3.1) either (1) or (2) hold, or there exist at most two conjugates of Λ containing A. But in the second case (3) holds by (3.3).   □

(3.5) Notation

If Λ is a TI-subset of Σ, then by (2.11)

$$T \cap T^g = \begin{cases} T \\ 1 \end{cases} \quad \text{for all } g \in G.$$

Hence T is a TI-subgroup of G. In this case we denote for $D \in \Sigma$ by T_D the unique conjugate of T containing D.

In the following we will speak of hyperplanes of T (and of conjugates of T). If $m > 2$ it is clear what this means, since T is a K-vectorspace. In case $m = 2$ hyperplanes are just points, i.e. elements of Λ.

(3.6) Lemma

Suppose Λ is a TI-subset of Σ. Then the following hold:

1. If $B \in \Sigma$ with $\Lambda \subseteq C_\Sigma(B) \cup \Psi_B$, then $B \in N_\Sigma(\Lambda)$.
2. If $B \in \Sigma - N_\Sigma(\Lambda)$ then $H = \langle \Lambda \cap \Psi_B \rangle$ is a hyperplane of T. Further $[H, B] = N_{T_B}(T) = C_{T_B}(H)$ is a hyperplane of T_B.
3. If $T^g \neq T$, then $N_T(T^g)$ resp. $N_{T^g}(T)$ are hyperplanes of T resp. T^g.

Proof. (1) If $B \notin N_\Sigma(\Lambda)$, then $\Lambda \subseteq \Psi_B$ and

$$[T, T^b] \leq T \cap T^b = 1 \text{ for each } b \in B^\#,$$

a contradiction to (2.12).

(2) Pick $C \in \Lambda \cap \Omega_B$ by (1). Then for each $C \neq D \in \Lambda$ there exists by II(2.3) and (2.11) a $D' \in CD \cap \Psi_B$. Hence $T = CH$ and H is a hyperplane of T, since C is a 1-subspace of T. Now we have

$$
\begin{array}{ccc}
C & B & D' \\
\circ\!\!-\!\!\!-\!\!\!-\!\!\!-\!\!\circ\!\!=\!\!=\!\!=\!\!\circ
\end{array}
$$

and so CD' is conjugate to $B[B, D']$ in $R = \langle C, B, D' \rangle$ by II(2.19). Hence $[B, D'] \leq T_B$ and thus $[B, H] \leq T_B$. Since $[B, D'] \in C_\Sigma(D')$ we also have

$[B, H] \leq N_{T_B}(T)$. As $D' \in C_\Sigma([B, D'])$ also $D' \leq N_T(T_B)$ and thus $H \leq N_T(T_B)$. Now $[H, N_{T_B}(T)] \leq T \cap T_B = 1$ and so with symmetry $H = N_T(T_B)$ and $N_{T_B}(T)$ are hyperplanes of T resp. T_B. Finally for $E \in \Sigma \cap N_{T_B}(T)$ we have $E \in \Psi_C$ and thus by II(2.19) $E = [E, C, B] \leq [H, B]$, which proves (2). (3) is a special case of (2). Indeed by (1) and (2.12) there exists a $B \in \Sigma \cap T^g$ with $B \notin N_\Sigma(T)$. Thus, as $T^g = T_B$, (3) holds. $\qquad\square$

As a corollary to (3.6) and II(2.19) we obtain:

(3.7) Corollary

Suppose Λ is a TI-subset of Σ and let $T_1 = T^g \neq T$. Pick $C \in \Lambda - N_\Sigma(T_1), B \in T_1 \cap \Omega_C$ and set $Y = \langle B, C \rangle, R = \langle T_1, T \rangle$ and $N = N_T(T_1) \times N_{T_1}(T)$. Then the following hold:

1. $R = N \cdot Y, N \lhd R$ and $N' = 1$.
2. Y is a perfect central extension of $SL_2(K), K$ a division ring or a Cayley division algebra.
3. $T = CN_T(T_1), T_1 = BN_{T_1}(T)$.
4. $T^G \cap R = T^R$.
5. If $1 \neq u \in T \cap N$, then $\langle u^Y \rangle$ is a natural module for Y.

Proof. (1)–(3) are direct consequences of (3.7) and II(2.19). To prove (5) let $u \in U \in \Sigma$. Then we have

$$\begin{array}{ccc} C & B & U \\ \circ\!\!-\!\!\!-\!\!\!-\!\!\circ\!\!=\!\!\!=\!\!\circ & & \end{array}$$

and thus (5) follows from II(2.19).

So (4) remains to be proved. For this let $T^g \leq R$. Then by (2.12) and (3.6) $T^g \not\leq N$. Hence by (1) either $T^g \leq N(NC)$ or there exists an $E \in \Sigma \cap T^g$ such that $\langle E, C \rangle$ is not nilpotent. The first possibility contradicts (3.6). In the second case $\langle E, C \rangle$ is a rank one group, whence $E = C^h$ for some $h \in \langle E, C \rangle \leq R$. But then $T^g = T^h \in T^R$, since T is a TI-subgroup of G. $\qquad\square$

(3.8) Lemma

Suppose Λ is a TI-subset of Σ and let $T_1 = T^g \neq T$ and $T_2 = T^h \neq T$. Then the following hold:

1. Either $N_T(T_1) = N_T(T_2)$ or $T = N_T(T_1)N_T(T_2)$ and $N_T(T_1) \cap N_T(T_2)$ is a subspace of codimension 2.
2. Suppose $N_T(T_1) = N_T(T_2)$. Then also
$$N_{T_i}(T) = N_{T_i}(T_j) \text{ for } \{1, 2\} = \{i, j\}.$$
3. There exists a pair T_1, T_2 with $N_T(T_1) \neq N_T(T_2)$.

Proof. (1) follows from the fact that $N_T(T_i), i = 1, 2$ are hyperplanes of T. (3) is a consequence of (2.9), since otherwise $N_T(T_1)$ would be $\langle N_\Sigma(T) \rangle$ invariant. To prove (2) assume $N_T(T_1) = N_T(T_2)$ but $N_{T_1}(T) \neq N_{T_1}(T_2)$. Then $T_1 = N_{T_1}(T)N_{T_1}(T_2)$ and we may pick $B \in (\Sigma \cap N_{T_1}(T_2)) - N_{T_1}(T)$. Hence by (3.6)(2)

$$N_{T_1}(T) = [N_T(T_1), B] = [N_T(T_2), B] \leq N(T_2),$$

a contradiction. $\qquad\qquad\qquad\qquad\qquad\qquad\qquad\qquad\qquad\qquad\qquad\qquad\square$

(3.9) Lemma

Suppose Λ is a TI-subset of Σ and T_1, T_2, T_3 are pairwise different conjugates of T with $N_{T_1}(T_2) = N_{T_1}(T_3)$. Then $T_3 \in T_1^{T_2}$.

Proof. By (3.8)(2) we have $N_{T_i}(T_j) = N_{T_i}(T_k)$ for $\{i, j, k\} = \{1, 2, 3\}$. Let $R = \langle T_1, T_2, T_3 \rangle$ and $N = \Pi_i N_{T_i}(T_j)$. Because of $N_{T_i}(T_j) = N_{T_i}(T_k)$ we have $[N_{T_i}(T_j), T_k] = N_{T_k}(T_i)$ and thus $N \vartriangleleft R$. Let $\overline{R} = R/N$ and pick $A_i \in (\Sigma \cap T_i)$, $A_i \not\leq N$ for $i = 1, 2, 3$. Then by (3.7)(2) $\langle A_i, A_j \rangle$ is a perfect central extension of $SL_2(K)$.

Now suppose that (3.9) is false. If $\overline{T}_3 \leq \langle \overline{T}_1, \overline{T}_2 \rangle$ then $\overline{T}_3 = \overline{T}_1^{\overline{x}}$ for some $\overline{x} \in \langle \overline{T}_1, \overline{T}_2 \rangle$, since $\langle \overline{T}_1, \overline{T}_2 \rangle$ and $\langle \overline{T}_1, \overline{T}_3 \rangle$ are rank one groups. Hence $T_3 \leq NT_1^x$ and thus $T_3 = T_1^x$ by the same argument as in the proof of (3.7). But then, as $T_3 \neq T_2$, $T_3 \in T_1^{T_2}$ by (3.7)(4), a contradiction. This show $\overline{A}_k \not\leq \langle \overline{A}_i, \overline{A}_j \rangle$ for $\{i, j, k\} = \{1, 2, 3\}$.

Pick now by (3.8)(3) a T^g such that $N_{T_1}(T_2) \neq N_{T_1}(T^g)$. Then by (3.8)(2) $N_{T_i}(T_j) \neq N_{T_i}(T^g)$ for $i, j \leq 3$. Let $L_i = N_{T_i}(T_j) \cap N_{T_i}(T^g)$ for $i = 1, 2, 3$ and $j \neq i$. Then L_i is by (3.8)(1) of codimension 2 in T_i. Hence

$$N_0 = \Pi_{i=1}^3 L_i \neq N$$

and for each $i \leq 3$ there exists a $C_i \in T_i \cap N \cap \Sigma$ with $C_i \not\leq L_i$. Then $T_i \cap N = L_i C_i$ and thus $N = N_0 C_1 C_2 C_3$. Moreover, since $N_{T_i}(T^g)$ is a hyperplane of T_i different from $N_{T_i}(T_j)$, we may choose $A_i \leq N_{T_i}(T^g)$. Hence $[L_j, A_i] \leq T_i \cap N(T^g) = L_i$ for $i \neq j \leq 3$ and thus $N_0 \vartriangleleft R$, since N is by (3.7)(1) abelian.

Let $\widetilde{R} = R/N_0$. Then $\widetilde{C}_i \widetilde{C}_j$ is a natural module for $\langle A_i, A_j \rangle$ by II(2.19). Hence if $a \in A_3^\#$ we have

$$[\widetilde{C}_1, a] = \widetilde{C}_3 = [\widetilde{C}_2, a].$$

Thus $[\widetilde{c}_1 \widetilde{c}_2, a] = 1$ for $\widetilde{c}_1 \in \widetilde{C}_1^\#, \widetilde{c}_2 \in \widetilde{C}_2^\#$. Now $\widetilde{c}_1 \widetilde{c}_2 \in \widetilde{C}_1^y$ for some $y \in \langle A_1, A_2 \rangle$, since $\widetilde{C}_1 \widetilde{C}_2$ is a natural module. Hence there exists a coimage $c_1 c_2$ of $\widetilde{c}_1 \widetilde{c}_2$ with $c_1 c_2 \in T_1^y \cap N$. Moreover $[c_1 c_2, a] \in T_3 \cap N_0 = L_3$. Now by II(2.19) we have for $F \in \Sigma \cap L_3$:

$$[F, A_1, a] = F.$$

Hence $L_3 = [L_1, a]$ and thus also $L_3 = [L_1^y, a]$. This shows that there exists a $1 \neq \ell \in L_1^y$ with $[c_1 c_2, a] = [\ell, a]$ and thus $1 \neq h = \ell^{-1} c_1 c_2 \in C(a)$. But then $a \in N_{T_3}(T_1^y) = T_3 \cap N$, a contradiction to $a \in A_3^{\#}$. This proves (3.9). $\qquad\square$

(3.10) Corollary

Suppose Λ is a TI-subset of Σ and let T^g, T^h be conjugates of T with $T^h \not\leq \langle T, T^g \rangle$. Then $T = N_T(T^g) N_T(T^h)$.

This is a consequence of (3.8), (3.9).

(3.11) Notation

Suppose that Λ is a TI-subset of Σ. Then we set $\mathcal{P} = \{T^g \mid g \in G\} = \{T_D \mid D \in \Sigma\}$ and $\mathcal{L} = \{\langle T^g, T^h \rangle \cap \mathcal{P} \mid g, h \in G \text{ with } T^g \neq T^h\}$. Call the elements of $\mathcal{P}$ *points* and the elements of $\mathcal{L}$ *lines* and let $\mathcal{G} = (\mathcal{P}, \mathcal{L}, \in)$ be the corresponding point-line geometry.

We assume for the rest of this section that Λ is a TI-subset of Σ, which implies by (2.11) that T is a TI-subgroup of G. A *triangle* is a triple T_B, T_C, T_D of points of $\mathcal{P}$, which do not lie on a common line. We have:

(3.12) Corollary

The following hold:

1. A line is spanned by each pair of points on it.
2. Two different lines intersect in at most one point.

(1) is a consequence of (3.7)(4) and II(2.1). (2) follows from (1).

(3.13) Lemma

Let $B, C, D \in \Sigma$ such that $\{T_B, T_C, T_D\}$ is a triangle. Then

$$(*) \qquad T_D = \bigcup_{g \in N_{T_C}(T_D)} N_{T_D}(T_B^g) \cup N_{T_D}(T_C).$$

Proof. By (3.10) $T_D = N_{T_D}(T_B) N_{T_D}(T_C)$. Pick $D_0 \in (\Sigma \cap N_{T_D}(T_B)) - N_{T_D}(T_C)$. Then by (3.6) $T_D = N_{T_D}(T_C) \times D_0$. Hence by (3.7)(5) and the description of the natural $SL_2(K)$-module in I(1.8) we have for each $u \in N_{T_D}(T_c)$:

$$D_0 u \subseteq D(D_0^{N_{T_C}(T_D)}).$$

Hence

$$T_D - N_{T_D}(T_C) = \bigcup_{g \in N_{T_C}(T_D)} D_0^g$$

which proves (3.13).   □

(3.14) Proposition

$\mathcal{G}$ is a projective space.

Proof. By (3.12) it suffices, to prove (3.14), to verify the so-called axiom of Pasch.

For this let $\{T_D, T_B, T_C\}$ be a triangle, $\ell_1 = \langle T_D, T_B \rangle \cap \mathcal{P}, \ell_2 = \langle T_B, T_C \rangle \cap \mathcal{P}$ and $\ell_3 = \langle T_D, T_C \rangle \cap \mathcal{P}$. Then we have to show: If T_1, T_2 are different points on ℓ_1 and T_3, T_4 different points on ℓ_3, then the lines $s = \langle T_1, T_3 \rangle \cap \mathcal{P}$ and $r = \langle T_2, T_4 \rangle \cap \mathcal{P}$ intersect in a point.

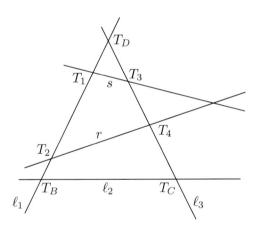

Now by (3.10) $T_D = N_{T_D}(T_4) N_{T_D}(T_B)$. Hence by (3.7)(4) there exist $d_1 \in N_{T_D}(T_4)$ and $d_2 \in N_{T_D}(T_B)$ with $T_2^{d_1} = T_B$ and $T_4^{d_2} = T_C$. Thus, if $d = d_1 d_2$, then $r^d = \ell_2$. Since s and r intersect in a point if and only if s^d and r^d intersect, this shows that we may assume $T_2 = T_B, T_4 = T_C$ and $r = \ell_2$.

By the same argument $s = \ell_2^{\bar{d}}$ for some $\bar{d} \in T_D$. But by (3.13)(∗) $\bar{d} \in N_{T_D}(T')$ for some point $T' \in \ell_2$. Hence

$$T' \in \ell_2 \cap \ell_2^{\bar{d}} = \ell_2 \cap s,$$

which proves (3.14).   □

(3.15)   Lemma

The following hold:

(1) $G \leq \mathrm{Aut}(\mathcal{G})$.

(2) For $D \in \Sigma$ let $\mathcal{H}_D = \{T^g \mid D \leq N(T^g)\}$. Then $\mathcal{H}_D$ is, with the lines contained in it, a hyperplane of $\mathcal{G}$.

(3) The elements of Σ act as elation subgroups corresponding to point hyperplane pairs on $\mathcal{G}$.

Proof. Since $Z(G) \leq R(G) = 1$, G is by II(4.13) simple, whence (1) holds. To prove (2) notice that if $D \in \Sigma$ normalizes T^g and T^h, then it normalizes by (3.6)(1) all conjugates of T contained in $\langle T^g, T^h \rangle$. This shows that, if two points of $\mathcal{P}$ are contained in $\mathcal{H}_D$, the line through these points is also contained in $\mathcal{H}_D$. Hence $\mathcal{H}_D$ is a subspace of $\mathcal{G}$. Thus, to prove (2), it remains to show that each line in $\mathcal{L}$ intersects $\mathcal{H}_D$ non-trivially. For this let $\{T_D, T_B, T_C\}$ be a triangle with $\ell = \langle T_B, T_C \rangle \cap \mathcal{P}$ not contained in $\mathcal{H}_D$. Then (3.13) shows that D normalizes some point on ℓ, whence (2) holds.

Finally, to prove (3), D fixes all points on $\mathcal{H}_D$ and all lines through T_D, whence D is contained in the elation subgroup corresponding to $(T_D, \mathcal{H}_D)$. Let now s be a line through T_D not in $\mathcal{H}_D$. Then for $T_D \neq T^g \in s$, there exists by (3.6) a $B \in T^g \cap \Omega_D$. Hence (3.7) implies that D is transitive on the points of s different from T_D. This shows that D is a full elation subgroup.   □

(3.16)   Corollary

Suppose $\dim \mathcal{G} = 2$. Then the following hold:

(1) $\mathcal{G}$ is a Moufang plane.

(2) Σ is the class of elation subgroups, corresponding to incident point line pairs.

(3) Either $G \simeq PSL_3(K)$, K a division ring, or to E_6^K, K a Cayley division algebra.

Proof. By (3.6)(3) and (3.15)(3) Σ contains an elation subgroup corresponding to each incident point line pair. By definition of Moufang planes, this shows (2) and (1).

Now by exercise (2.31)(4) the root subgroups corresponding to the roots of an apartment of $\mathcal{G}$ are such elation subgroups. Hence G contains and is generated by all root subgroups. Now (3) is just a matter of terminology, since in case $\mathcal{G}$ a Moufang plane coordinized by a Cayley division algebra K, we called the subgroup of $\mathrm{Aut}(\mathcal{G})$ generated by all root subgroups E_6^K in exercise II(5.24)(5).   □

Remark. Notice that we know by exercise (2.31)(4) and II(5.24)(5) that $\langle A, B \rangle \simeq SL_2(K)$, K a division ring or a Cayley division algebra, for $A \in \Sigma$ and $B \in \Omega_A$, without using the coordinatization of Moufang planes. But we do not know yet, that $\mathcal{G}$ is uniquely determined by K. (I.e. there could exist non-isomorphic Moufang planes belonging to the same K.)

That $\mathcal{G}$ is uniquely determined by K is part of theorem 2 of [Tim94a], which will be treated in the exercises.

(3.17) Notation

For the rest of this section we assume that $\dim \mathcal{G} > 2$. Then $\mathcal{G} \simeq \mathcal{P}(V)$, V a vector space over some division ring K by the main theorem of projective geometry. Since G is by II(4.13) simple we have $G \leq PSL(V)$, where $PSL(V)$ is the normal subgroup of $\mathrm{Aut}(\mathcal{P}(V))$ generated by all elations. Thus $\widehat{G} \leq SL(V)$, where $\widehat{G}$ is a covering group of G generated by $\widehat{\Sigma}$ and $\widehat{\Sigma}$ is a set of transvection subgroups of $SL(V)$ corresponding to point hyperplane pairs. Moreover $\widehat{\Sigma} \to \Sigma$ by the natural homomorphism $SL(V) \to PSL(V)$. To simplify notation we identify, being slightly incorrect, G with $\widehat{G}$ and Σ with $\widehat{\Sigma}$.

Now by II(1.3) an element of Σ is of the form $T_{P,\Phi}$, P a point of V and Φ a point of V^*, V^* the dual space of V. Let

$$W^* = \langle \Phi \mid T_{P,\Phi} \in \Sigma \text{ for some point } P \text{ of } V \rangle.$$

Then W^* is a subspace of V^* with $\mathrm{Ann}_V(W^*) = 0$, since a point in $\mathrm{Ann}_V(W^*)$ would be fixed by all elements of Σ, a contradiction since Σ is a conjugacy class and $\mathcal{P} = T^G$ contains all points of $\mathcal{P}(V)$. So, to identify G, it remains to show that $\Sigma = \Sigma(W^*)$ in the notation of II(1.3). Indeed, if this is the case, then $G = T(V, W^*)$ as defined in II(1.3). (In fact $G = PT(V, W^*)$, where $PT(V, W^*)$ is the image of $T(V, W^*)$ under $SL(V) \to PSL(V)$.)

Identifying T with a point of $\mathcal{P}(V)$ let

$$T^* := \langle \Phi \leq W^* \mid T_{T,\Phi} \in \Lambda \rangle.$$

Then $T^* \subseteq T^0 \cap W^*$, $T^0 = \{\sigma \in V^* \mid \sigma(T) = 0\}$, and $\mathrm{Ann}_V(T^*) = T$, since T is the only element of $\mathcal{P}$ normalized by T.

We show $\Sigma = \Sigma(W^*)$ with several lemmata.

(3.18) Lemma

Let $D = T_{Q,\Psi} \in N_\Sigma(T)$. Then $\Psi \subseteq T^*$ and $T_{T,\Psi} \in \Lambda$.

Proof. If $D \in \Lambda$ then $T = Q$ and (3.18) holds. So assume $D \notin \Lambda$. Then, as $C_\Sigma(\Lambda) = \Lambda$, there exists a $B = T_{T,\Phi} \in \Lambda \cap \Psi_D$. Since $[D, B] \in \Lambda$, II(1.3)(2)(b) implies $T \subseteq \text{Kern } \Psi$ and

$$T_{T,\Psi} = C = [D, B] \in \Lambda,$$

whence $\Psi \subseteq T^*$.   □

(3.19) Lemma

Let $T_1 = T^g \neq T$ and T_1^* as defined in (3.17). Then the following hold:

1. $T^* \cap T_1^*$ is a hyperplane of T^* and T_1^*.
2. $T^* + T_1^* = \langle (T^h)^* \mid T^h \leq \langle T, T_1 \rangle \rangle$.
3. $W^* = T^* + T_1^*$.

Proof. By (3.6) $T_1 = N_{T_1}(T) \times F, F \in \Sigma$. Thus, if $F = T_{Q,\Psi}$, then (3.18) implies $T^* + T_1^* = T^* + \Psi$. Hence T^* is a hyperplane of $T^* + T_1^*$ and symmetry implies (1). (2) is a consequence of (1) and (3.7)(4).

Finally, to prove (3), let $D = T_{R,\Theta} \in \Sigma$ be arbitrary. Then D normalizes by (3.13) some $T^h \in \mathcal{P} \cap \langle T, T_1 \rangle$. Hence $\Theta \subseteq T^* + T_1^*$ by (3.18), which implies (3).
                                                                                                                □

(3.20) Lemma

The following hold:

(1) $T^* = T^0 \cap W^*$.
(2) $\Lambda = \{ T_{T,\Phi} \mid \Phi \text{ a point of } T^* \}$.

Proof. (3.19) implies that T^* is a hyperplane of W^*. Since $T^* \subseteq T^0 \cap W^*$ this shows (1). Let now $T_{T,\Psi}$ and $T_{T,\Phi} \in \Lambda$ and $N = T_{T,\Psi} T_{T,\Phi}$. Then by (2.11) $\Sigma \cap N = \Lambda \cap N$ is a partition of N and by II(2.23) $\langle N_\Sigma(N) \rangle$ induces the $SL_2(K)$ on N. Hence $\langle N_\Sigma(N) \rangle$ induces also the $SL_2(K)$ on $\Phi + \Psi$ and thus

$$\Lambda \cap N = \{ T_{T,\Theta} \mid \Theta \text{ a point of } \Psi + \Phi \}.$$

Since $T^* = \langle \Phi \mid T_{T,\Phi} \in \Lambda \rangle$ this implies that

$$\{ \Phi \mid \Phi \text{ a point of } T^* \} = \{ \Phi \mid T_{T,\Phi} \in \Lambda \}$$

which proves (2).   □

(3.21) Theorem

We have

$$\Sigma = \Sigma(W^*) = \{T_{P,\Phi} \mid P \text{ a point of } V \text{ and } \Phi \text{ a point of } W^*\}.$$

Further $G = T(V, W^*)$.

Proof. Let $T_{P,\Phi}$ be an arbitrary element of $\Sigma(W^*)$. Then P is a point of V and Φ a point of W^* with $P \subseteq \text{Kern } \Phi$. Hence by definition Φ is a point of $P^0 \cap W^* = P^*$ by (3.20)(1). If now $P \longleftrightarrow T^g$, then (3.20)(2) implies $T_{P,\Phi} \in \Sigma \cap T^g \subseteq \Sigma$. This proves $\Sigma = \Sigma(W^*)$ and thus (3.21). □

Putting (3.16) and (3.21) together we have

(3.22) Corollary

Suppose $R(G) = 1$ and $N_G(A)$ is not maximal in G for $A \in \Sigma$. Then one of the following holds:

1. There exists a Cayley division algebra K such that $G \simeq E_6^K$, $\mathcal{G}$ is the Cayley plane over K and Σ is the class of elation subgroups on $\mathcal{G}$.

2. There exists a division ring K, a K-vector space V of dimension ≥ 3, a subspace W^* of the dual space V^* of V with $\text{Ann}_V(W^*) = 0$ such that $G \simeq PT(V, W^*)$, $\mathcal{G} \simeq \mathcal{P}(V)$ and Σ is the class of elation subgroups on $\mathcal{G}$, which are images of the transvection subgroups in $\Sigma(W^*)$.

(3.23) Exercises

Exercise (1)–(3) are extracted from [Tim94a].

(1) Let K be a division ring or a Cayley division algebra, $\mathcal{S}$ in the first case the set of 1-dimensional subspaces of K^2 and in the second case the partition of K^2 of I(1.8)(3). Let $\mathcal{A}(K^2)$ be the point line geometry with:

point set	–	elements of K^2
line set	–	cosets of elements of $\mathcal{S}$,

a point lying on a line iff the element is contained in the corresponding coset. Show that $\mathcal{A}(K^2)$ is an affine plane.

(2) Let $\mathcal{P}$ be a (projective) Moufang plane, α and $-\alpha$ opposite roots in some apartments $\mathcal{A}$ of $\mathcal{P}$ such that

$$X_\alpha = \langle A_\alpha, A_{-\alpha} \rangle \simeq SL_2(K)$$

for some division ring or Cayley division algebra K. (Such a K exists
by exercise II(5.24)(5)(c)!) For a line ℓ of $\mathcal{P}$ let $\mathcal{P}_\ell$ be the affine plane
obtained by removing ℓ and all points on it.
Show that $\mathcal{P}_\ell \simeq \mathcal{A}(K^2)$ for each line ℓ of $\mathcal{P}$.

Hint. Let T be the set of all elations corresponding to ℓ (see exercise
(2.31)(4).) Then we may assume that $T = A_\beta A_{\alpha+\beta}$; where $\{\alpha, \beta, \alpha+\beta\}$
are the roots of $\mathcal{A}$ containing a fixed chamber (i.e. positive roots!), since
$\text{Aut}(\mathcal{P})$ is transitive on the lines of $\mathcal{P}$. Hence by Theorem I(3.2) T is
isomorphic to the natural $\mathbb{Z}X_\alpha$-module. In particular T may be identified
with K^2 and $A_\beta^{X_\alpha}$ with the partition $\mathcal{S}$ of exercise (1).

(3) Let $\mathcal{P}$ and $\overline{\mathcal{P}}$ be Moufang planes over the same division ring or Cayley
division algebra K. (I.e. with $X_\alpha \simeq SL_2(K) \simeq X_{\overline{\alpha}}$ for roots α of $\mathcal{P}$ resp.
$\overline{\alpha}$ of $\overline{\mathcal{P}}$!) Then $\mathcal{P} \simeq \overline{\mathcal{P}}$.
Hint. We have $\mathcal{P}_\ell \simeq \mathcal{A}(K^2) \simeq \overline{\mathcal{P}}_{\overline{\ell}}$ for lines $\ell, \overline{\ell}$ of $\mathcal{P}$ resp. $\overline{\mathcal{P}}$.

The following exercises are extracted from [Tim75b].

(4) Let T be an elementary abelian 2-subgroup of the group G which is a
TI-subgroup of G. (I.e. $T \cap T^g = 1$ or T for each $g \in G$.) Let $T_1 = T^g \neq T$
with $N_T(T_1) \neq 1$. Set $N = N_T(T_1) \times N_{T_1}(T)$ and $Y = \langle T, T_1 \rangle$. Show

   (a) $[T, N_{T_1}(T)] \leq N_T(T_1)$ and $[T_1, N_T(T_1)] \leq N_{T_1}(T)$.

   (b) $N \trianglelefteq Y$.

   (c) Let $t \in T - N_T(T_1)$. Then $C_N(t) = T \cap N = N_T(T_1)$ and $[N, t] \leq N_T(T_1)$.

(5) Let T, T_1 be as in (4) and $t \in T - N_T(T_1), \tau \in T_1 - N_{T_1}(T)$.

   (a) Show that $\langle t, \tau \rangle \simeq D_{2n}, 1 \neq n-$ odd.
       Hint. Suppose $\langle t, \tau \rangle \simeq D_{2n}$ with $2|n$ and let z be the central involu-
       tion in $\langle t, \tau \rangle$. Then $zt \sim \tau \in \langle t, \tau \rangle$. Let T_2 be the conjugate of T con-
       taining zt. Then $T_2 \sim T_1$ in Y. But t centralizes $T_2 \cap N \leq N_{T_2}(T)$,
       a contradiction to $C_N(t) = T \cap N$.

   (b) Conclude from (a) that $T \sim T_1$ in Y.

   (c) Show $[N, T] = T \cap N$ for each $t \in T - (T \cap N)$.

A set D of involutions of the group R is called a set of *root involutions*
of R, if the following holds:

   (I) $R = \langle D \rangle$ and $D^g \subseteq D$ for each $y \in R$.

   (II) For each pair $d, e \in D$ one of the following holds:
        (a) $[d, e] = 1$.

  (b) $o(de) = 4$ and $[d, e] \in D$.

  (c) $\langle d, e \rangle \simeq D_{2n}, n$ odd.

The connection between root involutions and abstract root subgroups will be explored in chapter IV.

  (6) Let T and G be as in exercise (4) and let N be an abelian subgroup of G with $T \cap N \neq 1$. Let $R = \langle T^g \mid g \in G$ and $T^g \cap N \neq 1 \rangle$ and $D = \{d \mid d \in T^g$ where $T^g \cap N \neq 1\}$. Show that D is a set of root involutions of R.

  Hint. Apply (5) to the pairs T^g, T^h with $T^g \cap N \neq 1 \neq T^h \cap N$.

§ 4 Moufang hexagons

We assume in this section that Σ is a non-degenerate class of abstract root subgroups of the group G, $R(G) = 1$ and $\mathcal{F}(\Sigma)$ is connected. We fix again $A \in \Sigma$, $A_0 \in \Omega_A$ and $X = \langle A, A_0 \rangle$. Suppose further that $\Sigma_A \subseteq M_A$, but $N_G(A)$ is maximal in G. (The groups with $N_G(A)$ not maximal have been classified in §3.) Then we have:

(4.1) Lemma

Let $D \in \Sigma_A$. Then $AD \cap \Sigma$ is a weak TI-subset of Σ, but not a TI-subset.

Proof. As $N_G(A)$ is maximal, G is primitive on Σ, whence no TI-subsets exist. On the other hand, by hypothesis and (2.6) $D \in \Lambda_A$, so that by II(2.23) $R = \langle N_\Sigma(AD) \rangle > \langle C_\Sigma(A) \rangle$. Hence by (2.8) $AD \cap \Sigma = A^R$ is a weak TI-subset. $\qquad \square$

(4.2) Notation

Let $\mathcal{G}$ be the point line geometry with Σ as point set and

$$\mathcal{L} = \{BD \cap \Sigma \mid B \in \Sigma, D \in \Lambda_B\}$$

as line set, a point $E \in \Sigma$ lying on the line $\ell = BD \cap \Sigma$ if and only if $E \in BD \cap \Sigma$.

Then we have:

(4.3) Lemma

The following hold for $\mathcal{G}$:

  1. There exists at most one line through two different points.

  2. The lines are thick. (I.e. there exist at least three points on each line!)

3. Two different lines intersect in at most one point.

4. If $B \in \Sigma$ and $C \in \Psi_B$, then there exists exactly one point adjacent to B and C, namely $[B, C]$.

5. Let $\ell = AB \cap \Sigma$ and $g = CD \cap \Sigma$ be different lines with $\ell \cap g = \emptyset$ and $N_{AB}(CD) \neq 1$. Then there exists exactly one line s with $\ell \cap s \neq \emptyset \neq s \cap g$.

Proof. (1)–(3) are obvious. In (4) let D be a point adjacent to B and C. Then $\{B, C\} \subseteq \Lambda_D$, whence $[B, C] \leq D$ by (2.6). This proves (4).

Finally, to prove (5), let $1 \neq e \in N_{AB}(CD)$ and $e \in E \in \Sigma$. ($\Sigma \cap AB$ is a partition of AB!) Then $E \in \Psi_C$ or $E \in \Psi_D$ by (2.12)(2). Hence there exists an $F \in CD \cap \Lambda_E$. Thus $s = FE \cap \Sigma$ is a line which satisfies the requirements of (5).

Suppose $s' = F'E' \cap \Sigma$ is another such line with $E' \in \ell \cap s'$, $F' = g \cap s'$. Then, if $s \neq s'$, either $E \neq E'$ or $F \neq F'$ and we may without loss assume that the first holds. But then $AB = EE' \leq N(CD)$. Since by (2.12)(2) $[E, CD] = F$ and $[E', CD] = F'$ II(2.23) implies that $F = F'$ and so $AB \cap \Sigma \subseteq \Lambda_F$ by (2.6). But then, applying (2.1) to $N = ABF$, we obtain a contradiction to $\Sigma_A \subseteq M_A$. Hence $s = s'$, which proves (5). □

(4.4) Proposition

$\mathcal{G}$ is a thick generalized hexagon.

Proof. By (4.3) the lines of $\mathcal{G}$ are thick. Since by (4.1) the lines are no TI-subsets, there exist by (3.3) at least 3 lines through each point. Hence $\mathcal{G}$ is thick.

If now B, C are points with $C \in \Sigma_B \cup \Psi_B$, (resp. ℓ and g) are lines with $N_\ell(g) \neq \emptyset$, then it follows from (4.3)(4) and (5) that B and C resp. ℓ and g are of distance at most 4 in the incidence graph of $\mathcal{G}$. Moreover the chain from B to C resp. (ℓ to g) is uniquely determined.

If now $\ell = BC \cap \Sigma$ is a line and D is a point with $D \not\leq N(BC)$, then $BC \cap \Omega_D \neq \emptyset$ and thus we may by II(2.11) assume $B \in \Omega_D$ and $C \in \Psi_D$. But then

$$\ell, C, C[C, D] \cap \Sigma, [C, D], [C, D]D \cap \Sigma, D$$

is by II(2.11) the unique chain of length 5 from ℓ to D in the incidence graph of $\mathcal{G}$.

The same argument also shows that each pair of points is connected by a chain of length at most 6. Thus let finally $\ell = AB \cap \Sigma$ and $g = CD \cap \Sigma$ be two lines with $N_{AB}(CD) = 1$. Then, as above we may assume, $D \in \Omega_A \cap \Psi_B$. Let $E = [B, D]$. Then II(2.19) shows that

$$\ell, B, BE \cap \Sigma, E, ED \cap \Sigma, D, g$$

is a chain of length 6 from ℓ to g. □

(4.5) Lemma

Let $B \in \Lambda_A$ and $\ell = AB \cap \Sigma$. Then $M_B \cap C(A)$ is transitive on the lines through A different from ℓ. In particular $C(A)$ is doubly transitive on the lines through A.

Proof. Let $R = \langle N_\Sigma(AB) \rangle$. Then $R = \langle M_A, M_B \rangle$ by II(2.23). By (2.6) there exist $E \in \Lambda_A$ and $F \in \Lambda_B$ such that $M_A = (M_A \cap C(B))E$ and $M_B = (M_B \cap C(A))F$. Hence if $N = (M_A \cap C(B))(M_B \cap C(A))$ and $Y = \langle E, F \rangle$, then $R = N \cdot Y$ with N a nilpotent normal subgroup and Y acting naturally on AB.

Let now $s = AE \cap \Sigma$ and $r = AL \cap \Sigma$ with $L \in \Lambda_A$ and $s \neq r \neq \ell$. Then there exists by (2.12)(2) a $g \in R$ with $E^g = L$. Since $NE = NM_A = NL$ we have $g \in N_R(NE) = NN_Y(E)$, since Y is a rank one group and thus $N_Y(E)$ is maximal in Y. This shows that we may assume $g \in N$ and thus $g \in M_B \cap C(A)$, since M_A stabilizes all lines through A.

We have shown that there exist a $g \in M_B \cap C(A)$ with $s^g = r$ for each pair r, s of lines through A different from ℓ. This implies the first part of (4.5). The second is a consequence of the first, since by (3.3) there exist at least tree lines through A. □

(4.6) Lemma

The following hold:

(1) Let $B \in \Lambda_A$. Then $N = (M_A \cap C(B))(M_B \cap C(A))$ is transitive on $\Omega_A \cap \Psi_B$.

(2) Σ is a class of k-root subgroups, k a fixed commutative field.

Proof. Let $E, F \in \Omega_A \cap \Psi_B$. Then $B[B, E] \cap \Sigma$ and $[B, F]B \cap \Sigma$ are lines through B different from $AB \cap C$. Hence by (4.5) there exists a $g \in N$ with $[B, E]^g = [B, F]$. Thus, to prove (1), we may assume $[B, E] = [B, F]$. But then $[B, E]E \cap \Sigma$ and $[B, F]F \cap \Sigma$ are lines through $[B, E]$ different from $B[B, E] \cap \Sigma$. Hence again by (4.5) there exists a $g \in M_B \cap C([B, E])$ with $E^g = F$, since B is already transitive on the points of $[B, F]F \cap \Sigma$ different from $[B, F]$. By the same reason there exists an $h \in [B, E]$ with $A^{gh} = A$, whence $gh \in C(A) \cap M_B$ and $E^{gh} = F$. This proves (1).

To prove (2) consider $R = \langle A, E, B \rangle$. By II(2.19) $R = L \cdot Y$ with $Y = \langle A, E \rangle$ and $L = B[B, E]$ a natural module for $Y/C_Y(L) \simeq SL_2(K)$, K a division ring or a Cayley division algebra. In particular if K is not commutative, there exists an $h \in N_Y(A) \cap N_Y(E)$ with $[B, h] = 1$ but $[A, h] \neq 1$. (If K is a division ring, h corresponds to a matrix of the form $\begin{pmatrix} 1 & \\ & c \end{pmatrix}$ with $1 \neq c \in (K^*)'$. If K is a Cayley division algebra one may construct h using the action of the elements $h(t), t \in K^*$ on K^2 described in I(1.8)(11). Indeed for example $h = h(t)h(\tau)h((t\tau)^{-1})$ with t, τ non-commuting elements of K^* satisfies the requirements.)

Let by (4.5) $g \in C(A)$ with $[B, B^g] = A$. Then, conjugating with some element of M_A, we may assume $E^g \in \Psi_B$, whence by (1) $E^g = E$. Thus g centralizes Y, since it centralizes A and normalizes E. This implies

$$[B, h] = 1 = [B^g, h],$$

a contradiction to $A = [B, B^g]$ and $[A, h] \neq 1$.

This shows that K is commutative and thus that K is a field. Since by II(2.19) $C_Y(L) \leq Z(Y)$, II(4.8) and $M_A = \langle \Lambda_A^* \rangle$ by (2.13) imply $C_Y(L) = 1$. Thus $Y \simeq SL_2(k)$ for $k = K$ a field. That we obtain always the same field follows from (4.5), (1) and I(5.7).   □

(4.7)   Notation

A *central elation* (corresponding to the point A) of the generalized hexagon $\mathcal{G}$ is an automorphism which fixes all lines through A pointwise. Such a central elation is called *strong*, if it fixes all lines through points D, which are adjacent to A (i.e. lie on some line through A).

(4.8)   Lemma

Let $B \in \Lambda_A$, $C \in \Lambda_A \cap \Psi_B$ and $\ell = AB \cap \Sigma$, $s = AC \cap \Sigma$ and g a line through B different from ℓ and h a line through C different from g. Then the following holds:

(1) $\mathcal{A}$ is the group of all possible strong central elations corresponding to the point A in $\mathrm{Aut}(\mathcal{G})$.

(2) $\mathcal{A}$ is a root subgroup of $\mathrm{Aut}(\mathcal{G})$ corresponding to the half apartment

$$(g, B, \ell, A, s, C, h) = r.$$

Proof. Clearly A fixes each line through A pointwise. Since $A \leq M_B$ for $B \in$ Λ_A, A fixes also all lines through points in Λ_A. This shows that A consists of strong central elation. That A is the group of all possible strong central elations follows from I(4.7)(1) and the fact that A is already transitive on $g - B$ and $h - C$. (A proof in terms of the theory of abstract root subgroups may be obtained as follows: Let α be a central elation of $\operatorname{Aut}(\mathcal{G})$ corresponding to A. Then, as $g - B = E^A$ for $E \in g - B$, αa fixes E for some $a \in A$. Hence αa fixes the unique $F \in h \cap \Psi_E$ by II(2.11). Since $\langle E, C \rangle$ (resp. $\langle B, F \rangle$) act as $PSL_2(k)$ on ℓ (resp. s) and αa fixes all points of ℓ and s, it follows that αa centralizes E, C, B and F. The same argument shows that αa centralizes each point in Λ_A and also $D = [E, F] \in \Omega_A$ and thus $G = \langle \Lambda_A, D \rangle$.)

This proves (1). (2) is a direct consequence of (1), since A is transitive on $g - B$ and $h - c$ and thus on the set of apartments of $\mathcal{G}$ containing r. □

(4.9) Remark

It follows from (4.8) and a theorem of Ronan [Ron80], see [VM98, Th. (6.3.9)], applied to the dual hexagon of $\mathcal{G}$, that $\mathcal{G}$ is a Moufang hexagon. Hence by II(5.20) normal subgroups of automorphism groups of Moufang hexagons $\mathcal{G}$ generated by root subgroups and groups generated by a class Σ of abstract root subgroups with $\Sigma_A \subseteq M_A$ and $N(A)$ maximal in G are the same.

The proof of Ronan (in case $|k| > 2$) works by constructing a $G_2(k)$ sub-hexagon, which in our situation means constructing a subgroup $\langle \Delta \rangle \simeq G_2(k)$, $\Delta \subseteq \Sigma$. A proof in terms of the theory of abstract root subgroups could be given as follows:

Let $A, B, C, D, E, F \in \Sigma$ satisfying:

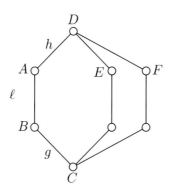

where $\ell = AB \cap \Sigma, g = BC \cap \Sigma, h = AD \cap \Sigma$ and so on are lines of $\mathcal{G}$. Then it suffices to show that there exists a $\sigma \in M_A \cap M_B \cap C(C) \cap C(D) = N$ with

$E^\sigma = F$. Indeed N is contained in the root subgroup of $\mathrm{Aut}(\mathcal{G})$ corresponding to $(D, h, A, \ell, B, g, C) = r$, since it fixes all points on ℓ linewise and it fixes h and g pointwise. $E^\sigma = F$ shows that σ maps the first apartment (of the picture) containing r onto the second. Hence if this holds for all such E, F, then N is transitive on the apartments containing r, which is the Moufang property for r.

Now consider $Y = \langle B, E, F \rangle$. Then $D = [E, F]$ as $E, F \in \Lambda_D$. Further $A = [D, B]$ and by II(2.19) $[A, E] = [A, F] = D$. This implies that $AD \lhd Y$ and Y induces the $SL_2(k)$ on AD. Let $\widetilde{Y} = Y/AD$. Then $\widetilde{B}^{\widetilde{Y}}$ is a class of k-transvections of $\widetilde{Y}$ and $\widetilde{Y}$ is a triangle group in the sense of II §3. Let $\widetilde{M} = R(\widetilde{Y})$. Then it follows from exercise II(3.17)(2) that $\widetilde{M}/C_{\widetilde{M}}(\widetilde{Y})$ is the natural module for $\widetilde{Y}/\widetilde{M} \simeq SL_2(k)$. Let $X = \langle B, E \rangle$, $H = N_X(B) \cap N_X(E)$ and $N_0 = [M \cap M_A \cap C(B), H]$. Then, if $H \neq 1$ (i.e. $|k| > 2!$), the structure of the group Y implies $M = N_0 C_M(E)$. Now, as H centralizes $(M_A \cap C(B)) M_B/M_B$, we have $N_0 \leq M \cap M_A \cap M_B$ and thus $N_0 \leq C(D)$. As shown in (4.5) there exists a $\sigma \in M$ and thence a $\sigma \in N_0$ with $E^\sigma = F$. Now $C = g \cap \Psi_E = g \cap \Psi_F$ and so $C^\sigma = C$, since $\sigma \in M_B$ and thus normalizes g. Since $\langle C, D \rangle$ is a rank one group, this shows $\sigma \in C(D)$ and thus $\sigma \in N$, which was to be shown.

(4.10)   Exercises

Let in the following exercises $\mathcal{G}$ be a Moufang hexagon, G the normal subgroup of $\mathrm{Aut}(\mathcal{G})$ generated by the root subgroups on $\mathcal{G}$ and Σ the class of k-root subgroups of G. (See II(5.20). We assume $G = \langle \Sigma \rangle$, which is by (4.9) the case when $|A| > 2$ for $A \in \Sigma$.) Let

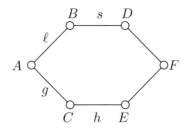

be an apartment of $\mathcal{G}$ (in self-explanatory notation), and $A_\ell = M_A \cap M_B \cap C(C) \cap C(D)$ and similar A_g, A_h, A_s. Then A_ℓ, A_g, A_h, A_s are by (4.9) root subgroups of G. Show:

(1)   (a) A_h acts regularly on the lines through A different from g.

 (b) $R_A = \langle A_h, A_s \rangle$ acts doubly transitively on the lines through A.

 (c) R_A is a rank one group with abelian unipotent subgroups A_h, A_s.

(2) Show that $ABA_\ell A_g C$ acts transitively on Ω_A.

Hint. Pick $F, L \in \Omega_A$. Conjugating with C one may assume $F, L \in \Psi_B$. Hence conjugating with AA_g one may by (1) assume that $F, L \in \Lambda_D$. But then F and L are conjugate by an element of $A_\ell B$.

(3) Show:

    (a) $M_A = ABA_\ell A_g C$.

    (b) Let $\widetilde{M}_A = M_A/A$. Then $\widetilde{M}_A = \tilde{B} \oplus \tilde{A}_\ell \oplus \tilde{A}_g \oplus \tilde{C}$.

    Hint. Show as in (2.16) that M_A acts regularly on Ω_A.

(4)* Show that, if Char $k \neq 3$, then:

    (a) $[\widetilde{M}_A, A_s] = \tilde{B} \oplus \tilde{A}_\ell \oplus \tilde{A}_g$.

    (b) $[\widetilde{M}_A, A_s, A_s] = \tilde{B} \oplus \tilde{A}_\ell$.

    (c) $[\widetilde{M}_A, A_s, A_s, A_s] = \tilde{B}$.

    Hint. If (4) does not hold, then $[A_\ell, A_s] = 1$. Thus if $U = A_s B A_\ell A A_g$, then $A_\ell \leq Z(U)$ and, if $\tilde{U} = U/A_\ell$, then $\tilde{U}$ involves a non-split extension of 2 natural $\mathbb{Z}SL_2(k)$-modules, where the $SL_2(k)$-action is given by the action of $\langle C, D \rangle$ on $\tilde{U}$. But such a non-split extension does only exist if Char $k = 3$.

Remark. $\widetilde{M}_A$ is a kR_A-module, where the scalar action is given by the action of the diagonal subgroup of $\langle A, F \rangle$ normalizing A and F. (See (2.25)). If now $\dim_k \widetilde{M}_A < \infty$ and $R_A \simeq SL_2(\ell), \ell$ some field, then it follows from (4), using representation theory of $SL_2(\ell)$, that $\ell = k$ or ℓ is a cubic extension of k. Further $\widetilde{M}_A$ is uniquely determined as a $kSL_2(\ell)$-module. This shows that $\widetilde{M}_A A_s$ and then that the unipotent subgroup $M_A A_s$ is uniquely determined.

§ 5   The orthogonal groups

We assume in this section that the group G is generated by the non-degenerate class Σ of abstract root subgroups and that $R(G) = 1$. We use the notation introduced in §2. In particular we fix $A \in \Sigma, A_0 \in \Omega_A, X = \langle A, A_0 \rangle, \Delta_0 = C_\Sigma(X)$ and $V = \langle \Delta_0 \rangle$. In addition we will assume that one of the following two hypotheses holds:

(A) $C_A = \langle C_\Sigma(A) \rangle$ is Σ-maximal, but Δ_0 is not a conjugacy class in V. (I.e. $\Delta_0 = \cup \Delta_i, i \in I$ and $|I| \geq 2$ in (2.17).)

(B) C_A is not Σ-maximal, but $N_G(A)$ is maximal in G. Further $\Sigma_A \not\subseteq M_A$.

Notice that, if G satisfies either of the hypotheses (A) or (B), then G satisfies hypothesis (M) of (2.13). Since by (3.3) there exist in case (B) at least three weak TI-subsets of Σ containing A, the hypothesis of (2.23) is in any case satisfied. Hence Σ is a class of k-root subgroups, k a fixed (commutative) field. Further $\widetilde{M_A}$ is, with the k-action defined in (2.24), by (2.25) a kV-module.

Before we start to obtain consequences of either (A) or (B), we will show that a non-degenerate (in the sense of II(1.5)(7)) orthogonal group of Witt index ≥ 3 satisfies (A) or (B). We have:

(5.1) Proposition

Suppose V is a non-degenerate orthogonal space of Witt index ≥ 3 over the field k with quadratic form $q : V \to k$ (Nondegenerate means $V^\perp \cap q^{-1}(0) = 0$. The Witt index might be infinite!). Let $G = PO(V, q)$ and Σ be the class of images under the homomorphism $P : O(V, q) \to PO(V, q)$ of the Siegel transvection groups (as defined in II(1.5).). Then one of the following holds:

   (1) If the Witt index of q is ≥ 4, then hypothesis (A) holds.
   (2) If the Witt index of q is 3, then hypothesis (B) holds.

Proof. If the Witt index of q is 3, then there exists by exercise II(1.6)(3)(b) a weak TI-subset Λ of Σ, which is not a TI-set. Hence, by §3, C_A is not Σ-maximal, but $N_G(A)$ is maximal. Now

$$\Lambda = \{T_\ell \mid \ell \text{ a singular line in } E\},$$

where E is a fixed 3-dimensional totally singular subspace of the underlying vector space V. Hence $\dim_k \langle \Lambda \rangle = 3$ by (3.4). (I.e. clearly $\langle \Lambda \rangle \neq D \cdot B; D, B \in \Sigma$!) But then by (2.1) $\Sigma_A \not\subseteq M_A$, whence (B) holds.

So assume that the Witt index of q is ≥ 4. Then again by exercise II(1.6)(3) no weak TI-subsets exist, whence C_A is Σ-maximal. Thus it remains to show that $\Delta_0 = \cup \Delta_i, i \in I$ with $|I| \geq 2$. Using the notation of II(1.5) we may set:

$$A = T_\ell, A_0 = T_s \text{ with } s \cap \ell^\perp = 0,$$

$\Delta_1 = \{T_g \mid g \subseteq \ell^\perp \cap s^\perp\}$ and $\Delta_2 = \{T_g \mid \ell \cap g \neq 0 \neq g \cap s\}$. Then by II(1.5)(4), $\Delta_0 = \Delta_1 \cup \Delta_2$ and obviously no element of Δ_1 is connected to an element of Δ_2 in $\mathcal{F}(\Delta_0)$. (We have not shown that $\mathcal{F}(\Delta_i)$ are connected for $i = 1, 2$, since this is not necessary for the proof of (5.1). In fact in $D_4(k)$ this is not the case for $\mathcal{F}(\Delta_1)$!)   □

(5.2) Lemma

Suppose that hypothesis (B) holds. Then the following hold:

(1) $\Lambda_A \cup \{A\} = \bigcup_{g \in N(A)} \Lambda^g$, where Λ is a weak TI-subset of Σ containing
A. Further $\dim \langle \Lambda \rangle = 3$ (in the notation of §3).

(2) $V \simeq SL_2(k)$.

Proof. (1) is a direct consequence of (2.21) and (3.4). By (2.12) $C_{\Delta_0}(\Lambda) = \emptyset$.
Hence by (2.1) and (2.12) $\mathcal{F}(\Delta_0)$ is connected. Also by (2.1) V induces an
$SL_2(k)$ on $\langle \Lambda \cap \Lambda_A^* \rangle$. Now, as V normalizes all $\Lambda^g, g \in N(A)$, we have

$$[\Lambda^g \cap \Lambda_A^*, C_V(\Lambda)] \leq \langle \Lambda^g \cap \Lambda_A^* \rangle \cap C(\Lambda) = 1$$

by (2.12). Hence, as $\langle \Lambda_A^* \rangle = M_A$ we have

$$C_V(\Lambda) \leq C(\langle M_A, X \rangle) \leq Z(G) = 1$$

by II(4.8). Hence $C_V(\Lambda) = 1$ and $V \simeq SL_2(k)$. □

(5.3) Notation

Suppose that hypothesis (B) holds. Then we set $J = k \cup \infty$ and $\Delta_0 = \{B_j \mid j \in J\}$. For $j \in J$ let:

$$\Xi_j = (\Lambda_A \cap \Lambda_{B_j})^{M_A} \cup (\Lambda_A \cap \Lambda_{B_j})^{M_{B_j}} \cup B_j^{M_A} \cup A^{M_{B_j}}.$$

Then the definition of Ξ_j is symmetric in A and B_j. As $[B_j, C] \in \Lambda_A \cap \Lambda_{B_j}$ for
each $C \in \Lambda_A$, we have $\langle \Xi_j \rangle = A \langle \Lambda_A \cap \Lambda_{B_j} \rangle B_j$. In particular $\langle \Xi_j \rangle$ is abelian.
Since $\langle \Lambda \cap \Lambda_A^* \rangle$ is the natural V-module, we have

(∗) $\Lambda \cap \Lambda_A^* = \bigcup_{j \in J} D_j$ where $\{D_j\} = \Lambda \cap \Lambda_A^* \cap \Lambda_{B_j}$.

Since (∗) also holds for $\Lambda^g, g \in N(A)$ we obtain by (5.2)(1):

(∗∗) $\Lambda_A^* = \bigcup_{j \in J} (\Lambda_A^* \cap \Lambda_{B_j})$.

Since $\Lambda_A = (\Lambda_A^*)^{M_A}$ and $\Sigma_A = \Lambda_A \cup \bigcup_{j \in J} B_j^{M_A}$ by (2.16) we obtain:

(+) $\Sigma_A \cup \{A\} = \bigcup_{j \in J} \Xi_j$ and $\Xi_j \cap \Xi_i = \{A\}$ for $i \neq j$.

We show next:

(5.4) Lemma

Suppose that hypothesis (B) holds. Then the following hold:

(1) $\langle N_\Sigma(\Xi_j) \rangle$ is transitive on Ξ_j.
(2) $M_E \leq \langle N_\Sigma(\Xi_j) \rangle$ for each $E \in \Xi_j$.
(3) $C_\Sigma(\Xi_j) = \Xi_j = \Sigma \cap \langle \Xi_j \rangle$.
(4) $\Xi_j = (\Lambda_A \cap \Lambda_{B_j}) \cup B_j^{M_A} \cup A^{M_{B_j}}$.

Proof. Set $\Xi = \Xi_j$. We show first:

$$(*) \qquad\qquad M_A \leq N(\Xi).$$

For this pick $E \in \Xi$ and show $E^{M_A} \subseteq \Xi$. If $E \in (\Lambda_A \cap \Lambda_{B_j})^{M_A} \cup B_j^{M_A}$ this is obvious. Suppose next $E \in (\Lambda_A \cap \Lambda_{B_j})^{M_{B_j}}$. Then $E^m \in \Lambda_A \cap \Lambda_{B_j}$ for some $m \in M_{B_j}$. Hence $E^m \in \Lambda_A^*$ and we may without loss of generality assume $E^m \in \Lambda$. (Otherwise replacing Λ by some $\Lambda^g, g \in N(A)$.) Hence $E^m = D_j$ by (5.3)$(*)$ and thus $E \in D_j B_j \cap \Sigma$. Since $D_j B_j \cap \Sigma \subseteq D_j \cup B_j^{M_A}$ we obtain $E^{M_A} \subseteq \Xi$ by definition of Ξ.

Finally assume $E \in A^{M_{B_j}}$. If $E \in \Lambda_A$, then since $EA \cap \Sigma$ is a partition of EA, we have $D = EA \cap M_{B_j} \in \Sigma$ and $D \in \Lambda_A \cap \Lambda_{B_j}$ by (2.6). Since $E = D^m$ for some $m \in M_A$ we clearly have $E^{M_A} \subseteq \Xi$. So we may assume $E \in \Sigma_A - \Lambda_A$. Since $[E, B_j] = 1$ we get

$$E \in B_j M_A \cap \Sigma \subseteq B_j^{M_A} \cup \Lambda_A \cup \{A\}.$$

Hence $E \in B_j^{M_A}$ and $(*)$ holds.

Let $R = \langle M_A, M_{B_j} \rangle$. Then by $(*)$ and symmetry in A and B_j in the definition of Ξ we obtain $R \leq N(\Xi)$. We claim that R is already transitive on Ξ. If $C \in \Lambda_A \cap \Lambda_{B_j}$ we may without loss of generality assume $C = D_j$. Hence $N_R(CB_j)$ acts already transitive on $CB_j \cap \Sigma$ and thus, by symmetry between A and B_j, also $N_R(CA)$ is transitive on $CA \cap \Sigma$. Hence each element of $(\Lambda_A \cap \Lambda_{B_j})^{M_A} \cup (\Lambda_A \cap \Lambda_{B_j})^{M_{B_j}}$ is conjugate to A in R. Moreover B_j is conjugate to C and C to A in R, which proves our claim.

Since $R \leq \langle N_\Sigma(\Xi) \rangle$ this implies (1) and (2). Now $C_\Sigma(\Xi) \subseteq C_\Sigma(A) \cap C_\Sigma(B_j)$. Thus

$$C_\Sigma(\Xi) - M_A \subseteq B_j^{M_A} \subseteq \Xi.$$

Further

$$
\begin{aligned}
C_\Sigma(\Xi) \cap M_A &\subseteq A \cup C_{\Lambda_A}(B_j) \subseteq A \cup C_{\Lambda^*_A}(B_j)^{M_A} \\
&= A \cup (\Lambda_A^* \cap \Lambda_{B_j})^{M_A} \subseteq \Xi
\end{aligned}
$$

by (2.28)(2). This proves (3). Finally, as $\Xi - M_A \subseteq B_j^{M_A}$ and

$$\Xi \cap M_A = A \cup (\Lambda_A \cap \Lambda_{B_j})^{M_A} \subseteq A \cup (\Lambda_A \cap \Lambda_{B_j}) \cup A^{M_{B_j}},$$

since $\Lambda_C \cap \Lambda_{B_j} \not\subseteq C_\Sigma(A)$ for $C \in \Lambda_A \cap \Lambda_{B_j}$, and since $CB_j \cap \Sigma$ is contained in some weak TI-subset of Σ, also (4) holds. □

(5.5) Lemma

Suppose that hypothesis (B) holds. Then

$$\Xi_j \cap \Xi_j^g = \begin{cases} E \in \Sigma \\ \emptyset \end{cases}$$

for each $j \in J$ and $g \in G$.

Proof. By (5.3)(+) we have $\Xi_j \cap \Xi_k = A$ for $j \neq k$. So, to prove (5.5) set $\Xi_j = \Xi$ and assume $\Xi \cap \Xi^g \neq \emptyset$. Then we may by (5.4)(1) assume $A^g \in \Xi \cap \Xi^g$. Hence there exists an $h \in N(\Xi)$ with

$$A = A^{gh} \in \Xi \cap \Xi^{gh}.$$

Now $gh \in N(A)$ and thus $B_j^{gh} \in B_k^{M_A}$ for some $k \in J$. By (5.4)(4) this implies $\Xi^{gh} = \Xi_k$, whence $A = \Xi \cap \Xi^{gh}$ and thus $A^g = \Xi \cap \Xi^g$. □

(5.6) Notation

Suppose that hypothesis (B) holds and set $\Xi = \Xi_j$ for some $j \in J$. (The Ξ_j are all conjugate in V!). We define a geometry $\mathcal{G}$ with point set $\mathcal{P} = \{\Xi^g \mid g \in G\}$ and line set Σ, a point Ξ^g lying on the line $D \in \Sigma$ if and only if $D \in \Xi^g$. Then we have, using the definition of a polar space of (1.3):

(5.7) Theorem

Suppose hypothesis (B) holds. Then $\mathcal{G}$ is a thick, non-degenerate polar space of rank 3, the planes of which are projective planes over k.

Proof. By (5.5) and (5.3)(+) we have $\{\Xi_j \mid j \in J = k \cup \infty\} = \{\Xi^g \mid A \in \Xi^g\}$. Hence the line A of $\mathcal{G}$ is a projective line over k. In particular, lines are thick.

To show that $\mathcal{G}$ is non-degenerate (if it is a polar space) it suffices to show that there exist Ξ^g, Ξ^h with $\Xi^g \cap \Xi^h = \emptyset$. For this assume $\{A, B_j\} \subseteq \Xi$ and pick $g, h \in G$ with $\Xi \cap \Xi^g = B_j$, $\Xi \cap \Xi^h = A$. Then

$$\Xi^g \cap \Xi^h \subseteq C_\Sigma(A) \cap C_\Sigma(B_j) \subseteq \Xi$$

as shown in the proof of (5.4)(3). But then $\Xi^g \cap \Xi^h = \emptyset$, which was to be shown.

So, to prove (5.7), let $E \in \Sigma$ be a line and Ξ^g a point not on E. We then have to show that Ξ^g is collinear with a unique or with all points on E. That is $E \notin \Xi^g$ and we have to show that there exists a unique Ξ^h with $E \in \Xi^h$ and $\Xi^g \cap \Xi^h \neq \emptyset$. To show this we may without loss of generality assume $\Xi^g = \Xi$. We first show the existence of Ξ^h.

If $E \in \Sigma_B$ for some $B \in \Xi$, then the existence of Ξ^h follows from (5.3)(+). Thus we may assume $E \notin \bigcup \Sigma_B, B \in \Xi$. If now $E \in \Omega_A$, then by (5.4) (2) and (2.15) we may assume $E = A_0$. But then $E \in \Sigma_{B_j}, B_j \in \Xi$, a contradiction to our assumption. Hence by (5.4)(1) we may assume

$$ E \notin \bigcup_{B \in \Xi} (\Sigma_B \cup \Omega_B). $$

Hence $\Xi \subseteq \Psi_E$ and (5.4) shows that Ξ^e and Ξ normalize each other for $e \in E^{\#}$. If $\Xi \cap \Xi^e \neq \emptyset$ this is by (5.5) and (5.3)($*$) impossible. (Since Ξ_i and Ξ_j do not normalize each other for $i \neq j \in J$!) Hence $\Xi \cap \Xi^e = \emptyset$ and thus $\Xi^e \subseteq C_\Sigma(\Xi) = \Xi$, which is together with (5.4)(3) a contradiction to $\Xi \subseteq \Psi_E$. This shows existence of Ξ^h.

Suppose now there exist Ξ^h, Ξ^y with $\Xi^h \cap \Xi^y = E$ and $\Xi^h \cap \Xi \neq \emptyset \neq \Xi^y \cap \Xi$. Then we may by (5.4)(1) without loss of generality assume $A = \Xi^h \cap \Xi$, whence $\Xi^h = \Xi_k$ for some $k \in J$. Let $B = \Xi \cap \Xi^y$. If $B \in B_j^{M_A}$ we get as before

$$ E \in C_\Sigma(A) \cap C_\Sigma(B) \subseteq \Xi, $$

a contradiction. Similarly, if $E \in B_k^{M_A}$, then $B \in C_\Sigma(A) \cap C_\Sigma(E) = \Xi_k$, a contradiction to $\Xi \neq \Xi_k$. Since $A^{M_{B_j}} \subseteq (A \cup \Lambda_A) \cup B_j^{M_A}$, this shows that we may assume $\{E, B\} \subseteq \Lambda_A$. Applying the same argument to Σ_B and Ξ^y we may also assume $E \in \Lambda_B$. Hence by (2.6) $\Theta = ABE \cap \Sigma$ is a partition of ABE and thus by (2.1), (2.12) and (5.2)(1) $\Theta = \Lambda^g$ for some $g \in N(A)$ where Λ is as in (5.2). This shows that, if $E \in \Xi^x, x \in G$, then $\Xi^x \cap AB \neq \emptyset$ and thence $\Xi^x \cap \Xi \neq \emptyset$. By definition of incidence in $\mathcal{G}$ this means that Ξ is collinear with all points on E.

This shows that $\mathcal{G}$ is a thick non-degenerate polar space. Moreover the above argument shows that if Ξ, Ξ^h and Ξ^y is a triple of pairwise collinear points and $A = \Xi \cap \Xi^h, B = \Xi \cap \Xi^y, E = \Xi^h \cap \Xi^y$, then $ABE \cap \Sigma = \Lambda^g$ for some $g \in N(A)$. Let $\mathcal{H}$ be the projective subspace of $\mathcal{G}$ spanned by this triple of points. Then $\dim_k \langle \Lambda \rangle = 3$ and the fact that $\langle \Lambda \rangle$ is by (2.12) a maximal abelian subgroup of G partitioned by elements of Σ implies that $\mathcal{H}$ is a maximal projective subspace of $\mathcal{G}$ and that $\mathcal{H}$ is a plane. Hence Rank $\mathcal{G} = 3$. Moreover visibly $\mathcal{H}$ is isomorphic to the dual of $\langle \Lambda^g \rangle$, the isomorphism given by

$$ \Xi^x \to \langle \Xi^x \cap \Lambda^g \rangle, \quad \Xi^x \text{ a point of } \mathcal{H}. $$

(I.e. $\langle \Xi \cap \Lambda^g \rangle = AB$, whence points of $\mathcal{H}$ are mapped onto planes of $\langle \Lambda^g \rangle$!) $\square$

The next difficult reduction theorem is crucial for the treatment of hypothesis (A).

(5.8)   Theorem

Suppose that hypothesis (A) holds and V_1 is not isomorphic to $SL_2(k)$. ($V_1 = \langle \Delta_1 \rangle$!) Then $|I| = 2, V_2 \simeq SL_2(k)$ and if $\Delta_2 = \{B_j \mid j \in J = k \cup \infty\}$ and

$$\Xi_j = (\Lambda_A \cap \Lambda_{B_j}) \cup B_j^{M_A} \cup A^{M_{B_j}}, j \in J,$$

then the following hold:

(a) $C_\Sigma(A) - \{C^m \mid C \in \Delta_1, m \in M_A\} = \bigcup_{j \in J} \Xi_j$.
(b) $\Xi_i \cap \Xi_j = A$ for $i \neq j$.
(c) $\langle \Xi_j \rangle = A \langle \Lambda_A \cap \Lambda_{B_j} \rangle B_j$. In particular $\langle \Xi_j \rangle$ is abelian.
(d) V_2 acts in its natural permutation action on $\{\Xi_j \mid j \in J\}$.
(e) $\langle V_1, M_A, M_{B_j} \rangle \leq \langle N_\Sigma(\Xi_j) \rangle$.
(f) $\Xi_j = \Sigma \cap \langle \Xi_j \rangle = C_\Sigma(\Xi_j)$.

Proof. Since the proof is relatively lengthy, we will divide it into several steps. Pick $B \in \Lambda_A^*$ and set $\Lambda_1 = B^{V_1}$. Then $\langle \Lambda_1 \rangle$ is abelian by (2.27)(1). Let

$$\{\Lambda_k \mid k \in K\} \text{ be the set of conjugates of } \Lambda_1 \text{ under } V.$$

Then, setting $V^1 = \Pi V_i, i \in I$ and $i \neq 1$, we have $V = V_1 * V^1$ and $\{\Lambda_k \mid k \in K\} = \Lambda_1^{V^1}$. Further $V_1 \leq N(\Lambda_k)$ for each $k \in K$, since $[V^1, V_1] = 1$. We will restrict the structure of V^1 through its action on $\{\Lambda_k \mid k \in K\}$. We have

(1) $\Lambda_k = C^{V_1}$ for each $C \in \Lambda_k$ and $k \in K$.

This is clear since it holds for Λ_1 by definition.

(2) $\Lambda_i \cap \Lambda_k = \emptyset$ for $i \neq k \in K$.

This follows from (1).

(3) $\Lambda_B \cap \Lambda_k \neq \emptyset$ for $k \in K$.

This is the first serious step. If $k = 1$ then (3) follows from (2.27)(2). So assume $k \neq 1$. Again by (2.27)(2) there exists a $C \in \Psi_B \cap \Delta_1$. By (2.1) and (2.18) there exists a $D \in \Delta_0$ with $[B, C, D] = B$. Hence $\langle C, D \rangle \simeq SL_2(k)$ and $D \in \Delta_1 \cap \Lambda_B$. If now $E \in \Lambda_k \cap \Psi_D \cap \Sigma_B$, then by the above argument $[E, D] \in \Lambda_k \cap \Lambda_B$, which was to be shown. Thus we may assume $\Lambda_k \cap \Psi_D \cap \Sigma_B = \emptyset$. If now $F \in \Lambda_1 \cap \Lambda_B \cap \Lambda_D$, then $BF \cap \Sigma \subseteq \Lambda_D$ and each $E \in \Lambda_k$ centralizes some

$\overline{B} \in BF \cap \Sigma$. But since also $\Lambda_k \cap \Psi_D \cap \Sigma_{\overline{B}} = \emptyset$, we obtain $\Lambda_k \subseteq \Sigma_D$, whence V_1 centralizes Λ_k, which contradicts (2.27). We have shown

(∗) $\Lambda_1 \cap \Lambda_B \cap \Lambda_D = \emptyset.$

Also clearly

$$[\Lambda_1, \Lambda_k] \neq 1,$$

since otherwise $\Lambda_k \cap \Psi_D \cap \Sigma_B \neq \emptyset$.

Let $\Psi = \Delta_1 \cap \Lambda_B$. If $c \in C \in \Delta_1$, then $\Psi \cap \Psi^c \subseteq \Lambda_B \cap \Lambda_{B^c}$ and BB^c is partitioned by $BB^c \cap \Sigma$. Then (∗) applied to $D \in \Psi \cap \Psi^c$ implies either $B = B^c$ and $\Psi = \Psi^c$ or $\Psi \cap \Psi^c = \emptyset$. This shows that Ψ is a weak TI-subset of Δ_1 or $|\Psi| = 1$. Suppose first that Δ_1 is degenerate. Then for each $L \in \Psi$ we have:

$$C_{\Delta_1}(L) \subseteq N_{\Delta_1}(\Psi) \subseteq C_{\Delta_1}(B) = \Delta_1 \cap \Lambda_B = \Psi$$

by (2.28) (2). Since $\langle \Psi \rangle$ is by II(2.3) abelian, this implies $\Psi \subseteq \underline{L}$ for each $L \in \Psi$ in the notation of II(3.11). On the other hand, as $R(V_1) \leq Z(V_1)$ by (2.30), the definition of $R(V_1)$ in II(4.12) implies $\underline{L} = \{L\}$. Hence $\Psi = \{L\}$ and Δ_1 contains no different commuting elements. We show that this implies $V_1 \simeq SL_2(k)$, which contradicts the hypothesis of (5.8).

Since $\langle C, D \rangle \simeq SL_2(k)$ it suffices to show that there exists no $L \in \Delta_1 - (\langle C, D \rangle \cap \Delta_1)$. As shown $N = B[B,C]$ is a natural module for $\langle C, D \rangle$. Hence if L centralizes some element of $\Sigma \cap N$, then (2.28) (2) and (2.6) imply $[L, N] \leq N$, which is by II(2.23) a contradiction to the fact that Δ_1 contains no different commuting elements. We obtain that

$$[L, N] = [\ell, N] \text{ is partitioned by } \{[\ell, \overline{B}] \mid \overline{B} \in \Sigma \cap N\}$$

for $\ell \in L^\#$. Hence $[L, N] \cap \Sigma \subseteq \Lambda_L$, a contradiction to (∗) applied to some $\overline{B} \in [L, N] \cap \Sigma$ and L in place of D.

We obtain that Δ_1 is non-degenerate. Pick $T \in \Delta_1 \cap \Psi_D$ and $f \in T^\#$. By definition $D \in \Delta_1 \cap \Lambda_B = \Psi$. If $T \in \Sigma_B$, then $\{D, D^f\} \subseteq \Psi$ and if $T \in \Psi_B$, then $\{T, [T, D]\} \subseteq \Delta_1 \cap \Lambda_{[B,T]}$. As $[B, T] \in \Lambda_1$, as shown in the beginning of the proof of (3), $[B, T]$ is conjugate to B by an element of V_1. So in any case $|\Psi| \geq 2$ and Ψ is a weak TI-subset. Now (2.11) applied to Ψ and II(2.11) show that there exists a $D_1 \in \Psi \cap \Omega_C$. Let $h \in V^1$ with $\Lambda_1^h = \Lambda_k$. Then, as $[V_1, h] = 1$, $\Psi = \Delta_1 \cap \Lambda_{B^h}$. If $[N, N^h] = 1$, then as $D^e \in \Lambda_{B^{hc}}$ for $c \in C^\#$, we obtain $[D^e, B] \in \Lambda_{B^{hc}}$. Hence, conjugating with some element of $\langle C, D \rangle \leq V_1$, (3) holds.

So we may assume $[N, N^h] \neq 1$ and thus $[B, B^{hc}] = A$. (As $N = BB^c$!). We now obtain the final contradiction. Indeed $D_1^c \in \Lambda_{D_1}$ and thus $D_1^c \leq N(\Psi)$.

Now $(*)$ implies $D_1^c \in \Sigma_B$. But as $D_1 \in \Lambda_{B^h}$, also $D_1^c \in \Lambda_{B^{hc}}$. Hence (2.6) implies $D_1^c \in \Lambda_A$, which is not the case.

This shows that (3) holds. As an immediate consequence of (1) and (3) we obtain:

(4) $\Lambda_D \cap \Lambda_k \neq \emptyset$ for each $D \in \Lambda_j$ and $j \neq k$.

We show next:

(5) Suppose $j \neq k \in K$. Then there exists an

$$F \in \Delta^1 = \Delta_0 - \Delta_1 \text{ and } f \in F \text{ with } \Lambda_j^f = \Lambda_k.$$

Indeed pick by (4) $D \in \Lambda_j$ and $E \in \Lambda_k \cap \Lambda_D$. Then there exists by (2.1) and (2.18) an $F \in \Delta_0$ and $f \in F$ with $D^f = E$. Since $V_1 \leq N(\Lambda_j)$, (2) implies $F \in \Delta^1$. Thus $[V_1, F] = 1$ and so $\Lambda_j^f = \Lambda_k$.

(6) The Δ_j are degenerate for $1 \neq j \in I$.

To prove (6) we may without loss of generality assume that Δ_2 is non-degenerate. If $\Delta_2 \subseteq C(\Lambda_1)$, then as $\Lambda_A^* = \cup \Lambda_k, k \in K$, it follows that $\Delta_2 \subseteq C_\Sigma(M_A)$, contradicting (2.6). Hence, arguing as in the beginning of the proof of (5.8), there exists an $E \in \Delta_2 \cap \Lambda_B$. We claim

$(+)$ $\qquad\qquad\qquad\qquad \Delta_2 \cap \Lambda_E \subseteq \Lambda_B.$

To prove $(+)$ assume without loss of generality that $[\Lambda_1, \Lambda_2] = A$ and pick by (4) $C \in \Lambda_2 \cap \Lambda_B$ and $F \in \Delta_2 \cap \Lambda_E$ with $F \notin \Lambda_B$. Then $F \in \Sigma_B$, since otherwise by (2.6) $[F, B] = E$, which is impossible. Now $[E, C] = B$, since if $[E, C] = 1$ then as $\Lambda_1 = B^{V_1} \subseteq \Lambda_E$, we obtain $A = [\Lambda_1, C] \leq M_E$, a contradiction to (2.6). If now $F \in \Sigma_C$, then $B = [E, C] \in \Lambda_F$ and whence $F \in \Delta_2 \cap \Lambda_B$ which was to be shown. So, to prove $(+)$, we may assume $F \in \Psi_C$. Also, if $[F, C] \leq BC$, then $[F, C] = B$ and again $F \in \Lambda_B$. Hence $N = BC[C, F]$ is by (2.1) a three dimensional vector space over k with $\Sigma \cap N$ the set of 1-spaces. Since $C^F \subseteq \Lambda_A^*$ it is easy to see that $\Sigma \cap N \subseteq \Lambda_A^*$. Thus (2.1) (resp. exercise (2.31)(1)) and (2.18) show that $\langle N_{\Delta_0}(N) \rangle$ induces the $SL_3(k)$ on N. But since E and F act non-trivially on N, this then implies that $\langle N_{\Delta_2}(N) \rangle$ induces already the $SL_3(k)$ on N. In particular there exists an $L \in \Delta_2 \cap \Lambda_B \cap \Sigma_C$. (I.e. L induces the transvection group corresponding to the point B and hyperplane BC on L!) Then, as $\Lambda_1 = B^{V_1}$ and $\Lambda_2 = C^{V_1}$ we obtain $\Lambda_1 \subseteq \Lambda_L$ and $\Lambda_2 \subseteq \Sigma_L$ and thus

$$A = [\Lambda_1, \Lambda_2] \in \Lambda_L,$$

which is impossible.

This contradiction shows that no such F exists, whence $(+)$ holds. But then the connectedness of $\mathcal{E}(\Delta_2)$ implies $\Delta_2 \subseteq \Lambda_B$, which contradicts II(2.3). This proves (6).

Next we show:

(7) Each $E \in \Delta^1$ normalizes exactly one $\Lambda_k, k \in K$. Further $\Lambda_k \subseteq C_\Sigma(E)$.

It is easy to see that E centralizes Λ_k if it normalizes Λ_k, since $\Lambda_k = D^{V_1}$ for $D \in \Lambda_k$ and if $C \in \Lambda_k \cap \Psi_E$ then $D = [C, E] \in \Lambda_k$.

So assume $\Lambda_i, \Lambda_k \subseteq \Sigma_E$ for $i \neq k$. By (6) and (2.28)(2) this implies $\Lambda_i \cup \Lambda_k \subseteq \Lambda_E$. Hence $[\Lambda_i, \Lambda_k] = 1$. In particular since $M_A = \langle \Lambda_A^* \rangle$ is non-abelian, there exists an $\ell \in K$ such that Λ_ℓ is not normalized by E. Pick $C \in \Lambda_\ell$ and by (4) $D \in \Lambda_k \cap \Lambda_C$ and $F \in \Lambda_i \cap \Lambda_C$. Then $C = [D, E] = F$ as $E \in \Lambda_C \cap \Lambda_F$, a contradiction to $k \neq i$.

(8) $|I| = 2$ and $V_2 \simeq SL_2(k)$.

Pick by (7) $E \in \Delta_2$ such that $\Lambda_1 \subseteq C(E)$. Then by (6) and (2.8) $\Lambda_1 \subseteq \Lambda_E$. Further by (7) $C_{\Delta_0}(E) \subseteq N(\Lambda_1)$. Hence (6), (7) and (2.28) together imply $C_{\Delta^1}(E) \subseteq \Lambda_C$ for each $C \in \Lambda_1$. In particular $\langle C_{\Delta^1}(E) \rangle$ is by (6) and II(2.3) abelian, which implies $\Delta^1 = \Delta_2$ and $|I| = 2$. Moreover, since by (2.30) $R(V_2) \leq Z(V_2), \Delta_2$ contains no different commuting elements.

Now let $F \in \Delta_2 - \{E\}$ with $R = \langle E, F \rangle \simeq SL_2(k)$ and assume that there exists a $D \in \Delta_2 - (\Delta_2 \cap R)$. Then, as $B \in \Lambda_E, N = B[B, F]$ is by II(2.19) partitioned by $\Sigma \cap N$. Set $C = [B, F]$. Then $\{C, [B, D]\} \subseteq \Lambda_B \cap \Lambda_A$ and thus $[C, [B, D]] = 1$. This implies that $[N, D] = [N, d] = [B, d][C, d]$ is abelian for $d \in D^\#$ and that $[N, D]$ is also partitioned by $\Sigma \cap [N, D]$, where the partitioning groups are just the images of the elements of $\Sigma \cap N$ under the commutator map.

Let $B_1 = [B, D]$ and $C_1 = [C, D]$. Then by II(2.19) $\{B_1, C_1\} \subseteq B^{V_2}$. Thus there exists an $h \in V_2$ with $B_1^h = C_1$. On the other hand by (7) there exists a fixed ℓ with $B_1 C_1 \cap \Sigma \subseteq \Lambda_\ell$ and thus there exists a subgroup $Y = \langle Y \cap \Delta_1 \rangle \simeq SL_2(k)$ acting naturally on $B_1 C_1$. But $[Y, h] \leq [V_1, V_2] = 1$, a contradiction to $C_Y(B_1) \neq C_Y(C_1)$.

This shows that no such D exists, whence $V_2 = R \simeq SL_2(k)$. Now (7) and (8) together show $N_{V_2}(\Lambda_1) = N_{V_2}(E)$ and thus V_2 acts naturally on $\{\Lambda_k \mid k \in K\}$. In particular $K = J = k \cup \infty$. Let B_j be the element of Δ_2 normalizing Λ_j. Then (7) and (2.28) imply $\Lambda_j = \Lambda_A \cap \Lambda_{B_j} = \Lambda_A^* \cap \Lambda_{B_j}$, since by (2.20) $\Lambda_A^* = \bigcup \Lambda_j, j \in J$. Now

$$\{[B_j, C] \mid C \in \Lambda_A, [B_j, C] \neq 1\} \subseteq \Lambda_j,$$

so that $[B_j, M_A] \leq A\langle \Lambda_j \rangle$.

Let $j \neq \ell \in J$. Then

$$\Delta_1 \cup (\Sigma \cap \langle A, A_0 \rangle) \subseteq \Sigma_{B_j} \cap \Sigma_{B_\ell}.$$

Hence Δ_1 is contained in a connectivity component Δ_1' of $\mathcal{F}(\Sigma_{B_j} \cap \Sigma_{B_\ell})$. But then, since $|I| = 2$, either $\Delta_1 = \Delta_1'$ or $\Delta_1 \cup \{A, A_0\} \subseteq \Delta_1'$. Now the second case is impossible, since otherwise there would exist an element in Σ centralizing $\Delta_1 \cup \{A, A_0\} \cup \{B_j, B_\ell\}$. Hence $\Delta_1 = \Delta_1'$ and

$$\Sigma_{B_j} \cap \Sigma_{B_\ell} = \Delta_1 \cup (\Sigma \cap \langle A, A_0 \rangle).$$

This shows that Λ_j plays the same role in Σ_{B_j} as in Σ_A. In particular $[A, M_{B_j}] \leq B_j \langle \Lambda_j \rangle$. We claim

$$(*) \qquad\qquad \Xi_j = B_j \langle \Lambda_j \rangle A \cap \Sigma.$$

To prove $(*)$ let Ψ_j be the right side of $(*)$. Then by the above and definition of Ξ_j we have $\Xi_j \subseteq \Psi_j$. On the other hand, if $\overline{B} \in \Psi_j - (\Lambda_A \cup A)$, then $\overline{B} M_A = B_j M_A$ and thus $\overline{B} \in B_j^{M_A} \subseteq \Xi_j$. Hence $\Psi_j - (\Lambda_A \cup A) \subseteq \Xi_j$ and, by the same reason, $\Psi_j - (\Lambda_{B_j} \cup B_j) \subseteq \Xi_j$. As $\Psi_j \cap \Lambda_A \cap \Lambda_{B_j} = \Lambda_j$, this proves $(*)$.

Now (a)–(d) of (5.8) are immediate. (By $(*)$ $\Sigma \cap \langle \Lambda_j \rangle A \subseteq \Xi_j$.) To prove (e) it remains to show that $V_1 \subseteq N(\Xi_j), j \in J$. But this is obvious since $V_1 \leq C(A) \cap C(B_j)$. Finally $\Xi_j = \langle \Xi_j \rangle \cap \Sigma$ is a consequence of $(*)$. Moreover $C_{\Delta_1}(\Lambda_j) = \emptyset$ is a consequence of (2.6), since $\Lambda_A^* = \bigcup \Lambda_j^v, v \in V$ by (d). Since by (7) $C_{\Lambda_A^*}(B_j) = \Lambda_j$ this implies $C_\Sigma(\Xi_j) = \Xi_j$, which finally proves (5.8). $\qquad\square$

(5.9) Corollary

Assume that hypothesis (A) holds. Then, up to renumbering, one of the following holds:

(1) $|I| = 2, V_1 \not\simeq SL_2(k), V_2 \simeq SL_2(k)$ and (5.8) holds.
(2) $|I| = 3$ and $V_i \simeq SL_2(k)$ for $i \in I$.

Proof. If one $V_i \not\simeq SL_2(k)$ we may without loss of generality assume $V_1 \not\simeq SL_2(k)$. Hence (1) holds by (5.8).

So we may assume $V_i \simeq SL_2(k)$ for $i \in I$. It remains to show $|I| = 3$. Suppose $|I| \geq 3$. Then, taking $V_1 * V_2$ instead of V_1 in the proof of (5.8), it follows as in (5.8) that $|I| = 3$. So it remains to show that $|I| \neq 2$ when $V_i \simeq SL_2(k)$ for $i \in I$.

Suppose $|I| = 2$ and $V_1 = \langle E, F \rangle, V_2 = \langle C, D \rangle$ with $\{E, F, C, D\} \subseteq \Delta_0$. Pick $B \in \Lambda_A^* \cap \Lambda_E$ (exists!). Then $N = B[B, F]$ is the natural module for $V_1 \simeq$

$SL_2(k)$. Assume without loss of generality $C \in \Psi_B$. Then by (2.29)(1) $B[B,C]$ is the natural module for V_2. Since $\overline{B} = [B,F] \in B^{V_1}$, also $[\overline{B}, C]\overline{B}$ is the natural module for V_2. This implies that

$$N[N,C] = B[B,C]\overline{B}[\overline{B},C]$$

is $V_1 * V_2$ invariant. But on the other hand $N[N,C]$ is abelian. (For example $\{\overline{B}, [B,C]\} \subseteq \Lambda_A \cap \Lambda_B$, whence they commute.) This is a contradiction to $\Lambda_A^* = B^V$ by (2.20). $\qquad\square$

(5.10)   Lemma

Suppose that hypothesis (A) holds and $|I| \neq 3$. Let $\Xi = \Xi_j$ for some $j \in J$. Then the following hold:

(1) $\langle N_\Sigma(\Xi)\rangle$ is transitive on Ξ. Further

$$M_C \leq \langle N_\Sigma(\Xi)\rangle \text{ for each } C \in \Xi.$$

(2) $\Xi \cap \Xi^g = \begin{cases} \emptyset \\ C \in \Xi \end{cases}$ for each $g \in G$.

    Further, if $\Xi \cap \Xi^g = C$, then $\langle C_\Sigma(C)\rangle$ induces a $PSL_2(k)$ in its natural permutation action on $\{\Xi^h \mid h \in G \text{ with } C \in \Xi^h\}$.

(3) There exists a Ξ^h with $\Xi \cap \Xi^h = \emptyset$.

Proof. To prove (1) we show, as in the proof of (5.4), that $R = \langle M_A, M_{B_j}\rangle$ is already transitive on $\Xi = \Xi_j$. Then (1) follows from (5.8)(e). Let $E \in \Lambda_j$. Then there exists a $D \in \Lambda_{B_j}$ and a $C \in \Lambda_A$ such that $\langle D, C\rangle \simeq SL_2(k)$ and acts naturally on EB_j. Hence E is conjugate to B_j in R. Similarly E is conjugate to A in R. But then the definition of Ξ_j implies that R is transitive on Ξ_j.

To prove (2) we may assume $A \in \Xi \cap \Xi^g$. But then $A = A^{hg}$ for some $h \in N(\Xi)$ by (1). Now, as $V_1 \not\simeq SL_2(k)$, hg normalizes $M_A V_2$. Hence $B_j^{hg} \in B_\ell^{M_A}$ for some $\ell \in J$ and thus $\Xi^{gh} = \Xi_\ell$. Now (2) is a consequence of (5.8)(b).

To prove (3) pick $x \in M_A V$ with $\Xi^x = \Xi_\ell$ with $j \neq \ell \in J$ and $y \in \langle C_\Sigma(B_\ell)\rangle$ with $\Xi_\ell \cap \Xi_\ell^y = B_\ell$ and set $g = xy$. Then, as $C_\Xi(B_\ell) = A$ by (5.8), we obtain

$$\Xi \cap \Xi^g \subseteq \{A\}.$$

But if $\Xi \cap \Xi^g = \{A\}$, then by (2) $\Xi = \Xi_\ell$, a contradiction to the choice of g.$\square$

Remark. The proof of (2) depends on the fact that $V_1 \not\simeq SL_2(k)$. This is the reason why we must treat the case $|I| = 3$ and $V_i \simeq SL_2(k)$, which corresponds to $D_4(k)$, separately. In fact in this case there might exist elements in $N(X)$ interchanging the V_i.

(5.11) Theorem

Suppose that hypothesis (A) holds and $|I| \neq 3$. Let $\mathcal{G}$ be the point-line geometry with point set $\mathcal{P} = \Xi^G$ ($\Xi = \Xi_j$ for some $j \in J$) and line set Σ, a point Ξ^g lying on a line $E \in \Sigma$ if and only if $E \in \Xi^g$. Then $\mathcal{G}$ is a non-degenerate polar space of rank ≥ 3, the (singular) subspaces of which are projective spaces over k.

Proof. Since the proof is similar to the proof of (5.7) we just sketch it. First (5.10)(2) shows that there is at most one line through two different points. (5.10)(2) also shows that the lines are projective lines over k. That $\mathcal{G}$ is non-degenerate, i.e. there exists no point collinear with all the others, follows from (5.10)(3).

We need to show: Given a point Ξ^g and a line $E \in \Sigma$, then Ξ^g is collinear to exactly one or to all points of E. To prove this we may without loss of generality assume $\Xi = \Xi^g$. We first show that Ξ is collinear to some point on E. If $E \in C_\Xi(D)$ for some $D \in \Xi$, then we may by (5.10)(1) assume $D = A$. If $E \leq M_A B_j$ for some $j \in J$, then $E \in \Xi_j$ and Ξ is collinear to the point Ξ_j on E. So we may assume $E \in F^{M_A}$ for some $F \in \Delta_1$ and thus $E \in \Delta_1$. But then $E \in N_\Sigma(\Xi)$ by (5.8)(e) and thus $E \in \Lambda_D$ for some $D \in \Xi$ by (5.8)(f). But then there exists by (5.8)(a) a Ξ^h with $\{E, D\} \subseteq \Xi^h$, whence Ξ^h is a point on E collinear with Ξ.

So we may assume $E \notin \bigcup C_\Xi(D), D \in \Xi$. If $E \in \Omega_D$ for some $D \in \Xi$, then we may by (5.10)(1) assume $D = A$ and thus by (2.15) $E = A_0$. But then $E \in \Sigma_{B_j}, B_j \in \Xi$. Hence we obtain

$$E \in \bigcap \Psi_D, D \in \Xi.$$

But this is, as shown in the proof of (5.7), impossible.

Suppose next Ξ is collinear with two points Ξ^x and Ξ^y on E. Then $\Xi^x \cap \Xi^y = E, \Xi \cap \Xi^x = F \in \Sigma$ and $\Xi^y \cap \Xi = L \in \Sigma$. Now one shows as in the proof of (5.7) that $N = EFL$ is partitioned by $\Sigma \cap N$. Thus, if Ξ^h is any point on E, i.e. $E \in \Xi^h$, then it follows from (5.10)(2) that $\Xi^h \cap FL \neq \emptyset$. Since $FL \cap \Sigma \subseteq \Xi$ this implies $\Xi^h \cap \Xi \neq \emptyset$, which shows that Ξ is collinear to all points on E.

Moreover, as in (5.7), this argument shows that the plane spanned by Ξ, Ξ^x, Ξ^y is isomorphic to the dual of the plane with point set $\Sigma \cap N$ given by (2.1), whence is a projective plane over k. Since all subspaces of $\mathcal{G}$ either contain such a plane or are contained in it, it follows that the subspaces are projective spaces over k. Finally Rank $\mathcal{G} \geq 3$ since $\dim_k N = 3$. $\square$

The next lemma will be needed, together with II(1.5)(7), to show that the elements of Σ are images of Siegel transvection groups.

(5.12)   Lemma

Suppose that either hypothesis (B) or hypothesis (A) with $|I| \neq 3$ hold. Let $\mathcal{G}$ be the polar space constructed in (5.7) or (5.11). Then, identifying $A \in \Sigma$ with the line ℓ of $\mathcal{G}$ (i.e. $\ell = A!$) the following hold for each $\alpha \in A$:

(1) α fixes all points in $\ell^{\perp}$. (I.e. collinear to all points of ℓ!)

(2) If P is a point on ℓ, then α fixes all lines through P.

Proof. (2) is rather obvious. Indeed if $\Xi^g = P$ is a point on ℓ, then $A \in \Xi^g$, whence A fixes all lines g through P since this means $g = D \in \Xi^g$.

To prove (1) let Ξ, Ξ^h be two different points on ℓ. Then $\Xi \cap \Xi^h = A$. Let Ξ^g be a point on $\ell^{\perp} - \ell$. Then Ξ is collinear with Ξ^g and Ξ^h, whence $B = \Xi \cap \Xi^g \in \Sigma$ and $C = \Xi^g \cap \Xi^h \in \Sigma$. Hence A centralizes B and C and thus $A \leq N(\Xi^g)$ by (5.10)(2) resp. (5.5). $\qquad\qquad\square$

The proof of the next lemma is due to A. Steinbach.

(5.13)   Lemma

Let $\mathcal{G}$ be a classical polar space obtained from some vector space V together with a non-degenerate (σ, ϵ)-hermitian form f or pseudoquadratic form q. Suppose there exists an $1 \neq \alpha \in \mathrm{Aut}(\mathcal{G})$ satisfying:

(1) α fixes all points in $\ell^{\perp}$ for some line ℓ of $\mathcal{G}$.

(2) If P is a point on ℓ, then α fixes all lines through P.

Then $\mathcal{G}$ is the polar space obtained from a k-vector space W together with a quadratic form $q : W \to k$.

Proof. By [Tit74, (8.2.2)] we may assume that we are in one of the following cases:

(i) f symplectic in Char $\neq 2$.

(ii) q is quadratic.

(iii) q is pseudoquadratic with $\epsilon = -1$ and $1 \in \{c + c^{\sigma} \mid c \in L\}$, L the division ring over which V is defined and σ an involutory antiautomorphism of L.

(See also Exercise II(1.6)(4).) Assume now that (ii) does not hold. Then in any case $\epsilon = -1$. Let $\ell = \langle x_1, x_2 \rangle$. Then, since $\alpha \neq \mathrm{id}$, there exist $y_1, y_2 \in V$ such that $\langle x_1, y_1 \rangle \perp \langle x_2, y_2 \rangle \subseteq V$, $\langle x_i, y_i \rangle$ are hyperbolic lines and $\alpha|_{\langle y_i \rangle} \neq \mathrm{id}$ for $i = 1$ or 2. We claim that there exists a $0 \neq \lambda \in L$ such that:

$$\langle y_1 \rangle^{\alpha} = \langle \lambda^{\sigma} x_2 + y_1 \rangle, \quad \langle y_2 \rangle^{\alpha} = \langle \lambda x_1 + y_2 \rangle.$$

Indeed $\langle y_1 \rangle^\alpha = \langle ax_2 + y_1 \rangle, \langle y_2 \rangle^\alpha = \langle bx_1 + y_2 \rangle$ with $a, b \in L$. Moreover, since $\langle y_1 \rangle^\alpha \perp \langle y_2 \rangle^\alpha$, we have

$$0 = f(ax_2 + y_1, bx_1 + y_2) = a - b^\sigma \quad \text{as} \quad \epsilon = -1.$$

Setting $\lambda = b$ our claim holds. ($b \neq 0$ as $\alpha \mid_{\langle y_i \rangle} \neq$ id for $i = 1$ or $2!$)

Now for $\mu \in L$ arbitrary $g = \langle x_1 - \mu^\sigma x_2, \mu y_1 + y_2 \rangle$ is a singular line through the point $\langle x_1 - \mu^\sigma x_2 \rangle$ on ℓ. Hence

$$\langle \mu y_1 + y_2 \rangle^\alpha \subseteq \langle y_1, y_2 \rangle^\alpha = \langle \lambda^\sigma x_2 + y_1, \lambda x_1 + y_2 \rangle$$

and also $\langle \mu y_1 + y_2 \rangle^\alpha \subseteq g$ by hypothesis (2). Thus the intersection of these two lines is non-empty, from which we obtain $\mu \lambda^\sigma = -\lambda \mu^\sigma$ for arbitrary $\mu \in L$. Setting $x = \lambda \mu^\alpha$ we have $x^\sigma = -x$ for all $x \in L$. In particular $1 = -1$, whence Char $L = 2$, $\sigma =$ id and $\{c + c^\sigma \mid c \in L\} = 0$, a contradiction. Hence (ii) holds. $\qquad \square$

(5.14) Theorem

Assume that either hypothesis (B) or hypothesis (A) with $|I| \neq 3$ is satisfied. Then the following hold:

(1) $G = P\Omega(V, q)$, where V is a k-vector space and $q : V \rightarrow k$ a non-degenerate quadratic form of Witt index ≥ 3. (Nondegenerate means $V^\perp \cap q^{-1}(0) = \{0\}$. $P\Omega(V, q)$ is the image of $\Omega(V, q)$ under the homomorphism $P : GL(V) \rightarrow PGL(V)$.)

(2) Σ is the class of images of the Siegel transvection groups of $\Omega(V, q)$ under P.

(3) $G \neq D_4(k)$.

(The case $D_4(k)$ corresponds to hypothesis (A) and $|I| = 3$, which will be treated in the next section.)

Proof. Let $\mathcal{G}$ be the polar space constructed in (5.7) (resp. (5.11)). Then, since the subspaces of $\mathcal{G}$ are projective spaces over k, it follows from the classification of polar spaces, see [Coh95, (4.3.4)], that $\mathcal{G}$ is isomorphic to the polar space of some k-vector space V, together with some non-degenerate pseudoquadratic form q or (σ, ϵ)-hermitian form f. By (5.12) and (5.13), $\mathcal{G}$ is the polar space obtained from some non-degenerate quadratic form $q : V \rightarrow k$ of Witt index ≥ 3. (The latter holds, since there exist planes in $\mathcal{G}$!) Again by (5.12) each $\alpha \in A, A \in \Sigma$ satisfies the hypothesis of II(1.5)(9). Hence $A \leq PT_\ell$ for some singular line ℓ of V and similar $D \leq PT_g$ for each $D \in \Sigma$. Hence to prove (1) and (2) it remains to show $A = PT_\ell$. ($\Omega(V, q)$ is by definition in II(2.16) the subgroup of $O(V, q)$ generated by all Siegel transvection groups.)

Let g be the line with $A_0 \leq PT_g$. Then we have by (2.26) for $\widehat{X} = \langle PT_\ell, PT_g \rangle$ and $\Omega = \Sigma \cap \widehat{X}, Y = \langle \Omega \rangle$ that the hypothesis of (1.7) is satisfied, since for each singular line in $\ell \oplus g$ there exists an element of Σ corresponding to this line. Hence $Y = \widehat{X}$, since $\widehat{X}$ is isomorphic to $(P)SL_2(k)$. But then we have by (2.26)

$$\Sigma \cap Y = \Sigma \cap \widehat{X} \subseteq C_\Sigma(V) = \Sigma \cap X,$$

whence $\widehat{X} = Y = X$ and $A = PT_\ell$, since both are full unipotent subgroups of X (resp. $\widehat{X}$). Finally $G \neq D_4(k)$, since in the latter group, as will be shown in the exercises, hypothesis (A) with $|I| = 3$ holds. $\qquad \square$

(5.15) Exercises

(1) Show that, if G satisfies hypothesis (A) and $|I| \neq 3$, then $\operatorname{Rank} \mathcal{G} \geq 4$.

    Hint. Show that there exists a subgroup N of G with $\Sigma \cap N$ a partition of N and $\dim_k N \geq 3$, which corresponds to a pencil of lines through a given point P, such that for each pair of these lines ℓ, s one has $s \subseteq \ell^\perp$.

(2) Let $G = D_4(k) = \Omega^+(8, k)$ and Σ be the class of Siegel transvection groups of G. Show that $V = V_1 * V_2 * V_3$ with $V_i \simeq SL_3(k)$. (I.e. G satisfies hypothesis (A) with $|I| = 3$!)

    Hint. Let $A, A_0 \in \Sigma$ with $\langle A, A_0 \rangle \simeq SL_2(k)$ and ℓ, s be the lines corresponding to A and A_0. Then $W = \ell + s$ is a non-degenerate 4-dimensional orthogonal space and $U = W \perp W^\perp$, U the 8-dimensional space for G. Hence there is a Σ-subgroup $\Omega^+(4, k) * \Omega^+(4, k)$ induced on this decomposition.

(3) Let V be a vector space over the field k of Char 2 and $(,)$ a non-degenerate symplectic form on V of Witt index ≥ 2. For each isotropic line ℓ of V and basis v_1, v_2 of ℓ and $c \in k$ let

$$t_c : w \to w + (w, v_2)cv_1 - (w, cv_1)v_2.$$

Show that then the following hold:

    (a) t_c is an isometry of W.

    (b) $t_c t_d = t_{c+d}$ for $c, d \in k$. Hence $T_\ell = \{t_c \mid c \in k\} \simeq k^+$.

    (c) T_ℓ is independent of the chosen basis of ℓ.

    (d) $(v, v(t_c - \mathrm{id})) = 0$ for each $v \in V$.

(4) Assume the same hypothesis as in (3). Let ℓ, s be isotropic lines of V. Show

    (a) $[T_\ell, T_s] = 1$ if $s \subseteq \ell^\perp$.

(b) $[T_\ell, T_s] = 1$ if $\ell \cap s \neq 0$.

(c) $\langle T_\ell, T_s \rangle \simeq SL_2(k)$ if $s \cap \ell^\perp = 0$.

(d) If dim $s \cap \ell^\perp = 1$, but $\ell \cap s = 0$ let $g = (s \cap \ell^\perp) \oplus (\ell \cap s^\perp)$. Then g is an isotropic line and

$$[T_\ell, T_s] = [t, T_s] = [T_\ell, \tau] = T_g \text{ for all } t \in T_\ell^\# \text{ and } \tau \in T_s^\#.$$

Remark. (4) shows that $FSp(V)$ is generated by a non-degenerate class of abstract root subgroups, where F stands for finitary. For the corresponding projective group $PSp(V)$ we know this already by exercise II(1.6)(5) and II(1.5).

(5) Let $V, (,)$ and T_ℓ be as in (3), but Char k is 0 or odd. Show that if ℓ and s are isotropic lines with $\ell \cap s = P$ a point and $s \not\subseteq \ell^\perp$, then

$$[T_\ell, T_s] = T_P,$$

where T_P is the set of all isotropic transvections corresponding to P. (See II(1.4)).

(This shows that, if Char $k \neq 2$, $\{T_\ell \mid \ell$ and isotropic line of $V\}$ is not a class of abstract root subgroups.)

§6 $D_4(k)$

We assume in this section that hypothesis (A) of §5 holds and $V = V_1 * V_2 * V_3$. We will show that under this hypothesis $G \simeq D_4(k)$, using the theory of Tits chamber systems and corresponding parabolic systems developed in [Tit82] [Tim83]. It is not possible to determine G with the same construction as in §5, since it is not clear that the set of points constructed in §5 is G-invariant. To do this would amount to showing that the triality automorphism is not inner.

(6.1) Notation

A thick connected chamber system $\mathcal{C}$ over I as defined in I§4 will be called a *Tits chamber system*, if for each $c \in \mathcal{C}$ and $J \subseteq I$ with $|J| = 2, \Delta_J(c)$ is a generalized m-gon for some $m \geq 2$. (As defined in I(4.5)). If $M = M(I) = (m_{ij})$ is a Coxeter matrix, we say that $\mathcal{C}$ is a *Tits chamber system of type M*, if for each $J = \{i, j\} \subseteq I$ and each $c \in \mathcal{C}$, $\Delta_J(c)$ is a generalized m_{ij}-gon. This notation is motivated by the so called universal covering theorem of Tits [Tit82] or [Ron89, (4.9)]. Here a surjective morphism $\sigma : \mathcal{D} \to \mathcal{C}$ of chamber systems over I is called a *2-covering*, if for each $J \subseteq I$ with $|J| = 2$ and each $d \in \mathcal{D}$ the map

$$\sigma|_{\Delta_J(d)} : \Delta_J(d) \to \Delta_J(\sigma(d))$$

is an isomorphism (of chamber systems). Such a 2-covering $\sigma : \mathcal{D} \to \mathcal{C}$ is called *universal*, if for any other 2-covering $\tau : \mathcal{E} \to \mathcal{C}$, there exists a 2 covering $\varphi : \mathcal{D} \to \mathcal{E}$ making the diagram

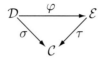

commutative.

Now the universal covering theorem of Tits (a little bit specialized for our purpose) is:

Let $\mathcal{C}$ be a Tits chamber system of type M over I, such that each rank 3 residue of $\mathcal{C}$ of type A_3 or C_3 is a building. Let $\sigma : \mathcal{D} \to \mathcal{C}$ be the universal 2-covering of $\mathcal{C}$. Then $\mathcal{D}$ is a building of type M.

This universal covering theorem will be applied in the following way in this section:

From our conditions, i.e. G satisfies hypothesis (A) with $V = V_1 * V_2 * V_3$ and $V_i \simeq SL_2(k)$ we will construct a subgroup B and subgroups P_i containing B for $i \in I = \{1,2,3,4\}$ such that $\mathcal{C} = \mathcal{C}(G, B, (P_i)_{i \in I})$ is a Tits chamber system of type D_4. We will then show that all rank 3-residues of type A_3 are projective 3-spaces over k, whence the hypothesis of the universal covering theorem is satisfied. Let

$$\sigma : \mathcal{D} \to \mathcal{C}$$

be the universal 2-covering of $\mathcal{C}$. Then $\mathcal{D}$ is a building of type D_4 all of whose rank 2-residues of type A_2 are projective planes over k, since σ restricted to the rank 2-residues is an isomorphism. Hence $\mathcal{D}$ is a building of some group $D_4(k)$ by the classification of spherical buildings [Tit74]. Now our chamber transitive group G is lifted to some chamber transitive subgroup $\widehat{G}$ of $\mathrm{Aut}(\mathcal{D})$ such that

$$\widehat{G}/N \simeq G,$$

where N is the so called group of deck transformations acting regularly on $\sigma^{-1}(c)$ for $c \in \mathcal{C}$. (See [Tit82] or the exercises of chapter 4 of [Ron89]).

Now in this situation we are able to show that $\widehat{G}$ is (essentially) the group of Lie type $\mathcal{D}$, whence $\widehat{G}$ is simple. But then $N = 1$, $\widehat{G} \simeq G$ and $\mathcal{D} \simeq \mathcal{C}$ which proves that $G \simeq D_4(k)$.

We now start with the details. For this we use the notation of §5. In addition we fix the following notation:

U_i is a full unipotent subgroup of $V_i \simeq SL_2(k), i = 1, 2, 3$.
Then, since $V_i = \langle \Delta_i \rangle, U_i \in \Sigma$.
$U = M_A U_1 U_2 U_3$.
H_i is a diagonal subgroup of V_i normalizing U_i.
$B_i = U_i H_i \qquad ; i = 1, 2, 3$ and $B_0 = B_1 B_2 B_3$.
$\widetilde{B} = M_A B_0, \qquad \widetilde{P}_i = \langle \widetilde{B}, V_i \rangle = M_A V_i B_0, \qquad i = 1, 2, 3$.

Then we have

(6.2) Lemma

The following hold:

(1) There exists a unique $C \in \Lambda_A^*$ normalized by $U_1 U_2 U_3$.

(2) $U_i \in \Lambda_C$ for $i = 1, 2, 3$.

(3) $B_0 \leq N(C)$.

(4) $\langle C^{V_i} \rangle$ is a natural module for $V_i \simeq SL_2(k)$, $i = 1, 2, 3$. Moreover $\langle C^{V_i} \rangle \cap \Sigma \subseteq \Lambda_A^*$.

Proof. It is clear that there exists a $C \in \Lambda_A^*$ centralized by $U_1 U_2 U_3$. Indeed, if $D \in \Lambda_A^* \cap \Psi_{U_i}$, then $[D, U_i] \in C_{\Lambda_A^*}(U_i)$. Now by $(2.28)(2)$ $C \in \Lambda_{U_i}$ for $i = 1, 2, 3$. Hence by II(2.19) $\langle C^{V_i} \rangle$ is a natural module for V_i. In particular (4) holds and $H_i \leq N(C)$. Hence also $B_0 \leq N(C)$.

It remains to prove the uniqueness of C. Suppose $C \neq C_1 \in \Lambda_A^*$ centralized by $U_1 U_2 U_3$. Then $C_1 = C^g$ for some $g \in V$ by (2.20). But then by II(2.3) $g \in N_V(U_1 U_2 U_3) = B_0 \leq N(C)$, a contradiction to $C_1 \neq C$. □

(6.3) Lemma

Let C be as in (6.2). Then there exists an $E \in \Lambda_A$ with $M_A = E \cdot C_{M_A}(C)$ and an $F \in \Lambda_C$ with $M_C = F C_{M_C}(A)$. Further $\langle E, F \rangle \simeq SL_2(k)$ and acts naturally on AC.

Proof. Pick $E \in \Lambda_A^* - C_\Sigma(C)$. Then, by the definition of abstract root groups, E acts transitively on $(AC \cap \Sigma) - A$. Hence for each $m \in M_A$ there exists an $e \in E$ with $C^{me} = C$. Thus

$$[C, me] \leq A \cap C = 1 \text{ and } m \in C_{M_A}(C)E.$$

This proves the first part of (6.3). The second follows from II(2.23) and (2.23). □

(6.4)   Lemma

Let $\widetilde{P_4} = \langle M_A, M_C \rangle$ and $Y = \langle E, F \rangle$. Then the following hold:

(1) $\widetilde{P_4} = NY$ with $N \triangleleft \widetilde{P_4}$ and $N \cap Y = 1$; where $N = C_{M_A}(C) C_{M_C}(A)$.

(2) N is nilpotent and $1 \triangleleft AC \triangleleft M_A \cap M_C \triangleleft N$ is a central series of N.

(3) $\overline{N} = N/M_A \cap M_C$ is a direct sum of natural $\mathbb{Z}Y$-modules. Further $\langle \overline{u}_i^Y \rangle$ is a natural Y-submodule of $\widetilde{N}$ for $i = 1, 2, 3$.

(4) $N M_A = M_A U_1 U_2 U_3 = U$.

Proof. We have $[C_{M_C}(A), E] \leq C_{M_A}(AC)$. Hence by symmetry $Y \leq N(N)$ and thus $N \triangleleft \widetilde{P_4}$ and $\widetilde{P_4} = N \cdot Y$. Since by (6.3) $C_Y(AC) = 1$, also $Y \cap N = 1$.

Since $[C_{M_A}(C), C_{M_C}(A)] \leq M_A \cap M_C$ it is obvious that (2) holds. Now by the above

$$[\overline{C_{M_A}(C)}, E] \leq \overline{A} = 1 \text{ and } [\overline{N}, E] \leq \overline{C_{M_A}(C)}.$$

Claim:

$(*)$ $\qquad\qquad C_{\overline{N}}(E) = C_{\overline{N}}(e) = \overline{C_{M_A}(C)}$ for each $e \in E^{\#}$.

Indeed, if $(*)$ is false, then

$$1 \neq \overline{C_{M_C}(A)} \cap C_{\overline{N}}(e) \leq C_{\overline{N}}(Y) \text{ for some } e \in E^{\#},$$

since $Y = \langle F, e \rangle$. Hence $\overline{C_{M_C}(A)} \cap \overline{C_{M_A}(C)} \neq 1$, since A is conjugate to C in Y, a contradiction to the definition of $\overline{N}$. Hence $(*)$ holds.

It is obvious that $(*)$ implies $C_{\overline{N}}(Y) = 1$. Hence Theorem I(3.4) implies that $\langle \overline{n}^Y \rangle$ is a natural Y-module for each $1 \neq \overline{n} \in \overline{C_{M_A}(C)}$, which shows, since $\overline{N} = \overline{C_{M_A}(C)} \oplus \overline{C_{M_C}(A)}$, that $\overline{N}$ is a direct sum of natural $\mathbb{Z}Y$-modules.

Now by (6.2) $\overline{U}_i \leq \overline{C_{M_C}(A)}$. Hence by the above $\langle \overline{u}_i^Y \rangle$ is a natural Y-module for each $1 \neq \overline{u}_i \in \overline{U}_i$. On the other hand, if $u_i \in U_i$ is a coimage of $\overline{u}_i$, then $[u_i, E] = [U_i, E]$ by the definition of abstract root groups. Hence also $[\overline{u}_i, E] = [\overline{U}_i, E]$ and thus $\langle \overline{u}_i^Y \rangle = \langle \overline{U}_i^Y \rangle$. This proves (3).

To prove (4) let $L = M_A C_{M_C}(A)$. Then $L = N \cdot M_A$ and by (2.16) and the Frattini argument $L = M_A C_L(A_0)$. Now $[U_i, C_{M_C}(A)] \leq C \leq M_A$ by (6.2) and thus

$$[C_L(A_0), U_i] \leq C_{M_A}(A_0) = 1 \text{ for } i = 1, 2, 3.$$

Since $C_L(A_0) \leq N_L(\langle C_\Sigma(X) \rangle) = N_L(V)$ and since $C_L(A_0) \leq C_L(U_1 U_2 U_3)$ by the above, the structure of the automorphism group of $SL_2(k)$, see I(5.8), implies that $C_L(A_0)$ induces inner automorphisms corresponding to elements of U_i on each V_i. Hence

$$(+) \qquad\qquad C_L(A_0) \le U_1 U_2 U_3 \times C_L(V).$$

Now $C_L(A_0)$ normalizes $AC \cap \Lambda_A^* = C$ and thus centralizes C, since $[C, L] \le A$. Since $\Lambda_A^* = C^V$ by (2.20), we obtain $C_L(A_0) \cap C_L(V) \le C(M_A)$. But $G = \langle M_A, A_0 \rangle$ by II(4.8). Hence $C_L(A_0) \cap C_L(V) \le Z(G) = 1$, which together with $(+)$ shows that $C_L(A_0) = U_1 U_2 U_3$. Thus $L = U$ and (4) holds. □

(6.5) Lemma

Let $H_4 = N_Y(E) \cap N_Y(F)$. Then the following hold:

(1) Without loss of generality $H_4 \le N(A_0)$. (I.e. we may choose E and F accordingly.)

(2) $[U_i, h] = U_i$ and $C_{U_i}(h) = 1$ for each $h \in H_4^{\#}$ and $i = 1, 2, 3$.

(3) Without loss of generality $[H_i, H_4] = 1$ for $i = 1, 2, 3$. (I.e. we may choose H_i in B_i accordingly.)

Proof. By (6.3) $H_4 \le N(A)$. We claim that there exists a $D \in \Omega_A \cap \Lambda_F \cap \Psi_E$. Indeed by II(2.23) there exists a $\overline{D} \in \Sigma$ with $[\overline{D}, C] = F$. Since $[A, F] = C$ we have $\overline{D} \in \Omega_A \cap \Lambda_F$. By II(2.3) $\overline{D} \notin \Sigma_E$. Suppose $\overline{D} \in \Omega_E$. Then there exists by II(2.11) a $D \in \overline{D}F \cap \Psi_E$. Clearly $D \ne F$. Since we have

$$\underset{\overline{D}}{\circ}\!\!-\!\!-\!\!-\!\!\underset{A}{\circ}\!\!=\!\!=\!\!=\!\!\underset{F}{\circ}\ ,$$

II(2.11) then implies $D \in \Omega_A$, which proves our claim.

Now we have

$$\underset{F}{\circ}\!\!-\!\!-\!\!-\!\!\underset{E}{\circ}\!\!=\!\!=\!\!=\!\!\underset{D}{\circ}\ ,$$

so that II(2.19) implies $H_4 \le N(D)$. But then there exists by (2.16) an $m \in M_A$ with $H_4^m \le N(A_0)$. Substituting H_4 by H_4^m (and E by E^m, F by F^m) (1) holds.

Now by (6.4)(3) $H_4 \le N(\overline{U}_i)$. Hence

$$[U_i, H_4] \le M_A U_i \cap C(A_0) \le U_i, \;\; i = 1, 2 \text{ and } 3.$$

Since $[\overline{U}_i, h] = \overline{U}_i$ and $C_{\overline{U}_i}(h) = 1$ for each $h \in H_4^{\#}$ this implies (2).

Now (1) and (2) imply $H_4 \le N(V_i)$ for $i = 1, 2, 3$ and no element of H_4 induces a field automorphism on V_i. Hence by the structure of the automorphism group of $SL_2(k)$, see I(5.8), we obtain $H_4 \leq U_i \widehat{H}_i C(V_i)$, where $\widehat{H}_i$ is the full group of diagonal automorphisms of V_i. Hence $H_4 \leq \widehat{H}_i^{u_i} C(V_i)$ for some $u_i \in U_i$ (see exercise $(6.18)(1)$). This shows $H_4 \leq C(H_i^{u_i})$ and, replacing H_i by $H_i^{u_i}$) (3) holds. $\qquad\square$

(6.6) Notation

Set $H = H_1 H_2 H_3 H_4$, $B = \widetilde{B} H_4$, $P_i = \widetilde{P}_i H_4$ for $i = 1, 2, 3$ and $P_4 = \widetilde{P}_4 H_1 H_2 H_3$.

Then we obtain:

(6.7) Lemma

The following hold:

(1) H is abelian.

(2) B is a maximal subgroup of each P_i. Further $P_i \cap P_j = B$ for $i \neq j \leq 4$.

(3) Let $Q_i = B_{P_i} = \cap B^g, g \in P_i$. Then $P_i / Q_i \simeq PSL_2(k)$ or $PGL_2(k)$ for each $i \leq 4$.

Proof. (1) follows from $(6.5)(3)$. For $i \leq 3$ it is obvious that B is maximal in P_i, since $\widetilde{B}$ is maximal in $\widetilde{P}_i$. Now $B_4 = N M_A H_4$ is maximal in $\widetilde{P}_4$ and each $H_i, i \leq 3$ normalizes B_4 and $\widetilde{P}_4 / N$. Since $B = B_4 H_1 H_2 H_3$ this implies that B is maximal in P_4. Now (2) follows from the fact that $B \leq P_i \cap P_j$ and $P_i \neq P_j$ for $i \neq j \leq 4$.

For $i \leq 3$ let $\widetilde{Q}_i = \widetilde{B}_{\widetilde{P}_i}$. Then $\widetilde{P}_i / \widetilde{Q}_i \simeq PSL_2(k)$ by definition of $\widetilde{P}_i$ for $i \leq 3$. Now by $(6.5)(2)$ and (3) H_4 induces diagonal automorphisms on $\widetilde{P}_i / \widetilde{Q}_i = \overline{P}_i$. Since $P_i / Q_i \simeq (\widetilde{P}_i H_4 / \widetilde{Q}_i H_4) / C_H(\widetilde{P}_i / \widetilde{Q}_i)$ this shows that (3) holds for $i \leq 3$.

Now $\widetilde{P}_4 / N \simeq Y \simeq SL_2(k)$. Since $[H_4, H_1 H_2 H_3] = 1$, $H_1 H_2 H_3$ also induces diagonal automorphisms on $\widetilde{P}_4 / N$. Hence it follows as before that (3) also holds for P_4 / Q_4. $\qquad\square$

(6.8) Lemma

Set $L_i = \langle P_i, P_4 \rangle$ for $i = 1, 2, 3$ and $W_i = \langle A^{L_i} \rangle$, $\overline{L}_i = L_i / C_{L_i}(W_i)$. Then the following hold:

(1) $SL_3(k) \leq \overline{L}_i \leq GL_3(k)$.

(2) W_i is the natural $k\overline{L}_i$-module and $\Sigma \cap W_i$ is the set of 1-subspaces of W_i.

Proof. Set $W_i' = \langle C^{V_i} \rangle$ and claim

(∗) $W_i = W_i' A.$

To prove (∗) let $\{i, j, \ell\} = \{1, 2, 3\}$. Then

$$[W_i', U_s] = 1 \text{ and } B_s \leq N(W_i') \text{ for } s \in \{j, \ell\}.$$

Since H_4 normalizes C and V_i we have $H_4 \leq N(W_i')$. Hence $H \leq N(W_i')$. Now $[W_i', M_A] \leq A$, as $W_i' \leq M_A$. As $P_i = M_A V_i U_j U_\ell H$ this implies $P_i \leq N(W_i' A)$. Now $W_i' A \leq M_A \cap M_C$. Thus $\widetilde{P}_4 \leq N(W_i' A)$ by definition of $\widetilde{P}_4$ and (2.6). Since $P_4 = \widetilde{P}_4 H$ this shows that $P_4 \leq N(W_i' A)$ and thus $L_i \leq N(W_i' A)$. Hence by definition of W_i, (∗) holds.

Now (∗) immediately implies that $\Sigma \cap W_i$ is a partition of W_i. Hence by (2.1) $\langle N_\Sigma(W_i) \rangle$ induces the $SL_3(k)$ on W_i and W_i is a 3-dimensional k-vector space with $\Sigma \cap W_i$ the set of 1-subspaces of W_i. Now (6.2)(4) and (6.3) imply that $M_D \leq N(W_i)$ for each $D \in \Sigma \cap W_i$. Hence $\overline{L}_i$ contains all transvections corresponding to each point of $\Sigma \cap W_i$. This implies $SL_3(k) \leq \overline{L}_i \leq \Gamma L_3(k)$, since $\overline{L}_i$ respects the partition $\Sigma \cap W_i$ of W_i and thus acts as a group of semilinear mappings on W_i. Now by (6.3)–(6.5) it is easy to see that H acts as a group of diagonal automorphisms on W_i. Since $L_i = \langle M_A, M_C, V_i \rangle H$ and since $\langle M_A, M_C, V_i \rangle \subseteq SL_3(k)$ since it is generated by transvections, we obtain $\overline{L}_i \leq GL_3(k)$.   □

(6.9) Lemma

Let $C_i \in \Sigma \cap W_i'$ with $W_i' = CC_i$ and $H \leq N(C_i)$ for $i = 1, 2, 3$ (such a C_i exists by (6.2)(4)!) Set

$$\begin{aligned} \widetilde{L}_i &= \langle M_A, M_C, M_{C_i} \rangle, \\ N_i &= C_{M_A}(W_i) C_{M_C}(W_i) C_{M_{C_i}}(W_i). \end{aligned}$$

Then the following hold:

(1) $N_i = C_{\widetilde{L}_i}(W_i)$ and $\widetilde{L}_i / N_i \simeq SL_3(k)$.

(2) $L_i = \widetilde{L}_i H$.

Proof. We have $[M_A, C_{M_C}(W_i)] \leq M_A \cap C(W_i) = C_{M_A}(W_i) \leq N_i$. This shows $[M_A, N_i] \leq N_i$ and with symmetry $N_i \lhd \widetilde{L}_i$. Since $SL_3(k)$ is generated by the sets of all transvections corresponding to the three points which generate W_i, (6.8)(1) implies $\widetilde{L}_i / C_{\widetilde{L}_i}(W_i) \simeq SL_3(k)$. Thus, to prove (1), it remains to show $N_i = C_{\widetilde{L}_i}(W_i)$. To simplify notation we do this for $i = 1$.

Let $F_1 = C_{\widetilde{L}_1}(W_1)$ and $L_1^* = \widetilde{L}_1/N_1$. We have $C_\Sigma(W_1) \subseteq C_{C_\Sigma(A)}(W_1) \subseteq M_1U_2U_3$. This implies $C_\Sigma(W_1) \subseteq C_{M_A}(W_1)C_{M_C}(W_1) \le N_1$. Hence II(2.21)(3) implies that L_1^* is generated by a non-degenerate class of k-root subgroups, namely the images of $\Sigma \cap \widetilde{L}_1$ under the $(*)$ homomorphism. This shows that L_1^* is perfect. In particular $F_1^* \le (L_1^*)'$. Moreover, since $[F_1, M_A] \le C_{M_A}(W_1) \le N_1$, we obtain $[F_1, \widetilde{L}_1] \le N_1$ and thus $F_1^* \le Z(L_1^*)$. Hence L_1^* is a perfect central extension of $SL_3(k)$ generated by k-root subgroups.

Suppose $F_1 > N_1$ and pick $x \in F_1 - N_1$. Then $xm \in N(A_0)$ for some $m \in M_A$ by (2.15). Hence $xm \in C(A_0)$, since $xm \in C(A)$ and $\langle A, A_0 \rangle$ is a rank one group. Now xm normalizes AC and AC_1 and thus normalizes C and C_1. Since $[W_1, M_A] \le A$ this implies $xm \in C(W_1)$ and thus $m \in C_{M_A}(W_1) \le N_1$. We obtain

$(*)$ $$F_1 = N_1 C_{F_1}(A_0).$$

Now $C_{F_1}(A_0)$ normalizes V and thus V_1, since $V \cap N(W_1) = V_1 B_2 B_3$. Hence $[V_1, C_{F_1}(A_0)] \le C_{V_1}(W_1) = 1$. Suppose now $1 \ne x \in C_{F_1}(A_0) - N_1$ normalizes V_2. Then x normalizes all $V_i, i = 1, 2$ and 3 and, since x centralizes C, $V_i \langle x \rangle$ induces an $SL_2(k)$ on W_i'. Hence x induces an inner automorphism according to some element of U_i on V_i. This shows that there exists a $y \in U_2U_3$ such that xy centralizes V. Since $xy \in C(C)$, (2.20) implies that $xy \in C(M_A)$, since $M_A = \langle \Lambda_A^* \rangle$. But, as $G = \langle M_A, A_0 \rangle$ by II(4.8), this implies $xy \in Z(G) = 1$. This is a contradiction to $y \in C_{M_C}(W_1) \le N_1$.

Together with $(*)$ this shows that $|F_1 : N_1| = 2$ and that each element x of $C_{F_1}(A_0) - N_1$ interchanges V_2 and V_3. Hence $B_2^x = B_3$ and we may assume $H_2^x = H_3$. Hence, if $H_2 \ne H_3$, we obtain:

$$1 \ne [H_2, x] \le H_2H_3 \cap [F_1, H_2] \le H_2H_3 \cap N_1 = 1,$$

a contradiction. (We have $H_2H_3 \cap N_1 = 1$, since each element of $(H_2H_3)^\#$ acts non-trivially on C_2 or C_3. On the other hand $[W_2W_3, N_1] \le M_A \cap M_C \cap M_{C_1}$ and $C_k \cap M_A \cap M_C \cap M_{C_1} = 1$ for $k = 2$ or 3!) This implies $H_2 = H_3$ and thus $k = GF(2)$ or $GF(3)$. Hence L_1^* is a proper covering group of $SL_3(2)$ or $SL_3(3)$. This implies $k = GF(2)$ and $L_1^* \simeq SL_2(7)$, which is impossible since $SL_2(7)$ is not generated by $GF(2)$-root subgroups.

This final contradiction shows $F_1 = N_1$ which proves (1).

Since $V_i \le \langle M_C, M_{C_i} \rangle \le \widetilde{L}_i$ we have $P_i \le \widetilde{L}_iH$. Also $P_4 \le \langle M_A, M_C \rangle H \le \widetilde{L}_iH$. Hence $L_i \le \widetilde{L}_iH$, which implies (2). $\qquad\square$

(6.10) Lemma

The following hold for $i \le 3$.

(1) $N_i \leq U$ and $C_{L_i}(W_i) \leq B$.

(2) B is a "Borel-subgroup" of L_i and P_i, P_4 are maximal "parabolic subgroups" of L_i containing B.

(I.e. $C_{L_i}(W_i) \leq B$ and $\overline{B}$ is a Borel subgroup of $\overline{L}_i$ and $\overline{P}_i, \overline{P}_4$ are parabolic subgroups of $\overline{L}_i$, where $\overline{L}_i = L_i/C_{L_i}(W_i)$!)

Proof. Assume without loss of generality $i = 1$. Then we have by (6.4) (4)

$$C_{M_C}(W_1) \leq C_U(W_1) \leq M_A U_2 U_3.$$

Now C and C_1 are conjugate in V_1 and thus also $C_{M_{C_1}}(W_1) \leq M_A U_2 U_3$. By definition of N_1 and U this proves the first part of (1).

Let $\widehat{L}_i = L_i/N_i$ and $\widehat{x} \in C_{\widehat{L}_i}(W_1)$. Then by (6.9)(2) we have $\widehat{x} = \widehat{y}\widehat{h}$ with $\widehat{y} \in \widetilde{\widehat{L}_1}$ and $\widehat{h} \in \widehat{H}$. Since $\widehat{h}$ fixes A, C and C_1 this shows that $\widehat{y}$ fixes A, C and C_1. Thus $\widehat{y}$ is a diagonal-element of $\widetilde{L}_1 \simeq SL_3(k)$. Hence $\widehat{y} \in \widehat{H}_1\widehat{H}_4 \leq \widehat{H}$ and thus $\widehat{x} \in C_{\widehat{H}}(W_1)$. This shows that

$$C_{L_1}(W_1) \leq N_1 H \leq B$$

which proves (1).

Now it is clear that U/N_1 is a full unipotent subgroup of $\widetilde{L}_1/N_1 \simeq SL_3(k)$, since $M_A N_1/N_1$ is the set of all transvections corresponding to the point A on W_1 and $U_1 N_1/N_1$ is the transvection group corresponding to the point C and hyperplane AC on W_1. This shows that UH_1H_4/N_1 is a Borel subgroup of $\widetilde{L}_1/N_1$. Now (6.9) and the part (1) show that (2) holds. □

(6.11) Lemma

The following hold:

(1) For $i \neq j \leq 3$ we have $P_{ij} = \langle P_i, P_j \rangle = P_i P_j$ and $B = P_i \cap P_j$.

(2) $G = \langle P_i \mid i = 1, \ldots, 4 \rangle$

Proof. (1) follows from $P_{ij} = M_A(V_i * V_j)B$ and (6.7)(2). To prove (2) let $R = \langle P_i \mid i = 1, \ldots, 4 \rangle$ and $\Delta = A^R$. Then, since $V \leq R$ and since P_4 is transitive on $AC \cap \Sigma$, (2.20) implies $\Lambda_A \cup A \subseteq \Delta$. Hence $\Lambda_D \cup D \subseteq \Delta$ for each $D \in \Delta$. Now II(4.7) implies $\Delta = \Sigma$ and $R = G$. □

(6.12)   Notation

Let $I = \{1, 2, 3, 4\}$ and $\mathcal{C} = \mathcal{C}(G, B, (P_i)_{i \in I})$ be the chamber system as defined in I § 4 Then by I(4.1) $\mathcal{C}$ is connected. Moreover obviously by definition of the $P_i, i \in I$ it follows that $\mathcal{C}$ is thick. We claim that $\mathcal{C}$ is a classical Tits chamber system of type D_4:

as defined in (6.1); where classical just means that the residues $\Delta_{i,4}(c)$, $c = B$, are desarguesian projective planes (i.e. chamber systems of such planes).

Now for $i \neq j \leq 3$ we know by (6.11)(1) and exercise (6.18)(2) that $\Delta_{ij}(c)$ is a generalized digon. Moreover by (6.10), (6.8) and exercise (6.18)(3) and (4) $\Delta_{i,4}(c)$ is the chamber system of the projective plane over k. Hence by definition $\mathcal{C}$ is a classical Tits chamber system of type D_4. Thus, to show that $\mathcal{C}$ satisfies the hypothesis of the universal covering theorem of Tits, it remains to show that the residues $\Delta_J(c)$ for $J = \{i, j, 4\}$ and $i \neq j \leq 3$ are buildings of type A_3, that is a chamber system of projective 3-spaces over k. Unfortunately it is not true in general that Tits chamber systems of type A_3 are buildings of type A_3. For this we state a lemma, the proof of which we do not give, since it needs the transition from chamber system to geometry via the "fundamental condition" of [MT83].

(6.13)   Lemma

Suppose $\mathcal{D}$ is a Tits chamber system of type A_3 over $J = \{1, 2, 3\}$. Suppose for all $L, K \subseteq J$ and each chamber $d \in \mathcal{D}$ we have:

$$\Delta_L(d) \cap \Delta_K(d) = \Delta_{L \cap K}(d) \text{ (fund. condition)} .$$

Then $\mathcal{D}$ is a building of type A_3.

The proof of (6.13) is contained in [MT83], [Pas94, (12.19)]. It proceeds roughly as follows: The fundamental condition shows that the geometry $\Gamma(\mathcal{D})$ obtained in a natural way from $\mathcal{D}$ is a Tits geometry of type A_3. Now it is easy to show that such a Tits geometry is actually a projective 3-space and hence a building of type A_3. To give details of the easy proof, one would need to give an introduction to the theory of Tits geometries and Tits chamber systems, which is not the content of this book.   □

From (6.13) we obtain as a Corollary:

(6.14) Corollary

For all $J = \{i, j, 4\}$ and $i \neq j \leq 3$, $\Delta_J(c)$ is a building of type A_3 over k; i.e. the chamber system of a projective 3-space over k.

Proof. We show that for all $L, K \subseteq J$ we have

$$(*) \qquad\qquad P_L \cap P_K = P_{L \cap K}$$

where $P_L = \langle P_\ell \mid \ell \in L \rangle$ and $P_\emptyset = B$. If $L = J$ or $K = J$ this is obvious. So we may assume that $L \neq K$ and both are properly contained in J. Now $(*)$ is a consequence of $(6.7)(2)$ and $(6.10)(2)$, since $P_i \not\leq P_{j,4}$. Since $\Delta_L(c) = \{Bg \mid g \in P_L\}$ and $\Delta_K(c) = \{Bh \mid h \in P_K\}$ $(*)$ immediately implies

$$\Delta_L(c) \cap \Delta_K(c) = \Delta_{L \cap K}(c).$$

But then this "fundamental condition" holds for all $d \in \mathcal{C}$, since G is transitive on $\mathcal{C}$.

(6.15) Lemma

Let $J = \{i, j, 4\}$ and $i \neq j \leq 3$, $P_J = \langle P_j \mid j \in J \rangle$ and Q_J the kernel of the action of P_J on $\Delta_J(c)$. (I.e. $Q_J = \cap B^g$, $g \in P_J$). Then $PSL_4(k) \leq P_J/Q_J \leq PGL_4(k)$.

Proof. By (6.14) $\Delta_J(c)$ is the chamber system of a projective 3-space over k and P_J/Q_J is a chamber transitive automorphism group of $\Delta_J(c)$. Hence $P_J/Q_J \leq P\Gamma L_4(k)$. Now for $\ell = i$ or j we have by (6.8) and (6.10) $Q_{\ell,4} = C_{L_\ell}(W_\ell)$ and $PSL_3(k) \leq P_{\ell,4}/Q_{\ell,4} \leq PGL_3(k)$.
Claim:

$$(*) \qquad\qquad U \cap Q_{\ell,4} \not\leq Q_J.$$

We have $U \cap Q_{\ell,4} = U \cap C_{L_\ell}(W_1) = N_\ell$ by (6.9), (6.10). Since $Q_J \subseteq Q_{j,4}$, $U \cap Q_{i,4} \subseteq Q_J$ would imply $N_i = U \cap Q_{i,4} \subseteq U \cap Q_{j,4} = N_j$, which is obviously not the case. This proves $(*)$.

Now by the structure of $PSL_4(k)$ a Levi complement of a parabolic subgroup, which is the stabilizer of a point or hyperplane of the projective 3-space acts irreducibly on the unipotent radical. (The unipotent radical is just the natural module for the Levi complement which is isomorphic to $SL_3(k)$!) Hence by $(*)$ we obtain $k^3 SL_3(k) \leq P_{\ell,4}/Q_J$ for $\ell = i$ and j. But then, since $PSL_4(k)$ is generated by two different maximal parabolic subgroups we obtain:

$$P_J/Q_J = \langle P_{i,4}/Q_J, P_{j,4}/Q_J \rangle \geq PSL_4(k).$$

Similarly, as $P_{\ell,4}/Q_{\ell,4} \leq PGL_3(k)$ for $\ell = i$ or j, $P_J/Q_J \leq PGL_4(k)$. $\qquad\square$

(6.16)   Theorem

The following hold:

(1) $\mathcal{C}$ is a building of type D_4 (over k).

(2) $G \simeq D_4(k)$.

Proof. Let $\pi : \widetilde{\mathcal{C}} \to \mathcal{C}$ be the universal 2-covering of $\mathcal{C}$ as defined in (6.1). Then $\widetilde{\mathcal{C}}$ is by (6.14) and the universal covering theorem a building of type D_4. Let $\sigma : \widetilde{G} \to G$ be the corresponding covering of G and $N = \ker \sigma$. We will show that π and σ are isomorphisms.

Now for $J = \{i, j, 4\}$ with $1 \le i \ne j \le 3$, $\Delta_J(c)$ is by (6.14) a building of type A_3 and whence by [Ron89, (4.3)] simply connected. Thus, if $\widetilde{c}$ is a coimage of c the map

$$\pi : \Delta_J(\widetilde{c}) \to \Delta_J(c)$$

is an isomorphism. Hence we have $|\pi^{-1}(d) \cap \Delta_J(\widetilde{c})| = 1$ for each $d \in \Delta_J(c)$. Since N acts regularly on $\pi^{-1}(d)$ it follows that $N \cap \widetilde{P}_J = 1$, where P_J is the stabilizer of $\Delta_J(\widetilde{c})$ in $\widetilde{G}$. Hence $\sigma : \widetilde{P}_J \to P_J$ is an isomorphism.

Suppose now $\widetilde{U} \cap \widetilde{Q}_J = 1$, where $\widetilde{U} = \sigma^{-1}(U)$ and $\widetilde{Q}_J = \sigma^{-1}(Q_J)$. Then $U \cap Q_J = 1$ and U can by (6.15) be identified with a full unipotent subgroup of $PSL_4(k)$. Hence U is of nilpotency class 3, which is obviously not the case. Thus $\widetilde{U} \cap \widetilde{Q}_J \ne 1$. Now, since $\widetilde{\mathcal{C}}$ is a building of type D_4, $\widetilde{G}$ can be identified with a subgroup of $D_4(k)$. Now by the structure of $D_4(k)$ a Levi complement isomorphic to $D_3(k) \simeq PSL_4(k)$ of the stabilizer of $\Delta_J(\widetilde{c})$ acts irreducibly on the unipotent radical. (which is just a natural module for the Levi complement considered as $D_3(k) \simeq \Omega_6^+(k)$!) Since by (6.15) $\widetilde{P}_J$ covers the Levi complement, it follows from $\widetilde{U} \cap \widetilde{Q}_J \ne 1$ that $\widetilde{Q}_J$ contains the unipotent radical. This implies that $\widetilde{P}_J$ contains R'_J, where R_J is the stabilizer of $\Delta_J(\widetilde{c})$ in $D_4(k)$. But since this holds for all $i \ne j \le 3$ we obtain

$$D_4(k) \le \widetilde{G} \le D_4(k)\mathrm{Aut}(k) = \mathrm{Aut}(\widetilde{\mathcal{C}}).$$

But on the other hand a field automorphism of $\widetilde{\mathcal{C}}$ must induce a field automorphism on $\Delta_J(\widetilde{c})$. Since $\widetilde{P}_J/\widetilde{Q}_J \simeq P_J/Q_J \subseteq PGL_4(k)$ by (6.16), $\widetilde{G}$ does not induce field automorphisms on $\widetilde{\mathcal{C}}$. Hence $\widetilde{G} = D_4(k)$. In particular $\widetilde{G}$ is simple and thus $N = 1$. Hence σ and thus π are isomorphisms and $G \simeq \widetilde{G} \simeq D_4(k)$ and $\mathcal{C} \simeq \widetilde{\mathcal{C}}$ is a building of type D_4.   □

We end this section by stating:

(6.17) Theorem

Suppose that G satisfies hypothesis (A) or (B) of §5. Then the following hold:

(1) $G \simeq P\Omega(V, q)$, V a vectorspace over the field k and $q : V \to k$ a non-degenerate quadratic form of Witt index ≥ 3.

(2) Σ is the class of images of the "Siegel transvection groups" of $\Omega(V, q)$.

Proof. (1) is a consequence of (5.13) and (6.16) and the fact that $D_4(k) = P\Omega^+(8, k)$. Also by (5.13) it remains to show that Σ is the class of images of the "Siegel transvection groups" in case $G \simeq D_4(k)$. Now by the proof of (6.16) U is a full unipotent subgroup of $D_4(k)$ and, by definition of U, $A = Z(U)$. But it is clear from the definition in II(1.5) that a "Siegel transvection group" is also central in a unipotent subgroup of $\Omega^+(8, k)$, since it is normalized by the stabilizer of the corresponding singular line. Since H acts transitively on $A^\#$, this shows that A is a Siegel transvection group. □

(6.18) Exercises

(1) Let $Y = PGL_2(k)$, k a (commutative) field and consider P as a homomorphism $P : GL_2(k) \to PGL_2(k)$. Let U be the image of $\left\{ \begin{pmatrix} 1 & \\ a & 1 \end{pmatrix} \mid a \in k \right\}$, H the image of $\left\{ \begin{pmatrix} c_1 & \\ & c_2 \end{pmatrix} \mid c_i \in k^* \right\}$ and $B = UH$. Show: If $K \leq B$ with $K \cap U = 1$, then there exists a $u \in U$ with $K \leq H^u$.

   Hint. $B - U = \bigcup_{u \in U} H^u$. So there exists a $u \in U$ with $K \cap H^u \neq 1$. Now K and H are abelian, whence for $1 \neq x \in K \cap H^u$ we have $\langle K, H^u \rangle \subseteq C(x)$. But $C(y) \subseteq H$ for each $1 \neq y \in H$.

(2) Let $G = P_1 P_2 = P_2 P_1$ be a group with subgroups P_1, P_2 and $B = P_1 \cap P_2$. Show that $\mathcal{C}(G, B, P_1, P_2)$ is (the chamber system of) a generalized digon in the sense of I(4.5).

(3) Let k be a field, $G = GL_3(k)$, B the set of upper triangular matrices of G and

$$P_1 = \left\{ \left(\begin{array}{c|cc} c & * & * \\ \hline 0 & & A \\ 0 & & \end{array} \right) \mid c \in k^*, A \in GL_2(k) \right\}$$

$$P_2 = \left\{ \left(\begin{array}{cc|c} & A & * \\ & & * \\ \hline 0 & 0 & c \end{array} \right) \mid c \in k^*, A \in GL_2(k) \right\}$$

be the two parabolic subgroups of G containing B. Let $\mathcal{P} = \{ P_1 g \mid g \in G \}$, $\mathcal{L} = \{ P_2 g \mid g \in G \}$ and Γ be the point-line geometry with point set

$\mathcal{P}$ and line set $\mathcal{L}$ and incidence $*$ defined by:

$$P_1 g * P_2 h \Leftrightarrow P_1 g \cap P_2 h \neq \emptyset.$$

(a) Show that Γ is isomorphic to the projective plane over k.

(b) Let R be a subgroup of G with $SL_3(k) \le R$, $\overline{B} = B \cap R$, $\overline{P}_i = P_i \cap R$ for $i = 1, 2$ and define $\overline{\Gamma}$ in the same way as Γ. Show $\overline{\Gamma} \simeq \Gamma$.

(4) Assume the same hypothesis as in (3). Show $\mathcal{C} = \mathcal{C}(G, B, P_1, P_2) \simeq \mathcal{C}(R, \overline{B}, \overline{P}_1, \overline{P}_2)$ is isomorphic to the chamber system of the projective plane over k. (If one considers the projective plane as a generalized 3-gon in the sense of I(4.5).)

Hint. As $B = P_1 \cap P_2, \mathcal{C} \simeq \mathcal{C}(\Gamma)$ where $\mathcal{C}(\Gamma)$ is the flag complex of the incidence geometry Γ as defined in I(4.5).

§7 Metasymplectic spaces

We assume in this section that Σ is a non-degenerate class of abstract root subgroups of G and that $R(G) = Z(G) = 1$. Fix $A \in \Sigma$, $A_0 \in \Omega_A$, $X = \langle A, A_0 \rangle$, and $\Delta_0 = \Sigma_A \cap \Sigma_{A_0}$, $V = \langle \Delta_0 \rangle$. Further assume that $\langle C_\Sigma(A) \rangle$ is Σ-maximal and Δ_0 is a conjugacy class in V. We will often use the results of §2 without reference. In particular we have under the hypothesis of §7:

Σ is a class of k-root subgroups, k a fixed commutative field.

M_A acts regularly on Ω_A.

V acts transitively on $\Lambda_A^* = \Lambda_A \cap \Psi_{A_0}$.

$R(V) \le Z(V)$.

Most of the results of this section are devoted to the case Δ_0 is degenerate, since this is the case corresponding to the groups of type F_4. In that case we cannot use directly the results of §1, since if $|k| \le 3$ hypothesis (H) of §1 is not satisfied. Therefore we will show in this case directly that $\mathcal{P}(\Delta_0)$ (as defined in §1) is a thick, non-degenerate polar space of rank three. Using this fact, we then can show that the points and lines of Σ together with certain symplecta form a thick metasymplectic space. We then use the classification of such metasymplectic spaces of [Tit74, §10] to obtain a classification of G.

For $B \in \Lambda_A^*$ let $\Delta_B = \Delta_0 \cap \Lambda_B$. We first prove some results on Δ_B without the assumption that Δ_0 is degenerate.

(7.1) Lemma

The following hold:

(1) $\langle \Delta_B \rangle$ is abelian.

(2) If $\Delta_B \subseteq \Delta_C$ for $B, C \in \Lambda_A^*$, then $B = C$.

(3) $N_V(\Delta_B) = N_V(B)$ and $N_{\Delta_0}(\Delta_B) = C_{\Delta_0}(B)$.

Proof. (1) is obvious, since

$$\langle \Delta_B \rangle' \leq B \cap V \leq M_A \cap V = 1.$$

Let $D \in (\Sigma_A \cap \Lambda_B) - \Lambda_A$. Then $\overline{D} = M_A D \cap V \in \Delta_0$ by (2.18) and $\overline{D} = D^m$ for some $m \in M_A$. Hence $\overline{D} \in \Lambda_{B^m}$ and thus $B^m = B$ by II(2.3). This shows $\Delta_B \neq \emptyset$ for each $B \in \Lambda_A^*$. Suppose now $\Delta_B \subseteq \Delta_C$, but $B \neq C$. Then $[B, C] = 1$. (If not, then $A = [B, C] \in \Delta_B$.) If now $B \in \Lambda_C$, then there exists by (2.1) and (2.18) an $E \in \Delta_B$ with $[E, C] = B$, a contradiction to $\Delta_B \subseteq \Delta_C$. Thus $B \notin \Lambda_C$. Let now N be a subgroup of G, satisfying:

(α) $AB \leq N$ and $N' = 1$.

(β) $\Sigma \cap N$ is a partition of N.

(γ) N is maximal with (α) and (β).

(Such an N exists by Zorn's lemma!) By (2.31)(1) N is a vector space over k and $\Sigma \cap N$ is the set of all 1-subspaces. Moreover, since $\Sigma_A \not\subseteq M_A$ by the hypothesis of this section and (2.8), $\dim_k N \geq 3$. If now $D \in (\Sigma \cap N) - AB$ and $D \in \Sigma_C$, then there exists as before an $E \in \Delta_B$ with $[D, E] = B$. But as $E \in \Lambda_C$, $B = [D, E] \in \Lambda_C$, a contradiction to the above.

This shows $(\Sigma \cap N) - AB \subseteq \Psi_C$. Since $[N, C] \leq A$ this in particular implies $\dim_k N = 3$. Let $D \in \Sigma \cap N$ with $N = ABD$. Then, since $\Sigma \cap N$ is by the hypothesis of this section no weak TI-subset, there exists an $F \in \Sigma_B$ and $f \in F$ with $D^f \not\leq N$. The maximality of N implies $D^f \notin \Lambda_A$. On the other hand $D^f \in \Sigma_A$, since $D^f \in \Lambda_B \cap \Lambda_D$. Hence by (2.18) and (2.29) $D^{fm} \in \Delta_0$ for some $m \in M_A$. As above $D^{fm} \in \Delta_B \cap \Delta_D \subseteq \Delta_C$. Hence $[D, C] \leq A \cap D^{fm} = 1$, a contradiction to $D \in \Psi_C$.

This contradiction proves (2). (3) is an immediate consequence of (2). □

(7.2) Lemma

The following hold for $B \in \Lambda_A^*$:

(1) $C_{\Delta_0}(\Delta_B) = \Delta_B$.

(2) $\langle \Delta_B \rangle \cap \Delta_0 = \Delta_B$.

(3) If $E \in \Delta_B$, then $\Lambda_E \cap \Delta_0 \subseteq C_{\Delta_0}(B)$.

Proof. Notice that (2) is by (7.1)(1) a consequence of (1). To prove (3) pick $F \in \Lambda_E \cap \Delta_0$. Then, as $B \in \Lambda_E$ by definition of Δ_B, we obtain

$$[B, F] \leq E \cap M_A = 1.$$

So (1) remains to be proved. Suppose $\Delta_B \subset C_{\Delta_0}(\Delta_B)$. Then, since $\mathcal{F}(\Delta_0)$ is by the assumption of this section connected, there exists an $E \in C_{\Delta_0}(\Delta_B) - \Delta_B$ and an $F \in \Delta_0 \cap \Omega_E$. By (7.1) $E \in \Sigma_B$ and $F \in \Psi_B$. Hence (2.7) implies

$$C = [B, F, E] \in \Lambda_A^* \cap \Lambda_E \cap \Lambda_B.$$

Now, as in the proof of (7.1), there exists a $D \in \Delta_B$ with $[C, D] = B$. Since $[D, E] = 1$, (2.6) now implies $B \in \Lambda_E$, a contradiction to $E \notin \Delta_B$. □

(7.3) Lemma

Let $B, C \in \Lambda_A^*$. Then $[B, C] = 1$ if and only if $\Delta_B \cap \Delta_C \neq \emptyset$.

Proof. Suppose $E \in \Delta_B \cap \Delta_C$. Then $\{B, C\} \subseteq \Lambda_E$ and thus $[B, C] \leq A \cap E = 1$.

To prove the converse suppose first $C \in \Lambda_B$. If $\Delta_B \cap \Delta_C = \emptyset$ then, as $\Lambda_B \cap \Lambda_C \subseteq C_\Sigma(A)$, (2.18) implies $\Lambda_B \cap \Lambda_C \subseteq \Lambda_A \cup \{A\}$. Hence

$$(\Lambda_B \cap \Lambda_C) \cup \{B, C\} \subseteq (\Lambda_A \cap \Lambda_B) \cup \{A, B\}$$

and thus

(∗) $(\Lambda_B \cap \Lambda_C) \cup \{B, C\} = (\Lambda_A \cap \Lambda_B) \cup \{A, B\}$

since there exists by (2.1) an involution fixing B and interchanging A and C. Let $d \in D(\Delta_0)$ with $B^d \neq B$. Then by (∗) and (2.6) $\Sigma \cap ABCB^d$ is a partition of $ABCB^d$. Hence there exists by (2.31)(2) an element g centralizing AB with $C^g = B^d$. Thus by (∗)

$$(\Lambda_A \cap \Lambda_B) \cup \{A, B\} = (\Lambda_B \cap \Lambda_{B^d}) \cup \{B, B^d\} = (\Lambda_A \cap \Lambda_{B^d}) \cup \{A, B^d\}.$$

This implies that V normalizes $(\Lambda_A \cap \Lambda_B) \cup \{A, B\}$, a contradiction to the transitivity of V on Λ_A^*.

Hence we may assume $B \notin \Lambda_C$. As before $\Lambda_B \cap \Lambda_C \subseteq C_\Sigma(A)$, since $A \in \Lambda_B \cap \Lambda_C$. Suppose $D \in (\Lambda_B \cap \Lambda_C) - (\Lambda_A \cup A)$. Then by (2.18) $M_A D \cap V \in \Delta_B \cap \Delta_C$, whence $\Delta_B \cap \Delta_C \neq \emptyset$, which was to be shown. So we may assume $\Lambda_B \cap \Lambda_C \subseteq \Lambda_A \cup \{A\}$. Pick $E \in \Lambda_B \cap \Psi_A$. Then $B = [A, E]$ and thus $E \notin \Sigma_C$, since otherwise by (2.6) $B \in \Lambda_C$. Now $[C, E] \in \Lambda_C \cap \Lambda_B$ and thus $[C, E] \in \Lambda_A$. Hence if $N = AC[C, E]$, then $\Sigma \cap N$ is a partition of N. Now for $e \in E^\#$ we have $N^e = A^e C[C, E]$ and $B \leq AA^e$. Since $\{A, A^e\} \subseteq \Lambda_C$, (2.6) implies $B \in \Lambda_C$, contradicting our assumption. □

For the rest of this section we assume that Δ_0 is degenerate. By (2.31)(1) and (2) this implies that $\dim_k N \leq 3$, if N is an abelian subgroup of G which is partitioned by $\Sigma \cap N$. With this additional hypothesis we can show:

(7.4) Lemma

Let $B \neq C \in \Lambda_A^*$. Then

$$|\Delta_B \cap \Delta_C| \geq 2 \text{ if and only if } B \in \Lambda_C.$$

Proof. Suppose first $B \in \Lambda_C$ and set $N = ABC$. Then N is partitioned by $\Sigma \cap N$ and N is, since $\dim_k N = 3$, maximal with this property. We first show

(*)   There exist $e, f \in D(\Sigma_C)$ with $[N^e, N^f] \neq 1$.

To prove (*) pick $\alpha_1, \ldots, \alpha_n \in D(\Sigma_C)$ with $B^{\alpha_1 \cdots \alpha_n} \not\leq C(N)$ and n is minimal with this property. (Such elements exist by (2.19)!) Then $B^{\alpha_1 \cdots \alpha_{n-1}} \leq C(N)$. Hence by (2.28), (2.29) and (2.18) there exists an $F \in AB \cap \Sigma$ with $B^{\alpha_1 \cdots \alpha_{n-1}} \in \Lambda_F$. By (2.1) applied to $CFB^{\alpha_1 \cdots \alpha_{n-1}}$ there exists an $e \in D(\Sigma_C)$ with $F^e = B^{\alpha_1 \cdots \alpha_{n-1}}$. Setting $f = \alpha_n^{-1}$ we obtain $[N, N^{ef^{-1}}] \neq 1$, which proves (*).

Now $\{B^e, B^f\} \subseteq \Lambda_B \cap \Lambda_C$ and thus $[B^e, B^f] \leq B \cap C = 1$. Similarly $[A^e, A^f] = 1$. Hence $[A^e, B^f] \neq 1 \neq [A^f, B^e]$. The maximality of N implies $\{B^f, B^e\} \subseteq \Sigma_A - \Lambda_A$. ($[B^f, A] \leq B \cap C = 1$!) Further $M_A B^e \neq M_A B^f$ since $[AA^e, M_A B^e] = A$, but $[A^e, B^f] = C$. Hence by (2.18) and (2.29) there exist $m_1, m_2 \in M_A$ with $\{B^{em_1}, B^{fm_2}\} \subseteq \Delta_0$. By (2.28) $B^{m_1} = B = B^{m_2}$ and $C^{m_1} = C = C^{m_2}$. Hence $\{B^{em_1}, B^{fm_2}\} \subseteq \Delta_B \cap \Delta_C$ and $B^{em_1} \neq B^{fm_2}$.

Next suppose $\{E, F\} \subseteq \Delta_B \cap \Delta_C$ and $E \neq F$. Then $[B, C] = 1$. Moreover, as $R(V) \leq Z(V)$, we have $C_{\Delta_0}(F) \not\subseteq C_{\Delta_0}(E)$. Thus there exists a $D \in \Delta_0 \cap \Sigma_F \cap \Omega_E$. By II(2.3) $D \notin \Sigma_B$. Hence $[B, D] \in \Lambda_B \cap \Lambda_F$, since D centralizes F. Moreover

$$[B, D, C] \leq A \cap F = 1.$$

Thus, applying (2.28) to $C \in \Sigma_B$, this shows that there exists an $L \in B[B, D] \cap \Lambda_C$. If $L = B$ then (7.4) holds. So we may assume $L \neq B$. But then by II(2.19) $[L, E] = B$ and thus by (2.6) $C \in \Lambda_B$, which was to be shown. $\square$

(7.5) Lemma

Let $B \neq C \in \Lambda_A^*$ with $E \neq F \in \Delta_B \cap \Delta_C$. Then

$$\Delta_B \cap \Delta_C = Z(\langle C_{\Delta_0}(E) \cap C_{\Delta_0}(F) \rangle) \cap \Delta_0 = \ell_{E,F}.$$

(In the notation of §1 applied to Δ_0. Notice that, as $R(V) \leq Z(V)$, we have $\underline{L} = L$ for $L \in \Delta_0$.)

Proof. We have $\Delta_B \subseteq C_{\Delta_0}(E) \cap C_{\Delta_0}(F)$. Since Δ_B is by (7.2) self centralizing, this implies

$$Z(\langle C_{\Delta_0}(E) \cap C_{\Delta_0}(F) \rangle) \cap \Delta_0 \subseteq \Delta_B \cap \Delta_C.$$

On the other hand by (7.1) $\Delta_B \not\subseteq \Sigma_C$ and $\Delta_C \not\subseteq \Sigma_B$. Hence by II(2.23) $Y = \langle \Delta_B, \Delta_C \rangle$ induces the $SL_2(k)$ on BC and BC is a natural module for $Y/C_Y(BC)$. Thus we obtain

$$(*) \qquad\qquad \Delta_B \cap \Delta_C = \bigcap_{y \in Y} \Delta_B^y.$$

Assume now $L \in \Delta_B \cap \Delta_C$, but $L \notin Z(\langle C_{\Delta_0}(E) \cap C_{\Delta_0}(F) \rangle) \cap \Delta_0$. Then there exists a $D \in C_{\Delta_0}(E) \cap C_{\Delta_0}(F) \cap \Omega_L$. Pick $h \in \langle D, L \rangle$ with $D = L^h$. Then $\{E, F\} \subseteq \Delta_B^h = \Delta_{B^h}$. Hence by (7.4) $B^h \in \Lambda_B \cap \Lambda_C$ and thus $N = ABCB^h$ is by (2.6) partitioned by $\Sigma \cap N$. Now $\dim_k N \leq 3$ implies $B^h \leq ABC \cap \Psi_{A_0} \leq BC$. Hence by the above $\Delta_B^h = \Delta_B^y$ for some $y \in Y$. But then by $(*)$ $D \in \Delta_B^h$ centralizes L, a contradiction to $D \in \Omega_L$. So no such L exists, which proves (7.5). $\qquad\qquad\qquad\qquad\qquad\qquad\qquad\qquad\qquad\qquad\qquad\qquad\square$

(7.6) Lemma

Let $B \in \Lambda_A^*$. Then the following hold:

(1) If $L \neq T \in \Delta_B$, then there exists a $D \in \Lambda_A^* \cap \Lambda_B$ such that $\ell_{L,T} = \Delta_B \cap \Delta_D$.

(2) $\Delta_B = C_{\Delta_0}(\ell_{E,F}) \cap C_{\Delta_0}(D)$ for each line $\ell_{E,F} \subseteq \Delta_B$ and point $D \in \Delta_B - \ell_{E,F}$.

(We use the language of §1. Since $R(V) \leq Z(V)$ we have $D = \underline{D}$ for $D \in \Delta_0$. So the elements of Δ_0 are also points of the polar space $\mathcal{P}(\Delta_0)$.)

Proof. (1) If $|k| \geq 4$, then V acts by II(4.1) transitively on the commuting pairs of elements of Δ_0. (For example by (7.4) there exist commuting pairs!) Since by (7.5) there exists a pair E, F with $\ell_{E,F} = \Delta_B \cap \Delta_C$ this shows that (1) holds in that case.

So, to prove (1), we may assume $|k| \leq 3$. Fix a line $\ell_{E,F} = \Delta_B \cap \Delta_C$ with $C \in \Lambda_A^* \cap \Lambda_B$. Let $L \in \Delta_B - \ell_{E,F}$. Then it suffices to show that there exists a $D \in \Lambda_A^* \cap \Lambda_B$ with $\ell_{E,L} = \Delta_B \cap \Delta_D$. Clearly

$$\ell_{E,L} = Z(\langle C_{\Delta_0}(E) \cap C_{\Delta_0}(L) \rangle) \subseteq \Delta_B.$$

If $\Delta_B \subset C_{\Delta_0}(\ell_{E,L})$, then there exists by (7.2) an $h \in V$ with $\ell_{E,L} \subseteq \Delta_B \cap \Delta_B^h$ and $\Delta_B \neq \Delta_B^h = \Delta_{B^h}$. But then by (7.4) $B^h \in \Lambda_B$ and $\ell_{E,L} = \Delta_B \cap \Delta_{B^h}$. Hence, to prove (1), we may assume $\Delta_B = C_{\Delta_0}(\ell_{E,L}) = Z(\langle C_{\Delta_0}(\ell_{E,L}) \rangle) \cap \Delta_0 = \ell_{E,L}$. In particular $\ell_{E,F} \subset \ell_{E,L}$. Thus, if $T \in \Delta_0 \cap \Sigma_E \cap \Omega_F$, (which exists since $R(V) \leq Z(V)$!), then $T \in \Omega_L$. Hence (F, T, L) is a triangle and thus there exists by II(3.1) a $y \in Y = \langle F, T, L \rangle$ with $o(y) = 2$ or 3 and $L^y = F$.

Since $Y \leq C_V(E)$ we obtain $\ell_{E,L}^y = \ell_{E,F}$, a contradiction to $\ell_{E,F} \subset \ell_{E,L}$ and $o(y) < \infty$. This proves (1).

Assume that (2) is false for $\ell_{E,F} \subset \Delta_B$ and $D \in \Delta_B - \ell_{E,F}$. Then there exists an $h \in V$ with

$$\ell_{E,F} \cup \{D\} \subseteq \Delta_B \cap \Delta_B^h \text{ and } \Delta_B \neq \Delta_B^h.$$

But by (1) $\ell_{E,F} = \Delta_B \cap \Delta_C$ for some $C \in \Lambda_A^* \cap \Lambda_B$. Hence by (7.4) BCB^h is partitioned by $BCB^h \cap \Lambda_A^*$ and thus $B^h \leq BC$. (By $\dim_k N \leq 3$!) But then, as in (7.5),

$$\Delta_B \cap \Delta_{B^h} = \Delta_B \cap \Delta_C = \ell_{E,F},$$

a contradiction to $D \in \Delta_B \cap \Delta_{B^h}$.   □

(7.7) Lemma

The following hold:

(1) Let $B \in \Lambda_A^*$. Then Δ_B, with the lines contained in it, is a projective plane.

(2) $\mathcal{P}(\Delta_0)$ is a non-degenerate polar space of rank three. (With Δ_0 as point set and lines as defined in (7.5)!)

Proof. To prove (1) it suffices by (7.6) to show that if $\ell_{E,F}$ and $\ell_{L,T}$ are two different lines contained in Δ_B, then they intersect in a point. Now by (7.6)

$$\ell_{E,F} = \Delta_B \cap \Delta_C \,,\, \ell_{L,T} = \Delta_B \cap \Delta_D \text{ for } C, D \in \Lambda_A^* \cap \Lambda_B.$$

Hence $[C, D] \leq A \cap B = 1$ and thus there exists by (7.3) an $S \in \Delta_C \cap \Delta_D$. If $D \in \Lambda_C$ then by the maximality of ABC we obtain $D \in ABC \cap \Lambda_A^* = BC \cap \Sigma$. But then, as $D \neq B$, we obtain

$$\ell_{E,F} = \Delta_B \cap \Delta_C = \Delta_B \cap \Delta_D = \ell_{L,T},$$

a contradiction to the choice of lines. Hence $D \notin \Lambda_C$ and so by (7.4) $\Delta_C \cap \Delta_D = \{S\}$. Now S centralizes $\ell_{E,F}$ and $\ell_{L,T}$ and thus by (7.6)(2) $S \in \Delta_B$. Hence

$$\{S\} = (\Delta_B \cap \Delta_C) \cap (\Delta_B \cap \Delta_D), \text{ which proves (1) }.$$

(1) shows that even in case $|k| \leq 3$, two different lines cannot be contained in each other. To prove (2) let $\ell_{E,F}$ be a line of Δ_0 and $D \in \Delta_0$ a point not on $\ell_{E,F}$. Then $\ell_{E,F} \subseteq \Delta_B$ for some $B \in \Lambda_A^*$ by (2.28)(2). If $D \in \Sigma_B$, then $D \in N_{\Delta_0}(\Delta_B) = \Delta_B$ by (7.2). Hence D is incident with each point on $\ell_{E,F}$.

So suppose $D \in \Psi_B$. Then $C = [B, D] \in \Lambda_A^* \cap \Lambda_B \cap \Lambda_D$. Hence $D \in \Delta_C$ and $\Delta_B \cap \Delta_C$ is by (7.5) a line of Δ_B. Thus by (1) $\ell_{E,F}$ and $\Delta_B \cap \Delta_C$ intersect in a point L and D is incident with L. If D would be incident with another point on $\ell_{E,F}$, then it would be incident with all points on $\ell_{E,F}$, since (7.6)(1) together with (7.4) and the maximality of N with $\dim_k N = 3$ and N partitioned by $\Sigma \cap N$ show that each line is determined by two points on it.

This shows that each point $D \in \Delta_0$ is incident (i.e. commutes) with all or exactly one point on $\ell_{E,F}$, which implies that $\mathcal{P}(\Delta_0)$ is a polar space. Clearly $\mathcal{P}(\Delta_0)$ is non-degenerate, since $\mathcal{F}(\Delta_0)$ is connected. Since Δ_B is by (7.2) a maximal subspace of $\mathcal{P}(\Delta_0)$, (1) implies rank $(\mathcal{P}(\Delta_0)) = 3$. $\qquad\square$

(7.8) Lemma

$\mathcal{P}(\Delta_0)$ is thick.

Proof. We first show that the lines of $\mathcal{P}(\Delta_0)$ are thick. Suppose $\ell_{E,F} = \{E, F\}$. Then by (7.7)

$$\Delta_0 = C_{\Delta_0}(E) \cap C_{\Delta_0}(F),$$

a contradiction to I(2.13)(7).

So it suffices to show that each line is contained in at least three planes. If now Δ_B would be the only plane containing $\ell_{E,F}$ then

$$\Delta_B = C_{\Delta_0}(\ell_{E,F}) = Z(\langle C_{\Delta_0}(\ell_{E,F})\rangle) \cap \Delta_0 = \ell_{E,F},$$

a contradiction to (7.6). Hence $\ell_{E,F}$ is contained in two planes Δ_B, Δ_C, whence in at least three planes, since Δ_B and Δ_C do not commute. $\qquad\square$

(7.9) Notation

We call the elements of Σ points and the elements of

$$\mathcal{L} = \{DC \cap \Sigma \mid D \in \Sigma \text{ and } C \in \Lambda_D\}$$

lines. If $D \in \Sigma$ and $B \in \Sigma_D - \Lambda_D$ let

$$S(D, B) = D\langle \Lambda_D \cap \Lambda_B \rangle B \cap \Sigma.$$

Then clearly $S(D, B) = S(B, D)$. We call such an $S(D, B)$, together with the points and lines contained in it, a *symplecton*. Let

$$\mathcal{S} = \{S(D, B) \mid D \in \Sigma \text{ and } B \in \Sigma_D - \Lambda_D\}$$

and consider the elements of $\mathcal{S}$ as point-line geometries.

(7.10) Lemma

Let $B \in \Sigma_A - \Lambda_A$. Then the following hold:

(1) $\langle M_A, M_B \rangle$ acts transitively on the points of $S(A, B)$.

(2) $S(A, B)$ is a thick, non-degenerate, polar space of rank three, the planes of which are projective planes over k.

Proof. (1) If $D \in \Lambda_A$, then $[D, B] = 1$ or $[D, B] \in \Lambda_A \cap \Lambda_B$.
Hence $[S(A, B), M_A] \leq A\langle \Lambda_A \cap \Lambda_B \rangle$. Moreover by (2.29) M_A acts transitively on $(S(A, B) \cap \Sigma) - (A\langle \Lambda_A \cap \Lambda_B \rangle \cap \Sigma)$. Clearly $B^m \notin \Lambda_B$ for some $m \in M_A$. Hence by symmetry in A and B we have $B^{mn} = A$ for some $n \in M_B$.

Let $R = \langle M_A, M_B \rangle$. Then the above argument shows that all elements of

$$[(\Sigma \cap S(A, B)) - \Lambda_A] \cup [(\Sigma \cap S(A, B)) - \Lambda_B]$$

are conjugate in R. If now $D \in \Lambda_A \cap \Lambda_B$, then $D^m \notin \Lambda_B$ for some $m \in M_A$, whence D^m is conjugate to A by M_B. This proves (1).

To prove (2) we show first that each point of $S(A, B)$ is incident with all or exactly one point of each line of $S(A, B)$. For this we may by (1) assume that the line is of the form $AC \cap \Sigma$ with $C \in \Lambda_A^*$ and the point D is in $S(A, B) - (\Lambda_A \cup \{A\})$. Then by (2.29) we may assume $D \in \Delta_0$. Hence C is by (2.28) the only point incident with D on $AC \cap \Sigma$.

This shows that $S(A, B)$ is a polar space, which is by (1) obviously non-degenerate. Clearly the lines of $S(A, B)$ are thick and each point is contained in at least three lines. Moreover, by $\dim_k N \leq 3$, where N is a subgroup of G partitioned by $\Sigma \cap N$, maximal subspaces of $S(A, B)$ have (projective) dimension at most 2. If now $B \in \Delta_0$, then $B \in \ell_{B,F}$ for some line $\ell_{B,F}$ of $\mathcal{P}(\Delta_0)$.

Now by (7.6) $\ell_{B,F} = \Delta_C \cap \Delta_D$, where $D, C \in \Lambda_A^*$ with $|\Delta_D \cap \Delta_C| \geq 2$. Then by (2.28) $D, C \in \Lambda_B$ and thus D, C are points of $S(A, B)$ and by (7.4) $D \in \Lambda_C$. Hence $ADC \cap \Sigma$ and $DCB \cap \Sigma$ are planes of $S(A, B)$. In particular rank $S(A, B) \geq 3$ and thus rank $S(A, B) = 3$.

Now by (7.7) there are at least three lines of $\mathcal{P}(\Delta_0)$ contained in Δ_D, i.e. at least three sets of the form $\Delta_D \cap \Delta_C$, $C \in \Lambda_A^*$. Hence the line $AD \cap \Sigma$ of $S(A, B)$ is contained in at least three planes. This proves (2). $\qquad \square$

(7.11) Lemma

Let E, F be non-collinear points of $S(A, B)$. Then $S(A, B) = S(E, F)$.

Proof. If $E \in (S(A, B) \cap \Sigma) - (\Lambda_A \cup \{A\})$, then by (7.10)(1) $S(A, B) = S(A, E) = S(E, F)$. Thus, to prove (7.11) we may assume $\{E, F\} \subseteq \Lambda_A \cap \Lambda_B$. But then there exists an $m \in M_B$ with $A^m \notin \Lambda_E$. Hence

$$S(A, B) = S(A^m, B) = S(A^m, E) = S(E, F). \qquad \square$$

(7.12)   Theorem

The geometry of points, lines and symplecta (as defined in (7.9)) satisfies the following axioms:

(M0) The graph of points and lines is non-empty and connected.

(M1) The intersection of two distinct symplecta is empty, a point, a line or a plane.

(M2) A symplecton is, together with the points and lines contained in it, a non-degenerate polar space of rank three.

(M3) Let

$$\mathcal{P}_A = \{S(A, E) \mid E \in \Sigma_A - \Lambda_A\}$$

for the point A of Σ and for $C \in \Lambda_A^*$ and $D \in \Lambda_A^* \cap \Lambda_C$ let

$$\ell_{C,D}^* = \{S(A, E) \in \mathcal{P}_A \mid CD \leq S(A, E)\}$$

and

$$\mathcal{B}_A = \{\ell_{C,D}^* \mid C \in \Lambda_A^* \text{ and } D \in \Lambda_A^* \cap \Lambda_C\}.$$

Then $\mathcal{P}_A$ together with the lines of $\mathcal{B}_A$ is also a non-degenerate polar space of rank three.

Proof. (M0) is just the connectedness of $\mathcal{E}(\Sigma)$. To prove (M1) let $S(A, B)$ and $S(C, D)$ be a pair of symplecta, such that $S(A, B) \cap S(C, D) \cap \Sigma$ contains the non-collinear pair of points E, F. Then by (7.11)

$$S(A, B) = S(E, F) = S(C, D).$$

This proves (2). (M2) is just (7.7). So (M3) remains to be proved.

Now each $S(A, E) \in \mathcal{P}_A$ is equal to $S(A, \overline{E})$, where $\overline{E} = M_A E \cap V \in \Delta_0$ by (2.18). Thus $\mathcal{P}_A$ is in one-to-one correspondence with Δ_0. Moreover, a line $\ell_{C,D}^*$ is in one-to-one correspondence with CD, which in turn is in one-to-one correspondence to $\Delta_C \cap \Delta_D$ and the latter is by (7.6) a line of $\mathcal{P}(\Delta_0)$. Finally we have by (2.29) $S(A, \overline{E}) \in \ell_{C,D}^*$ if and only if $\overline{E} \in \Delta_0 \cap \Delta_C \cap \Delta_D = \Delta_C \cap \Delta_D$. Hence the geometry consisting of the points in $\mathcal{P}_A$ and lines in $\mathcal{B}_A$ is isomorphic to the polar space $\mathcal{P}(\Delta_0)$, which proves (M3). $\qquad \square$

(7.13) Notation

A geometry consisting of points, lines and certain point sets called symplecta, satisfying the axioms (M0)–(M3) of (7.13) is called in [Tit74, (10.13)] a metasymplectic space. Such a metasymplectic space is called thick if the symplecta are thick polar spaces and if every plane is contained in at least three symplecta.

By [Tit74, p.216] the flag complex of such a thick metasymplectic space is a building of type F_4. Hence we have

(7.14) Corollary

The geometry $\mathcal{S}(\Sigma) = (\Sigma, \mathcal{L}, \mathcal{S})$ with point set Σ, line set $\mathcal{L}$ and set of "symplecta" $\mathcal{S}$ is a thick metasymplectic space, the planes of which are projective planes over k.

Proof. By (7.11), (7.12) it just remains to be shown that each plane ABC, $B \in \Lambda_A^*$ and $C \in \Lambda_A^* \cap \Lambda_B$ is contained in at least three symplecta. Now by (7.6) $\Delta_B \cap \Delta_C = \ell_{E,F}$ for $E \neq F \in \Delta_0$ and by (7.8) $|\ell_{E,F}| \geq 3$. But for each $D \in \ell_{E,F}$ we have $ABC \leq S(A, D)$, which proves (7.14).   □

(7.15) Notation

Let $\mathcal{B}$ be the flag complex of the metasymplectic space $\mathcal{S}(\Sigma)$ constructed in (7.14). Then $\mathcal{B}$ is by the theorem of [Tit74, p.216] a (thick) building of type:

$$F_4 : \overset{1}{\circ}\!\!-\!\!-\!\!-\!\!\overset{2}{\circ}\!\!=\!\!=\!\!\overset{3}{\circ}\!\!-\!\!-\!\!-\!\!\overset{4}{\circ}$$

where the vertices of type 1 are the points, of type 2 the lines, of type 3 the planes and of type 4 the symplecta of $\mathcal{S}(\Sigma)$. Since G respects the metasymplectic space structure, it is clear that $G \leq \operatorname{Aut}(\mathcal{B})$, since as $R(G) = Z(G) = 1$, G acts faithfully on Σ.

Now by construction Σ is also the set of all points of $\mathcal{S}(\Sigma)$. Hence $\operatorname{Aut}(\mathcal{B}) \subseteq \operatorname{Sym}(\Sigma)$, considering Σ as a point set of $\mathcal{S}(\Sigma)$. To distinguish this action from the conjugation action of the elements of G, we write $D\alpha$ for the image of the point $D \in \Sigma$ under the automorphism $\alpha \in \operatorname{Aut}(\mathcal{B})$. Finally let

$$\mathcal{S}(\Delta_0) := \{D \in \Sigma \mid \{D, B\} \text{ is contained}$$
$$\text{in some symplecton of } \mathcal{S} \text{ for each } B \in \Delta_0\}.$$

We claim that

(∗) $\mathcal{S}(\Delta_0) = \Sigma \cap X.$

Indeed, if $D \in \Sigma \cap X$, then $D \in \Sigma_B - \Lambda_B$ for each $B \in \Delta_0$. Hence $\{D, B\} \subseteq D\langle \Lambda_D \cap \Lambda_B \rangle B \cap \Sigma = \mathcal{S}(D, B)$. This shows $\Sigma \cap X \subseteq \mathcal{S}(\Delta_0)$.

On the other hand, if $D \in \mathcal{S}(\Delta_0)$, then $D \in C_\Sigma(\Delta_0)$. But by (2.26) $C_\Sigma(\Delta_0) = \Sigma \cap X$, so that also $\mathcal{S}(\Delta_0) \subseteq \Sigma \cap X$. With this notation we obtain:

(7.16) Lemma

A is the set of all automorphisms $\alpha \in \mathrm{Aut}(\mathcal{B})$ satisfying:

$(+)$ α fixes all lines, points and symplecta containing the point A, pointwise.

Proof. It is clear that the elements of A satisfy $(+)$.
Now let

$$\overline{A} = \{\alpha \in \mathrm{Aut}(\mathcal{B}) \mid \alpha \text{ satisfies } (+)\}$$

and more generally, for each $D \in \Sigma$, let $\overline{D}$ be the corresponding subgroup of $\mathrm{Aut}(\mathcal{B})$. Then $D \leq \overline{D}$. Moreover, if $\sigma \in \mathrm{Aut}(\mathcal{B})$ fixes the point D, i.e. $D\sigma = D$, then $D^\sigma \leq \overline{D}$ since all elements of D^σ satisfy $(+)$ (with respect to D).

Now let Φ be the root system of some apartment of $\mathcal{B}$ with positive system Φ^+ and let $h \in \Phi^+$ be the highest root of Φ^+. (For definition see II§5!) Let A_h be the root subgroup corresponding to the root h. Then A_h corresponds to some point P of $\mathcal{S}(\Sigma)$ (since by the proof of II(5.20) $N_{\overline{G}}(A_h)$ is the maximal parabolic of $\overline{G}$, which is the stabilizer of P, where $\overline{G} = \langle A_\alpha \mid \alpha \in \Phi \rangle$)! Thus, if we assume without loss of generality that $P = A$ (as a point of $\mathcal{S}(\Sigma)$), then $A_h = \overline{A}$. In particular by II(5.20) and II(5.22)

$$\overline{\Sigma} = \{\overline{D} \mid D \in \Sigma\}$$

is a class of abstract root subgroups of $\overline{G} = \langle \overline{\Sigma} \rangle$ and thus $[\overline{E}, \overline{F}] = 1$, if $[E, F] = 1$.

Let now $\alpha \in \overline{A}$. Then α centralizes each element of $C_\Sigma(A)$ and thus $C_A = \langle C_\Sigma(A) \rangle$. In particular α fixes each element of Δ_0 and thus permutes $\mathcal{S}(\Delta_0)$. This implies $A_0\alpha \in \Sigma \cap X$, considering this set as a point set of $\mathcal{S}(\Sigma)$. Since A is transitive on $(\Sigma \cap X) - \{A\}$, there exists an $a \in A^{\#}$ with $A_0\alpha = A_0 a$. Thus $\alpha a^{-1} \in \overline{A}$ and fixes the point $\overline{A}_0$. Since $\langle \overline{A}, \overline{A}_0 \rangle$ is by I(4.12) a rank one group, this shows that αa^{-1} centralizes $\overline{A}_0$. Hence αa^{-1} centralizes $G = \langle C_\Sigma(A), A_0 \rangle$ and thus fixes all points of $\mathcal{S}(\Sigma)$. Hence $\alpha a^{-1} = 1$ (as element of $\mathrm{Aut}(\mathcal{B})$) and $\alpha = a \in A$. $\square$

(7.17)   Corollary

G is the group of Lie type $\mathcal{B}$ and Σ the class of long-root subgroups of Aut$\mathcal{B}$.

Indeed we have shown in (7.16) that Σ is the class of long root subgroups of Aut($\mathcal{B}$). Hence (7.17) follows from II(5.22)(2).

For the convenience of the reader we list the possibilities for G and $\overline{V} = V/Z(V)$. The lists for G follow from [Tit74, (10.4)] and the lists for $\overline{V}$ follow from the description of the buildings of type F_4 in [Tit74, (10.2)] and (7.7), (7.8).

(7.18)

For G and $\overline{V}$ one of the following holds:

(1) $\overline{V} \simeq PSp(6,k)$ and $G \simeq F_4(k)$.

(2) $\overline{V} \simeq PSU(6,\overline{k})$ and $G \simeq {}^2E_6(\overline{k})$, $\overline{k}$ a quadratic extension of k.

(3) $\overline{V} \simeq PSU_6(\overline{k}, f)$ resp. $P\Omega_6(\overline{k}, q)$; where $\overline{k}$ is a quaternion division algebra over k and Char $k = 0$ or odd in the first resp. Char $k = 2$ in the second case. Further f resp. q are σ-hermitian resp. σ-quadratic forms of Witt index 3 and $G \simeq E_{7,4}^9(k)$ in the notation of [Tit66].

(4) $\overline{V} \simeq E_7^K = E_{7,3}^{28}(k)$ in the notation of [Tit74, (9.2)] resp. [Tit66], K a Cayley division algebra over k and $G \simeq E_{8,4}^{28}(k)$.

(5) $\overline{k}^2 \subseteq k \subseteq \overline{k}$, $\overline{k}$ a non-perfect field of char 2. $\overline{V} \simeq P\Omega(W,q)$; $q : W \to \overline{k}$ a non-degenerate quadratic form of Witt index 3 with $q(W^\perp) = k$ and
$$\dim_{\overline{k}} W = 6 + \dim_k W^\perp = 6 + \dim_{\overline{k}^2} k.$$

Moreover $G \simeq F_4(k, \overline{k})$ in the notation of [Tit74, (10.3.2)].

(7.19)   Exercises

(1) Let $G = PSp(6,k)$ and Σ the class of symplectic transvection groups of G. Let Δ be a maximal commuting subset of Σ and
$$\mathcal{L} = \{\Delta^g \cap \Delta^h \mid \Delta^g \neq \Delta^h \text{ and } |\Delta^g \cap \Delta^h| \geq 2\}.$$
Show that $(\Sigma, \mathcal{L}, \subseteq)$ is the polar space of G.

(2) Let G be a group generated by the non-degenerate class Σ of abstract root subgroups satisfying $R(G) = Z(G) = 1$. Suppose, that if N is an abelian subgroup of G partioned by $\Sigma \cap N$, then $\dim N \leq 3$; N a vector space over
$$K = \{\sigma \in \text{End}N \mid A^\sigma \subseteq A \text{ for all } A \in \Sigma \cap N\}$$
(See (2.31)(1) !). Show that $\Delta_0 = \Sigma_A \cap \Sigma_{A_0}$ is degenerate for $A_0 \in \Omega_A$.

Hint. Let $B \in \Lambda_A^*$ and $C \in \Delta_0 \cap \Lambda_B$. Show that $\Delta_0 \cap \Lambda_C \subseteq \Lambda_B$. (Hence if Δ_0 is non-degenerate, then the connectivity of $\mathcal{E}(\Delta_0)$ would imply $\Delta_0 \subseteq \Lambda_B$.)

(3) Assume that the hypothesis of this section is satisfied, i.e. $\langle C_\Sigma(A) \rangle$ is Σ-maximal and Δ_0 is degenerate. For $B \in \Sigma_A - \Lambda_A$ let $S(A, B)$ be the symplecton containing A and B as defined in (7.9) and let $S = \langle S(A, B) \rangle = A \langle \Lambda_A \cap \Lambda_B \rangle B$. Show:

   (i) $[S, M_A] \leq A \langle \Lambda_A \cap \Lambda_B \rangle$ and $[S, M_A, M_A] \leq A$.

   (ii) Let $\overline{M}_A = M_A / C_{M_A}(S)$. Show that $\overline{M}_A$ acts regularly on $S(A, B) - (\Lambda_A \cup \{A\})$.

(4) Assume the same hypothesis and notation as in (3). Show that $S(A, B)$ is the polar space of a non-degenerate orthogonal space of Witt index 3 over k. (I.e. there exists a k-vector space W, a non-degenerate quadratic form $q : W \to k$ of Witt index 3 such that $S(A, B)$ is the polar space of (W, q).)

   Hint. Since the planes of $S(A, B)$ are projective planes over k, $S(A, B)$ is a classical polar space. Let $C \in \Lambda_A - \Sigma_B$ (exists)! Then $D = [C, B] \in \Lambda_A \cap \Lambda_C \cap \Lambda_B$ and $AD \cap \Sigma \subseteq \Lambda_C$. Thus for the line $\ell = AD \cap \Sigma$ each element $c \in C^\#$ satisfies the hypothesis of (5.13).

(5) Assume the same hypothesis and notation as in (3) and let $R = \langle M_A, M_B \rangle$ and $\overline{R} = R / C_R(S)$. Show that $\overline{R}$ is the group of Lie type $S(A, B)$ (i.e. a non-degenerate orthogonal group of Witt index 3 over k), considering the rank three polar space $S(A, B)$ as a building of type C_3.

   Hint. $\overline{R} \leq \mathrm{Aut}(S(A, B)) = Y$. $Y_A = N_Y(A)$ is the stabilizer of the "point" A in Y, whence a maximal parabolic subgroup of Y. Thus by (3)(i) $\overline{M}_A \leq U_A$, where U_A is the "unipotent radical" of Y_A. Now (3)(ii) implies $\overline{M}_A = U_A$.

   Now let $Y_0 \leq Y$ be the group of Lie type $S(A, B)$ as defined in II§5. Then Y_0 is generated by the unipotent radicals of two opposite maximal parabolic subgroups, see [Tim00b, (4.11)], whence $Y_0 = \langle \overline{M}_A, \overline{M}_B \rangle = \overline{R}$.

§ 8 $E_6(k)$, $E_7(k)$ and $E_8(k)$

In this section we assume the same hypothesis as in §7, i.e. Σ is a non-degenerate class of abstract root subgroups of G, $R(G) = Z(G) = 1$ and $\langle C_\Sigma(A) \rangle$ is Σ-maximal for $A \in \Sigma$. Fix $A_0 \in \Omega_A$, $\Delta_0 = \Sigma_A \cap \Sigma_{A_0}$ and $V = \langle \Delta_0 \rangle$, $X = \langle A, A_0 \rangle$. Contrary to most of §7, assume in addition that Δ_0 is a non-degenerate class of abstract root subgroups of V. We will show that under these conditions G is one of the groups of the title of this section. Contrary to §7 we will not construct a geometry of type E_6, E_7 or E_8 directly, since this would mean constructing six, seven resp. eight types of vertices. Instead we will use the classification of the root geometries of types E_6, E_7 and E_8 of Cohen-Cooperstein [CC83], see also [Coh95, (6.11)] as certain parapolar spaces.

(8.1)   Definition

A connected point line geometry $(\mathcal{P}, \mathcal{L})$ with thick lines (i.e. $|\ell| \geq 3$ for $\ell \in \mathcal{L}$) will be called a *parapolar space*, if it satisfies the following axioms:

(F1) If $P \in \mathcal{P}$ and $\ell \in \mathcal{L}$ such that P is collinear to two points of ℓ, then P is collinear to all points on ℓ.

(F2) If $P \neq Q \in \mathcal{P}$ are collinear, then the graph induced on $\{P, Q\}^\perp$ is not a clique. (Here $\{P, Q\}^\perp$ is the set of points collinear with P and Q!)

(F3) If $P, Q \in \mathcal{P}$ with $d(P, Q) = 2$ (d is the distance in the collinearity graph) and $|\{P, Q\}^\perp| > 1$, then $\{P, Q\}^\perp$ is, with the lines contained in it, a non-degenerate polar space of rank at least 2.

A subset $\mathcal{R}$ of $\mathcal{P}$ is called a *subspace* of the parapolar space $(\mathcal{P}, \mathcal{L})$, if whenever P, Q are collinear points of $\mathcal{R}$ and ℓ is the line through P and Q, then all points of ℓ lie in $\mathcal{R}$. Such a subspace $\mathcal{R}$ is called *singular*, if all pairs of points of $\mathcal{R}$ are collinear. The rank $\mathrm{rk}(\mathcal{R})$ of a singular subspace $\mathcal{R}$ is the length of a longest chain

$$\mathcal{R}_0 \subset \mathcal{R}_1 \subset \cdots \subset \mathcal{R}_s = \mathcal{R}$$

of singular subspaces of $\mathcal{R}$ if this length is restricted. Otherwise we set $\mathrm{rk}(\mathcal{R}) = \infty$. The *singular rank* s of $(\mathcal{P}, \mathcal{L})$ is the maximal number s (possibly ∞) for which singular subspaces of rank s of $(\mathcal{P}, \mathcal{L})$ exist. If $s < \infty$, then we say that $(\mathcal{P}, \mathcal{L})$ is of finite singular rank.

Let $(\mathcal{P}, \mathcal{L})$ be a parapolar space. Then we say that $(\mathcal{P}, \mathcal{L})$ satisfies:

(F3)$_n$ If $P, Q \in \mathcal{P}$ with $d(P, Q) = 2$ and $|\{P, Q\}^\perp| > 1$, then $\{P, Q\}^\perp$ is a non-degenerate polar space of rank n.

Such pairs P, Q with $d(P, Q) = 2$ and $|\{P, Q\}^\perp| > 1$ will be called *symplectic pairs*.

(F4) Suppose P, Q is a symplectic pair and ℓ is a line through Q with $P^\perp \cap \ell = \emptyset$. Then either $P^\perp \cap \ell^\perp$ is a point or $P^\perp \cap \ell^\perp$ is a maximal singular subspace of $\{P, Q\}^\perp$.

We will use the following classification theorem of [CC83, p.75], see also [Coh95, (6.11)].

Theorem. Suppose $(\mathcal{P}, \mathcal{L})$ is a parapolar space of finite singular rank and diameter at least 3 (of the collinearity graph of $(\mathcal{P}, \mathcal{L})$). Let $n \geq 3$. Then $(\mathcal{P}, \mathcal{L})$ satisfies $(F3)_n$ and F_4 if and only if $n = 3, 4$ or 6 and $(\mathcal{P}, \mathcal{L})$ is isomorphic to the root group geometry of a group of Lie type E_6, E_7 or E_8.

Here the root group geometry is defined as follows: If Σ is a non-degenerate class of abstract root subgroups of G, then the geometry with point set $\mathcal{P} = \Sigma$ and line set

$$\mathcal{L} = \{AB \cap \Sigma \mid A \in \Sigma, B \in \Lambda_A\}$$

is called the *root group geometry* of (G, Σ). In particular if G is a group of Lie type E_6, E_7 or E_8 and Σ is the class of long root subgroups of G, then Σ is by II(5.20) a non-degenerate class of abstract root subgroups of G and we will call the root group geometry of (G, Σ) the *root group geometry* of G.

Now suppose Σ satisfies the hypothesis of this section in G. Then we define the *singular rank* $rk(\Sigma)$ of Σ by

$$rk(\Sigma) \quad := \operatorname{Max}\{\dim_k N \mid N \text{ a subgroup of } G \text{ which is partitioned by } \Sigma \cap N\}$$

(respectively $rk(\Sigma) := \infty$ if this maximum does not exist). Finally, if $A \in \Sigma, B \in \Sigma_A - \Lambda_A$, then

$$S(A, B) := A\langle \Lambda_A \cap \Lambda_B \rangle B \cap \Sigma$$

$S(A, B)$ is called the *symplecton* spanned by A and B.

Our aim in this section is to show that the root group geometry of (G, Σ) satisfies the hypothesis of the theorem of Cohen-Cooperstein. For this we often use geometric language, i.e. we speak of lines of Σ and say that A and C are collinear, if $C \in \Lambda_A$.

(8.2) Lemma

Suppose $rk \, \Delta_0 \leq n$. Then $rk \, \Sigma \leq n + 2$.

Proof. Suppose $rk \, \Delta_0 = n$ but $rk \, \Sigma \geq n + 3$. Then there exists a subgroup N of G satisfying:

(i) $\Sigma \cap N$ is a partition of N and $A \in \Sigma \cap N$.
(ii) $\dim_k N = n + 3$.

Let $C \in \Sigma \cap N - \{A\}$. Then there exists by (2.31)(2) a subgroup M of G which is partitioned by $\Sigma \cap M$, such that $C_M(N) = 1$ and M induces the set of all transvections corresponding to the point C on N. Hence, by the structure of $SL_{n+3}(k)$ we have $\dim_k M = n+2$. Let $M_0 = N \cap C(A)$. Then $\dim_k M_0 = n+1$ and $M_0 \cap M_A = 1$. Thus, if $M_1 = M_A M_0 \cap V$, then M_1 is by (2.18) a singular subspace of Δ_0 with $\dim_k M_1 = n + 1$, a contradiction to $\mathrm{rk}\,\Delta_0 = n$. $\qquad\square$

(8.3) Lemma

The following hold for $S(A, B)$.

(1) $S(A, B)$, with the lines contained in it, is a non-degenerate polar space of rank ≥ 3, the planes of which are projective planes over k.

(2) $R = \langle M_A, M_B \rangle \leq N(S(A, B))$ acts transitively on the pairs of non-collinear point of $S(A, B)$. In particular $S(A, B) = S(E, F)$ for each pair $E \neq F \in S(A, B)$ with $F \notin \Lambda_E$.

Proof. Clearly $[M_A, S(A, B)] \leq A\langle \Lambda_A \cap \Lambda_B \rangle$ and $[M_B, S(A, B)] \leq B\langle \Lambda_A \cap \Lambda_B \rangle$. Hence $R \leq N(S(A, B))$. By (2.29) M_A acts transitively on $S(A, B) - (\Lambda_A \cup \{A\})$. Hence it follows as in (7.10) that R acts transitively on $S(A, B)$ and thus on the pairs of non-collinear points of $S(A, B)$. In particular, as in (7.11), (2) holds.

Now, if $\ell = CD \cap \Sigma$ is a line of $S(A, B)$, then clearly $\ell \cap \Lambda_A \cap S(A, B) \neq \emptyset$ and if $|\ell \cap (\Lambda_A \cup A)| \geq 2$, then $\ell \subseteq \Lambda_A \cup A$. This shows that $S(A, B)$ is a polar space. Thus it remains to show that $\mathrm{rk}(S(A, B) \geq 3$. For this we may assume $B \in \Delta_0$, since $\langle \Sigma_A \rangle$ acts transitively on $\Sigma_A - \Lambda_A$. Hence by (7.3) there exist $C, D \in \Lambda_A^*$ with $D \in \Lambda_C$ and $B \in \Delta_C \cap \Delta_D$. Thus $CD \cap \Sigma \subseteq \Lambda_A \cap \Lambda_B$ and $ACD \cap \Sigma$ is a singular plane of $S(A, B)$, which is a projective plane over k.$\square$

(8.4) Lemma

The following hold for $S(A, B)$:

(1) $S(A, B)$ is a (projective) non-degenerate orthogonal space over k of Witt index ≥ 3.

(2) If $C \in \Lambda_A - \Sigma_B$, $D = [C, B]$ and $\ell = AD \cap \Sigma$, then C induces the Siegel transvection group corresponding to the line ℓ on $S(A, B)$.

(3) $Y = \langle N_\Sigma(S(A, B)) \rangle$ induces the group $P\Omega(W, q)$ on $S(A, B)$; where $S(A, B)$ is obtained from the k-vector space W together with the quadratic form $q : W \to k$.

Proof. Clearly $S(A, B)$ is a classical polar space, since the planes of $S(A, B)$ are projective planes over k. Let C, D and ℓ be as in (2). Then C fixes all points in $\ell^{\perp}$, since for $E \in S(A, B) \cap \Lambda_A \cap \Lambda_D$ we have $[C, E] \le A \cap D = 1$. Further, if $F \in AD \cap \Sigma$, then $C \in \Lambda_F$ so that $[S(A, B) \cap \Lambda_F, C] \le F$. Hence C fixes all lines through F. Hence (5.13) implies that $S(A, B)$ is the polar space obtained from some k-vector space W, together with a non-degenerate quadratic form $q : W \to k$ of Witt index ≥ 3. Moreover by II(1.5)(7) C is contained in the image of the Siegel transvection group corresponding to ℓ. Since C is transitive on $(DB \cap \Sigma) - D$, this shows that C is (the image of) the Siegel transvection group corresponding to ℓ. In particular (1) and (2) hold.

To prove (3) it suffices by definition of $\Omega(W, q)$ in II(1.5) to show that $N_\Sigma(S(A, B))$ induces Siegel transvection groups corresponding to all lines of $S(A, B)$ on $S(A, B)$. For this let $D \in \Lambda_A \cap \Lambda_B$ and $F \in \Delta_0 \cap \Omega_B$. Then for $f \in F^{\#}$ we obtain by II(2.19)

$$C = D^f \in (\Lambda_D \cap \Lambda_A) - \Sigma_B.$$

Thus by (2) C induces the Siegel transvection group corresponding to $\ell = AD \cap \Sigma$. This shows that we get a Siegel transvection group for each line of $S(A, B)$ through A. But then (8.3)(2) shows that (3) holds. $\qquad \square$

(8.5) Lemma

The root group geometry of (G, Σ) is a parapolar space of diameter 3 satisfying F4.

Proof. It is clear that (F1) is satisfied, since if two points of the line $AC \cap \Sigma$ are collinear to D, then $AC \cap \Sigma \subseteq \Lambda_D \cup \{D\}$. If $A, B \in \Sigma$ with $d(A, B) = 2$, then either $B \in \Psi_A$ or $B \in \Sigma_A - \Lambda_A$. In the first case $\{[A, B]\} = \{A, B\}^{\perp}$, while in the second case $S(A, B)$ is a non-degenerate polar space of rank $n \ge 3$. Hence $\{A, B\}^{\perp} = \Lambda_A \cap \Lambda_B$ is also a non-degenerate polar space of rank $n - 1 \ge 2$. This shows (F3).

By (F3) it suffices to show (F2) for pairs A, D of points with $D \in \Lambda_A$. (If $C \in \Omega_A$, then $\{A, D\}^{\perp} = \Lambda_A \cap \Lambda_D = \emptyset$!) Hence by (2.19) we may assume $D \in \Lambda_A \cap \Lambda_B$. But then, as $\Lambda_A \cap \Lambda_B$ is a non-degenerate polar space, $\Lambda_A \cap \Lambda_B \cap \Lambda_D$ is not a clique, whence also $\Lambda_A \cap \Lambda_D = \{A, D\}^{\perp}$ is not a clique.

Finally, to prove (F4), we may by the connectivity of $\mathcal{F}(\Sigma_A - \Lambda_A)$ assume that $B = P$ and $A = Q$. Then the line ℓ is of the form $AC \cap \Sigma$ with $C \in \Lambda_A$ and $AC \cap \Lambda_B = \emptyset$. Hence

$$P^{\perp} \cap \ell^{\perp} = \Lambda_B \cap \Lambda_A \cap \Lambda_C \subseteq S(A, B) \cap \Lambda_A \cap \Lambda_B.$$

Suppose $E \ne F$ are non-collinear points of $\Lambda_A \cap \Lambda_B \cap \Lambda_C$. Then by (8.3)(2) $S(E, F) = S(A, B)$ and whence $C \in S(A, B)$. But then $AC \cap \Lambda_B \ne \emptyset$, since

$S(A, B)$ is a polar space. This shows that $P^\perp \cap \ell^\perp$ is totally singular. If now $C \in \Psi_B$, then clearly $\{[B, C]\} = \Lambda_B \cap \Lambda_A \cap \Lambda_C$. Thus, to prove (F4), we have to show that $P^\perp \cap \ell^\perp$ is a maximal totally singular subspace of $P^\perp \cap Q^\perp$ in case $C \in \Sigma_B$.

For this pick $E, L \in \Sigma_B$ with $[A, E] = C$ and $[C, L] = A$. Then $\langle E, L \rangle \simeq SL_2(k)$ and $\Lambda_B \cap \Lambda_C \cap C(L) = \Lambda_B \cap \Lambda_C \cap \Lambda_A$. Moreover for $D \in (\Lambda_B \cap \Lambda_C) - C(L)$ we have by (2.7) $[D, L] \in \Lambda_B \cap \Lambda_A$, $[D, L, E] \in \Lambda_B \cap \Lambda_C \cap \Lambda_E$ and $D \leq [D, L, E]D_0$, $D_0 \in \Lambda_B \cap \Lambda_C \cap C(\langle E, L \rangle)$. This implies

$$\langle \Lambda_B \cap \Lambda_C \rangle = \langle \Lambda_B \cap \Lambda_C \cap \Lambda_A \rangle \langle \Lambda_B \cap \Lambda_C \cap \Lambda_E \rangle.$$

Now by the above and (8.4) $\langle \Lambda_B \cap \Lambda_C \cap \Lambda_A \rangle$ and $\langle \Lambda_B \cap \Lambda_C \cap \Lambda_E \rangle$ are both totally singular subspaces of the orthogonal space $\langle \Lambda_B \cap \Lambda_C \rangle$ (subspace of $S(B, C)$). Hence they must be both maximal, which proves (F4). (In fact this shows that $S(B, C)$ is of type D!) Since $d(A, A_0) = 8$ this proves (8.5). □

Now by the hypothesis of this section G acts transitively on the pairs A, B with $A \in \Sigma$ and $B \in \Sigma_A - \Lambda_A$. Hence all symplecta $S(A, B)$ have the same rank and so also all the polar spaces $A^\perp \cap B^\perp = \Lambda_A \cap \Lambda_B$ have the same rank $n \geq 2$. Thus, to show that the hypothesis of the theorem of Cohen-Cooperstein holds, it suffices to show that $n \geq 3$ and that the singular rank of Σ is finite.

(8.6) Notation

Fix $B \in \Delta_0$ and let $B_0 \in \Omega_B \cap \Delta_0$, $M_B^0 = \langle \Lambda_B \cap \Delta^0 \rangle$, $\Delta^1 = \Sigma_B \cap \Sigma_{B^0} \cap \Delta_0$ and $V^1 = \langle \Delta^1 \rangle$. Since $R(V) \leq Z(V)$ by (2.30), the results of §2 hold for $\overline{V} = V/Z(V)$ and $\overline{\Delta}_0$. Being slightly incorrect we often omit, to simplify notation, the -homomorphism from $V \to \overline{V}$.

(8.7) Lemma

The following hold:

(1) $M_B^0 \leq C(S(A, B))$.
(2) V^1 induces the corresponding orthogonal group on $\langle \Lambda_A \cap \Lambda_B \rangle$; considering $\langle \Lambda_A \cap \Lambda_B \rangle$ as the orthogonal subspace $A^\perp \cap B^\perp$ of $S(A, B)$. (See (8.4).)

Proof. We have

$$[\langle S(A, B) \rangle \cap M_A, M_B^0] \leq M_A \cap B = 1,$$

whence (1) holds. Since $\langle C_{\Delta_0}(B) \rangle = M_B^0 V^1$, (1) shows that V_1 acts as $\langle C_\Sigma(A) \cap C_\Sigma(B) \rangle$ on $\Lambda_A \cap \Lambda_B$. Hence (2) is a consequence of (8.4)(3) and the fact that in an orthogonal group of Witt index $n \geq 3$ the stabilizer of a hyperbolic pair (A, B) induces the orthogonal group of Witt index $n - 1$ on $A^\perp \cap B^\perp$ (See II(1.5)!). □

(8.8) Lemma

The rank of $S(A,B)$ is greater than or equal to 4.

Proof. Suppose false. Then by (8.3) $\mathrm{rk}(S(A,B)) = 3$ and $\Lambda = \Lambda_A \cap \Lambda_B$ is a non-degenerate orthogonal space of Witt index 2 over k. Moreover, since by (8.4) the elements of Δ_1 act as Siegel transvection groups on Λ, II(1.5)(4) implies that $\widetilde{\Delta}^1$ is a degenerate class of abstract root subgroups of $\widetilde{V}^1$, where $\widetilde{V}^1 = V^1/C_{V^1}(\Lambda)$. Together with (5.9) applied to $\overline{V} = V/Z(V)$ this shows that one of the following holds:

(α) Δ^1 is a class of k-transvections of V^1.
(β) $\mathcal{F}(\Delta^1)$ is not connected, $V^1 = V_1 * V_2$ with $V_1 \simeq SL_2(k)$ and V_2 a non-degenerate orthogonal group of Witt index 2.

In case (α) it follows from (7.7) applied to $\overline{V}$ that $\mathcal{P}(\Delta^1)$ is a polar space of rank 3. But on the other hand, since Λ is an orthogonal space of Witt index 2, we have $\mathcal{P}(\Delta^1) \simeq \mathcal{P}(\widetilde{\Delta}^1)$ is isomorphic to the dual of Λ, whence $\mathcal{P}(\Delta^1)$ is of rank 2. This shows that case (α) is impossible.

In case (β) $\overline{V}$ is by (6.18) an orthogonal group of Witt index 4. Pick $C \in \Lambda$. Then $M_B^0 N_{V^1}(C) \le N_V(\Delta_C)$ by (7.1). Since $M_B^0 N_{V_1}(C)$ is by (8.7) a maximal parabolic subgroup of $N_V(B)$, it follows that $N_V(\Delta_C)$ is a maximal parabolic subgroup of V, since clearly $N_V(C) \not\le N_V(B)$.

If now $\overline{V}$ is of type B_4, then $N_V(B)$ corresponds to $\circ\!\!-\!\!\otimes\!\!-\!\!\circ\!\!=\!\!\circ$

and $N_V(B) \cap N_V(C)$ corresponds to $\circ\!\!-\!\!\otimes\!\!-\!\!\circ\!\!=\!\!\otimes$.

Thus $N_V(\Delta_C)$ corresponds to $\circ\!\!-\!\!\circ\!\!-\!\!\circ\!\!=\!\!\otimes$

and thus is the stabilizer of a maximal totally singular subspace U in the natural representation of V. The same holds (up to conjugation with a triality automorphism) if $\overline{V}$ is of type D_4, since $N_V(B)$ corresponds to

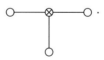

In any case, since $\dim_k U = 4$, there exist $x, y \in D(\Delta_0)$ with $U \cap U^{xy} = \{0\}$. Hence $\Delta_C \cap \Delta_C^{xy} = \emptyset$, since Δ_C consists of Siegel transvection groups corresponding to lines in U. But as $x, y \in D(\Delta_0)$ we have $C^x \in \Lambda_C \cap \Lambda_{C^{xy}}$ so that $[C, C^{xy}] \le C^x \cap A = 1$. Hence by (7.3) $\Delta_C \cap \Delta_{C^{xy}} \ne \emptyset$. This contradiction shows that (β) is also impossible, whence (8.8) holds. $\square$

Before we are able to show that the root group geometry of (G, Σ) is of finite singular rank, we need some additional properties of $N_\Delta(\Delta_C)$, $C \in \Lambda_A^*$.

(8.9) Lemma

Let $C \in \Lambda_A^*$ Then the following hold:

(1) $\mathcal{E}(\Delta_C)$ is connected and $N_{\Delta_0}(\Delta_C)$ is transitive on Δ_C.
(2) $\langle N_{\Delta_0}(\Delta_C) \rangle$ is Δ_0-maximal in V.

Proof. To prove (1) pick $E, F \in \Delta_C$ with $F \notin \Lambda_E$. Then by (2.6)(3) applied to Δ_0 there exists a $D \in \Delta_0 \cap \Delta_E$ with $[D, F] \neq 1$. Now by (7.2)(3)

$$[D, F] \in \Delta_C \cap \Lambda_E \cap \Lambda_F.$$

This proves the connectedness of $\mathcal{E}(\Delta_C)$. Now (1) follows immediately by (7.2)(3).

To prove (2) suppose that $V > R > \langle N_{\Delta_0}(\Delta_C) \rangle$ is a Δ_0-subgroup of V. Let $\Psi = R \cap \Delta_0$ and let Ω be the set of isolated vertices of $\mathcal{F}(\Psi)$. Then by part (1) and II(2.12) either $\Delta_C \subseteq \Delta_1$, where Δ_1 is a connectivity component of $\mathcal{F}(\Psi)$ or $\Delta_C \subseteq \Omega$. In the first case we have for an $E \in \Delta_C$ and $F \in \Delta_1 \cap \Omega_E$ by II(4.8) applied to V:

$$V = \langle \Lambda_E \cap \Delta_0, F \rangle \leq \langle N_{\Delta_0}(\Delta_C), F \rangle \leq R,$$

a contradiction. So $\Delta_C \subseteq \Psi$. Assume $L \in N_\Omega(\Delta_C) - \Delta_C$. Then $[L, F] \in \Delta_C$ for some $F \in \Delta_C$. But then there exists by (7.2) a $T \in \Delta_0 \cap \Lambda_F \subseteq N_{\Delta_0}(\Delta_C) \subseteq \Psi$ with $[L, F, T] = F$. Hence $T \in \Psi \cap \Omega_L$, a contradiction to the definition of Ω. We obtain $N_\Omega(\Delta_C) = \Delta_C$ and thus by (7.2) $\Delta_C = \Omega$. But then $R \leq N_V(\Delta_C)$, a contradiction to the choice of R. □

(8.10) Lemma

Suppose $\overline{V}$ is orthogonal of Witt index n and $\overline{\Delta}_0$ is the class of Siegel transvection groups of $\overline{V}$. Then $5 \leq n \leq 7$.

Proof. By (8.8) $\Lambda = \Lambda_A \cap \Lambda_B$ is an orthogonal space of Witt index ≥ 3 and V^1 induces the corresponding orthogonal group on Λ. Hence by the construction of the orthogonal groups in §5 and §6, $\overline{V}$ is of Witt index $n \geq 5$. Now pick $C \in \Lambda$. We show:

(*) $N_V(\Delta_C)$ is the stabilizer of a maximal totally
             singular subspace in the natural representation of V.

To prove (*) let W be the natural V-module. Then $[W, B]$ is a singular line. Moreover, if $E \in \Delta_0 \cap \Lambda_B$, then by II(1.5) $[W, E] \subseteq [W, B]^\perp$. Let now, $E \in \Delta_C - \Lambda_B$. We show that also in this case

(**) $[W, E] \subseteq [W, B]^\perp.$

Suppose $(**)$ does not hold. Then by II(1.5)(4) $\dim([W,B] \cap [W,E]) = 1$. Now $\langle C_{\Delta_0}(B) \rangle = M_B^0 \cdot V^1$ with $V^1 = V_2 * V_3$ and $SL_2(k) \simeq V_2 \le C_V(\Lambda)$ and V_3 acts as the orthogonal group on Λ. Moreover, it follows from (2.29) (2) and II(1.5) that

$$\{F^m \mid F \in \Delta_0 \cap V_2, m \in M_B^0\} = \{E \in C_{\Delta_0}(B) \mid [W,E] \not\subseteq [W,B]^\perp\}.$$

Hence, since by (8.7) $M_B^0 \subseteq N(\Delta_C)$, we may assume that $E \in V_2 \cap \Delta_C$. But then $\Lambda \subseteq \Lambda_C$, since $[V_2, \Lambda] = 1$ and $\Lambda = C^{V_3}$. Hence, picking a hyperbolic pair $L, T \in \Lambda$ we have by (8.3)

$$E \in S(L,T) = S(A,B),$$

a contradiction. This proves $(**)$. Now $(**)$ implies that $[W,E] \subseteq [W,F]^\perp$ for $E, F \in \Delta_C$, which shows that $U = [W, \Delta_C]$ is a singular subspace of W and $N_V(\Delta_C) \le N_V(U)$.

Suppose U is not a maximal totally singular subspace. Then there exists a $D \in \Delta_0$ with $[W,D] \subseteq U^\perp$ but $[W,D] \not\subseteq U$. Hence II(1.5) implies $D \in C_{\Delta_0}(\Lambda_C) = \Lambda_C$ a contradiction by (7.2). Thus U is a maximal totally singular subspace and

$$\Delta_C = \{D \in \Delta_0 \mid [W,D] \subseteq U\}.$$

Now $(*)$ is a consequence of the Δ_0-maximality of $\langle N_{\Delta_0}(\Delta_C) \rangle$ by (8.9).

If now $\dim U \ge 8$, then we have for each triple $x, y, z \in D(\Delta_0)$ that $\dim U \cap U^{xyz} \ge 2$, since $\operatorname{codim} C_W(x) = 2$. This implies $\Delta_C \cap \Delta_C^{xyz} \ne \emptyset$, since it contains the Siegel transvection group corresponding to a line in $U \cap U^{xyz}$. Now by (7.3) $[C, C^{xyz}] = 1$.

On the other hand, if $z \in D(\Delta_0)$ with $B^z = B_0$, then for some $F \in \Lambda^z$ we have $[\Lambda, F] \ne 1$. Indeed by II(2.3) we have $[B, F] \ne 1$, whence by (8.4) it must induce a Siegel transvection group on $S(A,B)$ and thus cannot centralize the hyperplane $A\langle \Lambda \rangle$. Now let $F = E^z, E \in \Lambda$ and $C \in \Lambda - \Sigma_F$. Then there exist by (8.7) $x, y \in D(\Delta^1)$ with $C^{xy} = E$, whence $C^{xyz} = F$ contradicting $[C, F] \ne 1$. This proves $n \le 7$. □

(8.11) Corollary

$\operatorname{rk}\Sigma \le 10$. (Notice that we only need $\operatorname{rk}\Sigma < \infty$ to quote the theorem of Cohen-Cooperstein!)

Proof. Suppose first $\overline{V}$ is orthogonal and Δ_0 is the class of Siegel transvection groups of $\overline{V}$. Then by (8.10) and exercise $(8.14)(2)$, $\operatorname{rk}\Delta_0 \le 6$. Hence $\operatorname{rk}\Sigma \le 8$ by (8.2).

Next suppose that V is not orthogonal. Then by §5 and §6 $\mathcal{F}(\Delta^1)$ is connected. Thus, if $\langle C_{\Delta_0}(B) \rangle$ is not Δ_0-maximal in V, then by (3.22), (6.18) and (8.7) $V_1 \simeq A_3(k) \simeq PSL_4(k)$ and $\overline{V} \simeq A_5(k) \simeq PSL_6(k)$, since by exercise (8.14)(1) $A_3(k)$ is the only linear group (i.e. satisfying the conclusion of (3.22)!) which is also an orthogonal group, such that the classes of transvection groups and Siegel transvection groups coincide. Hence $\mathrm{rk}\,\Delta_0 = 5$ and $\mathrm{rk}\,\Sigma \leq 7$ by (8.2).

Thus we may assume that $\langle C_{\Delta_0}(B) \rangle$ is Δ_0-maximal in V. Clearly Δ^1 is non-degenerate, since V^1 induces an orthogonal group of Witt index ≥ 3 on Λ by (8.8). Thus the hypothesis of this section is satisfied for (V, Δ_0) and (V^1, Δ^1). But then by (8.10) and exercise (8.14)(2) $\mathrm{rk}\,\Delta^1 \leq 6$, whence $\mathrm{rk}\,\Delta_0 \leq 8$ and $\mathrm{rk}\,\Sigma \leq 10$ by (8.2).   □

(8.12) Corollary

The root group geometry of (G, Σ) is isomorphic to the root group geometry of $E_6(k), E_7(k)$ or $E_8(k)$.

Indeed by (8.5), (8.8) and (8.11) the root group geometry of (G, Σ) satisfies the hypothesis of the theorem of Cohen-Cooperstein stated in (8.1), whence (8.12) holds.

(8.13) Theorem

G is isomorphic to $E_6(k), E_7(k)$ or $E_8(k)$ and Σ is the class of long root subgroups of G.

Proof. Let $\mathcal{G}$ be the root group geometry of (G, Σ). Then by (8.12) $\mathrm{Aut}(\mathcal{G})$ is isomorphic to the automorphism group of a building of type E_6, E_7 or E_8 over k, since as shown in [CC83] the building can be reconstructed from the root group geometry. Hence

$$G \leq \mathrm{Aut}(\mathcal{G}) = \widehat{G}\mathrm{Aut}(k); \qquad \widehat{G} \simeq E_6(k), E_7(k) \text{ or } E_8(k).$$

Now it can be shown exactly as in (7.15) that

(*) A is the set of all $\alpha \in \mathrm{Aut}(\mathcal{G})$, which fix all singular subspaces containing A and all symplecta $S(A, B), B \in \Sigma_A - \Lambda_A$, pointwise

Hence it follows from the proof of II(5.20) that A is a root subgroup of $\widehat{G}$. But then Σ is the class of root subgroups of $\widehat{G}$, since Σ contains a root subgroup corresponding to each point of $\mathcal{G}$. Together with II(5.22) this implies $G = \langle \Sigma \rangle = \widehat{G}$.   □

(8.14)   Exercises

(1) Suppose G is a group generated by the class Σ of abstract root subgroups, satisfying:

    (i) There exists an isomorphism $\varphi : G \to PT(V, W^*)$ mapping Σ on the class $\Sigma(W^*)$.

    (ii) There exists an isomorphism $\psi : G \to P\Omega(V, q)$ mapping Σ on the class of Siegel transvection groups of $P\Omega(V, q)$.

    Show that $G \simeq A_3(k) \simeq PSL_4(k)$.

    Hint. By (i) $\langle C_\Sigma(A) \rangle$ is not Σ-maximal for $A \in \Sigma$ and $\mathcal{F}(\Sigma_A - \Lambda_A)$ is connected. Thus the orthogonal group in (ii) must be of Witt index 3.

(2) Let G be an orthogonal group of Witt index n with $4 \leq n < \infty$ and Σ be the class of Siegel transvection groups of Σ. Show $\text{rk}(\Sigma) = n - 1$.

    Hint. Let V be the natural module for G. Show that a maximal singular subspace of Σ corresponds to a maximal pencil of singular lines of V contained in some maximal totally singular subspace.

(3) Show that in (8.10) $n \leq 6$.

    Hint. Suppose $n \geq 7$. For $x, y, z \in D(\Delta_o)$ we have $U \cap U^{xyz} \neq \{0\}$ and $U + U^{xyz} \subseteq (U \cap U^{xyz})^\perp$. Show that this implies $\Delta_C \cap N_{\Delta_0}(\Delta_C^{xyz}) \neq \emptyset$ for $C \in \Lambda$, which in turn implies $[C, C^{xyz}] = 1$ for all $x, y, z \in D(\Delta_0)$.

§9   The classification theorems

The purpose of this section is to describe the subdivision in the classification of groups generated by abstract root subgroups and state some "main theorems", which are consequences of the previous sections.

So, for the whole section, let G be a group generated by the set Σ of abstract root subgroups with $\mathcal{F}(\Sigma)$ connected (i.e. Σ is a conjugacy class). If Σ is degenerate assume that Σ satisfies the condition (H) of II §3; i.e. if $A \in \Sigma, B \in \Omega_A$, then there exist $A_0 \leq A, B_0 \leq B$ such that $\langle A_0, B_0 \rangle \simeq (P)SL_2(k)$, k a field with $|k| \geq 4$ and A_0, B_0 are full unipotent subgroups of $\langle A_0, B_0 \rangle$. We further assume that there exist commuting elements $C, D \in \Sigma$ with $\underline{C} \neq \underline{D}$. (For definition of $\underline{C}$ see the introduction of II §4.) Then we have by II(4.14):

(9.1)   Theorem

There exists a nilpotent normal subgroup $R(G)$, called radical, such that the following hold:

(1) $\widetilde{R} = R(G)/R(G) \cap Z(G)$ is abelian.

(2) Either $\overline{G} = G/R(G)$ is quasi-simple or Σ is degenerate, $o(a) \leq 3$ for each $a \in A \in \Sigma$ and $\overline{G}'$ is quasi-simple.

(3) $C_{\widetilde{R}}(\overline{G}) = 1$ and $\widetilde{R} = [\widetilde{R}, \overline{G}]$.

(4) $[\widetilde{R}, \overline{A}, \overline{C}] = 1$ for all commuting $A, C \in \Sigma$.

(5) $\underline{\overline{A}} = \{\overline{A}\}$ for each $A \in \Sigma$.

((5) means $Z(\langle C_{\overline{\Sigma}}(\overline{A}) \rangle) \cap \overline{\Sigma} = \{\overline{A}\}$!)

We will see at the end of this section that in most cases conditions (3) and (4) imply that $\widetilde{R} = 1$. Now one concentrates first on the easier case, when Σ is degenerate. Here we have:

(9.2) Theorem

Suppose Σ is degenerate and $R(G) = 1$. Then the following hold:

(1) $\mathcal{P}(\Sigma)$ is a thick non-degenerate polar space of (not necessarily finite) rank $r \geq 2$, satisfying the Moufang condition.

(2) G is the uniquely determined (normal) subgroup of $\mathrm{Aut}(\mathcal{P}(\Sigma))$ generated by all central elations. (If $r = 2$ up to duality)

(3) Let $T_A \leq \mathrm{Aut}(\mathcal{P}(\Sigma))$ be the group of all central elations corresponding to the point $A \in \Sigma$. Then $A \leq T_A$ and $A = T_A$ if $o(a) \neq 2$ for some $a \in A^{\#}$ and the polar space $\mathcal{P}(\Sigma)$ is classical.

(1) is (1.4) together with (1.7). (2) and the inequality $A \leq T_A$ are (1.10). Finally the equality $A = T_A$ if $\mathcal{P}(\Sigma)$ is classical and thence defined over a division ring of characteristic $\neq 2$ is exercise (1.14)(6).

Notice that by (1.12) the central elations are images of isotropic transvections as defined in II(1.4), if $\mathcal{P}(\Sigma)$ is classical. Hence our group G is some projective classical group. Moreover, as shown in II(1.4), all such groups generated by isotropic transvections must occur.

From now on we concentrate on the case when Σ is non-degenerate. Here we first treat the case when $N(A)$ is not a maximal subgroup of G for $A \in \Sigma$. In this case we have by (3.22):

(9.3) Theorem

Suppose $R(G) = 1$, Σ is non-degenerate and $N(A)$ is not maximal in G for $A \in \Sigma$. Then one of the following holds:

(1) There exists a Cayley division algebra K such that $G \simeq E_6^K$ and Σ is the class of elation subgroups of the Cayley plane over K.

(2) There exists a division ring K, a K-vector space V of dimension ≥ 3, a subspace W^* of the dual space V^* of V with $\mathrm{Ann}_V(W^*) = 0$ such that $G \simeq PT(V, W^*)$ and Σ is the class of elation subgroups of G, which are images of the transvection subgroups in $\Sigma(W^*)$.

For definition of $T(V, W^*)$ and $\Sigma(W^*)$ see II(1.3). Next we treat the case when $\langle C_\Sigma(A) \rangle = C_A$ is not Σ-maximal, but $N(A)$ is maximal in G. One subcase of this is the case when $\Sigma_A \subseteq M_A$. Here we have by §4:

(9.4) Theorem

Suppose $R(G) = 1$, Σ is non-degenerate and $\Sigma_A \subseteq M_A$ for $A \in \Sigma$, but $N(A)$ is maximal in G. Let $\mathcal{G}$ be the geometry with point-set Σ and line-set

$$\mathcal{L} = \{AB \cap \Sigma \mid A \in \Sigma \text{ and } B \in \Lambda_A\}.$$

Then the following hold:

(1) $\mathcal{G}$ is a thick Moufang hexagon.
(2) Σ is the class of long root subgroups on $\mathcal{G}$ (In the notation of II §5)
(3) Let $\widehat{G}$ be the group of Lie type $\mathcal{G}$ as defined in II §5. Then $G \trianglelefteq \widehat{G}$ and $\widehat{G} = G \cdot A_r$ for some short root subgroup A_r.

(1) and (2) are (4.4) and (4.8), while (3) is a consequence of II(5.22).

Now in the case when C_A is not Σ-maximal, it just remains to treat the subcase:

(B) C_A is not Σ-maximal, $\Sigma_A \not\subseteq M_A$, but $N(A)$ is maximal in G.

This case, together with the case

(A) C_A is Σ-maximal, $\mathcal{F}(\Delta_0)$ is not connected, where $\Delta_0 = \Sigma_A \cap \Sigma_{A_0}$ and $A_0 \in \Omega$.

is treated in §5 and §6. Here we have by (6.18) the following classification theorem:

(9.5) Theorem

Suppose Σ is non-degenerate, $R(G) = 1$ and either (A) or (B) holds. Then there exists a commutative field k, a k-vector space V and a non-degenerate quadratic form $q : V \to k$ of Witt index ≥ 3, such that the following hold:

(1) $G \simeq P\Omega(V, q)$.
(2) Σ is the class of "images" of the Siegel transvection groups of $\Omega(V, q)$.

Indeed, it was shown in §5 that (B) corresponds to the orthogonal groups of Witt index 3, while (A) corresponds to those of Witt index ≥ 4.

Now it just remains to treat the case when C_A is Σ-maximal and $\mathcal{F}(\Delta_0)$ connected. Here we have:

(9.6) Theorem

Suppose Σ is non-degenerate, $R(G) = 1$, C_A is Σ-maximal and $\Delta_0 = \Sigma_A \cap \Sigma_{A_0}$ is a degenerate class of abstract root subgroups of $V = \langle \Delta_0 \rangle$. Then there exists a thick building $\mathcal{B}$ of type F_4 such that G is the group of Lie type $\mathcal{B}$ (in the notation of II §5) and Σ is the class of long root subgroups on $\mathcal{B}$.

(9.6) is (7.16) restated. Notice that (9.6) as stated does not imply the uniqueness of Σ in the "abstract" group G.

Finally it remains to treat the case when Δ_0 is a non-degenerate class of abstract root subgroups of V. Here we have by (8.13):

(9.7) Theorem

Suppose Σ is non-degenerate, $R(G) = 1$, C_A is Σ-maximal and Δ_0 is a non-degenerate class of abstract root subgroups of V. Then there exists a field k, such that $G \simeq E_6(k), E_7(k)$ or $E_8(k)$ and Σ is the class of long root subgroups of G.

Notice that (9.2)–(9.7) together treat all possible cases for Σ, C_A and Δ_0. Hence together they give a complete classification of the groups G satisfying the hypothesis of this section with $R(G) = 1$.

Finally we want to show that, if Σ is a non-degenerate class of abstract root subgroups of G, then in most cases $R(G) \leq Z(G)$. We have:

(9.8) Corollary

Suppose Σ is non-degenerate and $\widetilde{R} = R(G)/R(G) \cap Z(G) \neq 1$. Let $\overline{G} = G/R(G)$ and $G^* = \overline{G}/Z(\overline{G})$. Then one of the following holds for G^* and Σ^*:

(1) $G^* \simeq PT(V, W^*)$ or E_6^K and Σ^* is the class of elation groups of G^*.

(2) G^* is of Lie type $\mathcal{G}$, $\mathcal{G}$ a thick Moufang hexagon and Σ^* is the class of long root subgroups of G^*.

(3) $G^* \simeq P\Omega(V, q)$, V a vector space over the field k and $q : V \to k$ a non-degenerate quadratic form of Witt index 3. Further Σ^* is the class of Siegel transvection groups of G^*.

Proof. By (9.1) $\widetilde{R}$ is a non-trivial $\mathbb{Z}G$-module satisfying $[\widetilde{R}, \overline{E}, \overline{F}] = 0$ for all commuting $\overline{E}, \overline{F} \in \overline{\Sigma}$. Hence by II(4.15) $\langle C_{\overline{\Sigma}}(\overline{A}) \rangle$ is not $\overline{\Sigma}$-maximal in $\overline{G}$.

Now by definition of the radical we have $R(\widetilde{G}) \leq Z(\widetilde{G})$ so that $R(G^*) = Z(G^*) = 1$ and $\langle C_{\Sigma^*}(A^*) \rangle$ is not Σ^* maximal in G^*. Now (9.3)–(9.5) show that one of the cases (1)–(3) holds. $\square$

(9.9) Exercise

Suppose Σ is non-degenerate, $R(G) = 1$ and $\mathrm{rk}\,\Sigma < \infty$, where $\mathrm{rk}\,\Sigma = \mathrm{Max}\{\dim_K N \mid N$ a subgroup of G partitioned by $\Sigma \cap N\}$ and $K = \{\sigma \in \mathrm{End}\,N \mid A^\sigma \leq A$ for all $A \in \Sigma \cap N\}$. Show that in this case G is of Lie type $\mathcal{B}$, $\mathcal{B}$ an irreducible spherical Moufang building of range ≥ 2.

Hint. Show that by the classification theorems (9.3)–(9.7) the only groups with $\mathrm{rk}\,\Sigma = \infty$ are the groups $PT(V, W^*)$ with $\dim V = \infty$ and orthogonal groups of infinite Witt index.

Chapter IV
Root involutions

In this chapter we will describe how the classification of finite groups generated by a class D of root involutions can be reduced to the classification of groups generated by abstract root subgroups. This root involution classification played an important role in the classification of finite simple groups and also for the determination of the subgroup structure of finite simple groups. Apart from its use for identifications of simple groups of characteristic 2 type, the main application of the root involution classification is through the theory of TI-subgroups, i.e. elementary abelian 2-subgroups T satisfying $T \cap T^g = T$ or 1 for all $g \in G$, see exercises III(3.23)(4)–(6). For a more detailed discussion of the role of TI-subgroups for the classification of finite simple groups see Aschbachers article [Asc80]. Also the special case of $\{3,4\}^+$-transpositions is of importance for the classification of groups with large extraspecial 2-subgroup, see [Tim78a] and [Smi80a], which was considered as one of the main open problems for the classification of finite simple groups, since most of the sporadic groups satisfy this condition.

To state the main result of this chapter, (4.11), we shortly repeat the definition of "root involutions" of III(3.23):

A class D of involutions generating the finite group G is called a class of *root involutions* of G, if for all $d, e \in D$ one of the following holds:

   (i) $de = ed$.
   (ii) $o(de)$ is odd.
   (iii) $o(de) = 4$ and $(de)^2 \in D$.

Let $\mathcal{D}(D)$ be the commuting graph of D, i.e. the graph with vertex set D and edges (d, e) with $d, e \in D$ and $1 \neq de = ed$. Further, for $d \in D$ let

$$E_d = \{e \in D \mid C_D(d) = C_D(e)\} \text{ and } E(d) := \langle E_d \rangle.$$

Then we will prove:

Root Involution Theorem (4.11): Let G be a finite group with $O_2(G) = Z(G) = 1$ generated by a class D of root involutions, satisfying $D \cap D^2 \neq \emptyset$, where $D^2 = \{ab \mid a, b \in D \text{ and } ab = ba \neq 1\}$. Then one of the following holds:

(1) $\mathcal{D}(D)$ is disconnected and $G \simeq L_2(q)$, $Sz(q)$ or $U_3(q)$, $q = 2^m > 2$.

(2) $\mathcal{D}(D)$ is connected, $E(d) = E_d \cup \{1\}$ for $d \in D$

and one of the following holds:

   (i) $\Sigma = E(d)^G$ is a class of k-root subgroups of G, where $k = GF(q)$ and $q = |E(d)|$.

   (ii) D is non-degenerate, $|E(d)| = 2$ and $|\langle C_D(d) \rangle| \leq 2^{10}$ for $d \in D$.

(In fact in case (ii) we get more information and we will also sketch the classification of groups satisfying (2)(ii). The groups are A_6, $U_3(3) \simeq G_2(2)'$, J_2 and $^3D_4(2)$. The arguments used to determine these groups in (2)(ii) are fairly standard local group theory.)

The proof of this so called "Root Involution Theorem" in this book will be largely self-contained. The only major results from finite group theory needed are Bender's classification of groups with a strongly embedded subgroup [Ben71], a theorem of Aschbacher [Asc73] giving a condition for the existence of a strongly embedded subgroup and a classical theorem of Suzuki classifying $L_3(2^n)$ [Suz65]. But of course, contrary to the other chapters, we assume a certain familiarity with finite group arguments and the structure of (the covering groups of) certain small finite simple groups, in particular $L_2(q)$, $Sz(q)$ and $U_3(q)$.

Unfortunately the theory of abstract root subgroups is not of much use for the classification of groups generated by a class D of odd transpositions (see (1.1) for definition) satisfying $D \cap D^2 = \emptyset$. (It is obvious that, if D is a class of 3-transpositions, always $D \cap D^2 = \emptyset$.) In fact the condition $D \cap D^2 \neq \emptyset$ splits the standard examples of odd transposition groups, i.e. symplectic and unitary groups over $GF(q)$, $q = 2^m > 2$ from the more exceptional examples, which correspond to "thin lines" in geometric terms. I believe it might be better to treat these groups (i.e. satisfying $D \cap D^2 = \emptyset$) together with 3-transposition groups.

Finally we will assume in this (and only this) chapter that all groups to be considered are finite. Most of the proofs in this chapter are either from [Asc72] or [Tim75a].

§ 1   General properties of groups generated by root involutions

(1.1)   Definitions

A set D of involutions of the group G is called a set of *root involutions* of G, if the following hold:

   (I) $G = \langle D \rangle$ and $D^g \subseteq D$ for all $g \in G$.
  (II) For all pairs $a, b \in D$ one of the following holds:

     (a) $ab = ba$.

     (b) $o(ab)$ is odd.

     (c) $o(ab) = 4$ and $(ab)^2 \in D$.

Since $(ab)^2 = [a, b]$ it is obvious that this definition is related to the definition of abstract root subgroups.

If in addition D is a conjugacy class in G, D is called a *class of root involutions* of G. There are certain specializations of the notion of root involutions, which are important for us, namely:

1. D is called a set of *odd transpositions* or also a *degenerate* set of *root involutions* of G, if II(c) never occurs.
   If all types of products occur, we say that D is a *non-degenerate set* of root involutions.

2. D is a set of *3-transpositions* of G, if it is a set of odd transpositions and $o(de) = 3$, whenever $o(de)$ is odd.

3. D is a set of $\{3, 4\}^+$-*transpositions* of G, if it is a set of root involutions of G with $o(de) = 3$, whenever $o(de)$ is odd.
   A set of 3-transpositions could also be called a degenerate set of $\{3, 4\}^+$-transpositions.

The notation $\{3, 4\}^+$-transpositions has historical reasons. Namely in the attempt to generalize Fischers 3-transposition classification in [Tim70] and [Tim73], I found it necessary to assume an additional condition, which is just II(c) above, which I called +-condition.

As in the case of abstract root subgroups, there are three graphs, all with vertex set D, connected with a set D of root involutions, namely:

$\mathcal{D}(D)$ (the commuting graph of D) has as edges the commuting pairs (e, d) (i.e. with $ed = de \neq 1$).

$\mathcal{F}(D)$ has as edges the pairs (e, d) with $o(dc)$ odd.

$\mathcal{E}(D)$ (which is a subgraph of $\mathcal{D}(D)$) has as edges the pairs (e, d) with $ed \in D$.

As in II §1 we also use pictorial notation. Namely for $d, e \in D$ with $d \neq e$ we write:

$$
\begin{array}{cc}
\overset{d}{\circ} & \overset{e}{\circ}
\end{array}
\qquad \text{if } de = ed,
$$

$$
\begin{array}{ccc}
\overset{d}{\circ} & \sim & \overset{e}{\circ}
\end{array}
\qquad \text{if } de = ed \in D,
$$

$$
\overset{d}{\circ}\!\!-\!\!-\!\!\overset{e}{\circ}
\qquad \text{if } o(de) = 3,
$$

$$
\overset{d}{\circ}\!\!\rightsquigarrow\!\!\overset{e}{\circ}
\qquad \text{if } o(de) \text{ is odd },
$$

$$
\overset{d}{\circ}\!\!=\!\!=\!\!\overset{e}{\circ}
\qquad \text{if } o(de) = 4.
$$

(This notation coincides with the standard notation for Coxeter groups in so far that, in case

$$
\begin{array}{cc}
\overset{d}{\circ} & \overset{e}{\circ}
\end{array}
$$

we have $\langle d, e \rangle \simeq D_4$. In case

$$
\overset{d}{\circ}\!\!-\!\!-\!\!\overset{e}{\circ}
$$

we have $\langle d, e \rangle \simeq D_6$ and in case

$$
\overset{d}{\circ}\!\!=\!\!=\!\!\overset{e}{\circ}
$$

we have $\langle d, c \rangle \simeq D_8$.)

Finally for $d \in D$ let:

$$
\begin{aligned}
D^2 &:= \{ de \mid d, e \in D \text{ and } de = ed \neq 1 \} \\
D_d &:= C_D(d) - \{d\} \\
X_d &:= \{ e \in D_d \mid ed \in D \} \\
Y_d &:= D_d - X_d \\
A_d &:= \{ e \in D \mid d \neq e \text{ and } o(de) \text{ is odd } \} \\
V_d &:= \{ e \in D \mid o(de) = 4 \} \\
E_d &:= \{ e \in D \mid C_D(d) = C_D(e) \}.
\end{aligned}
$$

Other types of natural notation, such as *D-subgroup* or $C_E(F)$ for $E, F \subseteq D$, the reader should translate from the corresponding notation for abstract root subgroups.

Notice that we will not discuss 3-transposition groups in this section in detail, since there is already a book of Aschbacher on 3-transposition groups [Asc97].

(1.2)   Examples

(a) Let G be a finite classical or Lie type group different from $^2F_4$ (in the sense of II(1.3)–(1.5) and II §5). Then some normal subgroup G_0 of G is by II(1.3)–(1.5) or II(5.20) generated by a class Σ of k-root subgroups of G_0, where k is a finite field. (This is so, since for example by the theorem of Kantor, Hering and Seitz [HKS72] a finite special rank one group is isomorphic to $(P)SL_2(k)$. But there are more direct proofs for that.) Suppose $k = GF(2^n), n \in \mathbb{N}$. Then $D(\Sigma)$ is a class of root involutions of G_0. Moreover, if $n = 1$, then $D(\Sigma)$ is a class of $\{3,4\}^+$-transpositions of G_0.

This is rather obvious to see. Indeed if $A, B \in \Sigma$ and $a \in A^\#$, $b \in B^\#$ we have:

$o(ab) = 4$ and $(ab)^2 \in D(\Sigma)$ if $\langle A, B \rangle$ is nilpotent of class 2 with $[A, B] \in \Sigma$.

$o(ab)$ is odd if $\langle A, B \rangle \simeq (P)SL_2(2^n)$.

Moreover, if $\langle A, B \rangle \simeq (P)SL_2(2)$, then $o(ab) = 3$.

The fact that $D(\Sigma)$ is a conjugacy class in G_0 follows immediately from the fact that $\mathcal{F}(\Sigma)$ is connected, which has been shown in II(1.3)–(1.5) (resp. II(5.20)).

This shows (of course modulo the results of II(1.3)–(1.5) and II(5.20)) that all finite simple groups of Lie type in characteristic 2, different from $^2F_4$, are generated by a class of root involutions.

(b) A_6. Then the only class D of involutions of A_6 is a class of root involutions of A_6 with $o(de) \in \{1, 2, 3, 4, 5\}$ for $d, e \in D$.

(c) The (simple) Hall-Janko group J_2 of order $2^7 \cdot 3^3 \cdot 5^2 \cdot 7$ with D the class of 2-central involutions of G (i.e. $d \in D$ is central in some 2-Sylow subgroup of G). Indeed it has been shown by M. Suzuki [Suz69] that $J_2 < G_2(4)$ with D contained in the set of long root elements of $G_2(4)$. Hence by (a)

$$o(de) \in \{1, 2, 3, 4, 5\} \text{ for all } d, e \in D.$$

If now $o(de) = 4$, then $(de)^2 \in D$. For example if $(de)^2 \notin D$, then the other class of involutions of J_2 would also be a set of long root elements of $G_2(4)$. But then these two classes would centralize each other $\mod O_2(J_1)$ by (1.6) below, which is nonsense. The same argument shows that $\mathcal{F}(D)$ is connected.

We now state the main result of this section, which will be proved at the very end of the section. To state it we provide another piece of notation. For any group R, let $Z_*(R)$ be the maximal coimage of $Z(R/O_2(R))$. We have:

(1.3) Theorem

Let D be a set of root involutions of G satisfying $D \cap D^2 \neq \emptyset$. Then one of the following holds:

(1) $\mathcal{D}(D)$ is connected   or

(2) $G/Z_*(G)$ is isomorphic to $L_2(q)$, $S_2(q)$ or $U_3(q)$, $q = 2^m > 2$.

In particular if D is non-degenerate, then $\mathcal{D}(D)$ is connected.

The proof uses Aschbacher's condition of the existence of a strongly embedded subgroup [Asc73] and of course Bender's theorem [Ben71]. Apart from that it is independant from deeper theorems on finite groups (of course we need properties of $L_2(q), U_3(q)$ and $S_2(q)$, which we will use without reference!).

Suppose from now on, that the group G is generated by a set D of root involutions.

(1.4) Lemma

The following hold:

(1) If U is a D-subgroup of G, then $U \cap D$ is a set of root involutions of U.

(2) If α is a homomorphism of G, then D^α is a set of root involutions of G^α.

(3) If $\mathcal{F}(D)$ is connected, then $|G : G'| \leq 2$.

(4) Let $S \in \mathrm{Syl}_2(G)$. Then $N_G(\langle S \cap D \rangle)$ controls fusion in $Z(\langle S \cap D \rangle)$.

(5) If $a, b, c \in D$ with

$$\underset{a}{\circ}\!=\!=\!=\!\underset{b}{\circ}\!-\!-\!-\!\underset{c}{\circ},$$

then $\langle a, b, c \rangle$ is an image of $(D_8 \times D_8)\langle b \rangle$. Moreover, if $ac \in D$, then $\langle a, b, c \rangle$ is already an image of $(D_8 * D_8)\langle b \rangle$.

(6) If $a, b \in D$ with $1 \neq o(ab)$ odd, then $X_a \cap D_b = \emptyset$.

Proof. (1) and (2) are obvious. (3) follows from (2) applied to $G \to G/G'$. The proof of (4) is the same as the proof of Burnside's fusion theorem, see [Gor68, Chap.7,(1.1)]. Finally, the proof of (5) is the same as the proof of II(2.10). (6) is completely obvious. (If $c \in X_a \cap D_b$, then $o(bac) = 2o(ab)$.)   □

(1.5) Lemma

Let $a, b, c \in D$ satisfying

$$\underset{a}{\circ}\!\!\sim\!\!\sim\!\!\underset{b}{\circ}\!=\!=\!=\!\underset{c}{\circ},$$

$X = \langle a, b \rangle$, $Y = \langle a, b, c \rangle$ and $Q = \langle c^X \rangle$. Then the following hold:

(1) $Y = QX$ with Q an elementary abelian normal 2-subgroup of Y and $Q \cap X = 1$.

(2) Let $Q_0 = \langle Q \cap X_a \rangle \langle Q \cap X_b \rangle$. Then Q_0 is normal in Y and $Q = Q_0 \langle c \rangle$.

(3) Suppose $ac \in D$. Then there exists an $x \in Q \cap X_b$ with $a^x = ac$.

(4) Let $\alpha \neq \beta \in c^X$. Then $\alpha\beta \in D$.

Proof. To prove (1) we show that each element of c^X commutes with c. Now any such element is of the form

$$c^{ba...b} \text{ or } c^{ba...ba}.$$

We will show that such an element commutes with c by induction on the number of a's and b's occuring in the exponent. Clearly c^b and c^{ba} commute with c. Let now $z = ba \dots b$. Then obviously c^z commutes with c if and only if c^{za} commutes with c. By the induction assumption, $[c, c^{zb}] = 1$. Now, as $X \simeq D_{2n}$, n odd, z is an involution of X. Hence $z \in D$. Assume first $o(zc)$ is odd. Then also $o(z^b c^b) \equiv 1(2)$ and $o(c^b c^{zb}) \equiv 1(2)$, since $c^{zb} = (c^b)^{z^b} \in \langle z^b, c^b \rangle$. As c commutes with c^b and c^{zb} and $cc^b = (cb)^2 \in D$, this implies that $o((cc^b)c^{zb}) = 2m$, where $1 \neq m$ is odd, a contradiction to the root involution hypothesis.

This shows $o(cz) = 2$ or 4 and thus in any case $[c, c^z] = 1$. This implies that each element of c^X commutes with c, whence Q is an X-invariant elementary abelian 2-subgroup. Since $c \in Q$ we obtain $Q \triangleleft Y$ and $Y = QX$. Clearly $Q \cap X = 1$, since $X \simeq D_{2n}$.

To prove (2) let $F_a = Q \cap X_a$, $F_b = Q \cap X_b$. Then $F_a \neq \emptyset \neq F_b$, since $(bc)^2 \in F_b$. If now $\alpha \in F_b \cap Y_a$, then $o(\alpha ba) = 2m$, $1 \neq m$ odd, a contradiction to $\alpha b \in D$. Hence $F_a \subseteq V_b$. In particular $[F_a, b] \subseteq \langle F_b \rangle$ and thus with symmetry $Q_0 \triangleleft Y$ and $Q = Q_0 \langle c \rangle$, since $[c, b] \in F_b$.

If now $ac \in D$, then $c \in F_a$ and, to prove (3), it suffices to show that there exists an $e \in F_b$ with $[e, a] = e$. (Since then $a^e = ac$!). Now by the above $F_b \subseteq V_a$. Moreover $|F_a| = |F_b|$, since a and b are conjugate in X. Thus to show that the map

$$e \to [e, a], \qquad c \in F_b$$

is a bijection from F_b on F_a, it suffices to show that there exists no $e \neq f \in F_b$ with $[e, a] = [f, a]$.

Suppose $[e, a] = [f, a]$. Then $ef \in \langle F_b \rangle \cap C(a) \leq Z(Y)$. Hence $o(a(bf)) = o((abe)(ef)) = 2m$, $1 \neq m$ odd, since clearly $be \in A_a$, a contradiction to $bf \in D$.

This proves (3). Finally to prove (4) notice that we have shown in the proof of (1)

$$c^X = \{c^y \mid y \in D \cap X\}.$$

In particular also $\alpha^X = \{\alpha^y \mid y \in D \cap X\}$, since $|\alpha^X| = |c^X| = n$ and $|C_{D \cap X}(\alpha)| = |C_{D \cap X}(c)| = 1$. As $D \cap X - (C_{D \cap X}(\alpha)) \subseteq V_\alpha$, (4) is now a consequence of the definition of root involutions. $\square$

(1.6) Lemma

Let $D_i, i = 1, \ldots, n$ be the connectivity components of $\mathcal{F}(D)$ of cardinality > 1 and W the set of isolated vertices of $\mathcal{F}(D)$. Then the following hold:

(1) $\langle W \rangle$ is a normal 2-subgroup of G and $W^g \subseteq W$ for all $g \in G$. Further $\langle W \rangle \cap D = W$.

(2) $[D_i, D_j] \le \langle W \rangle$ for $i \ne j$.

(3) If $a \in D_i$, then $X_a \subseteq D_i \cup W$.

Proof. $W^g \subseteq W$ follows from the definition of W. The well-known theorem of Baer, see [Asc86, (39.6)], implies that $\langle W \rangle$ is a 2-group. Hence (1) holds.

To prove (3) assume by way of contradiction $c \in D_j \cap X_a$ and $j \ne i$. Pick $b \in A_a \cap D_i$. (Exists by definition of D_i!) Then we have

since $c \notin A_b$ and since clearly $e \notin D_b$, since otherwise $o(acb) = 2m, 1 \ne m$ odd. Now (1.5)(3) implies $ac = a^e$ for some $e \in D_b$. As $b \in D_i$ we have $D_i^e \subseteq D_i$ and thus $ac \in D_i$.

Now, reversing the roles of a and c, pick $b' \in A_c \cap D_j$. Then $b' \in V_a$ and the same argument as above implies $ac \in D_j$, a contradiction to $D_i \cap D_j = \emptyset$.

This proves (3). Now (2) is a consequence of (3). Indeed if $a \in D_i$ and $b \in D_j$ with $o(ab) = 4$, then as $(ab)^2 \in X_a \cap X_b$, (3) implies $(ab)^2 \in W$. Hence D_i and D_j centralize each other mod $\langle W \rangle$. $\square$

We will apply (1.6) to *D-subgroups* of G, that is subgroups U with $U = \langle U \cap D \rangle$, since then $U \cap D$ is a set of root involutions of U.

(1.7) Proposition

Suppose $\mathcal{F}(D)$ is connected. Then $E_d = \{e \in D \mid C_D(d) \cup V_d = C_D(e) \cup V_e\}$ for $d \in D$. (By definition $E_d = \{e \in D \mid C_D(d) = C_D(e)!\}$)

Proof. Let $F_d = \{e \in D \mid C_D(d) \cup V_d = C_D(e) \cup V_e\}$. Then F_d is a TI-subset of D, whence $N(F_d)$ is transitive on F_d since by assumption D is a conjugacy class. Now by (1.5) $E_d \subseteq F_d$. Moreover, if D is degenerate, then (1.7) holds. Hence we may assume that D is non-degenerate. We show first:

(1) If $a, b \in F_d$ and $ab \in D$, then $ab \in F_d$. In particular $\langle F_d \rangle$ is elementary abelian.

Suppose $ab \in D - F_d$. Then there exists an $x \in A_{ab} \cap V_a$ since otherwise $C_D(a) \cup V_a \subseteq C_D(ab) \cup V_{ab}$. But then (1.5) implies $x \in A_a$, a contradiction.

This proves the first part of (1). Suppose $a, b \in F_d$ with $1 \neq o(ab)$ odd. Pick $c \in X_a$. (Exists as D is non-degenerate!) Then $c \in V_b$ and a is by (1.5) conjugate to ac in $C(b)$. Hence $ac \in F_d$ and thus $c \in F_d$ as shown above. But this is a contradiction to $b \in V_c \cap A_a$.

Thus $o(ab) = 2$ or 4 for all $a, b \in F_d$. Hence the theorem of Baer implies that $\langle F_d \rangle$ is a 2-group. If now $o(ab) = 4$, then by the above $(ab)^2 \in F_d$. This shows that $F_d \cap Z\langle F_d \rangle \neq \emptyset$. Hence the transitivity of $N(F_d)$ on F_d implies (1).

We show next:

(2) $C_D(F_d) = N_D(F_d)$.

To prove (2) assume $x \in N_D(F_d) - C_D(F_d)$. Then there exists an $a \in F_d \cap V_x$. By (1) $aa^x \in F_d$. If $x^a \in F_x$, then also by (1) $xx^a \in F_x$, a contradiction to $F_d \cap F_x = \emptyset$. Thus $x^a \notin F_x$ and so there exists a $y \in C_D(x^a) \cup V_{x^a}$ with $x \in A_y$. Then clearly $y \in V_{x^a}$ and thus by (1.5) $xx^a = x^e$ for some $e \in D$. Hence $(F_x)^e = F_d$ and, as $x \in V_e$ and $F_x \cap F_d = \emptyset$, $F_x \subseteq V_e$. But then F_x and F_d normalize each other and thus centralize each other by (1), a contradiction to $a \in F_d \cap V_x$.

Notice that (2) implies $F_d \subseteq E_d$, which proves (1.7). $\qquad\qquad\qquad\square$

It is obvious that (1.7) also implies:

$$E_d = \{e \in D \mid V_d = V_e\} = \{e \in D \mid A_d = A_e\}$$

if $\mathcal{F}(D)$ is connected.

(1.8) Corollary

Suppose $\mathcal{F}(D)$ is connected $a \in D$ and $b \in X_a - E_a$. Then there exists an $x \in V_b$ with $o(ax) = 3$.

Proof. By (1.7) there exists a $y \in A_a \cap (C_D(b) \cup V_b)$. Since $ab \in D$ clearly $y \in V_b$. Now $b^y b^{ya} \notin E_{b^y}$, since a centralizes $b^y b^{ya}$. Hence $b^{ya} \notin E_{b^y}$. Thus by (1.5) and (1.7) there exists an $x \in D$ with $b^{yx} = b^y b^{ya}$. Hence $\langle x, a \rangle$ acts on the 4-group $V = \langle b^y, b^{ya} \rangle$ and thus $\langle x, a \rangle / K \simeq \Sigma_3$, where K is the kernel of the action of $\langle x, a \rangle$ on V. In particular $a \in A_x$, whence $K \leq \langle xa \rangle$ and if $K \neq 1$, then there exists a $k \in K$ such that $x \neq x^k$. But then $\langle x, x^k \rangle \leq C(b^{ya})$, a contradiction to $xb^{ya} \in D$ and $1 \neq o(xx^k)$ odd. $\qquad \square$

In the following let $S(G)$ be the maximal solvable normal subgroup of G.

(1.9) Lemma

Suppose $\mathcal{F}(D)$ is connected and $D \cap D^2 \neq \emptyset$. Then $S(G) = Z_*(G)$. Moreover $G/Z_*(G)$ is simple.

Proof. For the proof of the first part of (1.9) we may assume $O_2(G) = 1$ and then show $S(G) = Z(G)$. For this let $M = O_p(G)$, p an odd prime and assume $M \not\leq Z(G)$. Let $\{a, b, ab\} \subseteq D$. Then there exists an $h \in C_M(b)$ with $a \neq a^h$. Hence $o((ab)b^h) = 2p^m$, $m \neq 0$, a contradiction. This shows that $N = F(G) \leq Z(G)$. If now Q is the coimage of $O_2(G/N)$, then $Q = N \times S$, $S \in \text{Syl}_2(Q)$ and thus $Q = N$, since we assume $O_2(G) = 1$. This proves the first part of (1.8).

For the second part assume $Z_*(G) = 1$ and let $N \neq 1$ be a minimal normal subgroup of G. Then, if $a \in N$, we have $G = N$ is simple. Thus we may assume $\{a, b, ab\} \cap N = \emptyset$. But then by (1.6) applied to $D \cap N \langle a, b \rangle$ we obtain a contradiction to (1.6)(3). $\qquad \square$

(1.10) Definition

Let D be a set of odd transpositions of G and $a, b, c \in D$. Then we call (a, b, c) a *triangle*, if $ac = ca \neq 1$ and $1 \neq o(ab)$ is odd and $1 \neq o(bc)$ is odd. If now (a, b, c) is a triangle, then $o(acb) = m$, 4 or $2m$, m odd and $o(acb) = m$ if and only if $ac \in D$. We call (a, c, b) a triangle of the *first* (resp. *second* resp. *third* kind,) if $o(acb) = m$ (resp. $o(acb) = 4$ resp. $o(acb) = 2m$).

Notice that

$$(*) \qquad\qquad C_D(ac) = C_D(a) \cap C_D(c),$$

since if $x \in C_D(ac)$ does not commute with a, we obtain $o(xc) = 2n$, n odd, a contradiction.

(1.11) Lemma

Suppose D is a set of odd transpositions of $G = \langle a, b, c \rangle$ and (a, b, c) is a triangle. Then the following hold:

(1) Suppose (a, b, c) is of the third kind and z is the central involution in $\langle ac, b \rangle$. Then $z \in Z(G)$.

(2) Suppose (a, b, c) is of the first or third kind. Then $S(G) = Z_*(G)$ and $G/Z_*(G)$ is simple.

(3) Suppose (a, b, c) is of the first kind. Then $\{a, c, ac\} \subseteq E_a$.

(4) If G is solvable, then all triangles in D are of the second kind.

Proof. (1) $z = (ac)d$ for $d \in D \cap \langle ac, b \rangle$. As $d \in C(ac)$ we have $d \in C_D(a) \cap C_D(b)$ and (1) holds.

(2) If (a, b, c) is of the first kind, then (2) is a consequence of (1.9). But if (a, b, c) is of the third kind and $\overline{G} = G/\langle z \rangle$, then $(\overline{a}, \overline{b}, \overline{c})$ is of the first kind in $\overline{D}$, whence (2) holds.

(3) is absolutely obvious since, as $C_D(ac) = C_D(a) \cap C_D(c)$ and a, b and c are conjugate in G, we have

$$C_D(a) = C_D(ac) = C_D(c).$$

(4) is a direct consequence of (2).   □

(1.12)   Lemma

Let D be a set of odd transpositions of $G = \langle a, b, c \rangle$ and (a, b, c) a triangle of the second kind. Suppose that for each $x \in D \cap \langle a, b \rangle$, (a, x, c) is not a triangle of the third kind. Let $X = \langle a, b \rangle$, $Q = \langle (ac)^X \rangle$. Then the following hold:

(1) $Q = O_2(G)$ is elementary abelian.

(2) $Q = [Q, a][Q, b]$ and $c = a^x$ for some $x \in [Q, b] \leq C_Q(b)$.

Proof. The proof is very similar to (1.5), so we just sketch it. To prove (1) it suffices to show that $(ac)^X \subseteq C(ac)$. Let $1 \neq x \in X$. Then $[ac, (ac)^x] = 1$ if and only if $[ac, (ac)^{ax}] = 1$. Hence we may assume $a \neq x \in D \cap X$. If $x \in C(b)$, then $\langle ax \rangle \triangleleft G$, a contradiction to a inverting ax and b centralizing it. Hence (a, x, c) is a triangle. Since $ab \notin D$, (a, x, b) is by hypothesis of the second type. Hence $o(acx) = 4$ and thus $[ac, (ac)^x] = 1$.

Clearly $Q_0 = [Q, a][Q, b]$ is X-invariant and thus $Q_0 \triangleleft G$ by (1). Moreover $Q/Q_0 \leq Z(G/Q_0)$, so that if $Q \neq Q_0$ and $\overline{G} = G/Q_0$, we have $\overline{G} = \overline{X} \times \langle \overline{ac} \rangle$. But then $o(\overline{bc}) = 2o(\overline{ba})$, a contradiction to (1.4).

Thus $Q = Q_0$ and, since $[Q, a] \leq C_Q(a)$ and a and c are conjugate under Q, (2) holds.   □

(1.13)   Lemma

Let D be a set of odd transpositions of G and $Q = O_2(G)$. For $a \in D$ let $Q_a = [Q, a]$. Then the following hold:

  (1) $C_D(a)$ centralizes Q_a.
  (2) Let $b \in A_a$. Then $C_Q(ab) = C_Q(a) \cap C_Q(b)$.
  (3) If $G = \langle C_D(a), b \rangle$, then $Q_a \cap Q_b \le Z(G)$ and $Q = C_Q(a)Q_b$. Moreover, if $c, d \in D \cap Qa$, then there exists a unique $e \in D$ with $acd \equiv e \bmod Z(G)$.

Proof. Let $f \in a^Q$. Then each $x \in D_a$ centralizes f, since otherwise $o(xa) \equiv 1 \bmod 2$, since $a \equiv f \bmod Q$. As $Q_a \le \langle a^Q \rangle$, this implies (1).

To prove (2) let $P = C_Q(ab)$. Then $\langle a, b \rangle$ acts on P. Hence if $a \ne a^g$ for $g \in P$, then $o(a^g b) = 2m$, $1 \ne m$ odd, a contradiction.

$Z = Q_a \cap Q_b \le Z(G)$ in (3) is a consequence of (1). If now $f \in D_a$, then by (1) $Q_f \le C_Q(a)$, so that $C_Q(a)Q_b$ is G invariant. Further $[Q_a, Q_b] \le Z$, whence Q_aQ_b/Z is abelian and Q_aQ_b is $\langle a, b \rangle$ invariant.

Now by (1.12)(2) applied to $\langle a, b, c \rangle$, there exists an $x_1 \in Q_b$ with $a^{x_1} = c$. Similarly $a^{x_2} = d$ for $x_2 \in Q_b$. Hence

$$
\begin{aligned}
acd &= aa^{x_1}a^{x_2} = ax_1(x_1x_2)^a x_2 \equiv a(x_1x_2)^a x_1 x_2 \\
    &\equiv a^{x_1 x_2} \bmod Z,
\end{aligned}
$$

since Q_b is elementary abelian. Since $aZ \cap D = a$ this proves (3). $\square$

(1.14)   Lemma

Let D be a class of odd transpositions of G, $Q = O_2(G)$ and $S \in \mathrm{Syl}_2(G)$. Let (a, b, c) be a triangle of the third kind and assume $\langle a, c, z \rangle \le S$, where z is the central involution of $\langle ac, b \rangle$. Suppose further that $G = \langle S \cap D, b \rangle$ and, that there exists no perfect central extension of G/Q by a 2-group, which is generated by odd transpositions. Then, if $d, e \in S \cap D$ and $f \in Qd \cap D$, there exist unique $u \in D$ and $z' \in C_Q(a) \cap C_Q(b) = Z(G) \cap Q$ such that

$$
def = uz'.
$$

Proof. Let $Q_1 = \Pi Q_v$ where v runs over $S \cap D$ and $Q_0 = Q_1 Q_b$. (In the notation of (1.13)). Then Q_0 is G-invariant. Moreover, as $[Q, b] = Q_b \le Q_0$, we have $Q/Q_0 \le Z(G/Q_0)$. Hence G/Q_0 is a central extension of G/Q generated by odd transpositions and thus by hypothesis either $Q = Q_0$ or the extension splits. In the second case $|G : G'| = 2$ and $Q_0 = Q \cap G'$. Now $zac \in S \cap D$ and thus $z = (zac)ca \in G - G'$. Since $G = \langle S \cap D, b \rangle$ we obtain $z \in Z(G)$ and either $Q = Q_0$ and $z \in Q_0$ or $G = G' \times \langle z \rangle$. Now, since $v \in A_b$ for $v \in S \cap D$, it follows from (1.13)(3) that $C_{Q_b}(v) = Q_b \cap Q_v$. Hence $C_{Q_0}(v) = Q_1$ and $C_Q(v) = Q_1 \langle z \rangle$. Therefore by (1.13) $C_Q(ab) = C_Q(a) \cap C_Q(b)$ is centralized by $G = \langle S \cap D, b \rangle$. Hence $Z(G) \cap Q = C_Q(ab)$.

Now Q_a is transitive on $Qb \cap D$. Hence, by the Frattini argument, $C_Q(a) = Q_a(C_Q(a) \cap C_Q(b)) = Q_a C_Q(ab)$. Further, if $a_i \in Qa \cap D$, for $i = 1, 2$, then by (1.13)(3)

$$a_1 a_2 = a(aa_1 a_2) \equiv aa_3 \quad \mod Z(G) \cap Q \text{ for some } a_3 \in Qa \cap D.$$

Hence

$$Q_a = \langle a_1 a_2 \mid a_i \in Qa \cap D \rangle \le a(Qa \cap D)C_Q(ab)$$

and thus $C_Q(a) = Q_a C_Q(ab) \le a(Qa \cap D)C_Q(ab)$.

Now, as $C_Q(a) = C_Q(v)$ for each $v \in S \cap D$, this implies $df \in C_Q(e) \le e(Qe \cap D)(Q \cap Z(G))$. Hence $def = uz'$ with $u \in Qe \cap D$ and $z' \in Q \cap Z(G)$. The uniqueness of u and z' follows from $uZ(G) \cap D = \{u\}$. □

(1.15) Corollary

Assume the same hypothesis and notation as in (1.14) and in addition $G/Z_*(G) \simeq L_2(q)$ or $S_2(q)$, $q = 2^m$, $(S \cap D) \cap (S \cap D)^2 \ne \emptyset$ and Q_0 is abelian. Then

$$\Omega_1(S) - Q \subseteq (S \cap D) \cup (S \cap D)^2.$$

Moreover, for each involution $s \in S - Q$, there exist $d \in S \cap D$, $z \in Z$ such that $s = dz$.

Proof. By hypothesis there exist $\{\alpha, \beta, \alpha\beta\} \subseteq S \cap D$, $\alpha \ne \beta$ such that $G = Q\langle \alpha, \beta, x \rangle$ for some $x \in D$. As $G = G'$, since $\alpha\beta \in D$, we have $Q = Q_0$. Thus $Q = Q_\alpha Q_\beta Q_x$, since the latter group is G-invariant. Since $xQ_x = x^Q \subseteq D$ we have $Z \cap Q_x = 1$ for $Z = Z(G) \cap Q$. Thus $Z \le C_Q(\alpha) = Q_\alpha Q_\beta$ and, as shown in (1.14), $C_Q(\alpha) = Q_\alpha Z$. Hence, to prove (1.15), it suffices to show $\alpha z \in D^2$ for each $z \in Z^\#$. Now $z = uv$ with $u \in Q_\beta$, $v \in Q_{\alpha\beta}$ and $\alpha u, \alpha\beta v \in D$. Hence

$$\alpha z = (\beta u)(\alpha\beta v) \in D^2,$$

which is to show. Further, if s is an involution in αQ, then $s \in \alpha C_Q(\alpha)$ and thus $sz \in \alpha Q_\alpha \subseteq \alpha^Q$ for some $z \in Z$. □

(1.16) Proposition

Let D be a set of odd transpositions of G and assume $D \cap D^2 \ne \emptyset$. Then one of the following holds:

(1) $\mathcal{D}(D)$ is connected.

(2) $G/Z_*(G)$ is isomorphic to $L_2(q)$, $Sz(q)$ or $U_3(q)$, $q = 2^m > 2$.

Proof. Let G be a minimal counterexample to (1.16). Then by (1.9) G is simple. Let $E \neq D$ be a connectivity component of $\mathcal{D}(D)$. If $ef \in D$ for all $e, f \in E$ with $ef = fe \neq 1$, then $S \cap D = \langle S \cap D \rangle^{\#}$ for some 2-Sylow subgroup S of G. Moreover for $d \in S \cap D$ we have $C_D(d) = S \cap D$. Hence $\langle S \cap D \rangle$ is strongly closed in the centralizer of each of its involutions and thus (2) holds by Aschbacher's condition for the existence of a strongly embedded subgroup [Asc73] and Bender's theorem [Ben71]. (See also exercise (2.14)(5)!)

Hence there exist different commuting elements a, b in E with $ab \notin D$. Suppose (a, x, b) is a triangle of the second kind for each $x \in D - E$. Then (1.12) implies that ab lies in the kernel of the action of G on the conjugates of E, a contradiction to the simplicity of G. Thus there exists an $x \in D - E$ such that (a, x, b) is of the third kind.

Let z be the central involution of $\langle ab, x \rangle$ and $S \in \mathrm{Syl}_2(G)$ containing a, b and z and $Y = \langle C_D(z) \rangle$. Then $S \cap D \subseteq Y$ and $\mathcal{D}(Y \cap D)$ is not connected. Since clearly $F = Y \cap D$ is a conjugacy class in Y with $F \cap F^2 \neq \emptyset$, Y satisfies the hypothesis of (1.16). Hence $Y/Z_*(Y) \simeq L_2(q)$, $S_z(q)$ or $U_3(q)$. Let $X = \langle S \cap D, x \rangle$. Then $X/Z_*(X) \simeq L_2(q)$ or $Sz(q)$.

Suppose $Q = O_2(X) \leq Z(X)$. Then, since $L_2(q)$ and $Sz(q)$ do not admit a perfect central extension by a 2-group, which is generated by odd transpositions, the extension of X by Q splits. ($L_2(q)$ does not admit any such extension except $SL_2(5)$. If $Sz(q)$ has such an extension, then $q = 8$ by [AG66]. But in this case it is easy to see that the extension is not generated by odd transpositions!) Hence $|X : X'| = 2$, a contradiction to $(S \cap D)^2 \cap S \cap D \neq \emptyset$. Thus $Q \not\leq Z(X)$.

Let $e \in aQ \cap D$ with $a \neq e$. As shown above, there exists a $y \in D - E$ such that (a, y, e) is a triangle of the third kind. Let t be the central involution of $\langle ae, y \rangle$ and $T \in \mathrm{Syl}_2(G)$ with $\{a, e, t\} \subseteq T$. Then $T \cap D$ and $S \cap D$ are conjugate in $C_G(a) \cap C_G(e)$, whence we may without loss of generality assume $t \in \langle S \cap D \rangle$.

Now, as shown above, X satisfies the hypothesis of (1.14). Hence $t = ae(aet) = dz'$ with $d \in S \cap D$ and $z' \in Z(X)$, since $aet \in D$. Further $te = a(aet) \in D^2$ and $d = e^h$ for some $h \in \langle C_D(t) \rangle$, since $\{d, e\} \subseteq E \cap C_D(t)$, $y \in C_D(t) - E$. Hence

$$z' = (td) = (te)^h = (a(aet))^h \in D^2$$

and thus $z' \in D^2$. But then (1.10)(*) implies that x and a are in the same connectivity component of $\mathcal{D}(D)$, a contradiction. □

The next lemma, [Asc72, (3.4)], is suggested by the D_d-theorem of Fischer [Fis71, (3.11)].

(1.17) Lemma

Let D be a class of odd transpositions of G with $\mathcal{D}(D)$ connected. Let $d \in D$, E an orbit of maximal length of $\langle D_d \rangle$ in D_d and $F = N_D(E) - E$. Then F is a maximal set of imprimitivity of the action of G on D.

Proof. Clearly $C_D(d) \subseteq E \cup F$ and $[E, F] = 1$ by (1.6). Suppose we have shown already that F is a set of imprimitivity of the action of G on D and $F \subseteq L$, L a maximal set of imprimitivity. Then, as $d \in F$, $E \subseteq N_D(L)$. Hence $E \subseteq L$ if $E \cap L \neq \emptyset$, since E is an orbit of $\langle D_d \rangle$. But then $C_D(d) \subseteq E \cup F \subseteq L$ and, since L is a set of imprimitivity, $N(L)$ is transitive on L. Hence $C_D(\ell) \subseteq L$ for each $\ell \in L$, a contradiction to the connectivity of $\mathcal{D}(D)$.

Thus $E \cap L = \emptyset$ and (1.6) implies $[E, L] = 1$. Hence $L \subseteq C_D(E) \subseteq N_D(E) - E = F$. ($\mathcal{F}(E)$ is connected, since E is an orbit of $\langle D_d \rangle$ length > 1!) This shows the maximality of F, when F is already a set of imprimitivity.

So suppose $g \in G$ and $f \in F \cap F^g \neq F$. Hence $E \subseteq D_f$ and, since $\mathcal{F}(E)$ is connected, E is contained in an orbit of $\langle D_f \rangle$ on D_f. The maximality of $|E|$ implies that E is an orbit of $\langle D_f \rangle$. Thus $C_D(f) \subseteq N_D(E)$. In particular $E^g \subseteq C_D(F^g) \subseteq C_D(f) \subseteq N_D(E)$. Since $\mathcal{F}(E^g)$ is connected this shows that either $E^g \subseteq F$ or $E = E^g$. In the second case $F = F^g$ by definition of F, a contradiction to the choice of g. Thus $E^g \subseteq F$ and by the same argument $E \subseteq F^g$.

We obtain $N_D(E) = E \cup F \subseteq F \cup F^g$. Moreover E^g centralizes $F^g \supseteq E$. Thus for each $e \in E^g$ we have $E \subseteq C_D(e)$. Hence the connectivity of $\mathcal{F}(E)$ and the maximality of $|E|$ imply that E is an orbit of $\langle D_e \rangle$ for all $e \in E^g$.

Since $F^g \subseteq D_e$ for such an e, we obtain $F^g \subseteq N_D(E)$. Thus $F \cup F^g \subseteq N_D(E) \subseteq F \cup F^g$ and $N_D(E) = (N_D(E))^g$.

Now, as shown above, $C_D(f) \subseteq N_D(E)$ for each $f \in F$. Thus $C_D(f^g) \subseteq (N_D(E))^g = N_D(E)$ for each $f^g \in F^g$. Hence $C_D(a) \subseteq N_D(E)$ for each $a \in N_D(E)$, a contradiction to the connectivity of $\mathcal{D}(D)$. This shows that F is a set of imprimitivity. $\square$

(1.18) Proposition

Let D be a class of odd transpositions of G with $\mathcal{D}(D)$ connected and $D \cap D^2 \neq \emptyset$. Then, for each $d \in D$ and $x \in A_d$, we have $C_D(\langle x, d \rangle) \neq \emptyset$.

Proof. Let G be a minimal counterexample to (1.18). Then $S(G) = Z_*(G)$ and $G/Z_*(G)$ is simple by (1.9). Let $\overline{G} = G/S(G)$. Then if $C_{\overline{D}}(\langle \overline{x}, \overline{d} \rangle) \neq \emptyset$, also $C_D(\langle x, d \rangle) \neq \emptyset$. Thus $G = \overline{G}$ is simple.

For $e, f \in D$ denote by $d(e, f)$ the distance from e to f in $\mathcal{D}(D)$ and let $n = d(d, x)$. We prove (1.18) by induction on n. Obviously $n \geq 2$ and if $n = 2$, then (1.18) holds. So suppose $n > 2$ and (1.18) holds for all $e \in A_x$ with $d(e, x) < n$. Let $d = d_0$ and

$$d_0, d_1, \ldots, d_n = x$$

a path of length n from d to x. If $dd_1 \in D$, then $d_1 \in E_d$ and whence $d(d, x) = d(d_1, x)$, a contradiction. Thus $dd_1 \notin D$ and (d, x, d_1) is a triangle. If now for all $y \in \langle d, x \rangle \cap D$ with $y \neq d$, (d, y, d_1) is not a triangle of the third kind, then $d_1^g = d$ for some $g \in C(x)$ by (1.12). Hence $d(d, x) = d(d_1, x)$, a contradiction.

Thus let (d, y, d_1) be a triangle of the third kind, $y \in \langle d, x \rangle \cap D$. Let z be the central involution in $\langle dd_1, y \rangle$ and $S \in \mathrm{Syl}_2(G)$ containing d, d_1 and z. Let $Y = \langle S \cap D, y \rangle$. Suppose $e \in S \cap D \cap C(y)$. Then e centralizes dy, but x inverts dy. ($dy \neq 1$ and $o(dy)$ is odd!) Hence x is not conjugate to e in $N(\langle dy \rangle)$ and thus $e \in C_D(x) \cap C_D(d)$, a contradiction to the assumption. This shows that $S \cap D \subseteq E = y^Y$ and E is a class of odd transpositions of Y. If now $\mathcal{D}(E)$ is connected, then the minimality of G implies $C_E(\langle d, y \rangle) \neq \emptyset$, which by the above argument implies $C_D(\langle d, x \rangle) \neq \emptyset$, a contradiction. Since clearly $(S \cap D)^2 \cap E \neq \emptyset$, now (1.16) implies $Y/Z_*(Y) \simeq L_2(q), Sz(q); q = 2^m > 2$.

Let $Q = O_2(Y)$ and assume first $Q \leq Z(Y)$. Then, as in (1.16), the extension of Y by Q splits. Hence $|Y : Y'| = 2$ since $z \in Q$, a contradiction to $E \cap E^2 \neq \emptyset$.

Thus $Q \not\leq Z(Y)$ and there exists a $d \neq a \in dQ \cap E$. Suppose that there exists an $r \in D$ such that (d, r, a) is a triangle of the third kind. (Clearly $da \notin D$!) We will lead this to a contradiction as in (1.16). Indeed let t be the central involution in $\langle da, r \rangle$ and $T \in \mathrm{Syl}_2(G)$ containing d, a and t. Then, since $S \cap D$ and $T \cap D$ are conjugate in $C(a) \cap C(d)$, we may assume $S \cap D = T \cap D$. Now by (1.14) $t = ad(adt) = fz'$, $f \in S \cap D$ and $z' \in Z(Y)$, since $adt \in S \cap D$. Further $td = a(adt) \in D^2$ and $f = d^h$ for some $h \in C(t)$, since r and f do not commute. Indeed, if $rf = fr$, then $R = \langle Y, r \rangle \leq C(z')$. Hence by minimality of G, $\mathcal{D}(D \cap R)$ is not connected and thus by (1.16) $R/Z_*(R) \simeq U_3(q)$ since $R \neq Y$. But this is a contradiction to $\{r, d\} \subseteq C_D(f)$.

We obtain $z' = tf = (td)^h = (a(adt))^h \in D^2$. Hence by (1.10)(*) $\mathcal{D}(C_D(z'))$ is connected and thus by minimality of G we obtain $C_D(\langle d, y \rangle) \neq \emptyset$. But then, as before also $C_D(\langle d, x \rangle) \neq \emptyset$.

Hence no such r exists. Let as in (1.17) E' be an orbit of $\langle D_d \rangle$ on D_d of maximal cardinality and $F' = N_D(E') - E'$. Suppose $d_1 \in F'$ and let $\ell \in C_D\langle d_1, x \rangle$. (Exists by induction assumption.) If also $\ell \in F'$, then $x \in F'$ by (1.17), since x is conjugate to d_1 in $C(\ell)$. But then $E' \subseteq C_D\langle x, d \rangle$. So $\ell \notin F'$ and thus $\ell \in C_D(d) - F' \subseteq E'$. But then ℓ centralizes $\langle d, x \rangle$ since $[E', F'] = 1$.

Hence $d_1 \in E'$. If also $a \in E'$ then such an r would exist, since d_1 is conjugate to a in $C(d)$. So $a \in F'$. Let $u \in D - N_D(F')$. Then (d, u, a) is a triangle and, since no triangle (d, r, a) of the third type exists, the hypothesis of (1.12) is satisfied. Let $N = O_2\langle d, u, a \rangle$ and $n \in N$. Since $da = (da)^n$ by (1.12), $E' \subseteq C_D(d^n a^n) \subseteq C_D(d^n)$. As $\mathcal{F}(E')$ is connected this shows $d^n \notin E'$. But as $d^n \in C_D(d)$, $d^n \in F'$. Hence $d^N \subseteq F'$ and $u^N \subseteq (F')^g$, where $g \in G$ with $u \in (F')^g$. Since $da \in N$ by (1.12) this shows that da is in the kernel of the permutation action of G on the conjugates of F', a contradiction to the simplicity of G. This finally proves (1.18).   □

(1.19) Lemma

Suppose D is a set of odd transpositions of G, $d \in D$ and $x \in A_d$. Suppose further that there exist no $a, b \in E_d$ such that (a, x, b) is a triangle of the third kind. Let $X = \langle E_d, x \rangle$, $Q_d = \langle ab \mid a, b \in E_d$ and $ab \notin D \rangle$, $Z = Q_d \cap Q_d^x$ and $\overline{X} = X/Z$. Then the following hold.

  (1) $Z \leq Z(X)$.
  (2) $\overline{Q}_d \times \overline{Q}_d^x \lhd \overline{X}$ and $[Q_d, E_d^x] \leq Q_d^x$, $[Q_d^x, E_d] \leq Q_d$.

Proof. Since $x \in \langle d, d^x \rangle$ (1) is obvious. Now $o(abx) = 4$ for $a \neq b \in E_d$ with $ab \notin D$. Let $y = (ab)^x$. Then, as $ab = (ab)^y$, we have

$$C_D(d^y) \subseteq C_D((ab)^y) = C_D(ab) \subseteq C_D(a) = C_D(d),$$

so that $d^y \in E_d$. This implies $E_d = E_d^y$ and $Q_d^x \subseteq N(E_d)$, $Q_d \subseteq N(E_d^x)$. In particular $[Q_d, Q_d^x] \leq Z$.

Suppose $ee^y \in D$ for some $e \in E_d$. Then $ee^y \in E_d$ and y centralizes ee^y. As

$$\langle ee^y, E_d^x \rangle = \langle E_d, E_d^x \rangle = X,$$

we obtain $y \in Z$, which is obviously impossible. Hence $[E_d, y] \leq Q_d$ and thus $[E_d, Q_d^x] \leq Q_d$, which proves (2).   □

(1.20) Lemma

Suppose D is a set of odd transpositions, $d \in D$ and $x \in A_d$. Let $X = \langle E_d, x \rangle$. Then $X/O_2(X)$ is isomorphic to $L_2(q)$, $Sz(q)$, $q = 2^m > 2$ or $X/O_2(X)$ is dihedral.

Proof. Let G be a minimal counterexample. Since for $\overline{G} = G/O_2(G)$ we have $\overline{E}_d \subseteq E_{\overline{d}}$, this implies $O_2(G) = 1$ and $X = G$. If now (a, x, b) is a triangle of the third kind for $a, b \in E_d$, then the central involution of $\langle ab, x \rangle$ lies in $Z(G)$, a contradiction to $O_2(G) = 1$. Hence (1.19) and $O_2(G) = 1$ imply $ab \in E_d$ for

$a \neq b \in E_d$. In particular either $|E_d| = 1$ and X is dihedral or $E(d) = E_d \cup 1$ is a subgroup with $E(d)^{\#} = E_d$ and $|E(d)| \geq 4$. Hence $D \cap D^2 \neq \emptyset$.

If now $\mathcal{D}(D)$ is not connected, then by (1.16) G is a perfect central extension of $L_2(q)$ or $Sz(q)$ by a group of odd order. But then $Z(G) = 1$ and (1.20) holds. So $\mathcal{D}(D)$ is connected. But then (1.18) implies $C_D(\langle E_d, x \rangle) \neq \emptyset$, a contradiction to $X = G$ and $Z(G) = 1$. $\qquad\square$

We finish this section with the proof of (1.3).

Proof of (1.3). Let G be a minimal counterexample to (1.3). Then by (1.6) $\mathcal{F}(D)$ is connected and by (1.9) G is simple. Let T be a connectivity component of $\mathcal{D}(D)$. Then $o(dx)$ is odd for all $d \in T, x \in D - T$. If now $ab \in T$ for all pairs of commuting involutions $a \neq b \in T$, then by a theorem of Aschbacher [Asc73] $N_G(T)$ is a strongly embedded subgroup and thus (2) holds by Bender's Theorem [Ben71].

Hence there exist $a \neq b \in T$ with $ab = ba \notin D$. If now $o(xab) \in \{2,4,8\}$ for all $x \in D - T$, then x, x^{ab} lie in the same connectivity component of $\mathcal{D}(D)$, whence ab lies in the kernel of the action of G on the conjugates of T, a contradiction to the simplicity of G. We obtain:

(1) There exist $a \neq b \in T$ with $ab = ba \notin T$ and $x \in D - T$ such that $o(xab) = 2n, 1 \neq n \equiv 1(2)$.

Let z be the central involution in $\langle x, ab \rangle$ and $c \in D$ with $z = cab$. We show next:

(2) $o(ca) = 2 = o(cb)$.

If $o(ca) = m \equiv 1(2)$, then as $c(ab) = (ab)c$, $o(cb) = 2m$, a contradiction. Thus, if (2) does not hold, we have $o(ca) = 4 = o(cb)$. As $(ab)^c = ab$, we obtain $d = aa^c \in \langle a, b, b^c \rangle \leq C(b)$. Thus $d \in C(ab) \cap C(c) \leq C(z)$. Now $\{d, c, dc\} \subseteq T$. Hence $o(xd), o(xc)$ and $o(xdc)$ are odd, since $x \in D - T$. This implies $d \sim c$ in $C(z)$. Thus

$$dz = (cz)^g = (ab)^g \in D^2, g \in C(z).$$

As $(dc)^a = c$ we obtain

$$z = cab = (dc)^a ab = (dz)^a \in D^2.$$

Let now $z = ef$; $e, f \in D$. Then, if $o(ce) = m \equiv 1(2)$, $o(cf) = o(cez) = 2m$, a contradiction. Thus $o(ce) = 2$ or 4 and $e \in T$. Similarly $o(ex) = 2$ or 4. Thus $e \in T^g$, where T^g is the connectivity component of $\mathcal{D}(D)$ containing x. But then $T = T^g$, a contradiction to $x \notin T$.

Now let $S^* \in \mathrm{Syl}_2(\langle C_D(z) \rangle)$ containing a, b and c, $L = S^* \cap D$ and $X = \langle L, x \rangle$, $E = D \cap X$ and $\overline{X} = X/\langle z \rangle$. Then $\{\overline{a}, \overline{b}, \overline{ab}\} \subseteq \overline{E}$ and $\mathcal{D}(\overline{E})$ is disconnected. Hence by minimality of G we obtain:

(3) $X/Z_*(X) \simeq L_2(q)$ or $Sz(q), q = 2^m > 2$.

If now D is degenerate (1.3) holds by (1.16). So D is non-degenerate. Let $S \in \mathrm{Syl}_2(G)$ containing S^*. Then $C_{S\cap D}(L) = L$, but by exercise (1.21)(1) $N_{S\cap D}(L) \neq L$. Let $F = N_D(L)$ and $F' = F - L$. In particular, since $N_X(L)$ is transitive on L, there exists for each $\ell \in L$ an $f \in F'$ with $(\ell f)^2 \in L$. Choose now $z' \in Z(X)$ such that $Y = \langle C_D(z') \rangle$ is maximal. Then, since $E \subseteq C_D(z')$ and $L \cap L^2 \neq \emptyset$, Y satisfies the hypothesis of (1.3). Hence $Y/Z_*(Y) \simeq L_2(2^k)$, $Sz(2^k)$ or $U_3(2^k)$, $k > 1$, by minimality of G. In particular $C_D(z')$ is degenerate. Hence, if $S' \in \mathrm{Syl}_2(Y)$ containing L, then $S' \cap D \subseteq C_D(z)$, whence $S' \cap D = L$. In particular this implies $C_Z(f) = 1$ for each $f \in F'$, where $Z = Z(X) \cap Q$ and $Q = O_2(X)$.

Let $M = \langle L \rangle \cap Q$. Since $X = X'$, as $L \cap L^2 \neq \emptyset$, the proof of (1.14) shows:

$$(*) \qquad Q = [Q, X] = (\Pi_{\ell \in L}[Q, \ell])[Q, x].$$

We show next:

(4) $F \subseteq N(M)$.

We have $\emptyset \neq Z(\langle S \cap D \rangle) \cap D \subseteq L$ for each $S \in \mathrm{Syl}_2(G)$ containing L. Indeed obviously $Z(\langle S \cap D \rangle) \cap D \neq \emptyset$ (See exercise (1.21)(1).) and $Z(\langle S \cap D \rangle) \cap D \subseteq C_D(L) = L$. Suppose that $Q' \neq 1$. Since $Q' \leq Z$ $(*)$ implies $Q' \leq [Q, a]$. Hence $[Q, a] \cap Z \neq 1$. Let $S \in \mathrm{Syl}_2(G)$ containing QL. Then, as $N_X(L)$ is transitive on L, without loss $a \in Z(\langle S \cap D \rangle) \cap L$. Hence $[Q, a] \cap Z \leq Z(\langle S \cap D \rangle)$, a contradiction to $C_Z(f) = 1$ for each $f \in F'$.

This shows that we can apply (1.15) to X. Clearly $D^2 \cap Z = \emptyset$, since otherwise a and x would be connected in $\mathcal{D}(D)$. Hence by (1.15) $Z^f \leq M$ for each $f \in F$. Let $L_1 = C_E(x)$. Then $M = N_{\langle L \rangle}(L_1)$ and $N_L(L_1) = \emptyset$. If now $m^f \notin M$ for some $m \in M$ and $f \in F$, then $m^f z \in L$ for some $z \in Z$ by (1.15). But, as $z \in M^f \leq N(L_1^f)$, we obtain $m^f z \in N_L(L_1^f)$, a contradiction.

We now come to the final contradiction.

Suppose $|\langle L \rangle / M| > 4$. Then $|C_{\langle L \rangle / M}(f)| \geq 4$ for each $f \in F'$.

Let $s \in \langle L \rangle - M$ with $[s, f] \in M$. Then, as $Q \cap D = \emptyset$, f centralizes $sQ \cap D$. Hence f centralizes $[s, Q]$, since $[s, Q] = [d, Q]$ for $d \in sQ \cap D$. Thus there exist $\alpha, \beta \in S \cap D$ with $\{\alpha, \beta, \alpha\beta\} \subseteq C(f)$ and $\alpha\beta \notin M$. Now there exists a $y \in E$ such that $X = Q\langle \alpha, \beta, y \rangle$ (by the structure of $L_2(q)$ or $Sz(q)$)! Hence, as shown in (1.15), $Q = [Q, \alpha][Q, \beta][Q, y]$ and $Z \leq [Q, \alpha][Q, \beta]$, a contradiction to $C_Z(f) = 1$.

This shows $|\langle L \rangle / M| = 4$ and $X/Q \simeq L_2(4)$. Next we show:

(5) $C_{S \cap D}(\langle a, b \rangle) = L$.

To prove (5) let $s \in C_{S \cap D}(\langle a, b \rangle)$ and assume by induction $s' \in L$ for all $s' \in C_{S \cap D}(\langle a, b \rangle)$ with $|C_L(s')| > |C_L(s)|$. Then, as $\ell \in C_L(\ell^s)$ for each $\ell \in L$ we have $\ell^s \in L$ by induction assumption for $s \in N_D(L) = F$. But then, as shown above, $s \in C_F(L) = L$, since s centralizes $\langle L \rangle / M$.

Let $d \in aM \cap D$, $d \neq a$. (d exists since $Q \not\leq Z(X)$ as $X = X'$!) Then $da \notin D$ and thus there exists an $x' \in D - T$ such that $o(dax') = 2k$, k odd. Let z' be the central involution in $\langle da, x' \rangle$ and $c' \in D$ with $z' = dac'$. Then by (2) $o(c'd) = o(c'a) = 2$. Conjugating in $C(\langle a, d \rangle)$ (and passing to some other 2-Sylow subgroup of G containing L if necessary) we may assume $c' \in S \cap D$. Suppose $c' \notin L$. Then by (5) $c' \in F$ and $(c'b')^2 = a$ for some $b' \in bM \cap D$. Hence $c'a \in D$ and $z' \in D^2$, a contradiction to $x' \in D - T$. Hence $c' \in L$ and by (1.14) $z' = dac' = uz$ for some $u \in L$, $z \in Z$. We now use the argument of (1.16). We have $z'd \in D^2$ and $d^g = u$ for some $g \in C(t)$, since $x' \in D - T$. Hence

$$z = z'u = z'd^g = (z'd)^g \in D^2,$$

which gives the final contradiction. □

(1.21) Exercises

(1) Let P be a p-group generated by the normal set Σ of subgroups. Then the following hold:

  (a) If $\Delta \subset \Sigma$, then $\Delta \subset N_\Sigma(\Delta)$.

  (b) If $[A, B] = 1$ or $[A, B] \in \Sigma$ for all $A, B \in \Sigma$, then

  $$(Z_i(P) \cap \Sigma) - Z_{i-1}(P) \neq \emptyset \text{ for all } 1 \leq i \leq cl(P).$$

(2) Let D be a set of root involutions of G and $a, b, c, d \in D$ satisfying

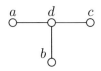

and $abc \in D$. ($a \neq b \neq c \neq a$!) Show that $X = \langle a, b, c, d \rangle \simeq W^*(D_4)$, the center-factor group of $W(D_4)$. In particular a is conjugate to b in $C_X(ac)$.

(3) Let $X \simeq SL_2(2^n)$, $n > 1$ and M be the set of 2×2 matrices over $GF(2^n)$ on which X acts by conjugation. Show

  (a) M is an indecomposable $\mathbb{Z}X$-module with composition series

  $$0 \subset D \subset M_1 \subset M$$

where D is the set of scalar matrices and M_1 the set of 2×2 matrices of trace zero.

(b) Let $N = M/D$ and $N_1 = M_1/D$. Then N is indecomposable, N_1 is isomorphic to the natural $\mathbb{Z}X$-module and N/N_1 is a trivial $\mathbb{Z}X$-module.

(4) Let X be as in (3), V the natural $\mathbb{Z}X$-module and $Y = V \cdot X$ the semidirect product of X with V. Show that Y contains 2^n conjugacy classes of complements to V.

Hint. Identify V with N_1 and embed $Y = VX = N_1X$ in NX. Then, as $[N, X] \leq N_1$, $Y \triangleleft NX$. Consider the conjugates of X in NX.

(5) Let Y be as in (4) and $R = \langle a, b, c \rangle$ be a subgroup of Y, a, b and c involutions satisfying $Y = VR$. Show $R \simeq L_2(2^n)$.

Hint. By (4) there exist at least 2^{3n} complements to V in Y. Hence the triple (a, b, c) is contained in some complement.

§ 2 Root subgroups

The purpose of this section is to show that $E(d) = E_d \cup 1$ for $d \in D$, if D is a class of root involutions of G, $O_2(G) = Z(G) = 1$, $D \cap D^2 \neq \emptyset$ and $\mathcal{D}(D)$ is connected. Moreover, we will also show that the root structure theorem, (2.5), holds in the degenerate case.

Before we start the main argument, we prove a special lemma.

(2.1) Lemma

Suppose $G = \langle x, a, b \rangle$ is generated by the set D of root involutions and $\{x, a, b\} \subseteq D$ satisfying:

(1) $\{a, b, ab\} \subseteq E_a$.
(2) $o(ax) = 3$.
(3) $G/S(G) \simeq L_2(q)$, $q = 2^n > 2$.

Then $G \simeq L_2(q)$.

Proof. Let G be a minimal counterexample to (2.1). Then by (1.9) $S(G) = Z_*(G)$. Since $L_2(q)$ does not have a proper perfect central extension generated by root involutions and since $G = G'$ by (1), we obtain

$$M = O_2(G) = S(G) = [M, G].$$

Let $N \lhd G$ such that M/N is a minimal normal subgroup of $\overline{G} = G/N$. We show that $\overline{M}$ is a natural module for G/M.

Let $S \in \mathrm{Syl}_2(G)$ containing E_a. If $\overline{D}$ is degenerate, then as shown in (1.14) $C_Q(a) = C_Q(S \cap D)$ and $C_Q(a) \cap C_Q(x) = 1$. Hence $\overline{M}$ is by I(3.4) the natural module for G/M. So suppose $\overline{D}$ is non-degenerate. Then $\overline{E} \neq \emptyset$ for $E = D \cap M$. Hence $\overline{M} = [\overline{E}, a][\overline{E}, b][\overline{E}, x]$. Now for $e \in E$ with $\overline{e} \in \overline{M} - C_{\overline{M}}(a)$, we have $o(ea) = 4$ and $[e, a] \in C_D(E_a)$. This shows that $[\overline{M}, a][\overline{M}, b] \leq C_{\overline{M}}(a) \cap C_{\overline{M}}(b)$.

If now $[\overline{E}, a] \neq [\overline{E}, b]$, then $[\overline{E}, a][\overline{E}, b] \cap C_{\overline{M}}(x) \neq 1$, a contradiction. Thus $\overline{M} = [\overline{E}, a] \times [\overline{E}, x]$ and

$$C_{\overline{M}}(a) = [\overline{E}, a] = [\overline{E}, b] = C_{\overline{M}}(b).$$

Hence exercise (2.14)(6) shows that $\overline{M}$ is the natural G/M module.

Let $\widetilde{G} = G/M$. Then, as $o(ax) = 3$, we can identify $\widetilde{a}$ with $\begin{pmatrix} 1 & \\ 1 & 1 \end{pmatrix}$ and $\widetilde{x}$ with $\begin{pmatrix} 1 & 1 \\ & 1 \end{pmatrix}$. Since $\langle \widetilde{a}, \widetilde{b}, \widetilde{x} \rangle \simeq L_2(q)$, $\widetilde{b}$ corresponds to the matrix $\begin{pmatrix} 1 & \\ \lambda & 1 \end{pmatrix}$ with $GF(q) = \mathbb{Z}_2(\lambda)$. Now there exist an $h \in G$ with $a^h = b$ such that $\widetilde{h}$ corresponds to a diagonal matrix. Hence $\widetilde{h}$ corresponds to $\begin{pmatrix} \mu & \\ & \mu^{-1} \end{pmatrix}$ with $\mu^2 = \lambda$. Let $H = \langle h \rangle$. Then $\widetilde{H}$ acts irreducibly on $\widetilde{S}$. On the other hand, as $a^h = b$, we have $H \leq N(E_a)$. Hence $\overline{S} = \overline{M} \langle \overline{E}_a \rangle$ and Gaschütz's theorem, [Hup67, I,17.4], implies that the extension of $\widetilde{G}$ by $\overline{M}$ splits. But then by exercise (1.21)(5) $\overline{G} \simeq L_2(q)$, a contradiction.

This shows that no such N exists. Hence $M = 1$ and $G \simeq L_2(q)$. □

For the rest of this section we assume that D is a class of root involutions of G and $D \cap D^2 \neq \emptyset$. We first start with the degenerate case.

(2.2) Lemma

Suppose D is degenerate, $X = \langle E_d, x \rangle$ with $d \in D$ and $x \in A_d$ and $X/Z_*(X) \simeq L_2(q)$, $q = 2^m > 2$. Let $Q = O_2(X)$, $E(d) = \langle E_d \rangle$, $E(x) = \langle E_x \rangle$ and $Z = Z(X) \cap Q$. Then one of the following holds:

(1) $X \simeq L_2(q)$, $E(d) = E_d \cup 1$ and $|E(d)| = q$.
(2) $Q = (E(d) \cap Q)(E(x) \cap Q)$, $Z = E(d) \cap E(x)$, Q/Z is the direct sum of natural $L_2(q)$-modules and $|E(d) : E(d) \cap Q| = q$. Further $Q \leq N(E_d)$ and $E_d = S \cap D$ for $S \in \mathrm{Syl}_2(X)$ containing E_d.

Proof. Since $D \cap D^2 \neq \emptyset$ and D is a conjugacy class, there exist $\{a, b, ab\} \subseteq E_d$. Let $S \in \mathrm{Syl}_2(X)$ containing E_d. Then the hypothesis of (1.13) is satisfied for X. Moreover, since there exists no proper perfect central extension of X/Q generated by odd transpositions, it follows, as shown in (1.14), that $Q = [Q, X]$ and $C_Q(a) \cap C_Q(x) = Z$. Let $\overline{X} = X/Z$. Since by (1.13) $C_Q(a) = C_Q(S \cap D)$, S acts quadratically on $\overline{Q}$ and $C_{\overline{Q}}(X) = 1$. Hence by I(3.7) $\overline{Q}$ is the direct sum of natural $\mathbb{Z}(X/Q)$-modules.

Now the structure of $L_2(q)$ implies that there exists a $y \in X \cap D$, that $X = QX_0$, and $X_0 = \langle a, b, y \rangle$. Since $\langle S \cap D \rangle$ is abelian and $S = Q\langle S \cap D \rangle$, Gaschütz's theorem, see [Hup67, I(17.4)], implies that the extension of $\overline{X}/\overline{Q}$ by $\overline{Q}$ splits. Hence by (1.21)(5) $\overline{X} = \overline{Q}\overline{X}_0$ with $\overline{Q} \cap \overline{X}_0 = 1$. This shows that X_0 is a central extension of $L_2(q)$ generated by odd transpositions. As $\{a, b, ab\} \subseteq X_0 \cap D$, we have $X_0 = X_0'$ and thus $X_0 \simeq L_2(q)$.

Let $S_0 \in \mathrm{Syl}_2(X_0)$ containing a. Then $S_0^{\#} \subset D$ and thus $S_0^{\#} \subset E_a = E_d$ by (1.10)(∗). Suppose $S_0^{\#} = E_d$. Then $|E_d| = q - 1$. Since there exist $\{e, f\} \subseteq S_0^{\#}$ with $X_1 = \langle e, f, x \rangle$ also supplementing Q in X, the above argument applied to X_1 implies $X_1 \simeq L_2(q)$. Hence $E_d \subseteq X_1$, as $|E_d| = q - 1$, and thus $X = X_1 \simeq L_2(q)$ and (1) holds.

Thus, to prove (2.2), we may assume $S_0^{\#} \neq E_d$. In particular $F_d = \{ef \mid e, f \in E_d \text{ and } 1 \neq ef \in Q\} \neq \emptyset$. Now by I(3.4) $\langle \overline{ef}^X \rangle = \overline{N}$ is a natural module for X/Q. Since $o(efy) = 2$ or 4 for each $y \in X \cap D$, the argument of (1.19) implies $N \leq N(E_d)$. In particular $[N, E_d]^{\#} \subseteq F_d Z$. Let $Q_0 = \langle F_d^X \rangle$ and $Z_0 = Z \cap Q_0$. Then, as shown, $Q_0 \subseteq N(E_d)$, $[Q_0, E_d]^{\#} \subseteq F_d Z_0$ and Q_0/Z_0 is the direct sum of natural X/Q-modules.

Suppose $Q = Q_0$. Then, as $S = QS_0$ and $S_0^{\#} \subset E_d$,

$$C_Q(d) = \Pi_{e \in S \cap D}[Q, e] \subseteq [Q, E_d] \subseteq E(d) \cap Q.$$

Hence $Z \leq E(d)$ and (2) holds.

Thus it suffices to show $Q = Q_0$. We show

(∗) $E(d) = (E_d \cup \{1\})\langle F_d \rangle = S_0 \langle F_d \rangle.$

For this let $\alpha, \beta \in E_d$ with $\alpha\beta \notin F_d$. Then there exist a $\beta' \in \beta Q \cap D$ with $\alpha\beta' \in D$. Hence $\{\alpha, \beta', \alpha\beta'\} \subseteq E_d$ and

$$\alpha\beta = \alpha\beta'(\beta\beta') \in E_d \langle F_d \rangle.$$

Thus $E_d^2 \subseteq E_d \langle F_d \rangle \cup F_d$ and an easy induction argument implies $E(d) \subseteq E_d \langle F_d \rangle \cup \langle F_d \rangle$. But since $E_d \subseteq S_0 \langle F_d \rangle$ this proves (∗).

Now we may choose $y \in E_x$, since by the above $E_x Q \cup Q \in \mathrm{Syl}_2(X)$. Hence

$$X = \langle E_d, E_x \rangle = \langle E_d, y \rangle \le Q_0 X_0,$$

which proves $Q = Q_0$. □

(2.3) Proposition

Suppose D is degenerate, $O_2(G) = 1$ and for all $d \in D$ and $x \in A_d$ we have $X/Z_*(X) \simeq L_2(2^m)$, $m > 1$, where $X = \langle E_d, x \rangle$. Then $E(d) = E_d \cup 1$ and if $q = |E(d)|$, then $X \simeq L_2(q)$ for all such X.

Proof. By (2.2) we may, to prove (2.3), assume $O_2(X) \ne 1$ for all such X. Let $x, y \in A_d$, $X = \langle E_d, x \rangle$, $Y = \langle E_d, y \rangle$ and $Q = O_2(X)$, $N = O_2(Y)$. We show

$$(*)\qquad\qquad\qquad E(d) \cap Q = E(d) \cap N.$$

Indeed $(*)$ implies that $E(d) \cap Q$ is in the kernel of the action of G on the conjugates of $E(d)$, a contradiction to $O_2(G) = 1$. Thus $E(d) \cap Q = 1 = Q$ and (2.3) holds.

So it remains to prove $(*)$. Assume first $z \notin N$ for $1 \ne z \in Z = E(d) \cap E(x)$. Then $y \in A_x$ and, applying (2.2) to Y and $\langle E(x), E(y) \rangle$ we obtain:

$$Y = \langle E(y), E(y)^z \rangle = \langle z, E(y) \rangle = \langle E(x), E(y) \rangle.$$

Hence, as $E(d) \le Y$, we obtain $X \le Y$. But since $E(d)$ is by (2.2) transitive on

$$\{ E(y) \mid y \in Y \cap D, y \notin E_d \}$$

this implies $X = Y$, a contradiction to $z \notin N$.

Thus $Z \le N$. Pick $h \in (E(d) \cap Q) - N$. Then, if $Z_1 = E(d) \cap E(y)$, there exist by (2.2) a $\bar{d} \in E_d$ and $\bar{z} \in Z_1$ with $h = \bar{d}\bar{z}$. (Since $\mathrm{mod} Z_1$ all elements of $E(d) - (E(d) \cap N)$ are in E_d!) Now with symmetry we have $Z_1 \le Q$. Since $\bar{d} \notin Q$ this implies $h \notin Q$, a contradiction to the choice of h. Hence $E(d) \cap Q \subseteq E(d) \cap N$ and $(*)$ holds with symmetry. □

Now suppose D is degenerate and $O_2(G) = 1$. Then, in view of (1.20), (2.3) and $E_d^2 \cap E_d \ne \emptyset$, it suffices to show that $\Sigma = E(d)^G$ is a class of k-transvection groups of G, where $k = GF(q)$ and $q = |E(d)|$, and to show that there exists no pair $d \in D$ and $x \in A_d$ with $X/Z_*(X) \simeq Sz(q)$ for $X = \langle E_d, x \rangle$. This will be shown in the next theorem. Before we are able to do this we need a technical lemma.

(2.4) Lemma

Suppose D is degenerate, $O_2(G) = 1$ and $\mathcal{D}(D)$ is connected. Let E and F be as in (1.17) and set $H = \langle E \rangle$ and $Q = O_2(H)$. Then $aQ \cap F^g = \{a\}$ for each $a \in E$ and F^g containing a. (Clearly $F^g \subseteq E!$)

Proof. Let $a \neq b \in aQ \cap F^g$. Then, since $C_D(a) \subseteq C_D(F^g) \cup F^g \subseteq C_D(F^g) \cup E$ and $C_E(a) \subseteq C_E(b)$, we have $b \in E_a$. We show that ab is in the kernel of the permutation action of G on the conjugates of F, contradicting $O_2(G) = 1$ and (1.9).

Since F is by (1.17) a maximal set of imprimitivity of the action of G on D, we have $E_d \subseteq F$ for $d \in F$. Let $a \in F^g$. Then, as $[F, a] = 1$, also $[F, F^g] = 1$. Hence $F^g \subseteq C_D(d)$ for $d \in F$ and thus $F^g \subseteq F \cup E$ by definition of E and F in (1.17). Thus $E_a \subseteq F^g \subseteq E$. Let $x \in E \cap A_a$. Then $\langle a, x, b \rangle \leq Q \langle a, x \rangle$ and the hypothesis of (1.12) is satisfied for (a, x, b). In particular for $y \in \langle a, x \rangle \cap D$ we have $[ab, (ab)^y] = 1$ and, as shown in (1.19), $(ab)^y \in N(E_a)$. Since by (1.12)

$$N = O_2(\langle a, x, b \rangle) = \langle (ab)^y \mid y \in \langle a, x \rangle \cap D \rangle$$

we obtain $x^N \subseteq E_x$. In particular $ab \in N(E_x) \subseteq N(F^h)$ where $E_x \subseteq F^h$.

This shows that ab normalizes all $F^h \subseteq E \cup F$. Now let $y \in A_d$ and $y \in F^u$, $u \in G$. If $y \in D_a$, then $ab \in N(F^u)$. So assume $y \in A_a$. Then there exists by (1.18) an $e \in C_D(y) \cap C_D(a)$. Hence $\{d, e\} \subseteq E^g$. Since $N_D(F^g) = E^g \cup F^g$ with $[E^g, F^g] = 1$ and since $\mathcal{F}(E^g)$ is connected, there exists a $w \in C_G(E_a)$ with $e^w = d$. Hence $F^{uw} \subseteq C_D(d)$ and thus $ab \in N(F^{uw})$. But then $ab = (ab)^{w^{-1}} \in N(F^u)$.

Since F is a set of imprimitivity this shows that ab normalizes all conjugates of F, proving (2.4). $\square$

(2.5) Theorem

Suppose D is degenerate, $O_2(G) = 1$ and $\mathcal{D}(D)$ is connected. Then $E(d) := \langle E_d \rangle = E_d \cup \{1\}$ and if $q = |E(d)|$, then $\langle E_d, x \rangle \simeq L_2(q)$ for all $d \in D$ and $x \in A_d$.

Proof. Let G be a minimal counterexample to (2.5). Then it suffices by (2.3) to show that there exists no $d \in D$ and $x \in A_d$ such that $X/Z_*(X) \simeq Sz(2^m)$ for $X = \langle E_d, x \rangle$. So suppose such a pair exists. Let E and F be as in (1.17) and $H = \langle E \rangle$. Then, since $F^g \subseteq E$ if $F^g \cap E \neq \emptyset$, (1.18) implies that there exists a pair $e, f \in E$ with $\langle E_e, f \rangle / Z_*(\langle E_e, f \rangle) \simeq Sz(2^m)$. Since $\mathcal{F}(E)$ is connected and $E \cap E^2 \neq \emptyset$ (as $E_e \subseteq E!$), the simplicity of $H/Z_*(H)$ by (1.9) and the minimality of G imply that $\mathcal{D}(E)$ is not connected. Hence by (1.16) $H/Z_*(H) \simeq Sz(q)$. This will lead to a contradiction.

Let $Q = O_2(H)$, $S \in \mathrm{Syl}_2(H)$ containing E_e and $E_e \subseteq F^g \subseteq E$. Suppose $\langle F \rangle$ is non-abelian. Then, since by (1.13) $Q = C_Q(S \cap E)[Q, f]$ we have $Q \subseteq N(F^g)$. But then by (2.4) $Q \leq Z(H)$ and thus $E = F^g$, since F^g is a set of imprimitivity. In particular $F \cup E = F \cup F^g = N_D(F^g) = N_D(F)$ and thus $C_D(d) \subseteq F \cup E$ for each $d \in F \cup E$, a contradiction to the connectivity of $\mathcal{D}(D)$. This implies that $\langle F \rangle$ is abelian and thus $F = E_d$ for $d \in F$, since $C_D(d) \subseteq E \cup F$.

Suppose next $Q \leq Z(H)$. Then $H/Z(H)$ is simple and thus $Q = 1$, since $Sz(q)$ does not have a proper perfect central extension generated by odd transpositions. Hence $\Omega_1(S)^\# \subset E$ and thus $E_e = \Omega_1(S)^\#$ and $\langle E_e \rangle = E(e) = E_e \cup \{1\}$. Suppose there exists an $x \in A_d \cap A_e$ such that (d, x, e) is a triangle of the third kind and let z be the central involution in $\langle de, x \rangle$. Then, as $dez \in D$ and centralizes d and e, $dez \in E_d \cup E_e$. Suppose without loss of generality that $dez \in E_d$. Then $ez \in E_d$ and $z = e(ez) \in D^2$, a contradiction to $(1.10)(*)$.

Thus all triangles of the form (d, x, e), $x \in A_d \cap A_e$, are of the second kind. (By I(2.13) (7) such an x exists!) Let $N = O_2(\langle d, x, e \rangle$. Then by (1.12) $d^N \subseteq E_d \cup E_e$ and thus $d^N = \{d, e\}$, since $E_d \cap N = \emptyset = E_e \cap N$. This implies $\langle d, x, e \rangle \simeq \Sigma_4$, a contradiction to (1.18) and $3 \nmid |Sz(q)|$.

Thus $Q \not\leq Z(H)$ and by (1.13) $Q = C_Q(e)[Q, f]$ and $C_Q(e) = \Pi_{a \in S \cap E}[Q, a]$, since $E \cap E^2 \neq \emptyset$ and thus $H = H'$. Let $T \in \mathrm{Syl}_2(G)$ containing $E(d)S$ and $\mathcal{L} = \{E_a \mid a \in T \cap D\}$. Then $\mathcal{L} = E_d \cup \{E_a \mid a \in S \cap E\}$, $|\mathcal{L}| > 2$ by (2.4) and $C_Q(e)$ fixes all elements of $\mathcal{L}$. Let $\overline{Q} = Q/Z$ with $Z = Z(H) \cap Q$. Then by the above $\overline{Q} = C_{\overline{Q}}(a) \times C_{\overline{Q}}(f)$ for each $a \in S \cap E$. Hence no element of $Q - C_Q(e)$ centralizes an element of $S \cap E$, whence by (2.3) $\tilde{Q} = Q/C_Q(e)$ acts fixed-point-freely on $\mathcal{L} - \{E_d\}$. Now $N_H(\mathcal{L})$ acts transitively on $\mathcal{L} - \{E_d\}$, since it acts transitively on $S \cap E$. Moreover by (1.4) $R = N_G(T \cap D)$ acts transitively on $\mathcal{L}$. Since $R \cap N_G(E_d)$ normalizes Q, R induces a doubly transitive permutation group on $\mathcal{L}$, such that the stabilizer of a point contains an elementary abelian normal 2-subgroup (i.e. $\tilde{Q}$) acting fixed-point-freely on the remaining points. Hence by exercise (2.14)(5) R/K contains a normal subgroup $R_0/K \simeq L_2(2^n)$, $Sz(2^n)$ or $U_3(2^n)$ acting naturally on $\mathcal{L}$, where $2^n = |\tilde{Q}|$ and K is the kernel of the action of R on $\mathcal{L}$. In particular

$$2^n = |\mathcal{L} - \{E_d\}| = |N_H(\mathcal{L}) : N_H(E_e)|.$$

Hence $\mathcal{L} = E_d \cup E_e^Q$ and a 2'-complement to S in $N_H(S)$ normalizes E_e. Together with (2.3) and the structure of $Sz(q)$ this implies $|E_e| = q - 1$ and $H = Q\langle E_e, f \rangle$. We claim

$(*)$ $|E(d)Z| = q.$

Suppose ($*$) holds. Then, as $\langle \mathcal{L} \rangle = E(d)C_Q(e)E(e)$ and $|C_Q(e)/Z| = 2^n$, we obtain $|\langle \mathcal{L} \rangle| \le q^2 2^n$. Now exercise (2.14)(1) implies $q^2 \ge 2^n$. Hence $|\overline{Q}| = 2^{2n} \le q^4$. But then exercise (2.14)(2) implies $|\overline{Q}| = q^4$ and $\overline{Q}$ is partioned by $\{C_{\overline{Q}}(x) \mid x \in E\}$. Since

$$Z \le C_Q(e) = \Pi_{a \in S \cap E}[Q, a],$$

$C_Q(e)$ is elementary abelian and thus each element of Q is an involution and so Q is elementary abelian. Hence $x[x, Q] \subseteq E$ for $x \in E$. We will follow this to a contradiction.

Let $\{e_1, e_2, e_1 e_2\} \subseteq E_e$. Then $[e_2, Q] \cap [e_1 e_2, Q] \ne 1$ since $|[e, Q]| \ge q^2$ and $|C_Q(e)| \le q^3$ by ($*$). Let $1 \ne n \in [e_2, Q] \cap [e_1 e_2, Q]$. Then $e_2 n \in E$ and $e_1(e_2 n) = (e_1 e_2)n \in Q$. Hence $\{e_2, e_2 n\} \subseteq E_e$, a contradiction to $E_d = F$ and (2.4).

Thus ($*$) remains to be proved. For this call the conjugates of $\mathcal{L}$ lines. Notice first that $\langle C_D(d) \rangle$ is D-maximal. Indeed, if $\langle C_D(d) \rangle < R < G$ and $R = \langle R \cap D \rangle$, then $\mathcal{F}(R \cap D)$ is connected, a contradiction to the connectivity of $\mathcal{D}(D)$. We need to show:

($**$) If $x \in A_d$, then x centralizes an element on each line through E_d.

To prove ($**$) we may by (1.18) assume $x \in A_d \cap D_e$, but x does not centralize any element of $\mathcal{L}^h$, where $\mathcal{L}^h$ is the line through E_d and E_f. Suppose first (d, x, f) is a triangle of the third kind and let z be the central involution in $\langle df, x \rangle$. Then, since $dfz \in D$ and centralizes d and f, $dfz \in E_{d'} \subseteq \mathcal{L}^h$ and $z \in \langle \mathcal{L}^h \rangle$. If now $z \notin E(d)Q$, then $H = Q\langle E(e), E(e)^z \rangle$ and thus x centralizes some element of $\mathcal{L}^h$. So $z \in E(d)Q$. But, as E_e is the only element of $\mathcal{L}$ centralized by x, we have $E_e^z = E_e$ and thence $z \in E(d)Z$. But then $R = \langle C_D(d), x \rangle \le C(z)$, a contradiction to the D-maximality of $\langle C_D(d) \rangle$.

So we may assume that (d, y, f) is of the second kind for all $d \ne y \in \langle d, x \rangle \cap D$ and for all $f \in E_f$. Let $N = O_2(\langle d, x, f \rangle)$. Then (1.12) implies $N \le N(\mathcal{L}^h)$. Let $\{f_1, f_2, f_1 f_2\} \subseteq E_f$. Then, as $(df_i)^x \in N(\mathcal{L}^h)$ for $i = 1, 2$, also

$$(f_1 f_2)^x = (df_1)^x (df_2)^x \in N_D(\mathcal{L}^h).$$

But then $E_{(f_1 f_2)^x} \in \mathcal{L}^h$, a contradiction to $x \in A_f$.

This proves ($**$). Now for the proof of ($*$) pick $x \in A_d \cap D_e \cap D_f$ (by ($**$))! Let $\mathcal{L}^g$ be the line containing E_x and E_e, $Y = \langle \mathcal{L}, \mathcal{L}^g \rangle$ and $M = O_2(Y)$. Then, if $|E(d)Z| > q$, $Z_0 = E(d)Z \cap M \ne 1$, since $Y/M \simeq Sz(q)$. Now $Z_0 \le N(\mathcal{L}^g)$ and thus $Z_0 \le N(E_x)$, since this is the only element of $\mathcal{L}^g$ centralized by f. But then $\langle E_x, C_D(d) \rangle \le C(Z_0)$, a contradiction to the D-maximality of $\langle C_D(d) \rangle$. This proves ($*$) and (2.5). $\qquad \square$

In the non-degenerate case the proof of the root structure theorem is much more complicated, mainly because the analog of (1.18) is false. In the rest of this section we will only show that, in the minimal counterexample to the root structure theorem, $E(d) = E_d \cup 1$.

(2.6)   Definition

A subset $T \subset D$ with $1 < |T| < D$ is called a *weak TI-subset* of D, if $T \cap T^d = T$ or $\emptyset$ for all $d \in D$. As in II(4.6) we have:

(2.7)   Lemma

Suppose D is non-degenrate and T a weak TI-subset of D. Then $\langle T \rangle$ is abelian.

Proof. Since the proof is similar to II(4.6) we just sketch it. Let G be a minimal counterexample and T a minimal weak TI-subset of D for which (2.7) is false. Then (1.6) and the minimality of T imply that either $\mathcal{F}(T)$ is connected or $\langle T \rangle$ is a 2-group. Let in the second case $a, b \in T$ with $o(ab) = 4$. Then by (1.7) $a^b \notin E_a$, whence by (1.8) there exists an $x \in A_a \cap V_{a^b}$ with $o(ax) = 3$. Hence $\langle a, x, a^b \rangle \simeq \Sigma_4$, since $aa^b \in D$, and there exists a $d \in C_D(a^b)$ with $a^d = aa^b$. But this is impossible, since if $a \in K_i(\langle T \rangle) - K_{i+1}(\langle T \rangle)$ then $aa^b = [a, b] \in K_{i+1}(\langle T \rangle)$. (The K_i are members of the descending central series!)

Hence in the second case $\langle T \rangle$ is abelian and thus we may assume that $\mathcal{F}(T)$ is connected. This implies $E_d \subset T$ for each $d \in T$. Suppose $d \in T$ and $a \in X_d - T$. Let $x \in A_d \cap T$. Then, as $a \in N_D(T) - T$ and $ad \in D$, $o(ax) = 4$. Hence by (1.4) there exists a $c \in C_D(x)$ with $d^c = ad$. Thus $ad \in T$ and arguing as above $a \in T$.

This shows $E_d \cup X_d \cup V_d \subseteq T$ for each $d \in T$. Hence we obtain $N_D(T) = T \cup C_D(T)$. Let $x \in C_D(T)$ and T^g a conjugate of T containing x. Then $T \subseteq N_D(T^g)$ and, if $T \cap T^g \neq \emptyset$, the connectedness of $\mathcal{F}(T)$ implies $T = T^g$, a contradiction to $x \notin T$. Thus $T^g \subseteq C_D(T)$. Suppose $T^g, T^h \subseteq C_D(T)$ with $[T^g, T^h] \neq 1$. Let $R = \langle T^g, T^h \rangle$ and $E = x^R$, $x \in T^g$. Then E is a non-degenerate class of root involutions of R. Hence the minimality of G and (1.9) imply $\langle T^g \rangle' \leq Z_*(R)$, a contradiction since $\mathcal{F}(T^g)$ is connected. We obtain:

$$N_D(T) = T_0 \cup T_1 \cup \cdots \cup T_n, \text{ where } T = T_0 \text{ and the } T_i \text{ are conjugate to } T.$$
$$\text{Further } [T_i, T_j] = 1 \text{ for } 0 \leq i \neq j \leq n.$$

In particular $N_D(T) = N_D(T_i)$ for $i \leq n$ and $C_D(x) \subseteq N_D(T)$ for each $x \in N_D(T)$. But this is a contradiction to (1.3). $\square$

(2.8) Corollary

Suppose D is non-degenerate. Then $G = \langle X_d, x \rangle$ for $x \in A_d$. In particular $\mathcal{E}(D)$ is connected.

Proof. For $a \in X_d$ we have $a \in A_x$ or $ad \in A_x$. If $a \in X_d - E_d$ then there exists by (1.8) a $c \in X_d$ with $a^c = ad$. This shows that, if $R = \langle X_d, x \rangle$ and $E = x^R$, then E is a non-degenerate class of root involutions of R with $X_e \subseteq E$ for each $e \in E$. Hence E is a connectivity component of $\mathcal{E}(D)$. But then (2.7) implies $E = D$ and $R = G$. It now follows from (1.8) that $\mathcal{E}(D)$ is connected.   □

Next a special lemma.

(2.9) Lemma

Let $d \in D$, $a, b \in E_d$ with $1 \neq ab \notin D$ and $x \in A_d$ such that $o(abx) = 2n$, n odd. Then

$$Z(\langle x, ab \rangle) \leq C(a) \cap C(b).$$

Proof. Let $c \in \langle x, ab \rangle \cap D$ such that $z = abc \in Z(\langle x, ab \rangle)$. Then, if $ac = ca$, $z \in C(a) \cap C(b)$. Hence we may assume $o(ac) = 4 = o(bc)$ and lead this to a contradiction. By (1.7) and (1.5) there exists a $y \in D$ with $a^y = aa^c$. Hence y centralizes $E(a^c)$. As $ab = (ab)^c \in E(a^c)$ we obtain $ab = (ab)^y = aa^c b^y \in E(aa^c)$. Thus $z = cab = c^a b^y \in D^2$. Now $b^y \in E_{aa^c}$. If $aa^c \in V_x$ then ab and $(ab)^x$ commute, a contradiction to $o(xab) = 2n$. Thus $b^y \in A_x$ and thence $o(xc^a) = 2o(xb^y)$, since $c^a = 2b^y$. This contradiction proves (2.9).   □

(2.10)

For the rest of this section we will consider the following hypothesis:

(A) G is a simple group of minimal order generated by the non-degenerate class D of root involutions, such that for $d \in D$ we have $|E(d)| > 2$ and $\Sigma = E(d)^G$ is <u>not</u> a class of abstract root subgroups of G.

We will show that under hypothesis (A) we have $E(d) = E_d \cup 1$.

(2.11) Lemma

Suppose that (A) holds. Let $d \in D$, $x \in A_d$ and assume that $X = \langle E_d, x \rangle < G$. Then $X/O_2(X)$ is isomorphic to $L_2(q)$, $Sz(q)$, $q = 2^m > 2$ or $X/O_2(X)$ is dihedral.

Proof. Let $E = x^X$, $Q = O_2(X)$ and $\overline{X} = X/Q$. Then $\overline{E}$ is by (1.6) a class of root involutions of $\overline{X}$. Let

$$
\begin{aligned}
E_{\overline{d}} &= \{\overline{e} \in \overline{E} \mid C_{\overline{E}}(\overline{e}) = C_{\overline{E}}(\overline{d})\}, \\
F_d &= \{e \in E \mid C_E(e) \cup (V_e \cap E) = C_E(d) \cup (V_d \cap E)\}, \\
\text{and } F_{\overline{d}} &= \{\overline{e} \in \overline{E} \mid \{\overline{x} \in \overline{E} \mid o(\overline{e}\,\overline{x}) = 2 \text{ or } 4\} = \{\overline{x} \in \overline{E} \mid o(\overline{x}\,\overline{d}) = 2 \text{ or } 4\}\}.
\end{aligned}
$$

Then by (1.7) $E_{\overline{d}} = F_{\overline{d}}$. By (1.5) $E_d \subseteq F_d$. Hence

$$\overline{E}_d \subseteq \overline{F}_d = F_{\overline{d}} = E_{\overline{d}},$$

since obviously $\overline{F}_d = F_{\overline{d}}$. This shows that $\overline{X} = \langle E_{\overline{d}}, \overline{x} \rangle$.

Suppose first that $\overline{E}$ is degenerate. Then, as $O_2(\overline{X}) = 1$, (1.19) implies that there exist $\overline{a} \ne \overline{b} \in E_{\overline{d}}$ such that $o(\overline{a}\,\overline{b}\,x) = n$ or $2n$, $1 \ne n$ odd or that $|E_{\overline{d}}| = 2$ and $\overline{X}$ is dihedral. In the second case (2.11) holds. In the first case we have $o(\overline{a}\,\overline{b}\,\overline{x}) = n$, since otherwise $1 \ne Z(\langle \overline{a}\,\overline{b}, \overline{x} \rangle \le Z(\overline{X})$, a contradiction to $O_2(\overline{X}) = 1$. Hence $\overline{E} \cap \overline{E}^2 \ne \emptyset$ and (2.11) holds by (1.20).

So we may assume that $\overline{E}$ is non-degenerate. Hence by (1.9) $\overline{X}/Z(\overline{X})$ is simple and thus by hypothesis (A) either $|E_{\overline{d}}| = 2$ or $\overline{X} \simeq L_2(q)$, $q = 2^m > 2$. But then in any case (2.11) holds. □

(2.12)   Lemma

Suppose that (A) holds. Let $d \in D$, $a \ne b \in E_d$ with $ab \notin D$ and $x \in A_d$ such that $o(ax) = 3$. Then also $o(xb) = 3$.

Proof. Obviously $o(xab) = 4, 8$ or $2n$, $1 \ne n$ odd. Assume first $o(xab) = 8$. Let $e = (xx^{ab})^2$. Then $e \in X_x \cap C(ab)$. Thus $o(ae) = 4$ and we have

$$\underset{\textstyle e}{\circ}\!\!=\!\!=\!\!\underset{\textstyle a}{\circ}\!\!-\!\!-\!\!\underset{\textstyle x}{\circ} \ .$$

Since $ex \in D$, this implies $\langle e, a, x \rangle \simeq \Sigma_4$ and $\langle a, x, b \rangle$ normalizes $\langle e, e^a \rangle$. Hence $o(xb) \equiv 3 \bmod C(\langle e, e^a \rangle)$. If now $o(xb) \ne 3$, then there exists an $h \in C(\langle e, e^a \rangle)$ with $1 \ne o(xx^h)$ odd, a contradiction to $x \in X_e$. Thus $o(xb) = 3$ and $o(xab) = 4$, as $\langle a, b, x \rangle \simeq \Sigma_4$.

This shows $o(xab) = 4$ or $2n$. In the first case we have

$$\underset{\textstyle a}{\circ}\!\!-\!\!-\!\!\underset{\textstyle x}{\circ}\!\!=\!\!=\!\!\underset{\textstyle ab}{\circ}$$

and thus by the structure of $W(B_3) \simeq \Sigma_4 \times \mathbb{Z}_2$ we obtain $\langle a, x, ab \rangle \simeq \Sigma_4$, since $b \in D$. Hence $o(xb) = 3$, which was to be shown.

So we may assume $o(xab) = 2n$. Let z be the central involution of $\langle ab, x \rangle$. Then by (2.9) $U = \langle x, a, b \rangle \leq C(z)$. Hence by (2.11) $U/O_2(U) \simeq L_2(q)$, since $o(xa) = 3$. Let $\overline{U} = U/\langle z \rangle$. Then $\{\overline{a}, \overline{b}, \overline{ab}\} \subseteq \overline{D \cap U}$ and thus the hypothesis of (2.1) holds for $\overline{U}$. Hence $U = X \times \langle z \rangle$, $X \simeq L_2(q)$, since there exists no proper perfect central extension of $L_2(q)$ generated by root involutions.

Now U contains exactly three classes of involutions, namely z, the involutions in X and the "diagonal" involutions. In particular all involutions in $U - (X \cup z)$ lie in D. By (2.8) $z \notin E(d)$. Let $A = E(d) \cap X$. Then, as $D \cap E(d) = E_d$, we obtain $at \in E_d$ for each $t \in A$. In particular $|E_d \cap U| > |A^{\#}|$.

Let $K = N_X(E_d \cap U)$ and $\overline{K} = K/C_K(E_d \cap U)$. Then by the structure of $L_2(q)$, $\overline{K}$ acts fixed-point-freely on $A^{\#}$ and transitively on $E_d \cap U$. Hence

$$|E_d \cap U| \leq |\overline{K}| \leq |A^{\#}|,$$

a contradiction to the above. $\qquad\qquad\square$

We finish this section with:

(2.13) Theorem

Suppose that hypothesis (A) holds. Then $E(d) = E_d \cup 1$ for $d \in D$.

Proof. Suppose (2.13) is false. Then there exist $a \neq b \in E_d$ with $ab \notin D$. Let $x \in D$ with $o(abx) = 8$. Then $o(xx^{ab}) = 4$. Hence for $y \in E_x$ and $h \in E(x^{ab})$ we have $o(hh^y) \leq 2$. Hence $[E(x), E(x)^h] = 1$ and thus $\langle E(x)^{E(x^{ab})} \rangle$ is elementary abelian. This shows that $\langle E(x), E(x^{ab}) \rangle$ is a 2-group.

This implies that, if $o(abx) = 2, 4$ or 8 for all $x \in D$, then $\langle ab, (ab)^g \rangle$ is a 2-group for all $g \in G$. Hence by a theorem of Baer [Asc86, (39.6)] $ab \in O_2(G)$, a contradiction to the simplicity of G.

Hence there exists an $x \in A_d$ such that $o(xab) = 2n$, $1 \neq n$ odd. Let $Y = \langle E_d, x \rangle$ and $\overline{Y} = Y/O_2(Y)$. Then (2.11) implies $\overline{Y} \simeq L_2(q)$ or $Sz(q)$, $q = 2^m > 2$, since by (2.9) $Z(\langle x, ab \rangle) \leq Z(Y)$. We show:

$(*)$ $\qquad\qquad$ Suppose $u, v, w \in E_d$ with $uv \notin D$. Then $uvw \in E_d$.

For the proof of $(*)$ pick by (1.9) $y \in D$ with $o(uy) = 3$. Then by (2.12) $o(vy) = 3$. Suppose $uw \notin D$ or $vw \notin D$. Then again by (2.12) $o(wy) = 3$ and $\langle u, v, w, y \rangle \simeq W(D_4)$ or $W^*(D_4)$ (where $W^*(D_4)$ is the center factor group of $W(D_4)$). Let $r = (uv)^y$. Then $u^r = v$, whence $s = w^r \in E_d$. Now either $uvws = 1$ or $1 \neq uvws \in Z(\langle u, v, w, y \rangle)$ (See exercise (2.14)(3)). Since $uvws \in E(d)$, the second case contradicts (2.8). Hence $s = uvw \in E_d$.

Thus, to prove $(*)$ we may assume $uw \in D$ and $vw \in D$. But if $uvw \notin E_d$ then by (2.12) $o(yvw) = 3 = o(yuw)$, whence $\langle u, y, uv \rangle \simeq \Sigma_4$, a contradiction to $v \in E_d$ and (1.7).

This proves $(*)$. Fix $y \in D$ with $o(ay) = 3$ and let

$$U = \langle ef | e, f \in E_d \text{ with } o(ey) = 3 = o(fy) \rangle.$$

Pick $u = u_1 u_2$ and $v = v_1 v_2$ in U with $u_i, v_i \in E_d$ and $o(u_i y) = 3 = o(v_i y)$. Then $\langle u, u^y, v, v^y \rangle$ is normalized by y and thus $uv \notin D$. (Otherwise $uv \in E_d$!) But by $(*)$ $u_1 u_2 v_1 \in E_d$, whence by (2.12) $o(y u_1 u_2 v_1) = 3$ and $uv = (u_1 u_2 v_1) v_2$. This shows:

$(**)$ Each element of $U^\#$ is of the form $e \cdot f$ with $e \neq f \in E_d$ and $o(ey) = 3 = o(fy)$.

Let now $e \neq f \in E_d$ with $ef \notin D$. Then by $(*)$ $aef \in E_d$, whence by (2.12) $o(yaef) = 3$ and $ef = a(aef) \in U$. This shows:

$(+)$ $$U = (E_d^2 - E_d) \cup 1 \quad \text{and} \quad E(d) - U = E_d.$$

In particular $|E_d| > |U^\#|$ and $N_G(E_d) \leq N_G(U)$.

Now clearly $E(d) \cap O_2(Y) \leq U$. But as $ab \notin O_2(Y)$, this shows $E(d) \cap O_2(Y) < U$. Let $K = N_Y(E_d)$. Then K acts transitively on E_d. Thus $\overline{K}/C_{\overline{K}}(\overline{E}_d)$ acts transitively on $\overline{E}_d$ and fixed-point-freely on $\overline{U}^\#$ by the structure of $L_2(q)$ and $Sz(q)$. Hence, as $|\overline{U}| > 1$,

$$|\overline{U}^\#| \geq |\overline{K}/C_{\overline{K}}(\overline{E}_d)| \geq \overline{E}_d,$$

a contradiction since by $(+)$ all elements of $\overline{E(d)} - \overline{U}$ lie in $\overline{E}_d$. This proves (2.13). $\square$

(2.14) Exercises

(1) Show that $L_2(2^n)$, $Sz(2^n)$ and $U_3(2^n)$ can not act non-trivially on a 2-group N with $|N| < 2^{2n}$.

     Hint. Count the orbits of the involutions of N.

(2) Let $X = Sz(q)$, $q = 2^{2m+1}$, $S \in \mathrm{Syl}_2(X)$ and $T = \Omega_1(S) = Z(S)$. Let V be a non-trivial $\mathbb{Z}_2 X$-module satisfying:

     (i) $|V| \leq q^4$.

     (ii) $[V, t_1] \leq C_V(t_2)$ for $t_1, t_2 \in T$.

Show that then the following hold:

(a) $[V, t] = C_V(t) = [V, T] = C_V(T)$ for $t \in T^{\#}$.

(b) $V = [V, T] \oplus [V, T^g]$ for all $T^g \neq T$.

(c) $|V| = q^4$ and V is partioned by the conjugates of $[V, T]$.

Such a $\mathbb{Z}_2 X$-module will be called the natural module for $Sz(q)$. It is uniquely determined up to equivalence as $\mathbb{Z}X$-module, a fact which we do not use.

(3) Let Y be a group generated by different involutions a, b, c, d satisfying:

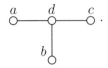

(Coxeter diagram). Let $N = O_2(Y)$ and $Z = Z(Y)$. Show

(a) $|Z| = 2$ or 1.

(b) If $|Z| = 1$, then $a^N = \{a, b, c, abc\}$, and if $1 \neq Z = \langle z \rangle$ then $a^N = \{a, b, c, abcz\}$.

(c) Let $y = (ab)^d$. Then $a^y = b$ and $c^y = abc$ resp. $abcz$.

Hint. Y is an image of $W(D_4)$, which is the split extension of Σ_4 by an elementary abelian group of order 2^3.

If $Z = 1$ then Y is isomorphic to the center factor group of $W(D_4)$, which is denoted by $W^*(D_4)$.

(4) Let G be a doubly transitive permutation group on the set Ω, such that for $\alpha \in \Omega$, G_α contains an elementary abelian normal 2-subgroup T acting fixed-point-freely on $\Omega - \{\alpha\}$. Show

(a) T is a TI-subgroup of G. (I.e. $T \cap T^g = T$ or 1 for all $g \in G$.)

(b) T is strongly closed in each 2-Sylow subgroup S of G containing T. (I.e. if $t^g \in S$ for $t \in T$, then $t^g \in T$.)

(c) $\bigcap_{g \in G} N_G(T^g) = 1$.

(5) Let G and T be as in (4) and assume in addition that $|T| \geq 4$. Let $Y = \langle T^G \rangle$. Show

(a) Let $D = \{t^g \mid t \in T^{\#}, g \in G\}$. Then D is a class of odd transpositions of Y with $D \cap D^2 \neq \emptyset$. Further $T^{\#}$ is a connectivity component of $\mathcal{D}(D)$.

(b) $O(G) = 1 = O_2(G) = C_G(Y)$. ($O(G)$ is the maximal normal subgroup of odd order.) Further Y is simple.

Remark. By the criterion for the existence of a strongly embedded subgroup of Aschbacher [Asc73] and Bender's theorem [Ben71] we have $Y \simeq L_2(q)$, $Sz(q)$ or $U_3(q)$, $|T| = q = 2^m > 2$.

(6) Let $X = \langle a, b, x \rangle \simeq L_2(q)$, $q = 2^m > 2$, where a, b and x are involutions of X satisfying $ab = ba$ and $o(ax) = 3$. Let V be an irreducible (non-trivial) $\mathbb{Z}_2 X$-module satisfying

(i) $[V, a] = C_V(a) = C_V(b) = [V, b]$.

(ii) $V = [V, a] \times [V, x]$.

Show that V is the natural $\mathbb{Z}X$-module.

Hint. Let $S \in \mathrm{Syl}_2(X)$ containing $\{a, b\}$, H a $2'$ complement to S in $N_X(S)$ and $h \in H$ with $a^h = b$. Then $H_0 = \langle h \rangle$ acts irreducibly on S, whence $C_V(a) = C_V(S)$.

§3 The Root Structure Theorem

We assume in this section that D is a non-degenerate class of root involutions of G, $O_2(G) = Z(G) = 1$ and that $E(d) = E_d \cup \{1\}$ for $d \in D$.

We fix the following notation (in addition to (1.1)):

$\Sigma := E(d)^G$.

$M_d := \langle X_d, d \rangle$.

W_d is the set of isolated vertices of $\mathcal{F}(C_D(d))$.

$N_d := \langle W_d \rangle$.

Then obviously $X_d \subseteq W_d$ and N_d is by (1.6) a normal 2-subgroup of $C(d)$. We first need to show that $W_d = X_d \cup \{d\}$ and $E(d) = N_d'$. If U is a nilpotent group denote by $cl(U)$ the nilpotence class of U. We have

(3.1) Lemma

$C_D(X_d) = E_d$. In particular $E_d = Z(M_d) \cap D = Z(N_d) \cap D = Z(\langle S \cap D \rangle) \cap D$ for each $S \in \mathrm{Syl}_2(G)$ containing M_d.

Proof. Suppose $c \in C_D(X_d) - E_d$. Then there exists by (1.7) an $x \in A_c \cap V_d$. Hence by (1.5) $c \notin C_D(d^x)$, a contradiction to $d^x \in X_d$. This proves the first part of (3.1). The second is a consequence of the first. □

(3.2) Lemma

Let $a, b \in N_d \cap D$, $ab \in D$, $a \in Z_2(N_d)$ and $x \in A_a \cap A_b \cap V_d$ such that $o(xa) = 3$. Then there exists a $y \in A_a \cap A_b \cap V_d$ with $o(yb) = 3$.

Proof. By (1.5) and (1.7) there exists a $y \in D$ satisfying $d^{xby} = d^x d^{xb}$. Hence $o(yb) = 3$, since $o(yb) \equiv 3 \bmod C_G(\langle d^x, x^{xb} \rangle)$. Clearly $y \in V_d$, since $y \in X_{d^x}$. As $[d^x, a] \in E_d$ by (3.1) we have $d^x d^{xb} \in C_D(a)$. (Since $[d^x, b] = [d^x E(d), b]$!). Assume $o(ya) = 4$. Then

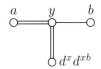

and $b(ba)(bd^x d^{xb}) = ab^{d^x}$. Hence (2.14)(3) implies $Y = \langle a, y, b, d^x d^{xb} \rangle \simeq W^*(D_4)$ or $W(D_4)$. In the second case $a \in Z_2(N_d)$ and the structure of $W(D_4)$ imply that the central involution of Y lies in E_d, a contradiction to $o(dy) = 4$. Thus $Y \simeq W^*(D_4)$ and $d^x = (d^x d^{xb})^{yb} \in C_D(a)$, a contradiction. This shows $y \in A_a$ and (3.2) holds.   □

(3.3) Lemma

Suppose $cl(N_d) \geq 3$. Then $Z_3(N_d) \cap X_d \not\subseteq Z_2(N_d)$.

Proof. By (1.21) (1) and (3.1) $E = (Z_3(N_d) \cap D) - Z_2(N_d) \neq \emptyset$ and $F = (Z_2(N_d) \cap D) - E_d \neq \emptyset$. Pick $f \in F$ and $x \in N_d \cap V_f$. Then $E_x \subseteq W_d$ by (1.7) and $[E_x, f] = E_d$ by (1.7) and (3.1). Let $E_d = \{d_1, \ldots, d_n\}$. Then, since $fd_i \in D$ for $i = 1, \ldots, n$ there exist by (1.5) $t_i \in D$ with $f^{t_i} = d_i$. Clearly $T = \langle t_i t_1 \mid i = 1, \ldots, n \rangle \leq N(E_d) \cap N(E_f)$ and $f^{t_i t_1} \neq f^{t_j t_1}$ for $i \neq j$. Hence $|f^T| \geq n$ and thus T is transitive on E_f. In particular $E_f \subseteq Z_2(N_d)$ and thus $[E(f), E(x)] = E(d) = [f', E(x)]$ for each $f' \in E_f$. We have shown:

(∗) $F \subseteq X_{\overline{d}}$ for each $\overline{d} \in E_d$ and $E_f \subseteq F$ for each $f \in F$.

Assume now that (3.3) is false. Then $E \cap X_d = \emptyset$. Pick $e \in E$ and $x \in N_d \cap D$ with $\alpha = [e, x] \notin Z(N_d)$. Then $\alpha \in E$. Let $y \in (N_d \cap D) - C(\alpha)$. (Exists by (3.1).) If $o(ey) = 2$, then $[e\alpha, y] = [\alpha, y] \in E_d$, a contradiction to $e\alpha = e^x \in E$. The same argument shows $[e, y] \notin E_d$. Hence $\beta = ee^y \in F$. Since $\alpha^y \in \alpha E(d)$, e^y centralizes α. Thus $o(\alpha\beta) = 2$ and

$$[y\beta, \alpha] = [y, \alpha] \in E_d \text{ and so } [y\beta, E_\alpha] = E_d \text{ by } (∗).$$

In particular $y\beta \in X_d$. By (1.8) there exists a $z \in A_\beta \cap V_d$ with $o(z\beta) = 3$. If $z \in V_y$ we have

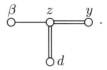

As $\beta(\beta y)(\beta d) = \beta y d \in D$, (2.14)(3) implies $\langle \beta, z, y, d \rangle \simeq W^*(D_4)$. Hence $\beta \sim \beta y$ in $C(d)$ and thus $y \in Z_2(N_d)$, a contradiction to $[e, y] \notin E(d)$.

Thus $z \in A_y$. Hence there exists by (3.2) a $\bar{z} \in A_\beta \cap A_y \cap V_d$ with $o(y\bar{z}) = 3$. Hence $d^{\bar{z}} \in X_d \cap V_\beta$, since otherwise $\langle \bar{z}, \beta \rangle \leq C(dd^{\bar{z}})$, which is impossible. By $(*)$ this implies $[d^{\bar{z}}, E_\beta] = E_d$. In particular there exists a $\bar{\beta} \in E_\beta$ with $[d^{\bar{z}}, \bar{\beta}] = d$. Then the action of $\langle \bar{\beta}, \bar{z} \rangle$ on $\langle d, d^{\bar{z}} \rangle$ implies $o(\bar{\beta}\bar{z}) = 3$ and we obtain:

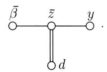

Now $y \in X_d$, since $[y, E_\alpha] = E_d$ by $(*)$. Further $\bar{\beta}y(yd) = \bar{\beta}d \in D$ by the above. Hence (2.14)(3) implies again $\langle \bar{\beta}, \bar{z}, y, d \rangle \simeq W^*(D_4)$ and thus $y \sim \bar{\beta}$ in $C(d)$. This implies $y \in Z_2(N_d)$, contradicting $[e, y] \notin E(d)$. $\square$

(3.4) Lemma

$cl(N_d) = 2$.

Proof. Suppose false. Then there exists by (3.3) an $x \in (X_d \cap Z_3(N_d)) - Z_2(N_d)$. Hence there exists a $y \in N_d \cap D$ such that $\alpha = (xy)^2 \in (Z_2(N_d) \cap D) - E_d$. By (1.8) there exists an $u \in V_d$ such that $o(u\alpha) = 3$. Assume first $u \in V_x$. Then we have

and as $\alpha d, \alpha x \in D$, $Y = \langle \alpha, u, d, x \rangle$ is by (2.14)(3) isomorphic to $W(D_4)$ or $W^*(D_4)$. In the first case $1 \neq t = (d^u x)^2 \in Z(Y) \cap D$. Since $x \in Z_3(N_d)$ and $t \notin E_d$, (3.1) implies $t \in Z_2(N_d) - E_d$. Moreover $xt = x^{d^u} \in X_d$. Let now

$v \in A_t \cap V_d$ with $o(vt) = 3$ and assume $o(vx) = 4$. Then

and $(2.14)(3)$ and $xtd \in D$ imply $\langle t, v, x, d \rangle \simeq W^*(D_4)$. Hence $t \sim tx$ in $C(d)$, a contradiction to $t \in Z_2(N_d)$ but $tx \in Z_3(N_d) - Z_2(N_d)$.

So we obtain one of the following possibilities:

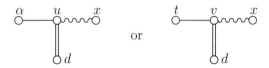

or

where $\alpha, t \in Z_2(N_d)$. Assume without loss of generality that we are in the first case. Then there exists by (3.2) an $\bar{u} \in A_\alpha \cap V_d$ with $o(\bar{u}x) = 3$. Since $d^{\bar{u}} \in X_d \cap V_\alpha$ we have by $(3.3)(*)$ $[E_\alpha, d^{\bar{u}}] = E_d$. In particular $[\bar{\alpha}, d^{\bar{u}}] = d$ for some $\bar{\alpha} \in E_\alpha$. Hence $o(\bar{\alpha}\,\bar{u}) = 3$ by the action on $\langle d, d^{\bar{u}} \rangle$ and we obtain

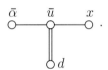

Since $\bar{\alpha} \in Z_2(N_d)$ and thus $\bar{\alpha} \in X_d$ by (3.3) $(*)$, $(2.14)(3)$ implies $\langle \bar{\alpha}, \bar{u}, x, d \rangle \simeq W^*(D_4)$. Hence $\bar{\alpha} \sim x$ in $C(d)$, a contradiction to $x \in Z_3(N_d) - Z_2(N_d)$. □

(3.5) Theorem

The following hold:

(1) $N_d = M_d$.

(2) $D \cap N_d = X_d \cup \{d\}$.

(3) $N'_d = E(d)$ and $E_d = Z(N_d) \cap D$.

Proof. Let $f \in (N_d \cap D) - E_d$. Then there exists by (3.4) and (3.1) an $x \in N_d \cap D$ with $[f, x] \in E_d$. By $(3.3)(*)$ $[f, E_x] = E_d$ and thus $f \in X_d$. Since $E_d = E(d)^{\#}$ by the hypothesis of this section, this proves (2) and (1). (3) is now a consequence of (3.1). □

(3.6)   Corollary

Let $x, y \in D$ such that $xy \in D$. Then $(E(x)E(y))^{\#} \subseteq D$, $\Sigma \cap E(x)E(y)$ is a partition of $E(x)E(y)$ and there exists a $z \in D$ such that

$$(*) \qquad E(x) = [E(y), E(z)] = [\bar{y}, E(z)] = [E(y), \bar{z}]$$

for each $\bar{y} \in E_y$, $\bar{z} \in E_z$.

Proof. As $E_y \subseteq N_x \cap D$, there exist by (3.5) a $z \in X_d$ such that $(*)$ holds. Hence $E(y)^{\bar{E}(z)} \cup E(x)$ is a partition of $E(y)E(x)$, which implies the first part of (3.6).   □

(3.7)   Corollary

Let $E \subset D$ with $\langle E \rangle^{\#} \subset D$. Let $\bar{E} = \bigcup_{e \in E} E_e$. Then the following hold:

(1) $\langle \bar{E} \rangle^{\#} \subset D$ and $\Sigma \cap \langle \bar{E} \rangle$ is a partition of $\langle \bar{E} \rangle$.
(2) One of the following holds:

   (a) $\langle \bar{E} \rangle = E(e)$.
   (b) $\langle \bar{E} \rangle = E(e)E(f)$; $e, f \in E$.
   (c) $\langle \bar{E} \rangle$ is a n-dimensional vector space over $GF(q)$, $q = |E(e)|$ and $n \geq 3$, with $\Sigma \cap \langle \bar{E} \rangle$ the set of 1-subspaces.

Moreover in (b) and (c) $\langle N_D(\bar{E}) \rangle$ induces the $SL_n(q)$ on $\langle \bar{E} \rangle$, where $n = 2$ in (b).

Proof. (1) follows from a repeated application of (3.6). Suppose that neither (a) nor (b) of (2) hold. Then (c) follows as in III(2.1) (See also exercise III(2.31)(1)). Moreover the proof of the same results also shows that $\langle N_D(\bar{E}) \rangle$ induces the $SL_n(q)$ on $\langle \bar{E} \rangle$, since for each $e \in \bar{E}$, M_e induces the set of all transvections corresponding to the point $E(e)$ on $\langle \bar{E} \rangle$.

So suppose that (2)(b) holds. Let $N = E(e)E(f) = \langle \bar{E} \rangle$. Pick $x \in X_e - C(f)$ and $y \in X_f - C(e)$. Then we obtain by (3.6):

$$(*) \qquad \begin{array}{l} E(e) = [\bar{x}, E(f)] = [E(x), \bar{f}] \text{ for all } \bar{x} \in E_x, \bar{f} \in E_f \\ E(f) = [\bar{y}, E(e)] = [E(y), \bar{e}] \text{ for all } \bar{y} \in E_y, \bar{e} \in E_e. \end{array}$$

Let now $\bar{x} \in E(x)^{\#}$. Then there exists by $(*)$ an $\bar{y} \in E(y)^{\#}$ with $E(f)^{\bar{x}} = E(e)^{\bar{y}}$, since the elements of Σ are trivial intersection subgroups of G. Hence $E(y)^{\overline{xy}^{-1}} \leq E(x)C(N)$, since for each $y_1 \in E(y)^{\overline{xy}^{-1}}$ there exists an $y_2 \in E(x)$ with $f^{y_1 y_2} = f$, whence $[E(f), y_1 y_2] \leq E(e) \cap E(f) = 1$. This implies that $E(y)^{\bar{x}} = E(x)^{\bar{y}} \mod C(N)$. Hence, if $X = \langle E(x), E(y) \rangle$, then $X/C_X(N) = \bar{X}$ is a rank one group. Now $(*)$ and I(3.2) imply $\bar{X} \simeq SL_2(q)$ and N is the natural $\bar{X}$-module. Moreover easily $\langle N_D(N) \rangle \leq XC(N)$.   □

(3.8) Lemma

Let $a \in A_d$ such that $X_d \cap V_a \neq \emptyset$. Then $\langle E(a), E(d) \rangle \simeq SL_2(q)$, $q = |E(d)|$.

Proof. Pick $e \in X_d \cap V_a$ and let $N = E(e)E(e)^a$. Then by (3.7) $N^\# \subset D$ and $\Sigma \cap N$ is a partition of N. Since $E(d) \leq N_e$ and $E(a) \leq N_{ee^a}$, (3.5) and (3.7) (2) imply that $X/C_X(N) \simeq SL_2(q)$ for $X = \langle E(d), E(a) \rangle$.

If now $|E(d)| = 2$, clearly $X \simeq D_6 \simeq SL_2(2)$. So assume $|E(d)| \geq 4$. Then (1.9) implies $C_X(N) = O_2(X)$. Pick now $d_1, d_2 \in E_d$ with $X = O_2(X)\langle d_1, d_2, a \rangle$ and $o(d_1 a) \equiv 3 \bmod O_2(X)$. Then $o(d_1 a) = 3$, since $o(d_1 a)$ is odd. Hence (2.1) implies $\langle d_1, d_2, a \rangle \simeq L_2(q)$. Moreover, if h is a diagonal element of $\langle d_1, d_2, a \rangle$ with $d_1^h = d_2$, then $d_1^{\langle h \rangle} \subseteq E_d$. Since $|\langle d_1^{\langle h \rangle} \rangle| = q$ as shown in (2.1), this shows $E(a) \subseteq \langle d_1, d_2, a \rangle$ and thus $X \simeq L_2(q)$. $\qquad\square$

(3.9) Lemma

Suppose T is a weak TI-subset of D. Then $\overline{T} = \bigcup_{d \in T} E_d$ is also a weak TI-subset.

Proof. We first show:

$(*)$ $\qquad$ If $\alpha \in T$ and $\beta \in T \cap X_\alpha$, then $(E(\alpha)E(\beta))^\# \subseteq T$.

Indeed by (3.7) each element of $E(\alpha)\beta$ is conjugate to β in $C_D(\alpha)$. Hence $E(\alpha)\beta \subseteq T$. Now again by (3.7) each element of $(E(\alpha)E(\beta))^\#$ is conjugate to some element of $E(\alpha)\beta$ in $C_D(\beta)$. This proves $(*)$.

Now suppose $\overline{T} \cap \overline{T}^x \neq \emptyset$ for $x \in D$. Then there exists an $a \in \overline{T}$ with $a^x \in \overline{T}$. If $a = a^x$ then $x \in C_D(E_a)$ and thus $x \in N_D(T)$, since $E_a \cap T \neq \emptyset$. So we may assume $a^x \in X_a$. Clearly $E_a \cap T \neq \emptyset \neq E_{a^x} \cap T$. Thus picking $\alpha \in E_a \cap T$, $\beta \in E_{a^x} \cap T$, (3.5) and $(*)$ imply $(E(\alpha)E(\beta))^\# \subseteq T$. But then $aa^x \in T$ and thence $T = T^x$ and $\overline{T} = \overline{T}^x$. $\qquad\square$

(3.10) Proposition

Let T be a maximal weak TI-subset of D. (In particular we assume weak TI-subsets exist!) Then $T = \langle T \rangle^\#$ and $\Sigma \cap \langle T \rangle$ is a partition of T.

Proof. By (3.9) we have $E_d \subseteq T$ for $d \in T$ by maximality of T.

By (3.7) $ab \in T$, if $a, b \in T$ and $b \in X_a$. Hence, to prove (3.10), it suffices to show that $ac \in T$ for all $a \neq c \in T$. By (2.7) $\langle T \rangle$ is abelian. Thus, if $a \neq c \in T$ with $ac \notin T$, then there exists a $b \in T \cap X_a \cap X_c$. Indeed by (3.1) there exists an $x \in X_a \cap V_c$, whence $b = (xc)^2 \in T \cap X_a \cap X_c$. Pick by (1.8) $x \in A_a \cap V_{ab}$

with $o(xa) = 3$. Then also $o(xb) = 3$. Clearly $o(xc)$ is 4 or odd. Suppose first $o(xc) = 4$. Then we have

and as $bc \in D$, (2.14)(3) implies that $Y = \langle a, b, c, x \rangle \simeq W(D_4)$ or $W^*(D_4)$. In the first case $z = [(ab)^x, c] \in T \cap Z(Y)$ since $(ab)^x \in C_D(ab)$, a contradiction. In the second case $b \sim bc$ in $C_D(a)$. Thus $bc \in N_a$ and thence $c \in X_a$ by (3.5), which was to be shown.

So we may assume $c \in A_x$. Now by (1.5) $ab(ab)^x \in X_x \cap V_c$. Hence (3.8) implies $\langle E(x), E(c) \rangle \simeq L_2(q)$, $q = |E(c)|$. In particular there exists a $\bar{c} \in E_c$ with $o(x\bar{c}) = 3$. We obtain:

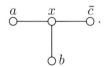

Since by (3.5) $b\bar{c} \in T$, (2.13)(3) and the same argument as above imply $\langle a, b, \bar{c}, x \rangle \simeq W^*(D_4)$. Hence $b \sim \bar{c}$ in $C_D(a)$ and thus $a\bar{c} \in T$. But then, as $E_c = E_{\bar{c}} \subseteq T$, (3.5) implies $ac \in T$, which was to shown. $\square$

(3.11)   Corollary

Let T be a weak TI-subset, satisfying $E_d \subset T$ for $d \in T$. Then T is maximal.

Proof. Embed T in a maximal weak TI-subset L. Pick $a \in T$, $b \in T - E_a$ and $c \in L - T$. Then by (3.7) and (3.10) $b \sim c$ in $C_D(a)$. Hence no such c exists and $T = L$. $\square$

(3.12)   Root Structure Theorem

One of the following holds:

(1) D contains a maximal weak TI-subset T such that $\Sigma \cap \langle T \rangle$ is a partition of $\langle T \rangle$ and $|\langle T \rangle| = q^2$, where $|E(d)| = q$ for $d \in D$.

(2) For each pair $E(a), E(b) \in \Sigma$ one of the following holds:

(i) $[E(a), E(b)] = 1$.

(ii) $E(c) = [E(a), E(b)] = [E(a), \bar{b}] = [\bar{a}, E(b)]$ for some $c \in D$ and each $\bar{a} \in E_a$, $\bar{b} \in E_b$.

(iii) $\langle E(a), E(b) \rangle \simeq SL_2(q)$, $q = |E(a)|$.

Proof. It is clear that, if $ab = ba$, (2)(i) holds and, if $c = (ab)^2$, (2)(ii) holds by (3.5). So we may assume that there exists no weak TI-subset T satisfying (1) and then show that (2)(iii) holds for each $a \in D$ and $b \in A_a$.

For this purpose let $_1A_a = \{b \in A_a \mid X_a \cap V_b \neq \emptyset\}$ and $_2A_a = A_a - {_1A_a}$. Then it suffices by (3.8) to show that $_2A_a = \emptyset$. Denote by $d(,)$ the distance in the graph $\mathcal{E}(D)$. ($\mathcal{E}(D)$ is connected by (2.8))!) and let $x \in {_2A_a}$ with $d(a, x)$ minimal. Let

$$a = a_0, a_1, \ldots, a_n = x \quad \text{and} \quad d = a_{n-1}$$

be a path of length $d(a, x)$ from a to x in $\mathcal{E}(D)$. Then by the minimality of $d(a, x)$ we have $d \in {_1A_a}$. Indeed, if $d \in V_a$, then by (1.5) $(ad)^2 \in X_a \cap V_x$, contradicting $x \in {_2A_a}$. Let $b \in X_a \cap V_d$. Then $c = (bd)^2 \in X_b \cap X_d$ and we may assume that our minimal path from a to x is

$$a, b, c, d, x \text{ and } x \in {_1A_b}.$$

Let $L = E(b)E(c)$. Then L is by (3.6) partioned by $\Sigma \cap L$. Since we assume that $L^{\#}$ is not a weak TI-subset of D, there exists by (3.7) a $y \in D_b$ with $E(c^y) \not\leq L$. Set $E = LE(c^y)$. Then, as $E(c^y) \leq N_b \cap N_c$, an application of (3.5) and (3.6) shows that $\Sigma \cap E$ is a partition of E. Hence by (3.7) $|E| = q^3$.

Let $W = E(c)E(d)$. Then by (3.5) $\langle E, E(x) \rangle \leq N(W)$. Hence (3.7) implies $|E_0| = q^2$ for $E_0 = E \cap C(W)$ and $E_0^{\#} \subset V_x$. But as $[E(a), E] \leq E(b)$ also $|C_E(a)| = q^2$. Hence there exists a $1 \neq u \in C_E(a) \cap V_x$. We have

But then (1.5) implies $X_a \cap V_x \neq \emptyset$, a contradiction to $x \in {_2A_a}$.   □

In view of (3.12) it suffices to show that also in the non-degenerate case Σ is a class of k root subgroups with $k = GF(q)$, to treat case (1) of (3.12). We will show in the next section, that if $q > 2$, also in case (1) Σ is a class of k-root subgroups. At the end of this section we will indicate how one treats case (1) of (3.12) when $q = 2$. In this case, as the examples (1.2)(b) and (c) show, Σ is not necessarily a class of $GF(2)$-root subgroups. Before we are able to do this, we need some additional facts about weak TI-subsets.

(3.13)   Lemma

Suppose T is a maximal weak TI-subset of D. Then $N_D(T) = V \dot{\cup} T$ with V a conjugacy class in $\langle N_D(T) \rangle$. In particular $C_D(T) = T$.

Proof. Decompose $N_D(T)$ according to (1.6) with W the set of isolated vertices of $\mathcal{F}(N_D(T))$. Then by (3.7) and (3.10) $W \subseteq C_D(T)$. Since $C_D(d) \subseteq N_D(T)$

for $d \in T$ we obtain $W \subseteq N_d \cap D = X_d \cup \{d\}$ for each $d \in T$ by (3.5). Suppose $a \in W - T$. Then (3.5) shows that $(E(a)\langle T \rangle)^{\#} \subseteq D$. But this is by (3.7) a contradiction to the fact that $T \neq E_d$ and T is a weak TI-subset.

So $W = T$. Suppose $V_1 \subseteq C_D(T)$ is a connectivity component of $\mathcal{F}(N_D(T))$. Then for $c \in V_1$ and $d \in T$ there exists by (1.7) an $x \in A_c \cap V_d$. Hence by (1.5) $y = (xd)^2 \in X_a \cap V_c$. Clearly $V_1 \cap X_a = \emptyset$ for each $a \in T$. If now $y \in T$, then as $x \in X_y$ we obtain $x \in N_D(T) - V_1$ a contradiction to $x \in A_c$. Thus $y \in N_D(T) - (T \cup V_1)$. But then $(yc)^2 \in T$ by (1.6), a contradiction to $T \cap X_c = \emptyset$.

Hence no such V_1 exists and $C_D(T) = T$. Now (1.6), (3.7) and (3.10) show that (3.13) holds. $\qquad \square$

(3.14) Lemma

Suppose (3.12)(1) holds and let $T = (E(a)E(d))^{\#}$. Pick $x \in X_a - C(d)$, $y \in X_d - C(a)$. Then the following hold:

(1) $M_a = E(x)(M_a \cap C(b))$ with $E(x) \cap C(b) = 1$.
(2) $X \simeq SL_2(q)$ for $X = \langle E(x), E(y) \rangle$.
(3) Set $N = (M_a \cap C(d))(M_d \cap C(a))$, $N_0 = M_a \cap M_d$ and $\widetilde{N} = N/N_0$. Then $\widetilde{N}$ is the direct sum of natural $\mathbf{Z}X$-modules. Further $NX = \langle N_D(T) \rangle$.
(4) $\widetilde{M_d \cap C(a)}$ acts regularly on the set of weak TI-subsets of D containing E_a and different from T.

Proof. (1) and (2) are direct consequences of (3.7) and (3.8). Now $E(x)$ centralizes $(\widetilde{M_a \cap C(d)})$ and $[\widetilde{M_d \cap C(a)}, E(x)] \leq \widetilde{M_a \cap C(d)}$ and the same holds for $E(y)$ with the roles of a and d reversed. Thus, if $C_{\widetilde{N}}(x) > \widetilde{M_a \cap C(d)}$, then $C_{\widetilde{M_d \cap C(a)}}(X) \neq 1$, a contradiction to $M_a \cap M_d = 1$. Hence the hypothesis of I(3.7) is satisfied for the action of X on $\widetilde{N}$, which implies the first part of (3). The second part is obvious, since by (3.13) and (3.5) $C_D(a) \subseteq M_a$.

Let T_1, T_2 be two different weak TI-subsets of D containing E_a different from T. Let $E_b \subseteq T_1 - E_a$ and $E_c \subseteq T_2 - E_a$. Then by (3.13) $[E(b), E(d)] = E(a) = [E(c), E(d)]$. Thus by (3.8) $X_1 = \langle E(y), E(b) \rangle \simeq SL_2(q) \simeq \langle E(y), E(c) \rangle = X_2$. Since $X_1 N = XN = X_2 N$, this implies $E(b)^n = E(c)$ for some $n \in N$. Since M_a normalizes T_1 and T_2 we may assume $n \in M_d \cap C(a)$.

This shows that $\widetilde{M_d \cap C(a)}$ acts transitively on the weak TI-subsets containing E_a different from T. Since $NE(b) = NE(x)$ we also have $C_{\widetilde{M_d \cap C(a)}}(E(b)) = 1$. This proves (4). $\qquad \square$

(3.15) Corollary

Suppose that (3.12)(1) holds and there exist more than two weak TI-subsets of D containing E_a. Let $R = \langle((M_d \cap C(a))M_a)^{C(a)}\rangle$ and $\overline{C(a)} = C(a)/M_a$. Then $\overline{R} \simeq L_2(q^n)$ and $\overline{R}$ acts triply transitively on the weak TI-subsets of D containing E_a.

Proof. By (3.14)(3) and (4) we have $|\overline{M_d \cap C(a)}| = q^n$ and $\overline{M_d \cap C(a)}$ acts regularly on the weak TI-subsets of D containing E_a different from T. Since by (3.14) $(M_d \cap C(a))M_d \lhd N_G(T) \cap C(a)$, reversing the roles of T and T_1, exercise (2.14)(5) implies $\overline{R}/\overline{K} \simeq L_2(q^n)$, where $\overline{K}$ is the kernel of the action of $\overline{R}$ on the weak TI-subsets containing E_a. By the above $\overline{K} \leq N(\overline{M_d \cap C(a)})$ so that $\overline{K} \leq Z(\overline{R})$. Since $\overline{M_d \cap C(a)}$ is elementary abelian, the extension of $\overline{R}/\overline{K}$ by $\overline{K}$ splits.

Now let $\overline{H} = N_{\overline{R}}(T)$. Then again by the above, $\overline{H} \leq N_{\overline{R}}(\overline{M_d \cap C(a)})$. Hence by the structure of $L_2(q^n)$,

$$\overline{M_d \cap C(a)} = [(\overline{M_d \cap C(a)}), \overline{H}] \leq \overline{R}'.$$

Hence $\overline{K} = 1$ and $\overline{R} \simeq L_2(q^n)$, which was to be shown. $\square$

(3.16) Lemma

Suppose (3.12)(1) holds and pick $b \in X_a - E_a$. Then $T_1 = (E(b)E(a))^{\#}$ is a maximal weak TI-subset of D.

Proof. Suppose $T = (E(d)E(a))^{\#}$. Then by (3.13) $C_D(b) \subseteq N_D(T_1)$. Hence (3.7) implies $C_D(e) \subseteq N_D(T_1)$ for each $e \in T_1$ and T_1 is a weak TI-subset. The maximality follows from (3.11). $\square$

(3.17) Proposition

Suppose (3.12) (1) holds and $|E(a)| = 2$ for $a \in D$. Let by (3.15) $2^n + 1$ be the number of weak TI-subsets of D containing a. Then one of the following holds:

(a) $n = 0$ and $M_a \simeq D_8$.
(b) $n = 1$ and $M_a \simeq Q_8 * \mathbb{Z}_4$ (i.e. $|M_a| = 16!$)
(c) $n = 2$ and $M_a \simeq (Q_8 * D_8) \times \langle t \rangle, t^2 = 1$.
(d) $n = 3$ and $M_a \simeq (D_8 * D_8 * D_8 * D_8) \times \langle t \rangle, t^2 = 1$.

Proof. By (3.5) we have $M_a' = \Phi(M_a) = \langle a \rangle \leq Z = Z(M_a)$. Let $\widetilde{M_a} = M_a/Z$. Then it is well known, see [Hup67, III(3.18)], that $\widetilde{M_a}$ is a non-degenerate symplectic space over $\mathbb{Z}_2$, with scalar product given by commutators. Moreover by

(3.13) and (3.16) the elements of $\widetilde{X}_a$ form a set of pairwise non-perpendicular singular vectors of $\widetilde{M}_a$.

Let $\widetilde{X}_a = \{\widetilde{d}_1, \ldots, \widetilde{d}_k\}$. Then, since $\{d_i, a, d_i a\}$ is by (3.16) a weak TI-subset of D, we have $k = 2^n + 1$ by (3.15). Now it is an easy exercise in linear algebra to show that the only linear relation among these $\widetilde{d}_i$ is of the form

$$\widetilde{d}_k = \widetilde{d}_1 + \cdots + \widetilde{d}_{k-1} \quad \text{(after reordering!)} \ .$$

Hence $\dim \widetilde{M}_a \geq 2^n$. On the other hand $\widetilde{M_a \cap M_{d_i}}$ is a totally singular subspace of $\widetilde{M}_a$, whence $\dim \widetilde{M_a \cap M_{d_i}} \leq \frac{1}{2} \dim \widetilde{M}_d$. This implies

$$\dim(\overline{M_{d_i} \cap C(a)}) \geq \frac{1}{2} \dim \widetilde{M}_a - 1 \geq 2^{n-1} - 1,$$

where $\overline{C(a)} = C(a)/M_a$ as in (3.15). Hence by (3.15) we obtain $2^n \geq 2^{2^{n-1}-1}$ which implies $n \leq 3$. (This argument is from [Tim78a, 4.2].)

If $n = 0$ clearly $M_a \simeq D_8$. If $n = 1$, then by the above $|\widetilde{M}_a| \geq 4$ and thus $|M_a| = 16$ and (b) holds. If $n = 2$ we have $|\widetilde{M}_a| = 16$, since $\widetilde{M}_a$ is non-degenerate. On the other hand M_a is generated by five elements of X_a. Hence $|M_a| \leq 2^6$. Now by a generator and relations argument, $D_8 * Q_8$ must be an epimorphic image of M_a, since $D_8 * Q_8$ possesses five involutions $t_1, \ldots, t_5$ satisfying $(t_i t_j)^2 = z$ for $1 \leq i \neq j \leq 5$, where z is the central involution. This shows that in case $n = 2$ (c) holds.

Finally if $n = 3$ we have $k = 9$ and thus $|\widetilde{M}_a| = 2^8$, since $\widetilde{M}_a$ is non-degenerate. Again $|M_a| \leq 2^{10}$ as $k = 9$. Since as before $D_8 * D_8 * D_8 * D_8$ is an image of M_a, this shows that (d) holds.

(3.18)   Remark

Here we will shortly describe how one classifies G in the cases (3.17) (a)–(d). For details the reader is referred to [Tim75a, §7].

If $n = 0$ one shows that $M_a \in \mathrm{Syl}_2(G)$ using transfer. Since M_d is a normal 2-complement in $C(d)$, a classical result of Suzuki [Suz65] implies $G \simeq A_6$ or $L_3(2)$.

If $n = 1$ one shows by (3.14) that $\langle N_D(T) \rangle \simeq (\mathbf{Z}_4 \times \mathbf{Z}_4)\Sigma_3$. Now, again using transfer, one obtains that $\langle N_D(T) \rangle$ contains a 2-Sylow subgroup of G, which is a wreath product of $\mathbf{Z}_4$ by $\mathbf{Z}_2$. Now a result of P. Fong [Fon67] implies $G \simeq U_3(3) \simeq G_2(2)'$.

In the cases $n = 2$ and 3 one first shows that, if $C = C_G(M_a) > \langle a \rangle$, then $C_0 = C_C(N_D(T)) \neq 1$. Now $\langle N_D(T) \rangle$ is D-maximal (i.e. maximal among subgroups generated by elements in D). Since it is easy to see that R centralizes C, this implies

$$C_0 \leq C(\langle N_D(T), R \rangle),$$

a contradiction to $G = \langle N_D(T), R \rangle$ and the simplicity of G. This shows that $C = \langle a \rangle$ and $M_a = F^*(C_G(a))$. We obtain with (3.15):

If $n = 2$,   then $C(a)/M_a \simeq A_5$ or Σ_5.
If $n = 3$,   then $C(a)/M_a \simeq PSL_2(8)$ or $P\Gamma L_2(8)$.

In case $n = 2$ one uses transfer again to show $C(a)/M_a \simeq A_5$. Now Janko's classification [Jan69] implies $G \simeq J_2$. In case $n = 3$ one uses the classification of Thomas [Thm70] to obtain $G \simeq {}^3D_4(2)$. (Notice that this sketch is slightly different from [Tim75a, §7], since there I computed $|_1A_a|$ and used the action of $C(a)$ on $_1A_a$, which is not really necessary.)   □

(3.19)   Exercises

(1) Suppose D is a non-degenerate class of root involutions of G and $O_2(G) = 1 = Z(G)$. Suppose $G > R = \langle R \cap D \rangle > \langle C_D(d) \rangle$, $d \in D$. Show:

(a) $T = d^R$ is a weak TI-subset of D.

(b) Show, that if L is a maximal weak TI-subset of D, then $\langle N_D(L) \rangle$ is D-maximal.

Hint. See III(2.8).

(2) (Thompson transfer.) Let G be a group with $2 \nmid |G : G'|$, $S \in \mathrm{Syl}_2(G)$ and $M \lhd S$ such that S/M is cyclic. Then each element $s \in S - M$ of minimal order (with the property $s \in S - M$) is conjugate to an element of M.

Hint. Consider the transfer of G in S/M.

(3) Let t be an involution of the group G, satisfying:

(a) $\langle t^G \cap C(t) \rangle = Q$ is a 2-group of class at most two.

(b) If $\tau \in t^G \cap C(t)$ with $\tau \neq t$, then $t\tau$ is conjugate to τ.

Show:

(i) $D = t^G$ is a class of root involutions of $R = \langle D \rangle \lhd G$.

(ii) $D = D^2$ or $D \cap C_G(t) = \{t\}$ and $R = O(R)\langle t \rangle$ by the **Z***-theorem of Glauberman[Asc86, p.261].

(iii) Show, using the classification of groups generated by abstract root
subgroups and the main theorem of the introduction to chapter IV
and (3.18), that if $D = D^2$ then $R/Z(R)$ is isomorphic to one of
the following groups: $L_2(q), Sz(q), U_3(q), L_3(q), G_2(q)$ (or $G_2(2)' = U_3(3)$) or $^3D_4(q)$, where $q = 2^m$, or to A_6 or J_2.

(4) Assume the same hypothesis as in (3)(a) and that Q is extra special.
Show:

(i) $D = t^G$ satisfies (3)(b).

(ii) Let $R = \langle D \rangle$. Then $R/Z(R)$ is isomorphic to $A_6, L_3(2), J_2$ or $^3D_4(2)$.

Hint. In (ii) $R/Z(R)$ is one of the groups of (3)(iii). By (3.18) $Q \simeq Q_8 * \mathbf{Z}_4$ if $R/Z(R) \simeq U_3(3)$ and thus is not extra special.

§4   The Rank Two Case

We assume in this section that D is a non-degenerate class of root involutions
of G, $O(G) = Z(G) = 1$ and $E(d) = E_d \cup \{1\}$ for $d \in D$. In addition, in view
of (3.12) and (3.17), (3.18) we assume:

(1) $q = |E(d)| > 2$.

(2) There exists a maximal weak TI-subset T of D such that $\Sigma \cap \langle T \rangle$ is a
partition of $\langle T \rangle$ and $|\langle T \rangle| = q^2$.

We carry on with the notation introduced in §3 and in addition use the following geometric language. Call the elements of Σ *points* and the maximal weak
TI-subsets *lines*, a point $E(a)$ lying on the line $T \subseteq D$ if and only if $E_a \subseteq T$.
Remember that by (3.16) $(E(a)E(d))^\#$ is a line, if $a \in X_d - E_d$. Moreover, by
(3.15), the number of lines through $E(a)$ is of the form $q^n + 1$ with $n \in \mathbb{N} \cup \{0\}$
and the group $\overline{R} = R/M_a \simeq L_2(q^n)$ and acts in its natural permutation action
on the lines through $E(a)$.

The aim of this section is to show that under these conditions Σ is also a class
of abstract root subgroups of G.

(4.1)   Proposition

Suppose $n = 0$. Then $G \simeq L_3(q)$. In particular Σ is a class of abstract root
subgroups of G.

Proof. Since large parts of the proof are standard finite groups theory we just
sketch it.

Let $T = (E(d)E(a))^{\#}$ be a line through $E(a)$. Then by (3.14) $\langle N_D(T)\rangle = \langle T\rangle X$, $X = \langle E(x), E(y)\rangle$ as in (3.14), since $M_a = E(d)E(a)E(x)$. Let $Q = \langle T\rangle$ and $h \in C(QX)$. Then h centralizes the line $E = (E(a)E(x))^{\#}$. Let

$$\langle N_D(E)\rangle = \langle E\rangle Y, Y = \langle Y \cap D\rangle \simeq L_2(q) \text{ and } L = \langle E, h\rangle.$$

Then we may assume $E(d) \le Y$. Hence $C_L(N_Y(E(d))) = \langle h\rangle$ and thus Y centralizes $\langle h\rangle$, since $\langle E\rangle$ is the natural Y-module. Hence by (3.13) (1) we obtain

$$G = \langle N_D(T), N_D(E)\rangle \le C(h)$$

and thus $C(QX) = 1$.

Let $C = C_G(Q)$. Then, since $[C, X] \le Q$, C is an elementary abelian 2-group. Now for $t \in C - Q$ there exists an $\overline{x} \in E(x)$ with $E(d)^{t\overline{x}} = E(d)$, since $\Sigma \cap \langle E\rangle$ is a partition of $\langle E\rangle$. This implies $t\overline{x} \in C(QE(d))$. Hence by Gaschütz's theorem [Hup67, I(17.4)] there exists an X-invariant complement to Q in $Q\langle t\rangle$. Hence $C(QX) = 1$ implies $C = Q$. Now, since $N_G(Q)$ normalizes $\Sigma \cap QX$, it is easy to see that this implies $N_G(Q) = QY$ with $X \le Y \le P\Gamma L(2, q)$.

Now let $S^* \in \mathrm{Syl}_2(N_G(Q))$ containing M_a and $S \in \mathrm{Syl}_2(G)$ containing S^*. Then, as $S \le N(M_a)$, we have $|S : S^*| \le 2$. Suppose first $S = S^* \ne M_a$. Then $S = M_a\langle y\rangle$, $M_a \cap \langle y\rangle = 1$ and $\langle y\rangle$ induces field automorphisms on Q. Let $\overline{y} \in \langle y\rangle$ with $\overline{y} \ne 1 = \overline{y}^2$. Then $\overline{y}$ is by (3.13)(2) conjugate to some element of M_a. Hence $\overline{y}$ centralizes some $E(f) \in \Sigma$. By (1.4)(5) there exists an $E_e \subseteq T \cap A_f$ with $C_{E(e)}(\overline{y}) \ne 1$. But then $E(e) \le C(\overline{y})$, a contradiction since $\overline{y}$ induces a field automorphism on Q.

Let H be a 2'-complement to $E(x)$ in $N_X(E(x))$. Then, as S normalizes M_aH, the Frattini argument implies $S = M_aN_S(H)$, $M_a \cap N_S(H) = 1$. Now $N_S(H)$ normalizes the Frobenius group $E(a)H$ and thus $N_S(H) = S_1 \times S_2$ with $S_1 = N_S(H) \cap S^*$ cyclic and $S_2 = C_S(H) \simeq \mathbb{Z}_2$. Hence $M_aS_2 = C_S(E(a))$. If now $S_1 \ne 1$, then as above some element of $S_1^{\#}$ fuses to M_aS_2 from which we get as above a contradiction to the fact that the elements of S_1 induce field automorphisms on Q. Hence $S_1 = 1$ and the involution z of S_2 fuses by (3.19)(2) to an element of M_a, a contradiction since all involutions in M_a are in D.

This shows $S = M_a$ and as $M_a \trianglelefteq C_G(a)$, the centralizer of each involution of G has a normal 2-Sylow subgroup. Now a classical theorem of Suzuki [Suz65] implies $G \simeq L_3(q)$. $\square$

(4.2) Notation

For the rest of this section we assume, in view of (4.1), that $n > 0$. For a fixed, $A = E(a) \in \Sigma$ let:

$$\Sigma_A = C_\Sigma(A) - \{A\}, \ M_A = M_a.$$
$$\Psi_A = \{B \in \Sigma \mid [B, A] \in \Sigma\}.$$
$$\Omega_A = \Sigma - (\Sigma_A \cup \Psi_A \cup \{A\}).$$
$$_1\Omega_A = \{B \in \Omega_A \mid \Sigma_A \cap \Psi_B \neq \emptyset\}.$$
$$_2\Omega_A = \Omega_A -_1 \Omega_A.$$

Notice that by (3.13) $\Sigma \cap AB$ is a partion of AB for each $B \in \Sigma_A$. Moreover, by (3.5), $B = E(b) \in \Psi_A$ if $o(ab) = 4$.

(4.3) Lemma

Let $A \in \Sigma$. Then $C_G(A)$ is transitive on $_1\Omega_A$.

Proof. Let $X, Y \in {}_1\Omega_A$. Since by (3.15) $C_G(A)$ is transitive on the lines through A and since M_A is transitive on $(\Sigma \cap AB) - \{A\}$ for $B \in \Sigma_A$, we may assume $B \in \Sigma_A \cap \Psi_X \cap \Psi_Y$. Let $N_1 = B[B, X]$ and $N_2 = B[B, Y]$. Then $N_1^\#$, $N_2^\#$ and $(AB)^\#$ are by (3.16) lines through B. Hence by (3.15) there exists a $g \in C(AB)$ with $N_1^g = N_2$. Thus we may assume $N_1 = N_2$ and also $C = [B, X] = [B, Y]$, since A is transitive on $(\Sigma \cap N_1) - \{B\}$. Now by (3.14)(4) there exists an $h \in M_B$ with $X^h = Y$. Since $M_B = (M_B \cap C(A))C$ we may choose $h \in M_B \cap C(A)$, which proves (4.3). □

(4.4) Lemma

Let $A \in \Sigma$. Then we have:

(1) $|\Sigma_A| = q(q^n + 1)$.

(2) $|\Psi_A| = q^{n+2}(q^n + 1)$.

(3) $|_1\Omega_A| = \frac{q^n+1}{k} q^{2n+3}$, where k is the number of lines T through A with $T \cap V_x \neq \emptyset$ for some fixed $E(x) \in {}_1\Omega_A$. Moreover

$$|\{E(x) \in {}_1\Omega_A | T \cap V_x \neq \emptyset\}| = q^{2n+3}$$

for a fixed line T through A.

Proof. (1) is a direct consequence of (3.16) and the fact that there are $q^n + 1$ lines through A. Now for $B \in \Psi_A$, there exists a unique $C \in \Sigma_A \cap \Sigma_B$, namely $C = [A, B]$. Hence

$$\begin{aligned}|\Psi_A| &= |\Sigma_A|(|\Sigma_C| - q) = q(q^n + 1)(q(q^n + 1) - q) \\ &= q^{n+2}(q^n + 1),\end{aligned}$$

where the q in $(|\Sigma_C| - q)$ stands for the number of elements in Σ_C centralizing A.

Now let $B \in {}_1\Omega_A$ and $C \in X_A \cap \Psi_B$. Then C is the only element of $AC \cap \Psi_B$. Hence we obtain by (4.3)

$$|{}_1\Omega_A| = \frac{(q^n + 1)q}{k}|{}_1\Omega_A \cap \Psi_C| \text{ for } C \in \Sigma_A.$$

Now by definition ${}_2\Omega_A \cap \Psi_C = \emptyset$. Hence we have

$$|{}_1\Omega_A \cap \Psi_C| = |\Psi_C| - |\Psi_C \cap \Sigma_A| - |\Psi_C \cap \Psi_A|.$$

Clearly $|\Psi_C \cap \Sigma_A| = q \cdot q^n = q^{n+1}$. Let $D \in \Psi_C \cap \Psi_A$ and $d \in D^{\#}$. Then AC and $(AC)^d$ normalize each other, since $(AC)^{\#}$ is a weak TI-subset. Hence, as $C_\Sigma(AC) = AC \cap \Sigma$ by (3.13), we obtain $AC \cap (AC)^d \neq 1$ and thus $d \in N(AC)$. This shows $D \in N_\Sigma(AC)$ for $D \in \Psi_A \cap \Psi_C$. Hence we obtain

$$|\Psi_A \cap \Psi_C| = (q - 1)q \cdot q^n,$$

where $q - 1$ stands for the points on AC different from A and C and q^n for the lines through such a point different from $(AC)^{\#}$. Hence we get

$$|{}_1\Omega_A \cap \Psi_C| = q^{n+2}(q^n + 1) - q^{n+1} - q^{n+1}(q - 1) = q^{2n+2},$$

which proves (3). $\square$

(4.5) Corollary

Let k be as in (4.4)(3). Then $k = 1, 2, q^n$ or $q^n + 1$.

Proof. The proof of (4.3) actually shows that R is already transitive on ${}_1\Omega_A$ for $A \in \Sigma$. Hence for $B \in {}_1\Omega_A$ we have by (4.4)

$$|R : N_R(B)| = \frac{q^n + 1}{k}q^{2n+3}.$$

Now $|R| = |M_A|q^n(q^n + 1)(q^n - 1)$ and M_A is a 2-group by (3.15). Thus $N_R(B)$ contains a subgroup H of order $q^n - 1$, stabilizing two lines through A and acting regularly on the other $q^n - 1$ lines through A. (Since by P. Hall's theorem [Hup67, VI,1.8] all such subgroups of R are conjugate.)

Let now $\mathcal{B}$ be the set of lines T through A with $\langle T \rangle \cap \Psi_B \neq \emptyset$. Then H acts on $\mathcal{B}$ and thus $k = |\mathcal{B}| = 1, 2, q^n - 1, q^n$ or $q^n + 1$. Since $k | (q^n + 1)(q^{2n+3})$, $k = q^n - 1$ is impossible. $\square$

(4.6) Lemma

If $k = q^n + 1$, then $_2\Omega_A = \emptyset$ for $A \in \Sigma$ and Σ is a class of $GF(q)$-root subgroups of G, where $q = |A|$.

Proof. Suppose $k = q^n + 1$. Then we have for each $B = E(b) \in {}_1\Omega_A$ and each line T through A, that $V_b \cap T \neq \emptyset$.

Let now $C = E(c) \in {}_2\Omega_A$ with the property that $d(a, c)$ is minimal, where $E(a) = A$ and d is the distance function in $\mathcal{E}(D)$. Then $C \in {}_1\Omega_E$ for some $E \in \Sigma_A$. But by (3.16) $(AE)^\#$ is a line through E and A so that $Y_c \cap AC \neq \emptyset$ as $k = q^n + 1$, contradicting $C \in {}_2\Omega_A$.

Thus $_2\Omega_A = \emptyset$ and $\langle A, B \rangle \simeq SL_2(q)$ for each $B \in \Omega_A$ by (3.8). This proves (4.6). $\qquad\qquad\qquad\qquad\qquad\qquad\qquad\qquad\qquad\qquad\qquad\qquad\quad\Box$

(4.7) Lemma

Let $B = E(b) \in {}_1\Omega_A$ and T be a line through A. Set $U = \langle B, T \rangle$, $\Delta = \Sigma \cap U$, $E = D \cap U$ and $\overline{U} = U/Z(U)$. Then one of the following holds:

(1) $\overline{E}$ is a non-degenerate class of root involutions of $\overline{U}$ and $(\overline{U}, \overline{E})$ satisfy the hypothesis of this section.

(2) $U \leq \langle N_D(L) \rangle$, L a line of D. In particular $T \cap V_b \neq \emptyset$.

(3) $U \simeq L_2(q^2)$, $\langle T \rangle \in \mathrm{Syl}_2(U)$ and if $X = \langle A, B \rangle$ and $H = N_X(A) \cap N_X(B)$, then H normalizes each element of $\Sigma \cap \langle T \rangle$.

Proof. First notice that, as $D = D^2$, we have $E = E^2$. Let L be the set of isolated vertices of $\mathcal{F}(E)$. Then by (1.6) $U \subseteq N(L)$. Suppose $L \neq \emptyset$ and pick $c \in L \cap Z(\langle L \rangle)$. If $L = E_c$, then $E \subseteq C_D(c)$, a contradiction since by (3.13) and (3.16) $\langle C_D(c) \rangle = N_c$ is a 2-group. Thus there exists an $E_f \subseteq L$ such that $L_0 = (E(c)E(f))^\#$ is a line. Now clearly $[L_0, L_0^e] \leq \langle L_0 \rangle \cap \langle L_0^e \rangle$ for each $e \in E$, so that by (3.13) $E \subseteq N_D(L_0)$. In particular by (3.7) and (3.13) $L = L_0 = C_E(L)$ and $U/C_U(L) \simeq SL_2(q)$. This implies

$$\emptyset \neq T \cap C_U(L) \subseteq T \cap V_b \text{ and (2) holds .}$$

So we may assume $L = \emptyset$. Then $E = E^2$ and (1.6)(3) imply that E is a conjugacy class in U. Hence, as $E = E^2$, $O_2(U) \leq Z(U)$ and thus $\overline{U}$ is simple by (1.9). If now E is non-degenerate, then $C_E(A) \supset T$, whence there exists a $C \in C_\Delta(A)$ with $[T, C] = A$. This shows that $(\overline{U}, \overline{E})$ satisfies the hypothesis of this section with the same q.

So we may assume that E is degenerate. But then, as $E = E^2$, T is a connectivity component of $\mathcal{D}(E)$. Hence, as $X = \langle A, B \rangle \simeq L_2(q)$, (1.16) implies

$\overline{U} \simeq L_2(2^n)$ with $2^n = |\langle T \rangle|$. Thus $2^n = q^2$. Since there exist no perfect central extensions of $L_2(q^2)$ generated by odd transpositions this shows that $U \simeq L_2(q^2)$.

Now $H \leq \widehat{H}$, $\widehat{H}$ a diagonal subgroup of U acting transitively on T. Hence $\widehat{H}$ acts transitively on $\Sigma \cap \langle T \rangle$ and thus H normalizes each element of $\Sigma \cap \langle T \rangle$, since it normalizes A.   □

(4.8) Lemma

Suppose G is a group of minimal order satisfying the hypothesis of this section, such that Σ is not a class of abstract root subgroups. Then $n = 1$.

Proof. Suppose $n \geq 2$ and by (4.6) $k \neq q^n + 1$. We show that there exists a line T through $A \in \Sigma$ and an $E(x) \in {}_1\Omega_A$ with $T \cap V_x = \emptyset$, such that $U = \langle T, E_x \rangle$ is a proper subgroup.

Indeed, if this is the case, then for U and $E = D \cap U$ one of the cases of (4.7) holds. If (1) holds, then by minimality of G we obtain that $\Delta = \Sigma \cap U$ is a class of abstract root subgroups of U. But then II(2.11) implies $T \cap V_x \neq \emptyset$.

By the same reason (4.7) (2) is impossible. Thus $U \simeq L_2(q^2)$. Let H be as in (4.7)(3). Then there exists a $y \in E$ inverting H. Now

$$[H, N_D(T)] \leq C_G(T) \cap \langle N_D(T) \rangle \leq O_2(\langle N_D(T) \rangle)$$

by (3.14). Since for $c \in N_D(T) - T$ we have $|cO_2(\langle N_D(T) \rangle) \cap D| = 2^n, n \in \mathbb{N}$, some subgroup $1 \neq H_0$ of H centralizes an element $c \in N_D(T)$. Since y inverts H_0 we obtain $o(cy) = 2$ or 4. In the first case we obtain $o(cyy^h) = 2n, 1 \neq n$ odd for some $h \in H_0$, a contradiction to $cy \in D$ as $D = D^2$. In the second case $(cy)^2 \in C_D(\langle H_0, y \rangle)$, a contradiction as before.

Hence the existence of T and $E(x)$ remains to be shown. Now by the proof of (4.5) there exists a subgroup H of R of order $q^n - 1$ fixing some element $E(x) \in {}_1\Omega_A$ and two lines through A and acting regularly on the remaining lines through A. Thus, if $k = 1$ or q^n, H normalizes a line T through A with $T \cap V_x = \emptyset$. Since $N_G(T)/C_G(T) \stackrel{\sim}{\subset} \Gamma L_2(q)$ by (3.7), we have $|H : C_H(T)| \mid q - 1$. Thus $C_H(T) \neq 1$, as $n \neq 1$, and so $U = \langle T, E(x) \rangle$ is a proper subgroup. (H centralizes $E(x)$, since it centralizes A!)

So we may assume $k = 2$. Then the number e of $E(x) \in {}_1\Omega_A$ with $T \cap V_x = \emptyset$ is

$$e = q^{2n+3}(q^n + 1)/2 - q^{2n+3} = q^{2n+3}(q^n - 1)/2.$$

Let $T = (AD)^{\#}$, $A = E(a)$, $D = E(d)$. Then by (3.14) and (3.15)

$$|M_a| = q^2 \cdot |M_a \cap M_d : AB| q^n \cdot q$$

and thus $|M_a(M_d \cap C(A)| = q^{2n+3} \cdot f$, $f = |M_a \cap M_d : AB|$. By (3.15) and (2.14)(1) $|M_a| \geq q^{2n+1}$, whence $|S| \geq q^{3n+1}$ for $S = M_a(M_d \cap C(A))$ and $|C_S(T)| \geq q^{3n}$. Hence if $n \geq 3$, then, as $3n \geq 2n + 3$, there exists a $1 \neq g \in C_S(T)$ normalizing and thus centralizing $E(x)$, a contradiction as before.

So we may assume $n = 2$. If $M_a \cap M_d = AB$ then (3.14) and exercise (4.12)(3) show that the group N of (3.14) is abelian. Hence by (4.12)(4) it is elementary abelian, which is obviously impossible. Thus $f > 1$. If now $f \geq q$, then by the above $|C_S(T)| \geq q^{2n+3}$, a contradiction as before.

So $1 < f < q$ and $q^4 < |\widetilde{M_a}| < q^5$ for $\widetilde{M_a} = M_a/A$. This will lead us to a contradiction. Now $\widetilde{M_a}$ is a non-trivial $\bar{R} = R/M_a \simeq L_2(q^2)$ module. Thus by (2.14)(1) there exists exactly one non-trivial $\bar{R}$-composition factor in $\widetilde{M_a}$. If $Z = Z(M_a) > A$, then $O^2(R) \leq C(Z)$. As $|Z : A| \leq f$, the structure of $\langle N_D(T) \rangle$ described in (3.14) implies $Z \leq M_a \cap M_d$. Since $ADZ \lhd \langle N_D(T) \rangle$ we obtain $ADZ \leq Z(O_2\langle N_D(T) \rangle)$. Hence $Z = A \times Z_0$, $1 \neq Z_0 = Z \cap Z(\langle N_D(T) \rangle)$, a contradiction to $G = \langle O^2(R), N_D(T) \rangle$ by (3.19)(1).

Thus $Z(M_a) = A$ and similarly $Z(O_2\langle N_D(T) \rangle) = \langle T \rangle$. In particular $[M_d \cap C(A), M_a \cap M_d] = E(d)$ and thus $\widetilde{E(d)} \leq [\widetilde{M_a}, R]$. Since by (3.15) $\widetilde{M_a} = \langle \widetilde{E(d)}^R \rangle$ this implies $\widetilde{M_a} = [\widetilde{M_a}, \bar{R}]$. Hence $M_a \leq O^2(R)$ and thus $R = O^2(R)$. If now C is the coimage of $C_{\widetilde{M_a}}(\bar{R})$ then, as $R = O^2(R)$, $C \leq Z(R)$. Hence by the above $C = A$ and $\widetilde{C} = 1$.

We obtain $C_{\widetilde{M_a}}(\bar{R}) = 1$, $\widetilde{M_a} = [\widetilde{M_a}, \bar{R}]$ and there exists exactly one non-trivial $\bar{R}$-composition factor in $\widetilde{M_a}$. Hence $\widetilde{M_a}$ is an irreducible $\mathbb{Z}_2\bar{R}$-module. Now exercise (4.12)(5) implies $|\widetilde{M_a}| = q^4$ as $|\widetilde{M_a}| < q^5$, a contradiction to $|\widetilde{M_a}| > q^4$. $\qquad \square$

(4.9) Lemma

Suppose $n = 1$ and let $d \in X_a - E_a$, $N = (M_a \cap C(d))(M_d \cap C(a))$, $N_0 = M_a \cap M_d$ and $X = \langle E(x), E(y) \rangle$ with $x \in X_a - C(d)$, $y \in X_d - C(a)$ be as in (3.14). Then $Z = C_{N_0}(X) \neq 1$.

Proof. Suppose $Z = 1$. Since $[E(x), N_0] = E(a)$ we have $|N_0 : C_{N_0}(E(x))| = q = |N_0 : C_{N_0}(E(y))|$. Hence $Z = 1$ implies $N_0 = E(a)E(d)$. Now by (3.14) N_0 and N/N_0 are natural $\mathbb{Z}X$-modules (for $X \simeq SL_2(q)$). Hence exercise (4.12)(3) shows that N is abelian. Now a well-known result of G. Higman [Hig68], see exercise (4.12)(4), shows that N is elementary abelian. But then $\langle x^N \rangle$ would also be abelian, which obviously contradicts (3.14)(4). $\qquad \square$

(4.10)   Theorem

Σ is a class of $GF(q)$-root subgroups of G, where $q = |A|$ for $A \in \Sigma$.

Proof. Let G be a minimal counterexample to (4.10). Then by (4.1) and (4.8) $n = 1$. Moreover by (4.5) and (4.6) $k = 1, 2$ or q. Let $A = E(a) \in \Sigma$. If there exists a line T through A and a $u \in {}_1A_a$ such that $T \cap V_u = \emptyset$ and $\langle T, E(u) \rangle$ is a proper subgroup of G, we obtain a contradiction to the minimality of G as in the proof of (4.8). So we may assume no such T and u exist.

We discuss the possibilities for k. If $k = q$, then $|{}_1\Omega_A| = (q+1)q^4$ by (4.4). Since by (4.9) $|M_a| > q^4$, there exists a $1 \neq t \in M_a$ normalizing and thus centralizing $E(u) \in {}_1\Omega_A$. Hence t centralizes $T_i \cap V_u$ for the q lines T_i through A with $T_i \cap V_u \neq \emptyset$. But if $R = \langle E(u), T_i | i = 1, \ldots, q \rangle$ and $E = a^R$, then E is a non-degenerate class of root involutions of R and $R/Z(R)$ satisfies the hypothesis of this section with $n > 0$. Hence there are already $q + 1$ lines through A in E and thus $R = G$ by (2.8), a contradiction to $t \in C_G(R)$.

Next assume $k = 1$. Let $T_1 \neq T_2$ be two fixed lines through A and

$$E_i = \{ E(x) \in {}_1\Omega_A \mid V_x \cap T_i \neq \emptyset \} \text{ for } i = 1, 2.$$

Then $E_1 \cap E_2 = \emptyset$ and $|E_i| = q^5$ by (4.4). Let H be a subgroup of R of order $q - 1$ normalizing T_1 and T_2. Then each element $1 \neq h \in H$ of prime order centralizes an element $E(x) \in E_1$ and $E(y) \in E_2$. Suppose $[E(x), E(y)] = 1$. Then $(E(x)E(y))^\# \subset D$ and there exists an $x \in E_x$, $y \in E_y$ with

$$\begin{array}{ccc} x & a & y \\ \circ\!\!\!-\!\!\!-\!\!\!-\!\!\!\circ\!\!\!-\!\!\!-\!\!\!-\!\!\!\circ \end{array}.$$

Hence $d = (xy)(xy)^a \in X_a$ and $E(x)^d = E(y)$, a contradiction since M_a normalizes E_1.

Let $U = \langle E(a), E(x), E(y) \rangle$ and $E = a^U$. Then E is a class of root involutions of U with $E = E^2$, since otherwise U normalizes by (1.6) a line, a contradiction to $k = 1$. Thus $O_2(U) \leq Z(U)$ and either E is degenerate or $U/Z(U)$ satisfies the hypothesis of this section. In the second case the minimality of G and $U \leq C(h)$ imply that $\Delta = E(a)^U$ is a class of $GF(q)$-root subgroups of U, a contradiction to $T_1 \cap V_y = \emptyset$.

Thus E is degenerate and $\mathcal{D}(E)$ is disconnected, as $E = E^2$. Now (1.16) shows that for $\overline{U} = U/Z(U)$ one of the following holds:

(1) $\overline{U} \simeq L_2(q^2)$.
(2) $\overline{U} \simeq U_3(q)$.
(3) $\overline{U} \simeq U_3(q^2)$.

In the first case there exists an $m \in M_a$ such that $E(x)$ and $E(y)^m$ commute, since $S^{\#} \subset D$ for $S \in \mathrm{Syl}_2(U)$ containing $E(a)$. But then, as above, $E(x)$ and $E(y)$ are conjugate in M_a, a contradiction. In case (2) there exists by the structure of $U_3(q)$ an element $g \in U$ of order $q + 1$ or $q + 1/3$ centralizing $E(a)$ and $E(x)$. Hence some element $\overline{g} \in \langle g \rangle$ of prime order p normalizes a line through A different from T_1, say T_2. Now, as $\langle T_2 \rangle$ is a projective line over $GF(q)$ and $\overline{g}$ centralizes $E(a)$ and $(q - 1, o(\overline{g})) = 1$, $\overline{g}$ centralizes T_2. Hence $\langle E(x), T_2 \rangle$ is the desired subgroup. Case (3) may be treated in the same way, since $\overline{E(a)}, \overline{E(x)}$ are contained in an $\overline{E}$-subgroup of $\overline{U}$ isomorphic to $U_3(q)$.

So we have $k = 2$, which is the most difficult case. Use the notation of (4.9) and let H_0 be a subgroup of R of order $q - 1$ normalizing the lines $(E(a)E(d))^{\#}$ and $(E(a)E(x))^{\#}$. Then H_0 normalizes $NX = \langle N_D(E(a)E(d)) \rangle$ and, since all conjugates of X in NX are by (3.14) conjugate under N, we may by the Frattini argument assume that $H_0 \leq N(X)$. Since $H_0 \cap X = 1$ as $H_0 \leq C(E(a))$, the action of $\widehat{X} = H_0 X$ on $E(a)E(d)$ implies $\widehat{X} \simeq GL_2(q)$. Let H be a diagonal subgroup of $\widehat{X}$ containing H_0 and assume without loss of generality H normalizes $E(a)$ and $E(d)$. (By choosing $E(d)$ appropriately!) Let $H_1 = C_H(X)$ and $H_2 = H \cap X$. Then $H = H_1 \times H_2 \simeq \mathbb{Z}_{q-1} \times \mathbb{Z}_{q-1}$. If now some $1 \neq h \in H_1$ of prime order centralizes an element of $(E(a)E(d))^{\#}$, then h centralizes $E(a)E(d)$ and N/N_0, since all squares of elements of $N - N_0$ are in $E(a)E(d)$. Hence h centralizes all $E(x)^n$, $n \in N$ since $N = C_N(h)Z$. But then by (3.14)(4) $h \in C(M_a)$. Since $|_1\Omega_{E(a)}| = q^5(q + 1)/2$, h normalizes and thus centralizes also an element of $_1\Omega_{E(a)}$, a contradiction to (2.8).

This shows that H_1 acts fixed-point-freely on $E(a)E(d)$ and N/N_0. Let $\overline{N} = N/E(a)E(d)$ and suppose some element $1 \neq h \in H_1$ centralizes $\overline{Z}$. Then $\overline{N} = [\overline{N}, h] \times \overline{Z}$ and $\overline{Z} = C_{\overline{N}}(h)$. Hence X acts on $[\overline{N}, h]$ and so $[N, h]$ is elementary abelian as shown in (4.9). But this is impossible since $\langle x^N \rangle$ is non-abelian. Hence H_1 acts faithfully on $\overline{Z}$ and thus by Schur's lemma $|\overline{Z}| = |Z| \geq q$.

Thus $|M_a| \geq q^5$ and $|S| \geq q^6$ for $S \in \mathrm{Syl}_2(R)$ normalizing $E(a)E(d)$. Now by (4.4)

$$|\{E(u) \in {}_1\Omega_{E(a)} | V_u \cap (E(a)E(d))^{\#} = \emptyset\}| = \frac{q^5(q + 1)}{2} - q^5 = \frac{q^5(q - 1)}{2}.$$

Hence a subgroup S_0 of order $\geq 2q$ normalizes and thus centralizes such an $E(u)$. Now, since $E(a)E(d)$ is a projective line over $GF(q)$ with $\Sigma \cap E(a)E(d)$ as point-set and since $S_0 \leq C(E(a))$, we have $S_1 = C_{S_0}(E(a)E(d)) \neq 1$. Hence $\langle E(a)E(d), E(u) \rangle$ is a proper subgroup, which by assumption does not exist. This final contradiction proves (4.10).   □

(4.10) is the final step in the problem of reducing the classification of groups generated by a class D of root involutions satisfying $D \cap D^2 \neq \emptyset$ to abstract root subgroups. For the convenience of the reader we state the results proved in this chapter in a single theorem:

(4.11) Root Involution Theorem

Let G be a (finite) group with $O_2(G) = Z(G) = 1$, generated by a class D of root involutions satisfying $D \cap D^2 \neq \emptyset$. Then one of the following holds:

(1) $\mathcal{D}(D)$ is disconnected and $G \simeq L_2(q)$, $Sz(q)$ or $U_3(q)$, $q = 2^m > 2$.

(2) $\mathcal{D}(D)$ is connected. For $d \in D$ we have $E(d) := \langle E_d \rangle = E_d \cup \{1\}$ and one of the following holds:

    (i) $\Sigma = E(d)^G$ is a class of $GF(q)$-root subgroups of G, where $q = |E(d)|$ or

    (ii) D is non-degenerate, $|E(d)| = 2$, $|\langle C_D(d) \rangle| \leq 2^{10}$ and one of the following holds:

        (a) $\langle C_D(d) \rangle \simeq D_8$.

        (b) $\langle C_D(d) \rangle \simeq Q_8 * \mathbb{Z}_4$.

        (c) $\langle C_D(d) \rangle \simeq (Q_8 * D_8) * \langle t \rangle$, $t^2 = 1$.

        (d) $\langle C_D(d) \rangle \simeq (D_8 * D_8 * D_8 * D_8) * \langle t \rangle$, $t^2 = 1$.

(For the classification of groups satisfying (ii) see (3.18)!)

Proof. If $\mathcal{D}(D)$ is disconnected (1) holds by (1.3). So assume that $\mathcal{D}(D)$ is connected. If now D is degenerate (2)(i) holds by (2.5). So suppose D is non-degenerate and G is a minimal counterexample to (2). Then, in particular, Σ is not a class of abstract root subgroups of G. Hence by (2.13) $E(d) = E_d \cup \{1\}$. Now by (3.12) there exists a maximal weak TI-subset T of D, such that $T = \langle T \rangle^\#$, $|\langle T \rangle| = q^2$ where $q = |E(d)|$ and $\Sigma \cap \langle T \rangle$ is a partition of $\langle T \rangle$. Moreover by (3.17) $q > 2$, since otherwise (ii) holds. But then the hypothesis of §4 is satisfied for G and D. Hence by (4.10) Σ is a class of $GF(q)$-root subgroups of G, a contradiction to the assumption. $\qquad\square$

(4.12) Exercises

In the following exercises let $X = SL_2(q)$, $q = 2^n > 2$ and $k = GF(q)$. The natural kX-module N is k^2 as an additive group on which the elements of kX act by matrix multiplication. If we forget about the scalar action of k, N is called the natural $\mathbb{Z}_2 X$-module. For $\sigma \in \mathrm{Aut}(k)$, the kX-module N^σ is N as additive group, while the element $(a_{ij}) \in X$ act by (a_{ij}^σ).

(1) (a) Show that for $\sigma \neq \tau \in \mathrm{Aut}(k)$, $N^\sigma \otimes_k N^\tau$ is an irreducible kX-module.

(b) Show that $N \otimes_k N$ is isomorphic to the (adjoint) kX-module of exercise (1.21)(3).

(c) Show that if $\varphi \in \mathrm{Hom}_{kX}(N^\sigma \otimes_k N^\tau, N)$ for $\sigma, \tau \in \mathrm{Aut}(k)$, then φ is the zero map.

Hint. Use the description of M of (1.21)(3).

(2) Consider N as a $\mathbb{Z}_2 X$-module. Show that if $\varphi \in \mathrm{Hom}_{\mathbb{Z}_2 X}(N \otimes_{\mathbb{Z}_2} N, N)$, then φ is the zero map.

Hint. Use the fact that $k \otimes_{\mathbb{Z}_2} N = \oplus_{\sigma \in \mathrm{Aut}(k)} N^\sigma$ and apply exercise (1).

(3) Let Q be a 2-group of order q^4 on which X acts, such that Q has an X-invariant normal subgroup Q_0 and Q_0 and Q/Q_0 are natural $\mathbb{Z}_2 X$-modules. Show that Q is abelian.

Hint. Let $\overline{Q} = Q/Q_0$. For $a, b \in Q$ let $(\overline{a}, \overline{b}) = [a, b]$. Show that $(,)$ is an X-invariant bilinear map from $\overline{Q} \times \overline{Q}$ in Q_0 and hence defines a map $\varphi \in \mathrm{Hom}_{\mathbb{Z}_2 X}(\overline{Q} \otimes_{\mathbb{Z}_2} \overline{Q}, Q_0)$.

(4) Assume the same hypothesis as in (3). Show that Q is elementary abelian. (See [Hig68, p.30]).

(5) Suppose $q = r^2$ and let V be an irreducible non-trivial $\mathbb{Z}_2 X$-module satisfying $|V| < r^5$. Show that $|V| = q^2 = r^4$.

Hint. Use the fact that each irreducible kX-module is the tensor product of algebraic conjugates of the natural module N [Hig68] and the usual field extension technique, i.e. the fact that $k \otimes_{\mathbb{Z}_2} V$ is a direct sum of algebraic conjugates of some irreducible kX-module [Hup67, V13.3].

Chapter V

Applications

In this chapter we will discuss some applications of the theory of abstract root subgroups. In §1 we will describe how the classification of certain groups, which are generated by elements acting quadratically on some finite dimensional vector space M (quadratic pairs), can be reduced to abstract root subgroups. In fact, this was the main motivation for developing the theory of abstract root subgroups.

The proofs in §1 are fairly complete. That is, apart from three very technical and lengthy proofs which consist mainly of matrix calculations and calculations in the endomorphism ring of M, we will describe completely how one reduces the determination of groups G, when (G, M) is a quadratic pair (for definition see §1), to abstract root subgroups.

In §2 we will describe some results on subgroups of Lie type groups generated by long root elements. Here we will concentrate on results holding over arbitrary fields. Finally in §3 we will briefly discuss further applications of the theory of abstract root subgroups. Since such results are still in progress, we will concentrate on the main ideas.

§1 Quadratic pairs

Let k be a field with Char $k \neq 2$ and $k \neq GF(3)$ and M a finite dimensional k-vector space. Then we say that the element $\sigma \in GL(M)$ acts *quadratically* on M if $\sigma = \mathrm{id} + \alpha$ with $\alpha^2 = 0 \neq \alpha$. Notice that, if Char $k = 2$, then all involutions in $GL(M)$ are quadratic. So in the characteristic 2 case this conception is meaningless.

(1.1) Notation

If now $\sigma = \mathrm{id} + \alpha \in GL(M)$ is quadratic and $c \in k$ let $c \circ \sigma := \mathrm{id} + c\alpha$. Then $(c \circ \sigma)(d \circ \sigma) = (c+d) \circ \sigma$, so that $k \circ \sigma := \{c \circ \sigma \mid c \in k\}$ is a subgroup of $GL(M)$

isomorphic to $(k, +)$. Further, if $k = \mathbb{Z}_p$, $p \geq 5$ a prime, then $k \circ \sigma = \langle \sigma \rangle$. If now G is a subgroup of $GL(M)$, then we call the pair (G, M) a *quadratic pair*, if:

(1) M is an irreducible kG-module.
(2) Let $\mathcal{Q} = \mathcal{Q}(G) := \{\sigma \in G \mid \sigma \text{ is quadratic on } M \text{ and } k \circ \sigma \leq G\}$. Then $G = \langle \mathcal{Q} \rangle$.

This notion was introduced by J. Thompson at the International Congress in Nice 1970 [Tho70] in the special case when $k = \mathbb{Z}_p$. (In that case one can, since $k \circ \sigma = \langle \sigma \rangle \leq G$ anyway, avoid this condition in the definition of $\mathcal{Q}(G)$. But if one avoids this condition also for arbitrary fields, one runs into the wide area of subgroups of $GL(M)$ defined over subrings of k.) J. Thompson also announced in Nice the classification of groups G, when (G, M) is a quadratic pair over $k = \mathbb{Z}_p$. The statement is, that G is a central product of finite Lie type groups in Char p, different from $E_8(q)$. ($E_8(q)$ does not admit a quadratic module!) But unfortunately his manuscript, although available in preprint form, still remains unpublished. In the case when $k = \mathbb{Z}_3$ some work on quadratic pairs was done by Ch. Ho[Ho76]. The quadratic modules M for arbitrary Chevalley groups and finite Lie type groups were determined by Premet and Suprunenko [PS83].

In this section we will describe how the classification of groups G, when (G, M) is a quadratic pair over arbitrary k, can be reduced to the classification of groups generated by abstract root subgroups. This reduction is based on the following two theorems:

(1.2) Theorem

(Bashkirov, Dickson) Let k be a field with Char $k \neq 2$ and $k \neq GF(3)$ and α be algebraic over k. Then $\langle \left(\begin{smallmatrix} 1 & c \\ & 1 \end{smallmatrix} \right), \left(\begin{smallmatrix} 1 & \\ c\alpha & 1 \end{smallmatrix} \right) \mid c \in k \rangle = SL_2(k(\alpha))$. (Considered as subgroup of $SL_2(\bar{k})$, $\bar{k}$ the algebraic closure of k!)

For finite fields this is just the well-known Dickson lemma, see [Gor68, Chap.2, (8.4)]. For infinite fields there is a proof of (1.1) due to Zalesskii included in the appendix of [Tim99].

(1.3) Theorem

(Aschbacher). Let $\mathcal{W}$ be a set of subgroups of the linear group G, satisfying:

(1) $\langle A, B \rangle$ is a unipotent subgroup of G for all $A, B \in \mathcal{W}$.
(2) $\mathcal{W}^g \subseteq \mathcal{W}$ for all $g \in G$.

Then $\langle \mathcal{W} \rangle$ is a unipotent normal subgroup of G.

The proof of (1.2), see [Asc90], is a generalization of a proof of Baer's theorem due to Alperin and Lyons.

(1.4) Definitions

Let (G, M) be a quadratic pair over k, where Char $k \neq 2$ and $k \neq GF(3)$. For $\sigma \in \mathcal{Q}(G)$ let $d(\sigma) := \dim M^{\sigma - \mathrm{id}}$ and

$$d := \min\{d(\sigma) \mid \sigma \in \mathcal{Q}(G)\},$$
$$\mathcal{Q}_d(G) = \mathcal{Q}_d := \{\sigma \in \mathcal{Q}(G) \mid d(\sigma) = d\}.$$

For $\sigma \in \mathcal{Q}_d$ let

$$E_\sigma := \{\tau \in \mathcal{Q}_d \mid C_M(\sigma) = C_M(\tau) \text{ and } M^{\sigma - \mathrm{id}} = M^{\tau - \mathrm{id}}\}.$$

Notice that $k \circ \sigma \subseteq E_\sigma \cup \{1\}$ for $\sigma \in \mathcal{Q}_d$. Let $\varphi, \rho \in E_\sigma$. Then $\varphi\rho$ acts quadratically on M with $C_M(\sigma) \subseteq C_M(\varphi\rho)$ and $M^{\varphi\rho - \mathrm{id}} \subseteq M^{\sigma - \mathrm{id}}$. If $\alpha = \varphi - \mathrm{id}$, $\beta = \rho - \mathrm{id}$ then $\alpha\beta = 0 = \beta\alpha$ since $\varphi, \rho \in E_\sigma$. Hence $\varphi\rho = \mathrm{id} + (\alpha + \beta)$ and

$$c \circ (\varphi\rho) = \mathrm{id} + c\alpha + c\beta = (c \circ \varphi)(c \circ \rho) \text{ for } c \in k.$$

This shows that $\varphi\rho \in \mathcal{Q}(G)$ and thus by definition of d

$$\varphi \cdot \rho \in E_\sigma \cup \{1\} \text{ if } \varphi, \rho \in E_\sigma.$$

This implies that $E(\sigma) := E_\sigma \cup \{1\}$ is an abelian subgroup of G, which is an elementary abelian p-group if Char $k = p$. (As $\sigma^m = m \circ \sigma$ for $m \in \mathbb{Z}$!) Let finally $\Sigma := \{E(\sigma) \mid \sigma \in \mathcal{Q}_d\}$.

We assume now for the rest of this section that (G, M) is a quadratic pair. We will show:

(1.5) Theorem

Σ is a set of abstract root subgroups of the normal subgroup $G_0 = \langle \Sigma \rangle$ of G. Further $\mathcal{F}(\Sigma)$ has no isolated vertices and each connectivity component Δ of $\mathcal{F}(\Sigma)$, which is a set of abstract transvection groups, satisfies the condition (H) of II§3.

Notice that by II(2.21) this implies that $\Sigma = \bigcup_{i \in I} \Sigma_i$, where the Σ_i are the connectivity components of $\mathcal{F}(\Sigma)$ and G_0 is a central product of groups $G_i = \langle \Sigma_i \rangle$. With the help of this we also show:

(1.6)   Theorem

Suppose (G, M) is a quadratic pair over the field k as above. Then the following hold:

(1) $G = G_1 * \cdots * G_s$ (central product), where the $G_i = \langle \mathcal{Q}(G) \cap G_i \rangle$ are quasi-simple normal subgroups of G.

   Further $\mathcal{Q}(G) = \bigcup_{i=1}^{s} (\mathcal{Q}(G) \cap G_i)$.

(2) M is homogeneous as kG_i-module for $i \leq s$.

(3) If N_i is an irreducible kG_i-module, then (G_i, N_i) is a quadratic pair.

(In general $G \neq G_0$, G_0 as defined in (1.5). $G_0 = G_1 * \cdots * G_r$ for some $r \leq s$!)

Notice that by (1.5) applied to (G_i, N_i) the set $\Delta_i := \{E(\sigma) \mid \sigma \in \mathcal{Q}_d(G_i)\}$ $(\mathcal{Q}_d(G_i)$ is defined with respect to $(G_i, N_i))$ is a class of abstract root subgroups of the quasi-simple group G_i, which satisfies the condition (H) of II§3 if Δ_i is degenerate. Hence the classification theory of Chapter III can be applied to (G_i, Δ_i). This shows that, if Δ_i contains commuting elements, the group G_i is completely determined. (I.e. it is essentially a quasi-simple classical or Lie type group over some division ring K of finite dimension over k.)

The proof of (1.5) and (1.6) will be given at the end of this section. We now start to discuss how one obtains abstract root subgroups from $\mathcal{Q}_d(G)$. In order not to confuse the reader with too many little lemmata, we often state simple facts together with definitions.

(1.7)   Notation

If $H \leq GL_k(M)$, then the *augmentation ideal* AH of kH is

$$AH := \{\Sigma_{g \in H} c_g g \mid \Sigma c_g = 0, c_g \in k\}.$$

It is easy to see that [Tim99, 20.1]:

(1) H acts unipotently on M if and only if AH is nilpotent.

For $\sigma = \mathrm{id} + \alpha$, $\tau = \mathrm{id} + \beta \in \mathcal{Q}(G)$ let

$$\Delta = \Delta(\sigma, \tau) = \alpha\beta + \beta\alpha \in \mathrm{End}_k M.$$

Then, if $H = \langle k \circ \sigma, k \circ \tau \rangle$, we have:

(2) (a) $\Delta(\sigma, \tau) \in AH$.

   (b) $\Delta(\sigma, \tau) = \Delta(\tau, \sigma)$.

   (c) $\Delta(c \circ \sigma, d \circ \tau) = (c \cdot d)\Delta(\sigma, \tau)$ for $c \in k$.

   (d) $\Delta(\sigma, \tau)$ centralizes H (in $\mathrm{End}_k M$!)

All these properties are very easy to compute. Crucial for us is the following little lemma [Tim99, (20.3)]:

(3) H acts unipotently on M if and only if $\Delta(\sigma, \tau)$ is nilpotent.

To prove (3) one shows that on each H composition factor of M on which Δ is zero one has $(AH)^3 = 0$. Thus, if $\Delta^n = 0$, then AH is nilpotent and thence H is unipotent on M by (1).

Now fix $\sigma \in \mathcal{Q}(G)$. Since the unipotent radical $R_u(G) = 1$, as G acts irreducibly on M, there exist by (1.2) a $\tau \in \mathcal{Q}(G)$ such that $H = \langle k \circ \sigma, k \circ \tau \rangle$ is not unipotent on M. Hence by (3) $\Delta = \Delta(\sigma, \tau)$ is not nilpotent. Let

$$f(x) = x^{n_0} f_1(x)^{n_1} \dots f_\ell(x)^{n_\ell}$$

be the minimal polynomial of Δ with distinct monic irreducible polynomials $f_i(x)$. Then $\ell \geq 1$ as Δ is not nilpotent. Let

$$M = M_0 \oplus M_1 \oplus \dots \oplus M_\ell \qquad (\Delta - \text{decomposition})$$

be the corresponding decomposition of M, i.e. $M_0 = \ker \Delta^{n_0}$ and $M_i = \ker f_i(\Delta)^{n_i}$ for $1 \leq i \leq \ell$. Then, since the minimal polynomial of $\Delta \mid_{M_i}$ is $f_i(x)^{n_i}$ resp. x^{n_0}, Δ is nilpotent on M_0 and invertible on each M_i for $i \geq 1$. By (2)(d) the M_i are H-invariant. Further by (3) applied to $H \mid_{M_i}$, H acts unipotently on M_0 but not unipotently on each M_i for $i \geq 1$. Since $\{m \in M_i \mid \Delta m = 0\} = \{0\}$ for $i \geq 1$, as $\Delta \mid_{M_i}$ is invertible, it follows that $C_{M_i}(\sigma) \cap C_{M_i}(\tau) = \{0\}$. Now, as $M_i^{\sigma - \mathrm{id}} \subseteq C_{M_i}(\sigma)$ by the quadratic action, this implies:

(4)
$$M_i = M_i^{\sigma - \mathrm{id}} \oplus M_i^{\tau - \mathrm{id}}$$

and $M_i^{\sigma - \mathrm{id}} = C_{M_i}(\sigma), M_i^{\tau - \mathrm{id}} = C_{M_i}(\tau)$ for $i \geq 1$.

Let $\mathcal{B}_i$ be a basis of $M_i^{\tau - \mathrm{id}}$. Then $\mathcal{B}_i^{\sigma - \mathrm{id}}$ is a basis of $M_i^{\sigma - \mathrm{id}}$ and $\mathcal{B}_i \cup \mathcal{B}_i^{\sigma - \mathrm{id}}$ is a basis of M_i. With respect to this basis σ is represented by:

$$\left[\begin{array}{c|c} I_n & I_n \\ \hline & I_n \end{array} \right] \quad \text{and } \tau \text{ by} \quad \left[\begin{array}{c|c} I_n & \\ \hline A_i & I_n \end{array} \right]$$

where $n = \dim M_i^{\sigma - \mathrm{id}} = \dim M_i^{\tau - \mathrm{id}}$. Hence Δ is represented by

$$\left[\begin{array}{c|c} A_i & \\ \hline & A_i \end{array} \right].$$

In particular A_i is an invertible $n \times n$-matrix.

We now bring A_i into Jordan canonical form over the algebraic closure $\overline{k}$ of k. For this let λ_i be a root of $f_i(x)$ in $\overline{k}$. Then λ_i^α is also a root of $f_i(x)$ for each $\alpha \in \mathrm{Gal}(\overline{k} : k)$. Moreover, since the roots of $f_i(x)$ are conjugate under the Galoisgroup of the splitting field of $f_i(x)$ and such automorphisms can be extended to elements of $\mathrm{Gal}(\overline{k} : k)$, we obtain:

(5) μ is a root of $f_i(x)$ if and only if $\mu = \lambda_i^\alpha$ for some $\alpha \in \mathrm{Gal}(\overline{k} : k)$.

Now let

$$
J_\nu(\lambda_i) = \begin{bmatrix} J_{n_1}(\lambda_i^{\alpha_1}) & & \\ & \ddots & \\ & & J_{n_s}(\lambda_i^{\alpha_s}) \end{bmatrix}
$$

be the Jordan-canonical form of A_i, where $n = n_1 + \cdots + n_s$ is a partition ν of n, $\{\alpha_1, \ldots, \alpha_s\} \subseteq \mathrm{Gal}(\overline{k} : k)$ and the $J_{n_k}(\lambda_i^{\alpha_k})$ are elementary Jordan-matrices. Let L_i be the kernel of the action of H on M_i for $i \geq 1$. Then we obtain:

(6) $H/L_i \simeq \left\langle \left[\begin{array}{c|c} I_n & cI_n \\ \hline & I_n \end{array}\right], \left[\begin{array}{c|c} I_n & \\ \hline dJ_\nu(\lambda_i) & I_n \end{array}\right] \mid c, d \in k \right\rangle.$

In fact, if $N \in GL_n(\overline{k})$ with $N^{-1}A_iN = J_\nu(\lambda_i)$, then H/L_i is conjugate to the group on the right side of (6) by

$$
\left[\begin{array}{c|c} N & \\ \hline & N \end{array}\right].
$$

Let $H_i = H/L_i$. Then we obtain:

(1.8) Theorem

One of the following holds:

(1) $H_i \simeq SL_2(k(\lambda_i))$. Moreover, if $\overline{M}_i = \overline{k} \oplus_k M_i$ and A, B are the unipotent subgroups of H_i containing the images of σ and τ, then

$$
\overline{M}_i = [\overline{M}_i, H_i] = [\overline{M}_i, A] \oplus [\overline{M}_i, B] \text{ and } [\overline{M}_i, A] = [\overline{M}_i, a] = C_{\overline{M}_i}(a)
$$
$$
\text{for each } a \in A^\#.
$$

(2) $R_u(H_i) \neq 1$ and $H_i/R_u(H_i) \simeq SL_2(k(\lambda_i))$. Moreover there exists some quadratic element $\rho \in R_u(H_i)$ with $k \circ \rho \leq R_u(H_i)$ and $\dim[M_i, \rho] < n$.

In both cases H_i is perfect.

(1.8) is (19.17) of [Tim99]. The proof occupies the whole section 19 of [Tim99] and consists of lengthy matrix calculations using the matrix representation of H_i given by (1.7)(6). The proof depends on (1.1), since in both cases one shows $H_i/R_u(H_i) \simeq SL_2(k(\lambda_i))$. Now, to prove the perfectness of H_i and the existence of ρ in case (2), one needs to determine the action of $H_i/R_u(H_i)$ on $R_u(H_i)$.

(1.9) Notation

Assume from now on that $\sigma, \tau \in \mathcal{Q}_d(G)$ such that $H = \langle k \cdot \sigma, k \circ \tau \rangle$ is not unipotent on M. By (1.8) there exist normal subgroups $L_i \leq N_i \lhd H$ such that $H/N_i \simeq SL_2(k(\lambda_i))$, λ_i algebraic over k.

Let $b(\sigma, \tau) = \dim M_0$ in the notation of (1.7) and

$$b = \max b(\sigma, \tau),$$

where σ, τ ranges over all elements of $\mathcal{Q}_d(G)$ such that $\langle k \circ \sigma, k \circ \tau \rangle$ is not unipotent.

(1.10) Lemma

$H = N_i N_j$ for $1 \leq i \neq j \leq \ell$.

Proof. Suppose false. Then, as $N_i N_j/N_i \lhd H/N_i$, we obtain $|N_i N_j : N_i| \leq 2 \geq |N_i N_j : N_j|$. Hence $\overline{N}_i = \overline{N}_j$ for the coimages $\overline{N}_i, \overline{N}_j$ of $Z(H/N_i)$ and $Z(H/N_j)$. This implies

$$PSL_2(k(\lambda_i)) \simeq H/\overline{N}_i = H/\overline{N}_j \simeq PSL_2(k(\lambda_j))$$

and this isomorphism maps the image of $\begin{pmatrix} 1 & c \\ & 1 \end{pmatrix}$ in $PSL_2(k(\lambda_i))$ onto the image of $\begin{pmatrix} 1 & c \\ & 1 \end{pmatrix}$ in $PSL_2(k(\lambda_j))$ and the image of $\begin{pmatrix} 1 & \\ \lambda_i & 1 \end{pmatrix}$ onto the image of $\begin{pmatrix} 1 & \\ \lambda_j & 1 \end{pmatrix}$, since these matrices are the images of $c \circ \sigma$ and τ in H/N_i (resp. H/N_j). Now I(5.7) shows that there exists an isomorphism $\alpha : k(\lambda_i) \to k(\lambda_j)$ with $\alpha|_k = \mathrm{id}$. Since α can be extended to an element of $\mathrm{Gal}(\overline{k} : k)$, this is a contradiction to (1.7)(5). $\qquad \square$

(1.11) Lemma

Let $\sigma, \tau \in \mathcal{Q}_d(G)$ such that $H = \langle k \circ \sigma, k \circ \tau \rangle$ is not unipotent on M and $b(\sigma, \tau) = b$. Then the following hold:

(1) $\ell = 1$.

(2) H induces the identity on M_0.

(3) $H \simeq SL_2(k(\lambda_1))$ and, if φ and ρ are in the same unipotent subgroup of H as σ resp. τ, then

$$[M_1, \varphi] = [M_1, \sigma] \text{ and } [M_1, \rho] = [M_1, \tau].$$

Proof. Suppose first $\ell \geq 2$. Let $K_i = \langle N_i, k \circ \sigma \rangle$ for $i = 1, \ldots, \ell$. By (1.8) N_i acts unipotently on M_i. Hence also K_i acts unipotently on M_i. Thus, if ρ is conjugate to σ in K_i, then $\langle k \circ \sigma, k \circ \rho \rangle$ acts unipotently on $M_i \oplus M_0$. By definition of b this implies that $\langle k \circ \sigma, k \circ \rho \rangle$ is unipotent on M. Hence (1.2) implies $k \circ \sigma \leq R_u(K_i)$ for $i = 1, \ldots, \ell$. Since $R_u(K_j)$ is N_j invariant, (1.10) implies $N_i R_u(K_j) \trianglelefteq H$ for $1 \leq i \neq j \leq \ell$. Since $H/N_i \simeq SL_2(k(\lambda_i))$ this shows that either $H = N_i R_u(K_j)$ or $k \circ \sigma \leq R_u(K_j) \leq N_i$. Now both cases are obviously impossible.

So we obtain $\ell = 1$. Now by definition L_1 is unipotent on M.
Claim: $H = L_1 H^{(n)}$ for some $n \in \mathbb{N}$ with $H^{(n)} = H^{(n+1)}$. Indeed, since H_1 is perfect, we have $H = L_1 H^{(m)}$ for all $m \in \mathbb{N}$. Now let $n \in \mathbb{N}$ so that $[M_0, H^{(n)}] = 1$ (H is unipotent on M_0!). Then $H^{(n)} \cap L_1 = 1$ and $H^{(n)} \simeq H/L_1$ is by (1.8) perfect. This proves our claim. This implies $\sigma = \rho \sigma_1$ with $\rho \in L_1$ and $\sigma_1 \in H^{(n)}$. Further $\sigma_1 \neq 1$ as $\sigma \notin L_1$. Now, as $H^{(n)}$ is perfect and acts unipotently on M_0, $H^{(n)}$ acts trivially on M_0. In particular

$$\sigma_1|_{M_0} = \text{id and } \sigma_1|_{M_1} = \sigma|_{M_1}.$$

This implies that σ_1 acts quadratically on M and $c \circ \sigma_1|_{M_1} = c \circ \sigma|_{M_1}$ for $c \in k$. On the other hand, if $c \circ \sigma = \rho' \sigma_1'$, $\rho' \in L_1$ and $\sigma_1' \in H^{(n)}$, then $\sigma_1'|_{M_0} = \text{id}$ and

$$\sigma_1'|_{M_1} = c \circ \sigma|_{M_1} = c \circ \sigma_1|_{M_1}.$$

Hence $c \circ \sigma_1 = \sigma_1' \in H^{(n)}$ for $c \in k$. We obtain $\sigma_1 \in \mathcal{Q}(G)$. Now the minimality of d implies $\sigma_1 \in \mathcal{Q}_d(G)$ and

$$[M_0, \sigma] = [M_0, \sigma_1] = 1.$$

By the same argument we also obtain $[M_0, \tau] = 1$ and (2) holds. In particular $L_1 = 1$. Thus (1.8) implies $H \simeq H_1 \simeq SL_2(k(\lambda_1))$. Now (3) follows from (1.8)(1). $\qquad\square$

(1.12) SL_2-lemma

Let $\sigma, \tau \in \mathcal{Q}_d(G)$ such that $H = \langle k \circ \sigma, k \circ \tau \rangle$ is not unipotent on M. Then the following hold:

(1) $b(\sigma, \tau) = b$.
(2) $H \simeq SL_2(k(\lambda))$ for some algebraic element λ over k.
(3) Let A (resp. B) be the full unipotent subgroups of H containing σ (resp. τ). Then $M = M_0 \oplus M_1$ with $[M_0, H] = 1$ and $M_1 = [M_1, A] \oplus [M_1, B]$ and

$$[M_1, A] = [M_1, \sigma] = [M_1, a] \text{ for each } a \in A^{\#}$$

(and similarly for $[M_1, B]$!).

(4) $A \le E(\sigma)$, $B \le E(\tau)$.

(5) M_1 is the direct sum of natural $\mathbb{Z}SL_2(k(\lambda))$-modules.

Proof. By definition of d we have

$$\operatorname{codim}(C_M(\sigma) \cap C_M(\tau)) \le 2d.$$

Hence $b \ge b(\sigma, \tau) \ge \dim M - 2d$. On the other hand, if $\bar\sigma, \bar\tau \in \mathcal{Q}_d(G)$ with $b(\bar\sigma, \bar\tau) = b$ and $\langle k \circ \bar\sigma, k \circ \bar\tau \rangle$ not unipotent on M, then (1.11) shows that $\dim M_1 = 2d$ and thus $b = \dim M - \dim M_1 = \dim M - 2d$. This implies $b(\sigma, \tau) = b$.

This proves (1). Now (2) and (3) are direct consequences of (1.11). (4) follows immediately from the definition of $E(\sigma)$ resp. $E(\tau)$ and (5) is a consequence of I(3.7). $\square$

Now we can derive the following important Proposition:

(1.13) Proposition

Let $\sigma, \tau \in \mathcal{Q}_d(G)$, $A = E(\sigma)$, $B = E(\tau)$ and $X = \langle A, B \rangle$. Then one of the following holds:

(1) X is unipotent on M.

(2) X is a special rank one group with unipotent subgroups A and B. Moreover the following hold:

    (a) $H = \langle k \circ \sigma', k \circ \tau' \rangle \simeq SL_2(k(\lambda'))$ for each $\sigma' \in A^{\#}$, $\tau' \in B^{\#}$. Further the unipotent subgroups of $\langle k \circ \sigma', k \circ \tau' \rangle$ containing σ' resp. τ' are contained in A (resp. B).

    (b) There exists a unique central involution i in X which inverts $M^X = [M, X]$ and centralizes $M_X = C_M(X)$.

    (c) $M = M_X \oplus M^X$ and M^X is the direct sum of natural $\mathbb{Z}H$-modules for each choice of $\sigma' \in A^{\#}$ and $\tau \in B^{\#}$.

Proof. Suppose first $\langle \sigma, \tau \rangle$ acts unipotently on M. Then either $[M, \sigma] \cap C_M(\tau) \ne 0$ or $[M, \tau] \cap C_M(\sigma) \ne 0$ and we may without loss of generality assume that $U = [M, \sigma] \cap C_M(\tau) \ne 0$. Hence X centralizes U and for each $\sigma' \in A^{\#}$ and $y \in X$ we obtain

$$[M, \sigma'] \cap [M, \sigma'^y] \ge [M, \sigma] \cap [M, \sigma]^y \ge U \ne 0.$$

Hence (1.12) implies that $\langle \sigma', \sigma'^y \rangle$ is unipotent. Now (1.2) implies that $\langle A^Y \rangle$ is unipotent on M and thus also X is unipotent on M, since $\langle A^X \rangle \trianglelefteq X$. This shows that in this case (1) holds.

So we may assume that $H = \langle k \circ \sigma', k \circ \tau' \rangle$ is not unipotent on M for each $\sigma' \in A^{\#}$ and $\tau' \in B^{\#}$. But then (1.12) shows that (2)(a) holds for each such H. Now the structure of SL_2, see I(1.5), implies that for $a \in A^{\#}$ there exists a $b \in B^{\#}$ with $a^b = b^{-a}$ (inside $\langle k \circ a, k \circ \tau \rangle$. Since obviously by definition $E(\sigma)$ is a TI-subgroup of G, we also obtain $A^b = B^a$, which shows that X is a special rank one group.

So (b) and (c) of (2) remain to be proved. Now by definition of $E(\sigma)$ we have $C_M(X) = C_M(H)$ and $[M, X] = [M, H]$, which proves (c). By (1.12)(5) there exists a unique involution i in H inverting M^X and centralizing M_X. Hence i is also unique in X, since the product of two such involutions centralizes M.

$\square$

(1.14)   Notation

We call a pair $A, B \in \Sigma$ such that $\langle A, B \rangle$ is not unipotent on M an *opposite*. By (1.2) there exists an opposite to each $A \in \Sigma$ since $R_u(G) = 1$ by the irreducibility of G on M. For $A \in \Sigma$ let

$$M_A := C_M(A), M^A := [M, A]$$

and

$$U(A) := \{g \in N_G(A) \mid M^{g-\mathrm{id}} \subseteq M_A, M_A^{g-\mathrm{id}} \subseteq M^A \text{ and } (M^A)^{g-\mathrm{id}} = 0\}.$$

Then $U(A)$ is a unipotent normal subgroup of $N(A)$. Let

$$\overline{A} := \{g \in N_G(A) \mid M^{g-\mathrm{id}} \subseteq M^A \text{ and } M_A^{g-\mathrm{id}} = 0\}.$$

Then the 3-subgroup lemma implies $\overline{A} \leq Z(U(A))$ and $U(A)' \leq \overline{A}$.

We fix for the rest of this section $A \in \Sigma$, an opposite B of A, $X = \langle A, B \rangle$ and the central involution i in X inverting M^X. In view of (1.12) we have to determine the structure of a unipotent subgroup $\langle E, F \rangle$, $E, F \in \Sigma$, to prove (1.5). This is done by complicated calculations in $\mathrm{End}_K(M)$, which we only can sketch. The most important tool is the following:

(1.15)   Theorem

Let $\varphi \in \mathcal{Q}_d(G) \cap U(A)$ with $\varphi \notin A$. Then there exists an $a \in A$ such that $\rho = \varphi a \in \mathcal{Q}_d(G)$ and is inverted by i. Moreover for $N = \langle \rho^X \rangle$ the following hold:

(a) $N' = 1$.

(b) $\overline{X} = X/C_X(N) \simeq SL_2(K)$, K a division ring or a Cayley division algebra (see I(1.8)). Further N is the natural $\mathbb{Z}X$-module (as defined in I(1.8)).

(c) $\Sigma \cap N$ is a partition of N.

(d) If $\rho \in F \in \Sigma \cap N$, then $F \leq U(A)$ and $NA \leq U(F)$.

This is (21.9) and (21.10) of [Tim99]. (The statements $F \leq U(A)$ and $A \leq U(F)$ are intermediate steps of the proof of (21.10), namely step (10) and (12).) The (technical) proof consists mainly of extensive calculations with the 3-subgroup lemma in $\mathrm{End}_K M$ and estimations of dimension, using the minimality of d. To identify N and $\overline{X}$ one finally applies I(3.2). As an application of (1.15) we prove:

(1.16) Corollary

Let $\sigma \in \mathcal{Q}_d(G) \cap U(A)$ with $\sigma \in F \in \Sigma$ and $F \neq A$. Then the following hold:

(a) $AF = AF' \leq U(A)$ with $F' \in \Sigma$ inverted by i.

(b) If $E \in \Sigma \cap U(A)$ and $[E, F] \neq 1$, then $[e, F] = [E, f] = [E, F] = A$ for all $e \in E^{\#}, f \in F^{\#}$.

Proof. By (1.15) there exists a $\sigma' \in \sigma A \cap \mathcal{Q}_d(G)$ inverted by i. Moreover $N = \langle \sigma'^X \rangle$ is the natural module for $\overline{X} = X/C_X(N)$ and $\sigma' \in F' \in \Sigma \cap N$. By the structure of the natural $SL_2(K)$-module, see I(1.8), $A^N \cup F'$ is a partition of AF'. In particular $\sigma \in F \in \Sigma \cap AF'$ and either $F = F'$ or $F = A^n$ for some $n \in N$ with $[A, n] = F'$. Hence in any case by (1.15) (d) $AF = AF' \leq U(A)$.

Now, since i inverts σ' it normalizes F'. Suppose $1 \neq f \in F'$ is centralized by i. Then, since $C_M(i) = C_M(X)$, we obtain

$$[C_M(i), f] = [M_X, f] \leq M^A \cap C_M(i) = 0.$$

Hence $[M, f] \leq M^A$ and $M_A = M^A \oplus C_M(i) \leq C_M(f)$. As $f \in \mathcal{Q}_d(G)$ this implies $f \in A$, a contradiction to $F' \cap A = 1$ since $F \neq A$. Thus i acts fixed-point-freely on F' and thence inverts F'. (Since F' is closed under the $\circ$ scalar multiplication with elements of k and Char $k \neq 2$!). This proves (a).

To prove (b) we first prove an independent statement which will be needed later on:

(∗) Suppose $\varphi, \rho \in \mathcal{Q}(G)$ such that $\varphi\rho = \rho\varphi$ is quadratic on M. Then $\varphi\rho \in \mathcal{Q}(G), [M, \varphi, \rho] = [M, \rho, \varphi] = 0$ and $c \circ (\varphi\rho) = (c \circ \varphi)(c \circ \rho)$ for $c \in k$.

To prove (∗) set $\varphi = \mathrm{id} + \alpha$, $\rho = \mathrm{id} + \beta$ with $\alpha^2 = 0 = \beta^2$. Then, as $\varphi\rho = \rho\varphi$, we have $\alpha\beta = \beta\alpha$. Since $\varphi\rho$ is quadratic on M we obtain:

$$(\alpha + \beta + \alpha\beta)^2 = (\varphi\rho - \mathrm{id})^2 = 0.$$

Together with $\alpha\beta = \beta\alpha$ this implies $\alpha\beta + \beta\alpha = 0$ and thence $\alpha\beta = 0$, since Char $k \neq 2$. This shows $\varphi\rho = \text{id} + \alpha + \beta$ and $c \circ (\varphi\rho) = \text{id} + c\alpha + c\beta = (c \circ \varphi)(c \circ \rho) \in G$, which proves $(*)$.

Now let $e \in E$, $f \in F$ with $[e, f] \neq 1$. Then by (1.14) $[e, f] \in \overline{A}$ and so $[e, f]$ commutes with E and F. Thus, applying $(*)$ above to $[e, f] = e^{-1}ef$, we obtain $[e, f] \in \overline{A} \cap \mathcal{Q}(G) = A^{\#}$. This shows $[E, F] \leq A$.

Let L be an opposite to F'. Then, since N is transitive on $(F'A \cap \Sigma) - F'$ and $A \leq U(F')$ by $(1.15)(\text{d})$, we may by part (a) choose L so that a central involution j of $Y = \langle L, F' \rangle$ inverts A. Hence by (1.15) $N_1 = \langle A^Y \rangle$ is a natural module for $\overline{Y} = Y/C_Y(N_1) \simeq SL_2(K)$. In particular, if $N_1 = AD$ with $D \in \Sigma$, then $D \leq U(A)$ by (1.15) and D acts regularly on $(AF \cap \Sigma) - A$.

Now $C_E(F) = N_E(F) = 1$, since otherwise F normalizes E (by the TI-set property of E) and thus $[F, E] \leq E \cap A = 1$, a contradiction to the choices of E and F. Thus for each $e \in E^{\#}$ there exists a $d \in D^{\#}$ such that $F \neq F^e = F^d$. Hence $F = F^{ed^{-1}}$ and, since $ed^{-1} \in U(A)$, $[F, ed^{-1}] = 1$. This shows $[F, e] = [F, d] = A$ by the structure of the natural $SL_2(K)$-module. Since we have symmetry between E and F this proves (b). $\square$

(1.17) Lemma

Let $e \in \mathcal{Q}_d(G) \cap C(A)$ such that $ea \in \mathcal{Q}_d(G)$ for some $e \in A^{\#}$. Then either $\langle e, B \rangle$ or $\langle ea, B \rangle$ is not unipotent.

Proof. Suppose $\langle e, B \rangle$ is unipotent on M and pick by (1.13) $b \in B$ with $a^b = b^{-a}$. Then by (1.12) $H = \langle k \circ a, k \circ b \rangle \simeq SL_2(k(\lambda))$ for some algebraic λ over k. Moreover M^X is the direct sum of natural $\mathbb{Z}H$-modules and by I(3.4) $\langle v^H \rangle$ is a natural $\mathbb{Z}H$-module for $0 \neq v \in V^B$. In particular $v = [v, a, b]$ for $0 \neq v \in V^B$. (If we set $a = \begin{pmatrix} 1 & \\ \alpha & 1 \end{pmatrix}$, then $b = \begin{pmatrix} 1 & \alpha^{-1} \\ & 1 \end{pmatrix}$ and for $v = (0, \beta)$ the above equation holds.)

Let now $b = \text{id} + \tau$, $a = \text{id} + \sigma$ and $e = \text{id} + \varphi$ with $\tau^2 = \sigma^2 = \varphi^2 = 0$. Then, as shown in $(1.16)(*)$, $ea = \text{id} + (\sigma + \varphi)$. Since $\langle e, B \rangle$ is unipotent, $\Delta = \Delta(b, e) = \tau\varphi + \varphi\tau$ is by $(1.7)(3)$ nilpotent. Now Δ centralizes b and thus stabilizes $M^B = [M, b]$. Since $\tau\varphi$ is 0 on M^B we obtain $(M^B)^{\varphi\tau} \subseteq M^B$ and $\varphi\tau|_{M^B} = \Delta|_{M^B}$ is nilpotent.

As shown in $(1.16)(*)$ we have $[M, a, e] = 0 = [M, e, a]$. Hence for $0 \neq v \in M^B$ we obtain:
$$[v, ea, b] = [v, e, b] + [v, a, b] = v + [v, e, b],$$
which is in additive notation:

$(*)$ $v^{(\sigma+\varphi)\tau} = v^{\sigma\tau} + v^{\varphi\tau} = v + v^{\varphi\tau}.$

Now let $v_1, \ldots, v_s$ be a basis of M^B and suppose that $v_1^{(\sigma+\varphi)\tau}, \ldots, v_s^{(\sigma+\varphi)\tau}$ are linearly dependent. Then by $(*)$ there exist $c_i \in k, i = 1, \ldots, s$ such that

$$\Sigma c_i v_i + \Sigma c_i v_i^{\varphi\tau} = 0 \text{ and some } c_j \neq 0.$$

Hence $\varphi\tau$ inverts $\Sigma c_i v_i \neq 0$, a contradiction as $\varphi\tau$ is nilpotent on M^B (and Char $k \neq 2$.)

This shows $(M^B)^{(\sigma+\varphi)\tau} = M^B$. In particular $\Delta(b, ea)$ is not unipotent on M^B and thus not unipotent on M. Hence $(1.7)(3)$ implies that $\langle ea, B \rangle \geq \langle k \circ (ea), k \circ b \rangle$ is not unipotent on M. □

(1.18) Lemma

Let $E, F \in \Sigma$. Then $[E, F] = 1$ if one of the following holds:

  (i) $[e, f] = 1$ for some $e \in E^\#, f \in F^\#$

    or

  (ii) $N_E(F) \neq 1$.

Proof. Suppose false. In both cases $\langle E, F \rangle$ is by (1.13) unipotent on M and thus nilpotent. Since $N_E(F)F$ is also nilpotent, we may in any case assume that (i) holds. Suppose $\bar{e} = [e', f] \neq 1$ for some $e' \in E$. Then $\bar{e} \in E$, since E is a TI-set. Let $E_0 = N_E(M^F) \cap N_E(M_F)$ and $F_0 = N_F(M^E) \cap N_F(M_E)$. Then $e \in E_0, f \in F_0$ and $[E_0, F_0] \leq E \cap F = 1$. Moreover

$$[M_F, e', f] \subseteq [M^E, f] \subseteq M^E \cap M^F,$$

so that by the 3-subgroup lemma $\bar{e} \in E_0$. On the other hand $\bar{e}f^{-1} = (f^{-1})^{e'} \in \mathcal{Q}_d(G)$, so that arguing as in (1.16) $(*)$ we obtain:

$$[M, \bar{e}, f^{-1}] = 0 = [M, f^{-1}, \bar{e}].$$

Thus $M^E = [M, \bar{e}] \subseteq M_F$ and $M^F \subseteq M_E$ and, again by the 3-subgroup lemma, $[E, F] = 1$.

This shows that no such e' exists. Hence $[E, f] = 1$ and with symmetry $[E, F] = 1$. □

The proof of the next theorem, (22.3) of [Tim99], is more complicated and so not given here.

(1.19) Theorem

Let $e \in \mathcal{Q}_d(G) \cap C(A)$ such that $ea \in \mathcal{Q}_d(G)$ for some $a \in A^\#$. Then $e \in U(A)$.

From (1.19) we obtain as a Corollary:

(1.20) Corollary

Let $E, F \in \Sigma$ such that $\langle E, F \rangle$ is nilpotent of class 2. Then

$$D = [E, F] = [e, F] = [E, f] \in \Sigma \text{ for each } e \in E^{\#}, f \in F^{\#}.$$

Further $\langle E, F \rangle \leq U(D)$.

Proof. Pick $f \in F^{\#}$ and let $D = [E, f]$ and $U = \langle E, F \rangle$. Then by (1.18) $1 \neq D \leq Z(U)$. Let $d = [e, f]$ for some $e \in E^{\#}$. (Again $d \neq 1$!) Then

$$(*) \qquad \begin{aligned} ee^{f^2} &= e^2 e^{-1} e^{f^2} = e^2 [e^2, f^2] = e^2 d^2 = e^2 [e^2, f] \\ &= (e^2)^f \in \mathcal{Q}_d(G). \end{aligned}$$

Thus $\{e, e^{f^2}, ee^{f^2}\} \subseteq \mathcal{Q}_d(G)$ and (1.19) implies $e^{f^2} \in U(E)$. Hence (1.15) and (1.16) imply that $\Sigma \cap E^{f^2} E$ is a partition of $E^{f^2} E$. By $(*)$ $E^f \cap EE^{f^2} \neq 1$, so that $E^f \leq EE^{f^2}$ and $EE^{f^2} = EE^f = ED$. Now $L \cap D \neq 1$ for some $L \in \Sigma \cap ED$. But then $D = C_{ED}(F) = L$ by (1.18).

Now by (1.19) and (1.16) $E \leq U(D)$. Further, since $f^{-1}d = (f^{-1})^e \in \mathcal{Q}_d(G)$, (1.19) and (1.16) also imply $F \leq U(D)$. Hence (1.20) is a consequence of (1.16)(b). □

(1.21) Lemma

The following hold:

(1) There exists a subgroup $X_0 = \langle X_0 \cap A, X_0 \cap B \rangle \simeq SL_2(k)$ of X with $X_0 \cap A = k \circ a$, $X_0 \cap B = k \circ b$ for some $a \in X_0 \cap A$, $b \in X_0 \cap B$.

(2) Let $H_0 = N_{X_0}(A) \cap N_{X_0}(B)$. Then we can label $H_0 = \{h(\lambda \mid \lambda \in k^*\}$ such that $h(\lambda)h(\mu) = h(\lambda\mu)$ and $h(\lambda)$ acts by multiplication with λ on M^A and by multiplication with λ^{-1} on M^B.

(3) $\alpha^{h(\lambda)} = \lambda^2 \circ \alpha$ for $\alpha \in \overline{A}$. In particular $\overline{A} = A$.

Proof. (1) is a direct consequence of (1.13), (1.14). By I(3.7) M^X is the direct sum of natural $\mathbb{Z}X_0$-modules M_i. Let $w \in M_i - C_{M_i}(a)$ and $v = w^{a-\mathrm{id}}$. Then $w = v^{coa-\mathrm{id}}$. This shows that each M_i is a k-subspace of M^X. Moreover the M_i are all equivalent as kX_0-modules, since if a acts by

$$\left[\begin{array}{c|c} I_n & I_n \\ \hline & I_n \end{array} \right]$$

on M^X, then $c \circ a$ acts by

$$\left[\begin{array}{c|c} I_n & cI_n \\ \hline & I_n \end{array} \right]$$

(see (1.7)!). This shows that we may choose notation so that (2) holds. (I. e. $h(\lambda) \in H_0$ acts by multiplication with

$$
\begin{bmatrix}
\lambda & & & & & & \\
& \lambda^{-1} & & & & & \\
& & \lambda & & & & \\
& & & \lambda^{-1} & & & \\
& & & & \ddots & & \\
& & & & & \lambda & \\
& & & & & & \lambda^{-1}
\end{bmatrix}
$$

on $M^X = \bigoplus M_i$.)

Pick now $\alpha \in \overline{A}^{\#}$ and $m \in M^B$. Then by (2)

$$
\begin{aligned}
[m, \alpha^{h(\lambda)}] &= [m^{h(\lambda^{-1})}, \alpha]^{h(\lambda)} = [\lambda m, \alpha]^{h(\lambda)} = \lambda^2 [m, \alpha] \\
&= [m, \lambda^2 \circ \alpha]
\end{aligned}
$$

since $[m, \alpha] \in M^A$. Since $\alpha^{h(\lambda)}$ and $\lambda^2 \circ \alpha$ centralize $M_X \oplus M^A$, this shows that $\alpha^{h(\lambda)} = \lambda^2 \circ \alpha$ (as elements of $\mathrm{End}_k(M)$). In particular $\lambda^2 \circ \alpha \in G$ for all $\lambda \in k^*$. Now we have

$$(*) \qquad\qquad k = \{\Sigma \lambda_i c_i^2 \mid c_i \in k \text{ and } \lambda_i \in \mathbb{Z} \text{ resp. } \mathbb{Z}_p\}.$$

Indeed, if F is the right side of equation $(*)$, then as $p/q = q \cdot p(1/q)^2$ for $p, q \in \mathbb{Z}$, the prime field k_0 of k is contained in F. Hence F is a k_0-vector space and F is closed under multiplication. Suppose $c \in k - F$. Then

$$1 + 2c + c^2 = (1 + c)^2 \in F$$

and thus $2c \in F$. But then $c \in F$, since Char $k \neq 2$ and F is a k_0-vector space. Hence $(*)$ holds.

Since $\lambda_i \circ \alpha = \alpha^{\lambda_i}$ for $\lambda_i \in \mathbb{Z}$ (resp. $\mathbb{Z}_p$), $(*)$ and $\lambda^2 \circ \alpha \in G$ for $\lambda \in k^*$ show that $k \circ \alpha \leq G$. But then by definition $\alpha \in \mathcal{Q}(G)$ and so $\alpha \in A$. This proves (3). $\qquad\qquad\qquad\qquad\qquad\qquad\qquad\qquad\qquad\qquad\qquad\qquad\qquad\qquad$ □

(1.22)   Corollary

Let k_o be the prime field of k and $\widetilde{U(A)} = U(A)/A$. Then $\widetilde{U(A)}$ is a $k_o C(A)$-module with scalar action given by:

$$\lambda \circ \tilde{u} := \tilde{u}^{h(\lambda)}; \lambda \in k_o^*, \tilde{u} \in \widetilde{U(A)}.$$

Proof. By (1.21)(3) and (1.14) $\widetilde{U(A)}$ is abelian. Hence $\widetilde{U(A)}$ is a $\mathbb{Z}C(A)$-module for the natural $\mathbb{Z}$-action, which we write exponential. We show:

$$(*) \qquad\qquad \widetilde{u}^\lambda = \widetilde{u}^{h(\lambda)} = \lambda \circ \widetilde{u} \text{ for } \lambda \in \mathbb{Z} \text{ resp. } \mathbb{Z}_p.$$

To prove $(*)$ let u be a coimage of $\widetilde{u}$ and $m \in M_X$. Then we have

$$[u^{h(\lambda)}, m] = [u, m]^{h(\lambda)} = [u, m]^\lambda = [u^\lambda, m]$$

by (1.21)(2) since $[u, m] \in M^A \leq C_M(u)$. Hence we obtain:

(i) $u^{h(\lambda)} \equiv u^\lambda \bmod C_{U(A)}(M_A)$.

Let now $\widehat{M} = M/M^A$ and $m \in M^B$. Then

$$\begin{aligned}[u^{h(\lambda)}, \widehat{m}] &= [u, \widehat{m}^\lambda]^{h(\lambda)} = [u, \widehat{m}^\lambda] = [u, \widehat{m}]^\lambda \\ &= [u^\lambda, \widehat{m}],\end{aligned}$$

since $[u, \widehat{M}] \leq \widehat{M}_A = \widehat{M}_X \leq C_{\widehat{M}}(u)$. This shows

(ii) $u^{h(\lambda)} \equiv u^\lambda \bmod C_{U(A)}(M/M^A)$.

Now (i) and (ii) together imply $u^{h(\lambda)} \equiv u^\lambda \bmod \overline{A}$, which with (1.21)(3) proves $(*)$.

If $k_\circ = \mathbb{Z}_p$, (1.22) is a direct consequence of $(*)$. So assume $k_\circ = \mathbb{Q}$. Then

$$\widetilde{U(A)} = \{\widetilde{u}^\lambda \mid \widetilde{u} \in U(A)\} \text{ for each fixed } \lambda \in \mathbb{Z},$$

since $\widetilde{U(A)}^{h(\lambda)} = \widetilde{U(A)}$ as $\widetilde{U(A)}^{h(\lambda^{-1})} \subseteq \widetilde{U(A)}$. Hence $\widetilde{U(A)}$ is divisible and it is an easy exercise to show that the natural $\mathbb{Q}$ action on $\widetilde{U(A)}$ coincides with the action by $h(\lambda), \lambda \in \mathbb{Q}$, since this holds by $(*)$ for $\lambda \in \mathbb{Z}$. $\qquad\square$

(1.23) Corollary

The following hold:

(1) $N(A) = U(A)(N(A) \cap C(i))$.
(2) $N(A) \cap C(i) = A(N(A) \cap N(B))$.
(3) $C(A) = U(A)C(X)$ and $C(A) \cap C(i) = AC(X)$.

Proof. Let $g \in N(A)$. Then i^g inverts M^A and M/M_A and centralizes M_A/M^A. Hence $ii^g \in U(A)$. Now, since $\widetilde{U(A)}$ is by (1.22) a $k_\circ$-vector space, $i^g \in iU(A)$ and both invert $\widetilde{U(A)}$, there exists an $u \in U(A)$ such that $i^g \equiv i^u \bmod A$. But then $i^g = i^u$, since i^g is the only involution in $\langle i^g \rangle \times A$. Hence the Frattini argument implies (1).

Let $x \in N(A) \cap C(i)$ and $h(t) \in H_0$ with $t^2 \neq 1$. Then $h(t)^x$ centralizes $M_X = C_M(i)$, acts by multiplication with t on M^A and by multiplication with t^{-1} on M^X/M^A ($M^X = [M, i]$!) Hence $h(t)^{-1}h(t)^x \in \overline{A} = A$ and $h(t)^x = h(t)a$ for some $a \in A$. Now A is with the $\circ$ action a k-vector space. Hence (1.21)(3) and $t^2 \neq 1$ imply that there exists an $\alpha \in A$ with $h(t)^x = h(t)^\alpha$. As $t \neq \pm 1$, M^A and M^B are by (1.21) the only eigenspaces of $h(t)$ on M^X. Hence $(M^B)^x = (M^B)^\alpha$, since x normalizes M^A. This shows that $x\alpha^{-1}$ normalizes M^B and $M_B = M^B + M_X$, whence $x\alpha^{-1} \in N(B)$ which proves (2).

(3) is an immediate consequence of (1) and (2), since $C(A) \cap N(B) = C(X)$, as X is by (1.13) a rank one group. $\square$

(1.24)   Lemma

Let $\tau \in \mathcal{Q}_d(G) \cap C(A)$ and $\sigma \in \mathcal{Q}_d(G) \cap U(A)$. Then either $[\sigma, \tau] = 1$ or $[\sigma, \sigma^\tau] = 1$.

Proof. By (1.15) we may, to prove (1.24), assume that σ is inverted by i. Now suppose that (1.24) is false. Then $1 \neq \rho = [\sigma, \sigma^\tau] \in A$. We will derive a contradiction from this in several steps. Let $m = \dim M^A$, $s = \dim[M^B, \sigma]$ and $\ell = \dim[M_A, \sigma]$. Then, as $\dim[M, \sigma] = m$ and $M = M_A \oplus M^B$ we have

(1) $m = s + \ell$.

Next, by the 3-subgroup lemma,

$$M^A = [M, \rho] = [M^B, \sigma, \sigma^\tau] + [M^B, \sigma^\tau, \sigma]$$

since $[M_A, \sigma, \sigma^\tau] = 0 = [M_A, \sigma^\tau, \sigma]$. Now clearly $\dim[M^B, \sigma, \sigma^\tau] \leq s$. As $[M^B, \sigma] \subseteq M_A$ and $\ell = \dim[M_A, \sigma^\tau]$ we obtain

$$\dim[M^B, \sigma, \sigma^\tau] =\leq \min(s, \ell)$$

and by the same argument also $\dim[M^B, \sigma^\tau, \sigma] \leq \min(s, \ell)$. Hence by (1):

(2) $s = \ell = \frac{m}{2}$ and $M^A = [M^B, \sigma, \sigma^\tau] \oplus [M^B, \sigma^\tau, \sigma]$,

since $[M^B, \sigma^\tau, \sigma] \subseteq [M_A, \sigma]$ and $[M^B, \sigma, \sigma^\tau] \subseteq [M_A, \sigma^\tau]$. As $\dim[M_A, \sigma] = \ell = \dim[M_A, \sigma^\tau]$ we obtain by (2):

(3) $M^A = [M_A, \sigma] \oplus [M_A, \sigma^\tau]$.

In particular, since τ interchanges $[M_A, \sigma]$ and $[M_A, \sigma^\tau]$, we have $\dim C_{M^A}(\tau) \leq m/2$. Hence, as τ acts quadratically on M^A, we obtain

$$\dim[M^A, \tau] = m/2 = \dim C_{M^A}(\tau).$$

Let by (1.23) $\tau = \tau' u$, $u \in U(A)$ and $\tau' \in C(X)$. Then $\tau|_{M^A} = \tau'|_{M^A}$ and $\tau|_{M/M_A} = \tau'|_{M/M_A}$. Since $M = M_A \oplus M^B$ and M^A and M^B are interchanged by X, we also have $\dim[M/M_A, \tau] = m/2$. As $\tau \in \mathcal{Q}_d(G)$ this implies $M_A \subseteq M^A + C_M(\tau)$. In particular $C_{M_A}(\sigma) = C_{M_A}(\sigma^\tau)$, which obviously contradicts (2) since $[M^B, \sigma] \subseteq C_{M_A}(\sigma)$. $\square$

(1.25)   Corollary

Let $E \in \Sigma \cap U(A)$ and $F \in \Sigma \cap C(A)$. Then either $[E, F] = 1$ or $L = [E, F] \in \Sigma$ and $\langle E, F \rangle \leq U(L)$.

Proof. Suppose $[E, F] \neq 1$. Then $T = \langle E^F \rangle$ is by (1.24) abelian. As $\langle E, F \rangle$ is by (1.13) unipotent on M, TF is nilpotent. Hence $T_0 = C_T(F) \neq 1$ and $T_1/T_0 = C_{T/T_0}(F) \neq 1$. Further, by (1.18), $T \cap F = 1$. Pick $t \in T_1 - T_0$. Then

$$F^t \times F = F \times [F, t] \text{ and } [F, t] \leq T_0.$$

Thus we have for $f \in F^\#$

$$[f, t^2] = [f, t]^2 = [f^2, t] \neq 1,$$

since $f^2 \neq 1$ and since $C_F(t) = 1$. (Otherwise $t \in N(F)$ and thus $1 \neq [F, t] \leq T \cap F \leq U(A)$, a contradiction to (1.18) and $[E, F] \neq 1$.) Hence, by the same reason, $F \neq F^{t^2} \neq F^t$ and $F^{t^2} \leq F \times [F, t] = F \times F^t$. Thus (1.19) implies $F^t \leq U(F)$ and so (1.15), (1.16) imply $\Sigma \cap FF^t$ is a partition of FF^t. Now

$$[F, t] = FF^t \cap T = FF^t \cap U(A).$$

Thus, if $L \in \Sigma \cap FF^t$ with $L \cap [F, t] \neq 1$, then (1.16) implies $L \leq FF^t \cap U(A) = [F, t]$. Since $\Sigma \cap LF$ is by (1.16) also a partition of LF, this shows that $[F, t] = L \in \Sigma$ and $F \leq U(L)$ by (1.19). Now, applying (1.24) to $F \leq U(L)$ and $E \leq C(L)$ we obtain $[F, F^e] = 1$ for each $e \in E$. Hence $[F, E] \leq Z(\langle E, F \rangle)$ and $\langle E, F \rangle$ is of class 2. Now (1.25) is a consequence of (1.20). $\square$

Before we can start with the proof of (1.5) we need a general lemma on p-groups.

(1.26)   Lemma

Suppose the non-abelian p-group P is generated by the abelian subgroups E and F. Then

$$cl(\langle E^P \rangle) < cl(P) \text{ and also } cl(\langle F^P \rangle) < cl(P).$$

($cl(P)$ is the nilpotency class of P!)

Proof. Let $N = [E, F]$. Then, as $P = \langle E, F \rangle$, we have $N \trianglelefteq P$. Since P/N is abelian we have $P' \leq N$ and thus $N = P'$. As $\langle E^P \rangle \leq EN$ we obtain

$$(NE)' = N'[N, E] \leq [N, P] \leq K_2(P),$$

where $K_i(P)$ is the i-th term of the descending central series of P. With induction this shows

$$K_i(NE) \leq K_{i+1}(P),$$

which shows $cl(\langle E^P \rangle) \leq cl(NE) \leq cl(P) - 1$. With symmetry this proves (1.26).   □

Now we can show:

(1.27) Theorem

Let $E, F \in \Sigma$ such that $\langle E, F \rangle$ is unipotent on M. Then one of the following holds:

(1) $[E, F] = 1$.
(2) $L = [E, F] = [e, F] = [E, f] \in \Sigma$ for each $e \in E^{\#}$, $f \in F^{\#}$. Further $\langle E, F \rangle \leq U(L)$ and $\langle E, F \rangle$ is of class 2.

(In theorem 2 of [Gla72] the same theorem has been proved under a slightly different hypothesis, i.e. G. Glauberman assumes that the quadratic elements of G with commutator space of minimal dimension are closed under the ∘ scalar multiplication, while we just start with any set $\mathcal{Q}$ of quadratic elements generating G, which are closed under ∘ multiplication and then let $\mathcal{Q}_d$ be the set of $\sigma \in \mathcal{Q}$ with minimal $d(\sigma)$.)

Proof. Let $P = \langle E, F \rangle$. Then we may assume that P is a non-abelian nilpotent group. We prove (1.27) by induction on $cl(P)$. If $cl(P) = 2$, (1.27) is just (1.20). So assume $cl(P) \geq 3$ and (1.27) holds for all subgroups Q of P generated by two elements of Σ with $cl(Q) < cl(P)$.

Let $P_1 = \langle E^P \rangle$ and $P_2 = \langle F^P \rangle$. Then $cl(P_i) < cl(P)$ by (1.26). If P_1 and P_2 are both abelian, then $cl(P) = 2$, a contradiction. So we may without loss of generality assume $P_1' \neq 1$. Let $E^P = \{E_i \mid i \in I\}$ and let $\{j_1, \ldots, j_s\} \subseteq I$ be a largest subset of I with

$$L = [E_{j_1}, E_{j_2}, \ldots, E_{j_s}] \neq 1$$

$(s \leq cl(P_1)!)$ Then by the maximality of s we have $[L, E_i] = 1$ for all $i \in I$ and whence $L \leq Z(P_1)$. Further, since $cl(P_1) < cl(P)$, we obtain by induction assumption

$$\{[E_{j_1}, E_{j_2}], [E_{j_1}, E_{j_2}, E_{j_3}], \ldots, L\} \subseteq \Sigma \cap P_1.$$

Moreover $E_{j_s} \leq U(L)$. Now we have

$$\langle L, F \rangle \leq (Z(P_1) \cap P_1')F \quad \text{and} \quad \langle L, F \rangle' \leq P_1''[P_1', F] \leq K_2(P).$$

Hence it follows with induction that $cl(\langle L, F \rangle) < cl(P)$ and so (1.27) holds for the pair L, F. Thus one of the following holds:

(1) $[L, F] = 1$.
(2) $T = [L, F] \in \Sigma$ and $\langle L, F \rangle \leq U(T)$.

In case (1) $L \leq Z(P)$ and, as $E_{j_s} \leq U(L)$, also $E \leq U(L)$. Since $F \leq C(L)$ we may apply (1.25). But then $cl(P) \leq 2$, a contradiction to the assumption. In case (2) we have $T \leq Z(P_1)$ and thus $F \leq U(T)$ and $E \leq C(T)$. Hence again $cl(P) \leq 2$ by (1.25), a contradiction. This proves (1.27).   □

Notice that (1.27) together with (1.13) prove Theorem (1.5). Indeed, both together show that Σ is a set of abstract root subgroups of $G_0 = \langle \Sigma \rangle$. Further, since the unipotent radical $R_u(G) = 1$ by the irreducibility of G on M, there exists an opposite to each $A \in \Sigma$ by (1.2). Hence $\mathcal{F}(\Sigma)$ has no isolated vertices. By (1.12) we have for $a \in A^\#$, $b \in B^\#$ with $A, B \in \Sigma$ opposite:

$$SL_2(k(\lambda)) \simeq \langle k \circ a, k \circ b \rangle \leq \langle A, B \rangle$$

and $A \cap \langle k \circ a, k \circ b \rangle$ is a full unipotent subgroup of $SL_2(k(\lambda))$. This shows that condition (H) of II§3 holds, since $|k(\lambda)| \geq |k| \geq 4$.

The rest of §1 is devoted to the proof of (1.6). First, as a Corollary to (1.5), we obtain:

(1.28) Corollary

$\Sigma = \Sigma_1 \dot{\cup} \cdots \dot{\cup} \Sigma_k$ such that for $G_i = \langle \Sigma_i \rangle$ the following hold:

(1) G_i is quasi-simple and Σ_i is a class of abstract root subgroups of G_i satisfying (H) of II§3.
(2) $[G_i, G_j] = 1$ for $1 \leq i \neq j \leq k$.

Proof. By (1.5) and II(2.21) we have

$$\Sigma = \dot{\cup}_{i \in I} \Sigma_i$$

such that Σ_i is a class of abstract root subgroups of $G_i = \langle \Sigma_i \rangle$ and $[G_i, G_j] = 1$ for $i \neq j$, since $\mathcal{F}(\Sigma)$ contains no isolated vertices. By (1.13) $\langle A, B \rangle$ is perfect for opposites $A, B \in \Sigma$. This implies that each G_i is perfect.

Now suppose $N_i \trianglelefteq G_i$. If $\langle A, A^n \rangle$ is a rank one group for some $A \in \Sigma$, $n \in N_i$, then as $\langle A, A^n \rangle \leq N_i A$ and $\langle A, A^n \rangle$ is perfect, we have $A \leq N_i$. Hence $N_i = G_i$

since Σ_i is a conjugacy class in G_i. Thus we may by (1.13) assume that $\langle A, A^n \rangle$ is unipotent for each $n \in N_i$. But then by (1.2) $\langle A^{N_i} \rangle$ is unipotent. Hence

$$[A, N_i] \leq R_u(N_i) \leq R_u(G_i) \leq R_u(G_0) \leq R_u(G) = 1$$

and thus $N_i \leq Z(G_i)$, again since Σ_i is a conjugacy class in G_i. This shows that each G_i is quasi-simple.

It remains to be shown that $|I| < \infty$. Suppose this is not the case. Then we may assume $\mathbb{N} \subseteq I$. Pick $A_i \in \Sigma_i$ for $i \in \mathbb{N}$ and claim

$$(*) \qquad\qquad [A_{i+1}, M] \not\subseteq \Sigma_{j=1}^i [A_j, M].$$

Indeed, if $(*)$ holds then the dimension of M is not restricted, a contradiction to $\dim M < \infty$.

Thus $(*)$ remains to be proved. Let B_i be opposite to A_i in Σ_i and $M_i = \Sigma_{j=1}^i [M, A_i]$. Reordering the A_i if necessary, we may assume $[A_{i+1}, M] \subseteq M_i$, but $[A_{i+1}, M] \not\subseteq M_k$ for $k < i$. Since $[M, A_i]$ and $[M, B_i]$ are conjugate in $X_i = \langle A_i, B_i \rangle$ and since X_i normalizes $[M, A_{i+1}]$ we obtain:

$$[A_{i+1}, M] \subseteq (M_{i-1} + [M, A_i]) \cap (M_{i-1} + [M, B_i]) \subseteq M_{i-1}$$

by (1.13). This contradiction proves (1.28).   □

(1.29) Lemma

$G = G_0 C_G(G_0)$.

Proof. It suffices to show $\mathcal{Q}(G) \subseteq G_0 C_G(G_0) = R$. Suppose this is false and pick $\sigma \in \mathcal{Q}(G) - R$ such that $d(\sigma)$ is minimal with this property. If $\langle \sigma, \sigma^g \rangle$ is unipotent on M for each $g \in G_0$, then by (1.2) $\sigma \in R_u(G_0 \langle \sigma \rangle)$ and thus

$$[\sigma, G_0] \leq G_0 \cap R_u(G_0 \langle \sigma \rangle) = R_u(G_0) = 1$$

by the irreducibility of G on M, a contradiction to $\sigma \notin R$. Thus we may pick a $g \in G_0$ such that $H = \langle k \circ \sigma, k \circ \sigma^g \rangle$ is not unipotent on M. Finally, among all such g, choose $g \in G_0$ so that $\dim[M, H]$ is minimal.

For the action of H on M we use the results of (1.7) and (1.8). Let $\Delta = \Delta(\sigma, \sigma^g)$ and $M = M_0 \oplus M_1 \oplus \cdots \oplus M_\ell$ be the corresponding Δ-decomposition of M. We show first:

(1) $\ell = 1$.

Suppose $\ell \geq 2$. Then by (1.10) $H = N_i N_j$ for $1 \leq i \neq j \leq \ell$, where $N_i \lhd H$ with $H/N_i \simeq SL_2(k(\lambda_i))$ in the notation of (1.9). In particular by (1.8) N_i acts unipotently on M_i (as $N_i/L_i = R_u(H_i)$!), whence $N_i \langle k \circ \sigma \rangle$ acts unipotently on M_i. Since $N_i N_j / N_j = H/N_j \simeq SL_2(k(\lambda_j))$, there exists a $y \in N_i \cap H'$ such that $H_1 = \langle k \circ \sigma, k \circ \sigma^y \rangle$ is not unipotent on M. But, as $H_1 \leq N_i \langle k \circ \sigma \rangle$, we have $M_i \not\subseteq [M_i, H_1]$ and whence $M_i \not\subseteq [M, H_1]$. Hence $\dim[M, H_1] < \dim[M, H]$, a contradiction to $y \in H' \leq (G_0(k \circ \sigma))' \leq G_0$ and the minimality of $\dim[M, H]$. Next we show:

(2) $[H, M_0] = [\sigma, M_0]$ and $H' \leq C(M_0)$.

Indeed, as $H|_{M_0}$ is unipotent, there exists an $n \in \mathbb{N}$ with $H^{(n)}|_{M_0} = 1$. Let L_1 be the kernel of the action of H on M_1 (as in (1.7)). Then H/L_i is by (1.8) perfect. Hence $H = H^{(n)} L_1$. As $H' \leq G_0$, there exists an $x \in H^{(n)} \cap G_0$ such that $H_0 = \langle k \circ \sigma, k \circ \sigma^x \rangle$, is not unipotent on M_1. The minimality of $\dim[M, H]$ implies $[M, H] = [M, H_0]$. Since $x \in H^{(n)} \leq C(M_0)$, this implies

$$[M_0, \sigma] = [M_0, H_0] = [M_0, H].$$

In particular $[M_0, \sigma, \sigma^g] = [M_0, \sigma, \sigma] = 0$ and thus $[k \circ \sigma, k \circ \sigma^g] \leq C(M_0)$, which proves (2).

Now we can obtain a final contradiction. By (2) $H' \cap L_1$ is the identity on $M_0 \oplus M_1 = M$ by (1), whence $H' \cap L_1 = 1$. Thus $H = L_1 \times H'$ and $H' \simeq H/L_1$ is perfect. Let $A = L_1(k \circ \sigma) \cap H'$. Then $L_1 A = L_1(k \circ \sigma)$, $A \cap L_1 = 1$ and $A \simeq k \circ \sigma$. Moreover

(*) $A|_{M_1} = (k \circ \sigma)|_{M_1}$ and $A|_{M_0} = \mathrm{id}$.

In particular A acts quadratically on M.

Now let $d \in k^*$ and $\alpha \in A$. Then by (*) there exists a $c \in k$ such that $\alpha = c \circ \sigma$ on M_1. Hence

$$d \circ \alpha = d \circ (c \circ \sigma) = (dc) \circ \sigma \text{ on } M_1.$$

Thus by (*) there exists an $\alpha' \in A$ such that $d \circ \alpha = \alpha'$ on M_1 and thence on M. This implies $d \circ \alpha \in A$ and $A^{\#} \subseteq \mathcal{Q}(G)$.

Pick $\alpha \in A^{\#}$ such that $\rho = \sigma \alpha^{-1}$ is the identity on M_1. Since $[A, k \circ \sigma] \leq L_1 \cap H' = 1$ (1.16) (*) implies $\rho \in \mathcal{Q}(G)$, as $\rho|_{M_0} = \sigma|_{M_0}$ and thus ρ is quadratic on M. Now the minimality of $d(\sigma)$ implies $\rho \in R$. Hence, as $A \leq H' \leq G_0$ we obtain $\sigma = \alpha \rho \in R$, a contradiction to the choice of σ. This proves (1.29). $\square$

As a Corollary to the proof of (1.29) we obtain:

(1.30)   Corollary

Let $\sigma \in \mathcal{Q}(G)$. Then the following hold:

  (a) $k \circ \sigma \le G_0$ or $k \circ \sigma \cap G_0 = 1$.
  (b) $k \circ \sigma \le G_i$ or $k \circ \sigma \cap G_i = 1$ for each $i \le k$, k as in (1.28).

Proof. To prove (a), choose σ among all $\sigma \in \mathcal{Q}(G)$ with $1 \ne k \circ \sigma \cap G_0 \ne k \circ \sigma$, such that $d(\sigma)$ is minimal. If now $\langle k \circ \sigma, k \circ \sigma^g \rangle$ is unipotent on M for all $g \in G_0$, then as in (1.29) $k \circ \sigma \le C_G(G_0)$, whence $k \circ \sigma \cap G_0 = 1$. Thus there exists a $g \in G_0$ such that $H = \langle k \circ \sigma, k \circ \sigma^g \rangle$ is not unipotent on M. Now choose g, among all such $g \in G_0$, such that $\dim[M, H]$ is minimal.

Then as in (1.29) $M = M_0 \oplus M_1$ is the $\Delta(\sigma, \sigma^g)$ decomposition of M, $[M_0, H] = [M_0, \sigma]$ and $H' \le C(M_0)$. Let as in (1.29) $A = L_1(k \circ \sigma) \cap H'$, where L_1 is the kernel of the action of H on M_1. Then

$$A|_{M_1} = k \circ \sigma|_{M_1}, A|_{M_0} = \mathrm{id} \text{ and } A^{\#} \subseteq \mathcal{Q}(G_0).$$

As $k \circ \sigma = k \circ \sigma'$ for each $1 \ne \sigma' \in k \circ \sigma$, we may assume $\sigma \notin G_0$. Pick again $\alpha \in A^{\#}$ such that $\rho = \sigma \alpha^{-1}$ is the identity on M_1. Then by (1.16) $(*)$ $\rho \in \mathcal{Q}(G)$ and obviously $d(\rho) < d(\sigma)$. Let $c \in k^*$ such that $c \circ \sigma \in G_0$. (Exists as $k \circ \sigma \cap G_0 \ne 1$!)

Then by (1.16) $(*)$

$$c \circ \rho = (c \circ \sigma)(c \circ \alpha^{-1}) \in G_0.$$

Hence by minimality of $d(\sigma)$ we obtain $k \circ \rho \le G_0$ and thus $\sigma = \rho \alpha \in G_0$, a contradiction to the choice of σ.

The proof of (b) is the same, since by (1.29) $G_i \trianglelefteq G$ for each $i \le k$.   □

(1.31)   Proposition

Let $G_0 = G_1 * \cdots * G_k$ as in (1.28). Then $\mathcal{Q}(G) = (q(G) \cap G_i) \dot{\cup} (\mathcal{Q}(G) \cap C_G(G_i))$ for $i \le k$.

Proof. Suppose false and let $\rho \in \mathcal{Q}(G)$ with $\rho \notin C = C_G(G_i)$ and $\rho \notin G_i$. Then by (1.28), (1.29) $G = G_i C$. Hence $\rho = \sigma_1 \sigma_2$ with $1 \ne \sigma_1 \in G_i$ and $1 \ne \sigma_2 \in C$. Let V be an irreducible kG_i-submodule of M. We first assume $V^{\sigma_2} \subseteq V$.

Then, by Schur's lemma, $\sigma_2 \in \mathrm{Hom}_{kG_i}(V, V) = K$, K a division ring. Thus V can be considered as a K-vector space. Hence σ_2 acts by multiplication with some $\lambda \in K^*$ on V. Let $\bar{k} = k(\lambda)$. Then clearly $\dim_k \bar{k} \le \dim_k K < \infty$. Now, since $\sigma_1 = \rho \sigma_2^{-1}$, we have

$$\sigma_1 = \lambda^{-1} \mathrm{id}_V + \alpha, \alpha^2 = 0 \text{ in } \mathrm{End}_{\bar{k}}(V).$$

By (1.28) $G_i = G_i'$. Hence $G_i \leq SL_{\overline{k}}(V)$ and thus $\det \sigma_1 = 1$. This implies $\lambda^s = 1$ for $s = \dim_{\overline{k}} V$. In particular we obtain $\lambda^m = 1$, where $m = \dim_k V$ if Char $k = 0$ (resp. $\dim_k V = m \cdot p^r$ with $(m, p) = 1$ when Char $k = p$).

Let now $\tau_i = \sigma_i^m, i = 1, 2$. Then $\tau_2 \neq 1$, since otherwise $1 \neq \rho^m = \sigma_1^m \in G_i$ and thus by (1.30) also $\rho \in G_i$ as $\rho \in k \circ \rho^m$. As $\lambda^m = 1$ we have $\tau_2|_V = \mathrm{id}$. Hence $\tau_1 = \rho^m|_V$ is quadratic on V. Since M is homogeneous as a kG_i-module, this shows that τ_1 is quadratic on M.

Let $U = V^g, g \in G$. If U is τ_2 invariant, τ_2 acts as shown above by scalar multiplication with some μ on U. On the other hand

$$\rho^m|_U = \tau_2|_U \cdot \tau_1|_U = (\mu \mathrm{id}_U)(\mathrm{id}_U + \alpha), \alpha^2 = 0$$

and ρ^m is quadratic. Hence $\mu = 1$. We have shown:

$$\tau_2|_U = \mathrm{id} \text{ for each } U = V^g \text{ which is } \tau_2\text{-invariant.}$$

Suppose now there exist a $U = V^g$ which is not τ_2-invariant. Pick $0 \neq u \in C_U(\tau_1)$. (τ_1 is quadratic on U!) Then $u^{\tau_2 - \mathrm{id}} = u^{\tau_1 \tau_2 - \mathrm{id}} = u^{\rho^m - \mathrm{id}} \in C_M(\rho^m) \cap C_M(\tau_1)$. Hence

$$u^{\tau_2 - \mathrm{id}} \in U^{\tau_2 - \mathrm{id}} \cap \ker(\tau_2 - \mathrm{id}) \neq 0,$$

since ρ^m is quadratic. In particular, as G_i acts irreducibly on $U^{\tau_2 - \mathrm{id}}$ we have $U^{\tau_2 - \mathrm{id}} \subseteq \ker(\tau_2 - \mathrm{id})$ and

$$W = U \oplus U^{\tau_2} = U \oplus U^{\tau_2 - \mathrm{id}} \text{ is } \tau_2\text{-invariant.}$$

Moreover τ_2 acts quadratically on W. Since $U \simeq U^{\tau_2 - \mathrm{id}} \simeq V$ as kG_i-module, this implies that τ_1, τ_2 and $\rho^m = \tau_1 \tau_2$ act quadratically on M. Hence by (1.16) $(*)$

$$U^{(\tau_2 - \mathrm{id})(\tau_1 - \mathrm{id})} = [W, \tau_2, \tau_1] = 0.$$

Since $U^{\tau_2 - \mathrm{id}} \simeq V$ as a kG_i-module, this implies $\tau_1|_V = \mathrm{id}$ and then $\tau_1 = \mathrm{id}$, since M is homogeneous as a kG_i-module. But then $\rho^m = \tau_2 \in C$, whence G_i centralizes $k \circ \rho^m$ and thus $\rho \in C$, a contradiction to the choice of ρ.

We have shown that, if V is σ_2 invariant, then $\tau_2|_U = \mathrm{id}$ for each $U = V^g, g \in C$. But then $\tau_2 = 1$, $\rho^m = \tau_1 \in G_i$ and thus $\rho \in G_0$ by (1.30), a contradiction to the choice of ρ.

So we are left with the case $V^{\sigma_2} \not\subseteq V$ for each irreducible kG_i-submodule V of M. This will lead to a contradiction. We first show

$(*)$ $W = V \oplus V^{\sigma_2}$ is σ_2-invariant.

Indeed, pick $0 \neq v' = v^{\sigma_1} \in V$. Then

$$v^{\rho-\mathrm{id}} = v'^{\sigma_2} - v \in C_W(\rho).$$

Hence

$$v'^{\sigma_2} - v = (v'^{\sigma_2} - v)^{\rho} = v'^{\sigma_1 \sigma_2^2} - v'^{\sigma_2}$$

and

$$v'^{\sigma_1 \sigma_2^2} = 2v'^{\sigma_2} - v \in W \cap V^{\sigma_2^2}.$$

Since $V^{\sigma_2^2} \simeq V$ as a kG_i-module, this implies $V^{\sigma_2^2} \subseteq W$ and either $V = V^{\sigma_2^2}$
or

$$W = V \oplus V^{\sigma_2} = V \oplus V^{\sigma_2^2},$$

which implies $(*)$.

If now $U \neq 0$ is a proper $kG_i\langle\sigma_2\rangle$-submodule of W, then $U \simeq V$ as a kG_i-module and thus is irreducible as a kG_i-module, a contradiction to the above. Hence W is an irreducible $kG_i\langle\sigma_2\rangle$-module. Now

$$\sigma_2 \in K = \mathrm{Hom}_{G_i\langle\sigma_2\rangle}(W, W), K \text{ a division ring}$$

and as before we obtain $\sigma_2^m = 1$ for some $m \in \mathbb{N}$ with $(\mathrm{Char}\, k, m) = 1$. Hence again $1 \neq \rho^m = \sigma_1^m \in G_i$ and (1.30) implies $\rho \in G_i$, the final contradiction. $\square$

(1.32) Proof of (1.6)

We prove (1.6) by induction on $\dim M$. If $G = G_0 = G_1$ in the notation of (1.28), then clearly (1.6) holds. So we may assume $G_1 < G$. By (1.31)

$$(*) \qquad \mathcal{Q}(G) = (\mathcal{Q}(G) \cap G_1) \dot\cup (\mathcal{Q}(G) \cap C_G(G_1)).$$

Hence, if we set $C_1 = \langle \mathcal{Q}(G) \cap C_G(G_1) \rangle$, then $G = G_1 * C_1$. Let now V be an irreducible kC_1 submodule of M. Then as $C_1 = \langle \mathcal{Q}(C_1) \rangle$ by definition, (V, C_1) is a quadratic pair over k. Further $V \neq M$, since C_1 normalizes $[M, \sigma]$ for $\sigma \in \mathcal{Q}(G) \cap G_1$. Hence by minimality of $\dim M$ we obtain:

$$C_1 = G_2 * \cdots * G_\ell, \, G_i \quad \text{quasi-simple and} \quad \mathcal{Q}(C_1) = \dot\cup_{i=2}^{\ell} (\mathcal{Q}(C_1) \cap G_i)$$

$(\mathcal{Q}(C_1)$ defined in the action of C_1 on V!) Now M is homogeneous as a kC_1 module since $M = \Sigma V^g$, $g \in G_1$. Hence $\mathcal{Q}(C_1) = \mathcal{Q}(G) \cap C_1$ and by $(*)$ above $(1.6)(1)$ holds. The proof of $(1.6)(2)$ and (3) is now completely obvious. $\square$

§2   Subgroups generated by root elements of groups generated by abstract root subgroups

This section is devoted to the following type of questions: Let G be a group generated by the set Σ of abstract root subgroups satisfying: $\mathcal{F}(\Sigma)$ is connected and $R(G) = 1$. Let U be a subgroup of G such that $U = \langle U \cap D(\Sigma)\rangle$. What can be said about the possible structure of U and its embedding in G? For example one such question is: For a pre-described isomorphism type of U and G, can U be such a subgroup of G and if so, is U uniquely determined up to conjugacy in $\mathrm{Aut}(G)$? In such a situation we call the elements of $D(\Sigma)$ *root elements* and say that U is generated by root elements. One, the easier possibility, is when U is already a Σ-subgroup, i.e.:

(i) If $a \in U \cap D(\Sigma)$ and $A \in \Sigma$ with $a \in A$, then $A \leq U$. In that case $U = \langle U \cap \Sigma\rangle$ and we may apply the theory of abstract root subgroups directly to U and $\Sigma \cap U$.

But in general we have:

(ii) $A \neq A \cap U \neq 1$ for $A \in \Sigma$ with $a \in A$ for some $a \in U \cap D(\Sigma)$.

It is clear that case (ii) is too general to say much about it. For example G might be a Lie type group defined over some field k and U the same Lie type group defined over some subring of k. Hence in case (ii) it is important to find conditions which guarantee that

$$\Sigma_0 := \{A_0 \mid A \in \Sigma, A_0 = A \cap U \neq 1\}$$

is a set of abstract root subgroups of U. One such condition is, see (2.3), that U is a fixed point subgroup under some automorphism group Y of G.

This section is organized as follows: First we prove some results showing that Σ_0 is, under some additional conditions, indeed a set of abstract root subgroups of U. Secondly we discuss examples when U and G are both Lie type groups. Here we will concentrate on those examples which occur over arbitrary fields. Perhaps the exceptional examples over small fields, in particular over $GF(2)$, are more interesting since they are sometimes connected with sporadic finite simple groups. (For example Fischer has shown that his smallest group $\mathrm{Fi}(22)$ is contained in $^2E_6(2^2)$ as a subgroup generated by long root elements.) But the discussion of such exceptional embeddings always requires special and lengthy arguments.

Then we will discuss some of the classification results of subgroups generated by long root elements. Normally such results say that a certain known type of subgroup is uniquely determined up to conjugacy in $\mathrm{Aut}(G)$. Here we will again concentrate on results which hold over arbitrary fields. Finally we will

give a proof of a theorem of H. Cuypers and A. Steinbach [CS99], which gives
a fairly uniform description of subgroups generated by transvections.

(2.1)–(2.6) are generalizations of the corresponding theorems for subgroups
of groups generated by k-root subgroups of [Tim94b], to the more general
situation of abstract root subgroups. (2.7) says that if U is finite, Σ_0 is always
a set of k-root subgroups of U if $|A_0| > 2$ for $A_0 \in \Sigma_0$.

First an elementary lemma.

(2.1) Lemma

Let $X = \langle A, B \rangle$ be a rank one group with unipotent subgroups A and B and
σ a homomorphism of X. Then the following hold:

(1) Suppose $A^\sigma \neq B^\sigma$. Then X^σ is also a rank one group with unipotent
subgroups A^σ and B^σ.

(2) Suppose X is special, $\sigma \in \mathrm{Aut}(X)$ and

$$A_0 = \{a \in A \mid a^\sigma = a\} \neq 1 \neq B_0 = \{b \in B \mid b^\sigma = b\}.$$

Then $X_0 = \langle A_0, B_0 \rangle$ is a special rank one group with unipotent subgroups A_0 and B_0. Moreover, if $A^\sigma \leq A$, then

$$X_0 \trianglelefteq X_\sigma = \{x \in X \mid x^\sigma = x\}.$$

(3) Suppose the hypothesis of (2) holds and $X \simeq (P)SL_2(k)$, k a field. Then
$X_0 \simeq (P)SL_2(L)$, L a subfield of k.

Proof. (1) is completely obvious, since for $U \leq X$ and $g \in X$ we have $(U^g)^\sigma = (U^\sigma)^{g^\sigma}$.

To prove (2) pick $a \in A_0^{\#}$ and claim $b = b(a) \in B_0$. Since X is special we have:

$$(*) \qquad a^{b^\sigma} = (a^\sigma)^{b^\sigma} = (a^b)^\sigma = (b^{-a})^\sigma = (b^\sigma)^{-a^\sigma} = (b^\sigma)^{-a}.$$

Hence for $\Omega = A^X$ we obtain $(b^\sigma)^{-1} = a^{b^\sigma a^{-1}} \in D(\Omega)$. On the other hand b^σ
centralizes $B_0 = B_0^\sigma$. Since $C_{D(\Omega)}(B_0) = B^{\#}$ by I(1.4) it follows that $b^\sigma \in B$.
But then the uniqueness of $b = b(a) \in B$ satisfying $(*)$, see I(1.2)(2), implies
$b^\sigma = b$ and $b \in B_0$. Now we obtain

$$B_0^{ab^{-1}} \leq B^{ab^{-1}} \cap X_\sigma = A \cap X_\sigma = A_0$$

and thus $B_0 \leq A_0^{ba^{-1}} = A^{ba^{-1}} \cap X_\sigma = B \cap X_\sigma = B_0$. Hence $A_0^b = B_0^a$ for each
$a \in A_0^{\#}$ and $b = b(a) \in B_0^{\#}$. Since with symmetry also $a(b) \in A_0$ for $b \in B_0^{\#}$,
this shows that X_0 is a special rank one group.

It remains to show $X_0 \trianglelefteq X_\sigma$ if $A^\sigma \leq A$. First notice that $A^\sigma \leq A$ implies $A^\sigma = A$. Indeed, if $A^\sigma < A$, then $A < A^{\sigma^{-1}}$ and $A_1 = A^{\sigma^{-1}} \cap N_X(B) \neq 1$. But since A_1 centralizes A it also centralizes B and thus $A_1 \leq Z(X)$. Hence

$$1 \neq A_1^\sigma \leq A \cap Z(X),$$

a contradiction. We obtain $A^\sigma = A$ and $\Omega^\sigma = \Omega$. Since $B_0 \neq 1$ also $B^\sigma = B$. Pick now $x \in X_\sigma$. Then either $A_0^x = A_0$ or $A_0^x \leq B^a$ for some $a \in A$. In the second case we have

$$A_0^x = (A_0^x)^\sigma \leq B^a \cap (B^a)^\sigma = B^a \cap B^{a^\sigma}.$$

Hence $B^a = B^{a^\sigma}$ since the elements of Ω are TI-subgroups and then $a = a^\sigma$ by I(1.2)(3). Thus $a \in A_0$ and

$$A_0^x \leq B^a \cap X_\sigma = B_0^a.$$

This shows $X_0^x \leq X_0$ for each $x \in X_\sigma$, which implies $X_0 \trianglelefteq X_\sigma$. This proves (2).

To prove (3) let $A = \{a(\lambda) \mid \lambda \in k\}$ and $B = \{b(\lambda) \mid \lambda \in k\}$ as in I(5.7). For $\lambda \in k$ define λ^σ by $a(\lambda^\sigma) = a(\lambda)^\sigma$ (clearly $A^\sigma = A$ since $A^\sigma \leq C_X(A_0)$ and $A_0 \neq 1$!) and set $L = \{\mu \in k \mid \mu^\sigma = \mu\}$. Then L is an additive subgroup of k and, since the group of diagonal automorphisms of X is transitive on $A^\#$, we may assume $1 \in L$. (We may label $A = \{a(\lambda)\}$, $B = \{b(\lambda)\}$ such that the Steinberg relations of I(5.1) (A) and (B) hold, starting with any element of $A^\#$ as a(1)!)

Then $b(1) = b(a(1)) \in B_0$ and $w = a(1)b(1)a(1) \in X_0$. Since $B_0 = A_0^w$ we obtain $B_0 = \{b(\mu) \mid \mu \in L\}$. On the other hand $b(a(\mu)) = b(\mu^{-1})$. (See I(1.5)!). Hence $L^\#$ is closed under inverses. Since $h(\mu) \in X_0$ for $\mu \in L$ now the argument of I(5.7) shows that L is a subfield of k. Hence $X_0 \simeq (P)SL_2(L)$. (In fact we have shown that it is conjugate by a diagonal automorphism to the natural $(P)SL_2(L)$ inside $(P)SL_2(k)$!)　　　　　$\square$

(2.2)　Lemma

Suppose the group G is generated by the set Σ of abstract root subgroups satisfying:

(a) $\mathcal{F}(\Sigma)$ is connected.

(b) $R(G) = 1$.

(c) Σ contains different commuting elements.

Let $Y \leq \mathrm{Aut}(G)$, $A \in \Sigma$ and $B \in \Omega_A$ satisfying:

$$A_0 = C_A(Y) \neq 1 \neq C_B(Y) = B_0.$$

Then the following hold:

(1) $X_0 = \langle A_0, B_0 \rangle$ is a special rank one group with unipotent subgroups A_0 and B_0.

(2) If Σ is a set of k-root groups of G, then $X_0 \simeq (P)SL_2(L)$, L a subfield of k.

Proof. (1) Pick $a \in A_0^{\#}$ and claim: $(*)$ $\quad b = b(a) \in B_0^{\#}$. To prove $(*)$ notice first, that by hypothesis and II(3.16) $X = \langle A, B \rangle$ is special. Hence $a^b = b^{-a}$. As in (2.1)$(*)$ this implies $a^{b^\sigma} = (b^\sigma)^{-a}$ for all $\sigma \in Y$. In particular

$$b^\sigma = (a^{b^\sigma a^{-1}})^{-1} \in D(\Sigma) \quad \text{for } \sigma \in Y.$$

Let $\overline{B} \in \Sigma$ with $b^\sigma \in \overline{B}$ for some $\sigma \in Y$ with $b^\sigma \neq b$ and let $T = a^G$. Then $T^\sigma = T$ and we obtain:

$$\begin{aligned}
C_T(B) &= C_T(B_0) = C_{T^\sigma}(B_0^\sigma) = C_T(B_0)^\sigma = C_T(b)^\sigma = C_T(b^\sigma) \\
&= C_T(\overline{B}),
\end{aligned}$$

since $C_\Sigma(B) = C_\Sigma(b)$. As $T \cap D \neq \emptyset$ for each $D \in \Sigma$ this implies $C_\Sigma(B) = C_\Sigma(\overline{B})$. But then $B = \overline{B}$ as $R(G) = 1$ by definition of $R(G)$ and II(4.12), II(4.10). Now the uniqueness of $b \in B$ with $a^b = b^{-a}$, see I(1.2), implies $b = b^\sigma$, a contradiction to the choice of σ.

This shows $b = b^\sigma$ for all $\sigma \in Y$ and $b \in B_0$ as claimed. With symmetry we obtain $B_0^{\#} = \{b(a) \mid a \in A_0^{\#}\}$ and vice versa. Now it follows as in the proof of (2.1)(2) that (1) holds. The proof of (2) is exactly the same as the proof of (2.1)(3). (One sets $A = \{a(\lambda) \mid \lambda \in k\}$, $A_0 = \{a(\lambda) \mid \lambda \in L\}$ with $L \subseteq k$. Then, assuming without loss pf generality that $1 \in L$, it follows that L is a field.) $\square$

Example. In this example we want to show that the condition $R(G) = 1$ in (2.2) is necessary.

Let $X = SL_2(4)$, N the natural $\mathbb{Z}_2X$-module and $Y = N \cdot X$. Then Y is generated by a class Σ of $GF(4)$ transvection groups. Let $\overline{N}$ be an indecomposable $\mathbb{Z}_2X$-module with $|\overline{N}| = 4^3$, containing N as a submodule. (See exercise IV(1.21)(3) and (4).) Let $\overline{Y} = \overline{N}X$. Then $Y \vartriangleleft \overline{Y}$, whence the elements of $\overline{N} - N$ induce automorphisms on Y. From the action of X on $\overline{N}$ it follows that there exists a $\sigma \in \overline{N} - N$ with:

$$C_X(\sigma) = \langle a, b \rangle \simeq D_{10}; \quad a, b \in D(\Sigma).$$

Thus, if $a \in A$, $b \in B$ and $A, B \in \Sigma$, then $b(a) \notin C_Y(\sigma)$. In particular $\langle C_A(\sigma), C_B(\sigma) \rangle = \langle a, b \rangle$ is not a rank one group.

(2.3)   Theorem

Suppose G is a group generated by the set Σ of abstract root subgroups satisfying:

(a) $\mathcal{F}(\Sigma)$ is connected.

(b) $R(G) = 1$.

(c) Σ contains different commuting elements.

Let $Y \leq \mathrm{Aut}(G)$ with $\Sigma^{\sigma} \subseteq \Sigma$ for $\sigma \in Y$ and let:

$$\Sigma_0 = \{A_0 \mid A \in \Sigma \text{ and } A_0 = C_A(Y) \neq 1\}.$$

Then Σ_0 is a set of abstract root subgroups of $U = \langle \Sigma_0 \rangle$. Moreover, if Σ is a set of k root subgroups and $\mathcal{F}(\Sigma_0)$ is connected, then Σ_0 is a class of L-root subgroups of U, L a subfield of k.

Proof. Let $A_0, B_0 \in \Sigma_0$ with $B \in \Omega_A$. Then by (2.2) $\langle A_0, B_0 \rangle$ is a rank one group with unipotent subgroups A_0 and B_0. Thus, to prove that Σ_0 is a set of abstract root subgroups of U, it suffices to show:

(∗)   Let $A_0, B_0 \in \Sigma_0$ with $B \in \Psi_A$ and let $C = [A, B]$.

Then

$$C_0 = C_C(Y) = [A_0, B_0] = [a, B_0] = [A_0, b]$$

for each $a \in A_0^{\#}, b \in B_0^{\#}$.

Fix $a \in A_0^{\#}$. Then, by definition of abstract root subgroups, the map: $\chi_a :$ $b \to [a, b]$ for $b \in B$ is an isomorphism from B onto C. Thus it suffices to show $\chi_a(B_0) = C_0 = C_C(Y)$.

Suppose this is not the case. Then $\chi_a(B_0) < C_0$ and there exists a $b \in B - B_0$ with $[a, b] = \chi_a(b) \in C_0$. Hence for each $\sigma \in Y$ we have:

$$[a, b] = [a, b]^{\sigma} = [a, b^{\sigma}].$$

Now by hypothesis $\Sigma^{\sigma} \subseteq \Sigma$, we have $B^{\sigma} \in \Sigma$ and thus $B = B^{\sigma}$, since $B_0 \neq 1$ and B is a TI-subgroup of G. (See II(2.7)). Hence $b^{\sigma} \in B$ and thus $b = b^{\sigma}$, since χ_a is an isomorphism from B onto C, a contradiction to $b \notin B_0$.

This shows that Σ_0 is a set of abstract root subgroups of U. Suppose that Σ is a set of k-root subgroup and $\mathcal{F}(\Sigma_0)$ is connected. Then, to show that Σ_0 is a class of L-root subgroups of U, it suffices by (2.2)(2) to show:

(+)   Let $A_0 \in \Sigma_0$ and $B_0, C_0 \in \Omega_{A_0}$. Then
         $\langle A_0, B_0 \rangle \simeq (P)SL_2(L) \simeq \langle A_0, C_0 \rangle$, L a subfield of k.

Indeed, the connectedness of $\mathcal{F}(\Sigma_0)$ together with $(+)$ then imply that we always get the same $(P)SL_2(L)$ for all $E_0 \in \Sigma_0$ and $F_0 \in \Omega_{E_0}$.

Let now by $(2.2)(2)$ L_i, $i = 1, 2$ be subfields of k with $\langle A_0, B_0 \rangle \simeq (P)SL_2(L_1)$ and $\langle A_0, C_0 \rangle \simeq (P)SL_2(L_2)$. Then, assuming with loss that $a(1) \in A_0$ in the notation of the proof of (2.2), we obtain as shown in (2.2):

$$L_1 = \{c \in k \mid a(c) \in A_0\} = L_2. \qquad \square$$

In the next theorem we study subgroups generated by root elements, which are not necessarily fixed point subgroups of some automorphism group. As the examples of subgroups defined over subrings show, we need an additional condition, to show that such a subgroup is generated again by abstract root subgroups. This condition is a kind of closure under the operation

$$a \to b(a).$$

(2.4) Notation

Let G be a group generated by the set Σ of abstract root subgroups such that $\mathcal{F}(\Sigma)$ is connected and $R(G) = 1$ and let U a subgroup of G satisfying $U = \langle U \cap D(\Sigma) \rangle$. Let $\Sigma_0 := \{A_0 \mid A \in \Sigma \text{ and } A_0 = A \cap U \neq 1\}$. Then we say the pair (U, Σ_0) satisfies hypothesis (S) (for subgroup) with respect to (G, Σ) if the following holds:

(S) If $A_0, B_0 \in \Sigma_0$ with $B \in \Omega_A$ and $a \in A_0^{\#}$, then $b(a) \in B_0$.

Suppose now that the pair (U, Σ_0) satisfies (S) and let $A_0 \in \Sigma_0$. Then we set:

$$
\begin{aligned}
\Omega_{A_0} &:= \{B_0 \in \Sigma_0 \mid B \in \Omega_A\}, \\
\Psi_{A_0} &:= \{B_0 \in \Sigma_0 \mid B \in \Psi_A\}, \\
\Lambda_{A_0} &:= \{B_0 \in \Sigma_0 \mid B \in \Lambda_A\}.
\end{aligned}
$$

Then for $B_0 \in \Sigma_0$ it is possible to decide from the structure of $\langle A_0, B_0 \rangle$ alone, if B_0 belongs to one of these sets. Indeed, $B_0 \in \Omega_{A_0}$ if and only if $\langle A_0, B_0 \rangle$ is not nilpotent. In fact, since we require (S) symmetrically in A_0 and B_0 we have for $a \in A_0^{\#}$ and $b = b(a) \in B_0^{\#}$:

$$A_0^b = A^b \cap U = B^a \cap U = B_0^a,$$

which shows that:

(1) $\langle A_0, B_0 \rangle$ is a rank one group if $\langle A, B \rangle$ is a rank one group.

In particular $\langle A_0, B_0 \rangle$ is by $I(2.1)$ not nilpotent. Next, $B_0 \in \Psi_{A_0}$ if and only if $\langle A_0, B_0 \rangle$ is nilpotent, but non-abelian. Finally $B_0 \in \Lambda_{A_0}$ if and only if $[A_0, B_0] = 1$ and

(2) $D(\Sigma_0) \cap A_0 B_0 \not\subseteq A_0^{\#} \cup B_0^{\#}$.

Indeed, if $B \in \Lambda_A$ then $(AB)^{\#} \subset D(\Sigma)$. Hence for $c \in (A_0 B_0)^{\#}$ with $c \notin A_0$ and $c \notin B_0$, there exists a $C \in \Sigma$ with $c \in C_0^{\#}$, whence $c \in D(\Sigma_0)$.

On the other hand, if there exists a $c \in D(\Sigma_0) \cap A_0 B_0$ with $c \notin A_0$ and $c \notin B_0$, then $c \in D(\Sigma) \cap AB$ and $c \notin A$, $c \notin B$. Let $c \in C \in \Sigma$. Then, as $c = ab$ for some $a \in A^{\#}$, $b \in B^{\#}$, B is by II(2.2) an isolated vertex of $\mathcal{F}(C_\Sigma(A))$. Hence by III(2.6) $B \in \Lambda_A$ and thus $B_0 \in \Lambda_{A_0}$.

We now define the graphs $\mathcal{E}(\Sigma_0)$ $\mathcal{D}(\Sigma_0)$ and $\mathcal{F}(\Sigma_0)$ in the natural way. Then the above shows that the definition of these graphs depends only on (U, Σ_0).

(2.5) Lemma

Suppose (U, Σ_0) satisfies hypothesis (S) with respect to (G, Σ) and that $\mathcal{F}(\Sigma_0)$ has no isolated vertices. Let $A_0 \in \Sigma_0$ and $C_0 \in \Lambda_{A_0}$. Then there exist $E_0, F_0 \in \Sigma_0$ such that $F_0 \in \Omega_{E_0}$ and $X_0 = \langle E_0, F_0 \rangle$ induces an $SL_2(L)$, L a division ring or a Cayley division algebra on $A_0 C_0$. Moreover $A_0 C_0$ is the natural module for X_0. (i.e. $\Sigma_0 \cap A_0 C_0$ is an X_0-invariant partition of $A_0 C_0$!)

Proof. First notice that the hypothesis $C_0 \in \Lambda_{A_0}$ implies that A_0 and C_0 are in the same connectivity component of $\mathcal{F}(\Sigma_0)$. Indeed, let $d \in (A_0 C_0)^{\#}$ with $d \notin A_0$ and $d \notin C_0$ and let $D_0 \in \Sigma_0$ with $d \in D_0$. (The elements of Σ_0 are TI-subgroups of U, so such a D_0 is uniquely determined!) Then, by the proof of II(2.12) D_0 is connected to A_0 and C_0 in $\mathcal{F}(\Sigma_0)$.

So we may for the rest of the proof assume that $\mathcal{F}(\Sigma_0)$ is connected. We first show:

(1) There exists (without loss of generality) a $B_0 \in \Omega_{A_0} \cap \Psi_{C_0}$.

Suppose this is not the case. Then $\Omega_{A_0} \subseteq \Omega_{C_0}$ by II(2.2). Pick $B_0 \in \Omega_{A_0}$. Then, since by (2.4)(1) $\langle A_0, B_0 \rangle$ and $\langle C_0, B_0 \rangle$ are rank one groups, we have by I(1.2):

$$A_0^{\#} = \{a(b) \mid b \in B_0^{\#}\}, \ C_0^{\#} = \{c(b) \mid b \in B_0^{\#}\}.$$

Let $d = a(b)^{-1} c(b)$ for some $b \in B_0^{\#}$. Then $d \in D(\Sigma_0)$ since $C_0 \in \Lambda_{A_0}$. Hence, if $D_0 \in \Sigma_0$ with $d \in D_0$, II(2.9) implies $D_0 \in \Psi_{B_0}$. On the other hand by II(2.11)

$$D_0 \supseteq \{a(b)^{-1} c(b) \mid b \in B_0^{\#}\},$$

whence $C_0 \leq A_0 D_0$. Now by (2.4)(2) $D_0 \in \Lambda_{A_0}$, so that the pair (A_0, D_0) satisfies the hypothesis of (2.5) with $B_0 \in \Omega_{A_0} \cap \Psi_{D_0}$. Hence, if (2.5) holds for $A_0 D_0$, then $A_0 D_0 = A_0 C_0$ and X_0 acts doubly transitively on $\Sigma_0 \cap A_0 D_0$. In particular $\Omega_{A_0} \cap \Psi_{C_0} \neq \emptyset$. Hence, replacing C_0 by D_0 we may assume that (1) holds.

Let $E_0 \in \Sigma_0$ with $[B_0, C_0] \leq E_0$. Then we show:

(2) $[E_0, A_0] = [E_0, a] = C_0$ for all $a \in A_0^{\#}$,
$[C_0, B_0] = [C_0, b] = E_0$ for all $b \in B_0^{\#}$.

Since $Y_0 = \langle A_0, B_0 \rangle$ is a rank one group, there exists a $w \in Y_0$ interchanging A_0 and B_0 and thus also A and B. Let $N = CE$. Then by II(2.19) N is a natural module for $Y = \langle A, B \rangle$. In particular $C = C_N(A)$, $E = C_N(B)$. Thus w interchanges C and E and whence also C_0 and E_0. Suppose now $[C_0, b] < E_0$ for $b \in B_0^{\#}$. Then $C_0[C_0, b] < C_0 E_0$ and $C_0[C_0, b]$ is $Y_0 = \langle A_0, b \rangle$ invariant, since $[E_0, A_0] \leq C_0$. But then

$$E_0 = C_0^w \leq C_0[C_0, b],$$

a contradiction. With symmetry this proves (2).
Let $M_0 = A_0 C_0$. Then (2) implies

$$[M_0, E_0] = C_0 = [m, E_0] \text{ for all } m \in M_0 - C_0.$$

Now, exchanging the roles of A_0 and C_0 we also find an $F_0 \in \Sigma_0$ such that we obtain:

(3) $C_0 = [M_0, E_0] = [m, E_0]$ for all $m \in M_0 - C_0$,
$A_0 = [M_0, F_0] = [m, F_0]$ for all $m \in M_0 - A_0$.

We need to show:

(4) $C_0 = [M_0, e] = [A_0, e]$ for all $e \in E_0^{\#}$,
$A_0 = [M_0, f] = [C_0, f]$ for all $f \in F_0^{\#}$.

Indeed, if (4) holds, then by I(3.2) and (3) $X_0/C_{X_0}(M_0) \simeq SL_2(L)$, L a division ring or a Cayley division algebra and M_0 is the natural $SL_2(L)$-module. In particular (2.5) holds.

So (4) remains to be proved. Now by II(2.23) $X = \langle E, F \rangle$ induces the $SL_2(K)$, K a division ring or a Cayley division algebra on $M = AC$. In particular X is a rank one group and thus by (2.4)(1) X_0 is a rank one group. Pick $w_0 \in X_0$ interchanging E_0 and F_0. Then as before w_0 interchanges E and F, A and C and thus also A_0 and C_0. If now $[A_0, e] < C_0$ for some $e \in E_0^{\#}$, then as $A_0[A_0, e]$ is $X_0 = \langle F_0, e \rangle$ invariant, we obtain a contradiction as before. Thus $[A_0, e] = C_0$ and $[C_0, f] = A_0$ for each $e \in E_0^{\#}$, $f \in F_0^{\#}$, which proves (4) and (2.4).   □

Now, from (2.5) we obtain:

(2.6)   Theorem

Suppose (U, Σ_0) satisfies (S) with respect to (G, Σ) and $\mathcal{F}(\Sigma_0)$ has no isolated vertices. Then the following holds:

(1) Σ_0 is a set of abstract root subgroups of U.
(2) Let $\Sigma_0 = \bigcup_{i \in I} \Sigma_i$, with Σ_i the connectivity components of $\mathcal{F}(\Sigma_0)$, and $U_i = \langle \Sigma_i \rangle, i \in I$. Then Σ_i is a class of abstract root subgroups of U_i and $[U_i, U_j] = 1$ for $i \neq j, i, j \in I$.
(3) Suppose Σ is a set of k-root subgroups of G, Char $k \neq 2$ and $\mathcal{F}(\Sigma_0)$ is connected. Then Σ_0 is a class of L-root subgroups of U, L a subfield of k.

Proof. (1) In view of (2.4)(1) it suffices to show:

$$(*) \qquad \begin{array}{l} \text{If } A_0 \in \Sigma_0,\ B_0 \in \Psi_{A_0} \text{ and } [A_0, B_0] \leq C_0 \in \Sigma_0, \\ \text{then } [a, B_0] = [A_0, b] = C_0 \text{ for each } a \in A_0^{\#}, b \in B_0^{\#}. \end{array}$$

Now by (2.4)(2) $C_0 \in \Lambda_{A_0}$. Hence by (2.4) there exist $E_0, F_0 \in \Sigma_0$ such that $X_0 = \langle E_0, F_0 \rangle$ induces an $SL_2(L)$ on $N_0 = A_0 C_0$ and N_0 is the natural module for $X_0 / C_{X_0}(N_0)$. In particular by the description of the natural $SL_2(L)$-module in I(1.8) we may assume $[N_0, E_0] = A_0$ and $[N_0, F_0] = C_0$.

Now by II(2.23) $N = AC$ is the natural module $X / C_X(N)$, where $X = \langle E, F \rangle$. Moreover, since $[N, B] = C$, we have $BC_G(N) = FC_G(N)$. This implies that $\langle E, B \rangle$ and whence $\langle E_0, B_0 \rangle$ are rank one groups, since $\langle E, B \rangle$ is not nilpotent. Now by I(1.2)

$$F_0^{\#} = \{f(e) \mid e \in E_0\} \text{ and } B_0^{\#} = \{b(e) \mid e \in E_0^{\#}\}.$$

Thus $F_0 C_G(N) = B_0 C_G(N)$ which with (2.5) implies $[A_0, b] = C_0$ for each $b \in B_0^{\#}$.

Since also $C_0 \in \Lambda_{B_0}$ this shows, exchanging the roles of A_0 and B_0, that $(*)$ and whence (1) holds. Now (2) follows from II(2.21) applied to Σ_0. (The set Ω of isolated vertices of $\mathcal{F}(\Sigma_0)$ is empty.)

To prove (3) pick $A_0 \in \Sigma_0$ and $B_0 \in \Omega_{A_0}$, then as $\langle A, B \rangle \simeq (P)SL_2(k)$, exercise I(1.13)(9) implies $\langle A_0, B_0 \rangle \simeq (P)SL_2(L)$, L a subfield of k. Thus it remains to show that we always get the same subfield L of k. But this follows from the connectivity of $\mathcal{F}(\Sigma_0)$ as in the proof of (2.3).   □

Example. In this example we show that the condition "$\mathcal{F}(\Sigma_0)$ has no isolated vertices" in (2.6) is necessary. Let $G = SL_3(k)$ and Σ the class of transvection subgroups of G. Then, if L is a subfield of k, there exists a subgroup $Y = N \cdot X$, with $N \simeq k^2$, $X \simeq SL_2(L)$ and $N = \langle \Sigma \cap N \rangle$, $X = \langle D(\Sigma) \cap X \rangle$. Thus, if $1 \neq A_0 = A \cap X$, $A \in \Sigma$, then $A_0 \simeq L$. Let $B \in \Sigma \cap N$ with $B \in \Psi_A$. Then, for $b \in B^\#$ we have

$$B_0 = [A_0, b] \simeq L \text{ and thus } B_0 \notin \Sigma \cap N.$$

In particular this shows that $\Sigma_0 = \{C_0 \mid C_0 = C \cap Y \neq 1, C \in \Sigma\}$ is not a set of abstract root subgroups of Y, nontheless $Y = \langle Y \cap D(\Sigma) \rangle$.

In the next proposition we show that in case U is finite and $|U \cap A| > 2$ for some $A \in \Sigma$, the condition (S) is superfluous. We have:

(2.7) Proposition

Suppose G is a group generated by the class Σ of k-root subgroups, such that $\mathcal{F}(\Sigma)$ is connected and $R(G) = 1$. Let U be a finite subgroup of G satisfying:

(1) $U = \langle U \cap D(\Sigma) \rangle$.
(2) Let $\Sigma_0 = \{A_0 \mid A \in \Sigma, A_0 = A \cap U \neq 1\}$. Then $\mathcal{F}(\Sigma_0)$ is connected. $((A_0, B_0)$ is an edge of $\mathcal{F}(\Sigma_0)$ if and only if $\langle A_0, B_0 \rangle$ is not nilpotent.)
(3) $|A_0| > 2$ for some $A_0 \in \Sigma_0$.
(4) Either $|A_0| > 3$ or there exist different commuting elements in Σ_0.

Then Σ_0 is a class of k_0-root subgroups of U, k_0 a finite subfield of k.

Proof. Let $A_0 \in \Sigma_0$ with $|A_0| > 2$ and $B_0 \in \Sigma_0$ such that (A_0, B_0) is an edge of $\mathcal{F}(\Sigma_0)$. We show (∗) $X_0 = \langle A_0, B_0 \rangle \simeq (P)SL_2(k_0)$, k_0 a subfield of k and A_0, B_0 are full unipotent subgroups of X_0.

In particular (∗) implies that A_0 and B_0 are conjugate in X_0 and whence $|B_0| > 2$. Now the connectedness of $\mathcal{F}(\Sigma)$ shows that (∗) holds for all $\langle E_0, F_0 \rangle$, where $E_0, F_0 \in \Sigma_0$ with $\langle E_0, F_0 \rangle$ not nilpotent. Moreover, as in the proof of (2.3), it follows that we always get the same field. This shows that condition (S) holds for Σ_0. Now (2.7) is a consequence of (2.6).

Thus (∗) remains to be shown. Let $X = \langle A, B \rangle$ and $A = \{a(\lambda) \mid \lambda \in k\}$, $B = \{b(\lambda) \mid \lambda \in k\}$ and assume without loss of generality that $a(1) \in A_0$. Let $\mathcal{W} = \{\lambda \in k \mid b(\lambda) \in B_0\}$. Then, since $o(a(1)b(\lambda)) < \infty$ for $\lambda \in \mathcal{W}$, λ satisfies a polynomal equation over $\mathbb{Z}_p$, $p = \text{Char } k$. Hence $|k(\lambda)| < \infty$ and thus $\mathcal{W} \subseteq k_1$, k_1 a finite subfield of k. Applying the same argument to $\{\mu \mid a(\mu) \in A_0\}$ we see that $X_0 \leq X_1 = \langle A_1, B_1 \rangle \simeq (P)SL_2(k_1)$, k_1 a finite subfield of k and $A_1 = A \cap X_1$, $B_1 = B \cap X_1$.

Suppose next $|A_0| > 3$. Then Dickson's theorem, see [Hup67, (8.27)], implies $X_0 \simeq (P)SL_2(k_0)$, k_0 a subfield of k_1 with $|k_0| \geq |A_0| > 3$. Now, since $X_0 \leq U$

and since a unipotent subgroup of X_0 containing A_0 lies in A, we obtain
$A_0 = A \cap X_0$ and $B_0 = B \cap X_0$. Hence $(*)$ holds.

So we may assume $|A_0| = 3$ and $(*)$ does not hold for X_0. Then Char $k = 3$
and $|D_0| = 3$ for each $D_0 \in \Sigma_0$. Moreover, if $k \simeq \mathbb{Z}_3$, then clearly $(*)$ holds.
Hence we may assume $|k| > 3$ and thus condition (H) of II§3 holds.

Now by condition (4) we may assume that there exists a $C_0 \in \Sigma_0$ with
$[A_0, C_0] = 1 \neq [B_0, C_0]$. Hence for $C \in \Sigma$ we obtain

$$(\alpha) \quad \overset{A}{\circ}\!\!\!\!\!\!\!\!\underline{}\!\!\!\overset{B}{\circ}\!\!\!=\!\!\!=\!\!\!\overset{C}{\circ} \qquad \text{or} \qquad (\beta) \quad \overset{A}{\circ}\!\!\!\!\!\!\!\!\underline{}\!\!\!\overset{B}{\circ}\!\!\!\!\!\!\!\!\underline{}\!\!\!\overset{C}{\circ} \; .$$

Now, in case (α), replacing C_0 by $[C_0, B_0, A_0]$, we may assume $C \in \Lambda_A$. Hence
$N = C[C, B]$ is by II(2.19) the natural $\mathbb{Z}X$-module. Now, since we assume
$(*)$ does not hold for X_0, Dickson's theorem implies $X_0 \simeq (P)SL_2(5)$. Hence
$[C_0, B_0, A_0] \not\leq C_0$, since $SL_2(5)$ can not act non-trivially on $\mathbb{Z}_3 \times \mathbb{Z}_3$. But, as N
is the natural X-module, we have $[C_0, B_0, A_0] \leq C \cap U$ and thus $|C \cap U| > 3$,
a contradiction.

So we are in case (β). If $C \in \Lambda_A$, then by II(2.11) the structure of $Y = \langle A, B, C \rangle$
is the same as in case (α). Let again $N \triangleleft Y$ with $Y = NX$, $N \cap X = 1$. If
$Y_0 \cap N = 1$, $Y_0 = \langle A_0, B_0, C_0 \rangle$, then Y_0 is conjugate to a subgroup of X, since
as Char $k \neq 2$ all complements to N in Y are conjugate. But this is obviously
impossible. Hence $N \cap Y_0 \neq 1$ and, since $N^{\#} \subset D(\Sigma)$, we may assume that we
are in case (α), a contradiction.

Thus $C \notin \Lambda_A$. In particular $\Omega_C \cap \Sigma_A \neq \emptyset$ by III(2.6). Hence, arguing as in
II(3.15) we see that $Z(Y) \neq 1$. Thus by II(3.17)(2) and (8) $Y \simeq k^{1+2}SL_2(k)$
and $Z(Y) \leq N'$, $N = O_3(Y)$. Now, since $X_0 \simeq SL_2(5)$, it is easy to see that
$Z_0 = Z(Y) \cap U$ has order $\geq 3^2$.

If now Σ is degenerate, then III(1.14)(10) implies $Z(Y) \leq E \in \Sigma$. Hence
$Z_0 \leq E_0 = E \cap U$, a contradiction to $|Z_0| \geq 9$. Thus we may assume Σ is non-
degenerate. Since $A^N \cap \Lambda_C = \emptyset$, the argument of the proof of II(3.15)(2) shows
$A^N \subseteq M_A C$. But by III(2.16) $C_{M_A C}(B) = D$ for $D \in C^{M_A}$. (Hypothesis (M)
of III(2.13) is satisfied since $\Sigma_A \not\subseteq M_A$, as $C \in \Sigma_A - \Lambda_A$!). Thus we obtain:

$$Z(Y) \leq \langle A^N \rangle \cap C(B) \leq C_{M_A C}(B) = D.$$

In particular $Z_0 \leq D_0 = D \cap U$, a contradiction to $|Z_0| \geq 9$. $\qquad\qquad\square$

We just sketched the proof in the final case $|A_0| = 3$ and $C \notin \Lambda_A$, since here
one needs many properties from the classification theory of groups generated
by abstract root subgroups. Without the condition $|A_0| > 2$ (2.7) is false. A
counterexample is $J_2 < G_2(4)$, see IV(1.2)(c). A counterexample that (2.7)

is false if condition (4) is not satisfied, is $SL_2(5) < SL_2(9)$ generated by 3-elements.

But if $|A_0| = 2$ for each $A \in \Sigma_0$ we have the following relatively obvious result:

(2.8)   Corollary

Assume the same hypothesis as in (2.7)(1) and (2), but $|A_0| = 2$ for each $A_0 \in \Sigma_0$. Then $D = D(\Sigma) \cap U$ is a class of root involutions of U.

Proof. This is fairly obvious. Indeed, if (A_0, B_0) is an edge of $\mathcal{F}(\Sigma_0)$ and $A_0 = \langle a \rangle$, $B_0 = \langle b \rangle$, then $\langle a, b \rangle$ is dihedral of finite order. Since $\langle a, b \rangle \leq \langle A, B \rangle \simeq (P)SL_2(k)$, k a field of characteristic two, it follows that $\langle a, b \rangle \simeq D_{2n}$, n odd. In particular the connectivity of $\mathcal{F}(\Sigma_0)$ implies that D is a conjugacy class.

If now $A_0, B_0 \in \Sigma_0$ such that $\langle A_0, B_0 \rangle$ is nilpotent but non-abelian and again $a \in A_0^{\#}$, $b \in B_0^{\#}$, then

$$(ab)^2 = [a, b] \in [A, B] \cap U \in \Sigma_0,$$

whence $(ab)^2 \in D$. □

(2.7) and (2.8) together show that we may apply in any case the classification results of this book to U, namely either the results of III or IV.

(2.9)   Examples

In (2.9) we will describe some examples of subgroups of a Chevalley group G, generated by elements of $D(\Sigma)$, where Σ is the class of long root subgroups. (See II§5). Most of these examples are well known and hence will be given without discussion. But we will discuss at some length the embedding

$$^3D_4(L) \subseteq F_4(K),$$

L a Galois extension of degree 3 of K as a subgroup generated by elements in $D(\Sigma)$. The same method also leads to an embedding $^2E_6(L) \subseteq E_7(K)$, L a Galois extension of degree 2. These embeddings were first constructed in the finite case in [Ste78] and then in the arbitrary field case in [Stb00].

Let for the rest of (2.9) $G = \langle x_r(t) \mid r \in \Phi, t \in K \rangle$ be a *Chevalley group* defined over the field K with irreducible root system Φ. Let h be the highest root of Φ^+, $A = \{x_h(t) \mid t \in K\}$ and $\Sigma = A^G$. Then, from the description of the root systems in [Car72, (3.6)], h is a long root and by II(5.20) Σ is a class of K-root subgroups of G (or possibly G', see II(5.22)!). We call Σ the class of *long root subgroups of G* and $D(\Sigma)$ the set of *long root elements of G*. (In case of a symplectic group the elements of $D(\Sigma)$ might not be conjugate.)

(1) *Subsystem groups*
Let Ψ be a root subsystem of Φ. Then $U = \langle x_r(t) \mid r \in \Psi, t \in K \rangle$ is a subgroup of U which is, if Ψ contains long roots, generated by $U \cap \Sigma$, since U is by II(5.22) (essentially) simple. Examples of such subgroups are:

$$D_\ell \subset B_\ell, \; D_8 \subset E_8, \; A_8 \subset E_8, \; A_7 \subset E_7, \; B_4 \subset F_4, \; A_2 \subset G_2.$$

(2) *Chevalley groups over subfields*
Let L be a subfield of K. Then $U = \langle x_r(t) \mid t \in L, r \in \Phi \rangle$ is a Chevalley-group of the same type over L. By (2.6) it is obvious that $\Sigma_0 = \{B_0 \mid B \in \Sigma, B_0 = B \cap U \neq 1\}$ is a class of L-root subgroups of U.

A special case of this example is, when $Y \leq \text{Aut}(K)$ and $L = \{c \in K \mid c^\sigma = c$ for all $\sigma \in Y\}$. In this case Y can be considered as a subgroup of $\text{Aut}(G)$ by:

$$\sigma : x_r(t) \rightarrow x_r(t^\sigma), \text{ for } \sigma \in Y, t \in K.$$

Hence $\Sigma_0 = \{B_0 \mid B \in \Sigma$ and $B_0 = C_B(Y) \neq 1\}$ and $U = \langle \Sigma_0 \rangle \trianglelefteq C_G(Y)$ as in (2.3). (It might be possible that $U \neq C_G(Y)$, i.e. $C_G(Y)$ is an extension of U by diagonal elements.)

(3) *Fixed point groups of diagram automorphisms.*
A symmetry σ of the Dynkin diagram of the root system Φ of G can be extended first to an isometry of the euclidian space spanned by Φ and then to an automorphism of G by

$$\sigma : x_r(t) \rightarrow x_{r^\sigma}(t), r \in \Phi, t \in K.$$

The subgroup $U = \langle A \in \Sigma \mid A \leq C_G(\sigma) \rangle$ is a normal subgroup of $C_G(\sigma)$ generated by the class $\Sigma \cap U$ of abstract root subgroups. Examples of such pairs (U, G) are:

$$C_m \subset A_{2m-1}, B_{m-1} \subseteq D_m, G_2 \subseteq D_4 \text{ and } F_4 \subset E_6.$$

Also the containment $G_2 \subset B_3$ can be obtained in this way. Indeed, let α, β be isometries of D_4:

Then α and β induce automorphisms of order 2 on the group $G = D_4(K)$ with fixed point groups U isomorphic to $B_3(K)$ as described above. $\alpha\beta$ is an automorphism of order 3 with fixed point group isomorphic to $G_2(K)$ and, from the action of α and β on the root system Φ of type D_4 one sees that

$$G_2(K) \simeq \langle A \in \Sigma \mid A \leq C_G(\alpha\beta) \rangle \leq \langle A \in \Sigma \mid A \leq C_G(\alpha) \rangle \simeq B_3(K).$$

(4) *Twisted groups*
Suppose the Dynkin diagram of Φ admits a symmetry α of order $e = 2$ or 3 and suppose K is a Galois extension of L of degree e. Let $\beta \in \text{Aut}(K)$ of order e with L the fixed field and extend α and β to graph (resp. field) automorphisms of order e of G as in (2) or (3). Let $\sigma = \alpha\beta$. Then $U = \langle A_0 \mid A \in \Sigma, A_0 = C_A(\sigma) \neq 1 \rangle$ is the corresponding "twisted" Chevalley-groups. It is generated by the class $\Sigma_0 = \{A_0 \mid A \in \Sigma, A_0 = C_A(\sigma) \neq 1\}$ of L-root subgroups. Examples of such pairs (U, G) are:

$$^2A_n(K) \subset A_n(K),\ ^2D_n(K) \subseteq D_n(K),\ ^2E_6(K) \subseteq E_6(K).$$

In these cases we have $|K : L| = 2$. Moreover

$$^3D_4(K) \subseteq D_4(K) \text{ in which case } |K : L| = 3.$$

(5) Here we describe at same length the embedding $^3D_4(K) \subset F_4(K)$, when K is a Galois extension of degree 3 of a field L, as a subgroup generated by long root subgroups. That is, if $G = F_4(K)$, Σ the class of long root subgroups of G, then $U = \langle U \cap \Sigma \rangle \simeq\ ^3D_4(K)$. With the same type of argument one also obtains the embedding $^2E_6(K) \subset E_7(K)$, when K is a Galois extension of degree 2 of L. These embeddings were originally discovered in the finite case by E. Stensholt [Ste78] and then constructed in the arbitrary field case by A. Steinbach [Stb97], from which we take the arguments.

Let $(e_1, \ldots, e_8)$ be an orthonomal basis of $\mathbb{R}^8$ and set:

$$
\begin{aligned}
\beta_2 &= e_1 + e_3 & \beta_3 &= e_2 - e_1 \\
\beta_4 &= e_3 - e_2 & \beta_5 &= e_4 - e_3 \\
\beta_6 &= e_5 - e_4 & &\text{and} \\
\beta_1 &= \tfrac{1}{2}(e_1 - e_2 - e_3 - e_4 - e_5 - e_6 - e_7 + e_8).
\end{aligned}
$$

Then $\Pi = \{\beta_1, \ldots, \beta_6\}$ forms a fundamental system for a root system Φ of type E_6. (Φ is as in [Car72, (3.6)(ix)]). Let $\beta^* = \beta_1 + 2\beta_2 + 2\beta_3 + 3\beta_4 + 2\beta_5 + \beta_6$

352   Chapter V. Applications

be the highest root of Φ. Then

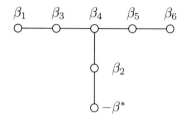

is the extended Dynkin diagram of Φ. The linear extension σ of the symmetry $(\beta_3\beta_5\beta_2)(\beta_1\beta_6 - \beta^*)$ of the Dynkin diagram of Φ is an isometry of the 6-dimensional euclidian space spanned by Φ with $\sigma(\Phi) = \Phi$. Choosing the structure constants $N_{r,s}$ of type E_6 as in [Spr98, (10.2)] one obtains $N_{r,s} = N_{r^\sigma,s^\sigma}$ for all $r, s \in \Phi$. Hence if K is a Galois extension of degree 3 of the field L and, if we denote by σ also the automorphism of order 3 with $\langle \sigma \rangle = \mathrm{Gal}(K : L)$, then we obtain:

(i) The map $n_\sigma : x_r(t) \to x_{r^\sigma}(t^\sigma), r \in \Phi, t \in K$ extends to an automorphism of order 3 of the universal Chevalley group G of type E_6 over K.

This is a direct consequence of the remarks above, since n_σ respects the Steinberg commutator relations defining G.

Let Σ be again the class of long root subgroups of G. Then one can show

(ii) There exists a symmetry of order two of Φ fixing $\beta_2, \beta_3, \beta_4$ and β_5, which extends to a graph automorphism τ of G such that:

(a) $U = \langle A \in \Sigma \mid A \leq C_G(\tau) \rangle \simeq F_4(K)$,

(b) τ commutes with n_σ.

(Indeed, each graph automorphism of order 2 of G has a fixed point group isomorphic to $F_4(K)$. The point is that one can conjugate τ under $W(\Phi)$ to a graph automorphism satisfying (b). This amounts to choosing the fundamental system of Φ, on which the symmetry is defined, appropriately.)

Clearly, by (ii)(b), n_σ fixes U. The next fact one can show is:

(iii) n_σ induces a field automorphism corresponding to σ with $\langle \sigma \rangle = \mathrm{Gal}(K : L)$ on U.

From (iii) it follows that:

$$U_0 = \langle A_0 \mid A \in \Sigma \cap U, A_0 = C_A(n_\sigma) \neq 1 \rangle \simeq F_4(L)$$

and
$$\Sigma_0 = \{A_0 \mid A \in \Sigma \cap U, A_0 = C_A(n_\sigma) \neq 1\}$$

is by (2.3) a class of L-root subgroups of U_0.

Let Ψ be the root subsystem of Φ spanned by the fundamental root β_2, β_3, β_4, β_5. Then Ψ is a root system of type D_4 and thus
$$R = \langle x_r(t) \mid r \in \Psi, t \in K \rangle \simeq D_4(K)$$

with $\Sigma \cap R$ a class of K-root subgroups of R. Now, by definition of σ, Ψ is σ-invariant. Hence n_σ induces a graph field automorphism on R and thus
$$R_0 = \langle A_0 \mid A \in \Sigma \cap R, A_0 = C_A(n_\sigma) \neq 1 \rangle \simeq {}^3D_4(K)$$

and $\Delta_0 = \{A_0 \mid A \in \Sigma \cap R, A_0 \neq 1\}$ is a class of L-root subgroups of R_0. Now by II(5.12)(3) R_0 is generated by the elements:
$$x_{\beta_4}(t), x_{-\beta_4}, t \in L$$

$$x_{\beta_2}(t) x_{\beta_3}(t^\sigma) x_{\beta_5}(t^{\sigma^2}), x_{-\beta_2}(t) x_{-\beta_3}(t^\sigma) x_{-\beta_5}(t^{\sigma^2}), t \in K.$$

Now by (ii) τ fixes all these elements. Hence we obtain $R_0 \leq U \cap C_G(n_\sigma)$ and $\Delta_0 \subseteq \Sigma_0$. Thus
$${}^3D_4(K) \simeq R_0 = \langle \Delta_0 \rangle \subseteq \langle \Sigma_0 \rangle = U_0 \simeq F_4(L)$$

as a subgroup generated by L-root subgroups.

Of course one can iterate the constructions (1)–(5) to obtain more subgroups of G generated by long root elements.

In the rest of this section we will describe some of the classifications of subgroups of Lie type groups generated by long root elements. Since the proofs of these results are lengthy and technical, the only result the proof of which we will discuss is a result of H. Cuypers and A. Steinbach [CS99], which gives a fairly uniform description of subgroups generated by transvections.

The first results in this area are two papers [McL67], [McL69] of Mc Laughlin, determining the irreducible subgroups of $GL_n(k)$ generated by full transvection groups. (I.e. root groups of $A_{n-1}(k)$.) This is split into two parts, first the case when $|k| > 2$ in which only canonical embeddings exist. Second, the case $|k| = 2$ in which exceptional cases exist like $\Sigma_n \subseteq GL_{n-1}(2)$, $O^\pm(2n, 2) \subseteq GL(2n, 2)$. His methods are matrix calculations.

The problem of determining nearly simple subgroups of finite Lie type groups generated by long root elements was settled by Kantor [Kan79] and Cooperstein [Coo79], [Coo81]. Here Kantor determined such subgroups of classical groups, while Cooperstein treated the case of finite exceptional Lie type groups. In particular in the case of exceptional groups the results are quite formidable, since there are many exceptional embeddings, in particular over $GF(2)$. Both, Kantor and Cooperstein, used in even characteristic the root involution classification and in odd characteristic an important theorem of Aschbacher [Asc77a], giving a uniform classification of finite Lie type groups in odd characteristic.

Liebeck and Seitz in [LS94] considered the problem from a different angle. In their paper they first determined simple, closed, connected subgroups X of a simple algebraic group G (over an algebraically closed field) satisfying:

(a) Either $X = \langle X \cap D(\Sigma) \rangle$ or

(b) There exists a $d \in D(\Sigma)$ inducing an outer automorphism on X.

(Here Σ is the class of long root subgroups of G.) Now the fact that X is also an algebraic group easily implies that for $A \in \Sigma$ with $A \cap X \neq 1$ either $A \leq X$ or $|A \cap X| = 2$, Char $G = 2$ and one obtains strong additional information. (See (2.2) and (3.3) of [LS94].) Now from the results for algebraic G they obtain partial results for finite Lie type groups G via Frobenius morphism. Although Liebeck and Seitz do not use deep classifcation results as Cooperstein and Kantor, they replace such results by the classification of simple algebraic groups and various deep results on the structure and representations of algebraic groups. So from that point of view it is a matter of taste, what type of quotations one prefers. I think the main advantage of Liebeck and Seitz's approach is, that they obtain for the case $|A \cap X| > 2$ a more conceptional result.

It is clear that Liebeck and Seitz's methods do not work for Lie type groups over arbitrary fields. Here A. Steinbachs work started. She considered (essentially simple) subgroups X of such a Lie type group G with $X = \langle \Sigma_0 \rangle$, $\Sigma_0 = \{A_0 \mid A_0 = A \cap X \neq 1, A \in \Sigma\}$ and Σ_0 a class of abstract root subgroups of X. In view of (2.3) and (2.6) such a hypothesis seems natural. Her first result in her thesis [Stb97] was a (partial) generalization of Kantor's theorem to arbitrary fields. (She assumed that the rank of X is ≥ 3 and that the groups are not defined over non-perfect fields of characteristic 2.) A slight generalization of this to subgroups of rank two is [Stb98b], where she allowed subgroups $X \simeq G_2(L)$. In [Stb00] she determined such subgroups $X = \langle \Sigma_0 \rangle$ of $F_4(k)$. But contrary to the above, she also allowed the case rank $X = 2$. This is of particular interest, since the new Moufang quadrangle discovered by R. Weiss and coming into the classification of Moufang polygons by Tits and Weiss [TW], corresponds to such a subgroup X of rank two of $F_4(k)$, when Char $k = 2$, which is in her list. The methods of both papers of A. Steinbach are a combination of an application of the theory of abstract root subgroups with certain geometric arguments. Most notably the theory of weak embeddings of polar spaces into projective spaces [SVM99], [SVM00].

At the end of this section we will describe a result of H. Cuypers and A. Steinbach [CS99], which gives a fairly uniform description of subgroups R of $GL(V)$ such that

$$\Sigma_0 = \{A_0 \mid A \in \Sigma, A_0 = A \cap R \neq 1\},$$
$$\Sigma \text{ the class of transvection groups of } GL(V),$$

is a class of abstract transvection groups of R.

A. Steinbach and H. Cuypers do not use the classification of groups generated by abstract transvection subgroups of III§1 in their proof (in particular not III(1.4) and III(1.10)), but just some of the ideas of the classification theory. Here we will use the results of II§3 and §4 and of III§1, to obtain a stream-lined construction of their embeddable polar space in case $C_V(R) = 0$. (This corresponds to theorem 1.5 of the introduction of [CS99]. They also consider in 1.6 of their paper the case $C_V(R) \neq 0$, which is more complicated.)

(2.10) Notation

Let V be a vector space over the division ring K, $G = SL(V)$ and Σ the class of transvection subgroups of G as defined in II(1.3). Let $R = \langle R \cap D(\Sigma) \rangle$ be a subgroup of G, satisfying:

(1) $R = \langle \Sigma_0 \rangle$, Σ_0 a class of abstract transvection subgroups of R and $D(\Sigma_0) \subseteq D(\Sigma)$. (The last condition says that if $1 \neq a \in A_0 \in \Sigma_0$, then a is a transvection on V.)

(2) $|A_0| \geq 4$ for $A_0 \in \Sigma_0$.

(3) There exist $A_0 \neq C_0 \in \Sigma_0$ with $[A_0, C_0] = 1$ and $C_{\Sigma_0}(A_0) \neq C_{\Sigma_0}(C_0)$.

(4) $V = [V, R]$ and $C_V(R) = 0$.

Then we say R satisfies (T) in $G = SL(V)$. Condition (3) will show that there exist different commuting points in $\underline{\Sigma}_0 = \{\underline{A}_0 \mid A_0 \in \Sigma_0\}$ in the language of III§1 and condition (4) shows that the radical $R(R) = 1$. We show:

(2.11) Theorem

Suppose R satisfies (T) in $G = SL(V)$. Then the following hold:

(1) $\dim[V, A_0] = 1 = \operatorname{codim} C_V(A_0)$ for $A_0 \in \Sigma$.

(2) $\underline{\Sigma}_0$ satisfies condition (H) of II§3 and R is quasi-simple.

(3) Let $\mathcal{P}(\Sigma_0)$ be the polar space obtained from III(1.4). Then $\mathcal{P}(\Sigma_0)$ is isomorphic to the incidence geometry with point-set

$$\mathcal{P} = \{[V, A_0] \mid A_0 \in \Sigma_0\}$$

and line-set

$$\mathcal{B} = \{g_{A_0, C_0} \mid A_0 \neq C_0 \in \Sigma_0 \text{ with } [A_0, C_0] = 1\},$$
$$\text{where } g_{A_0, C_0} = \mathcal{P} \cap ([V, A_0] + [V, C_0]).$$

Remark. Of course one obtains information about the structure of R already from (2) and III(1.4), III(1.7) and III(1.10). But (3) shows that the polar space $\mathcal{P}(\Sigma_0)$ is weakly embedded in the projective space $\mathcal{P}(V)$ in the sense of [SVM99], [SVM00]. It is this fact, from which H. Cuypers and A. Steinbach obtain precise information about the embedding of R in G, see (1.5) of [CS99]. (I don't state their theorem, since I have not introduced the Moufang quadrangles coming into the statement of (1.5) of [CS99].)

Proof of (2.11). We will prove (2.11) in a series of reductions. First notice that if $a_1, a_2 \in A_0^{\#}$ with $a_1 a_2 \neq 1$, then $a_1 a_2$ is a transvection again. Hence either $C_V(a_1) = C_V(a_2)$ or $[V, a_1] = [V, a_2]$ and, passing to the dual module V^* of V if necessary, we may assume that the first possibility holds. Thus, without loss of generality:

(1) codim $C_V(A_0) = 1$ for $A_0 \in \Sigma_0$.

Now let $A_0, B_0 \in \Sigma_0$ such that $X_0 = \langle A_0, B_0 \rangle$ is a rank one group. Let

$$T_A = \{\sigma \in D(\Sigma) \mid C_V(\sigma) = C_V(A_0)\} \cup \{1\}$$

and similarly T_B. Let $Y = \langle T_A, T_B \rangle$ and $N = N_{T_A}(T_B) N_{T_B}(T_A)$. Then II(1.6)(1) shows that $N \lhd Y$, N is abelian and $Y/N \simeq SL_2(K)$. Since X_0 is a rank one group, clearly $X_0 \cap N = 1$. Hence $X_0 \simeq X_0 N/N \leq Y/N \simeq SL_2(K)$ with $A_0 \overset{\sim}{\subset} T_A N/N$. Thus I(2.12) implies:

(2) X_0 is special and quasi-simple. Further Σ_0 satisfies condition (H) of II§3.

Now by II(1.3) we have for $A_0, C_0 \in \Sigma_0$:

$$[V, C_0] \subseteq C_V(A_0) \text{ if and only if } [A_0, C_0] = 1.$$

Since by II(4.4) and condition (3) of (S) we have $R = \langle C_{\Sigma_0}(A_0), B_0 \rangle = \langle C_{\Sigma_0}(A_0), b \rangle$ for $b \in B_0^{\#}$, this implies:

$$W = \Sigma_{C_0 \in C_{\Sigma_0}(A_0)}[V, C_0] \subseteq C_V(A_0) \text{ and } V = W + [V, b]$$

as $V = [V, R]$. In particular $W \cap [V, B_0]$ is centralized by $R = \langle C_{\Sigma_0}(B_0), A_0 \rangle$ and thus, as $C_V(R) = 0$, we obtain:

(3) $\dim[V, B_0] = 1$ for $B_0 \in \Sigma_0$ and $W = C_V(A_0)$.

Suppose next $R(R) \neq 1$. Then, by definition of $R(G)$ in II(4.12) we have $N_A = N_{\langle A_0 \rangle}(B_0) \neq 1$. Since $C_{\Sigma_0}(A_0) = C_{\Sigma_0}(A_0)$ we obtain $W = C_V(A_0)$ and $\dim[V, A_0] = 1$. (As $W \cap [V, B_0] = 1$.) Hence $[N_A, [V, B_0]] \leq [V, A_0] \cap [V, B_0] = 0$ and thus $N_A \leq C_R(V) = 1$. This contradiction shows:

(4) $R(R) = 1$.

In particular (2), (4) and II(4.14) imply R is quasi-simple. Let now $A_0 \neq C_0$ be commuting elements of Σ_0 and $E_0 \in \Sigma_0$ with $[V, E_0] \subseteq [V, A_0] + [V, C_0]$. (I. e. $[V, E_0] \in g_{A_0, C_0}$.) Then

$$C_{\Sigma_0}(A_0) \cap C_{\Sigma_0}(C_0) \subseteq C_{\Sigma_0}([V, E_0]) = C_{\Sigma_0}(E_0).$$

Hence

$$E_0 \in Z(\langle C_{\Sigma_0}(A_0) \cap C_{\Sigma_0}(C_0) \rangle) \cap \Sigma_0 = \ell_{A_0, C_0}$$

in the notation of III§1. This implies

(5) $\{E_0 \in \Sigma_0 \mid [V, E_0] \in g_{A_0, C_0}\} \subseteq \ell_{A_0, C_0}$

and, to prove (2.11), it suffices to show that we have equality in (5). To do this, pick $B_0 \in \Sigma_0$ such that (A_0, B_0, C_0) is a triangle in Σ_0 (B_0 exists by I(2.13)(7)!) and let $Y = \langle A_0, B_0, C_0 \rangle$. Then there exists by II(3.12) and II(3.15) a $y \in Y$ such that $D_0 = C_0^y \in \ell_{A_0, C_0}$ and $A_0 \neq D_0 \neq C_0$. Moreover by II(3.4), II(3.12) and II(3.15) we have $A_0 C_0 D_0 \cap Z(Y) \neq 1$, since by II(2.5) $D_0 \cap A_0 C_0 = 1$. Let $1 \neq z = acd \in Z(Y)$ with $a \in A_0^\#$, $c \in C_0^\#$ and $d \in D_0^\#$. Then

$$[V, z] \subseteq [V, Y] \cap C_V(Y),$$

since $A_0 \sim B_0$ in Y. Clearly $\dim([V, Y] \cap C_V(Y)) \leq 1$, since $\dim[V, Y] \leq 3$. Hence z is a transvection on V. In particular

$$[V, A_0] + [V, C_0] = [V, A_0] + [V, z] = [V, C_0] + [V, z].$$

Let F_0 be the unique point on ℓ_{A_0, C_0} commuting with B_0. We show:

(6) $[V, F_0] = [V, z]$.

Notice that (6) implies equality in (5) and thus (2.11). Indeed, if $F_0 \neq E_0 \in \ell_{A_0, C_0}$, then considering the triangle (A_0, B_0, E_0) we obtain with some $\bar{z} \in Z(\langle A_0, B_0, E_0 \rangle)$:

$$[V, A_0] + [V, E_0] = [V, A_0] + [V, \bar{z}] = [V, A_0] + [V, F_0] = [V, A_0] + [V, C_0],$$

which proves equality in (5).

To prove (6) pick $L_0 \in C_{\Sigma_0}(C_0) - C_{\Sigma_0}(A_0)$. Then (A_0, L_0, D_0) is a triangle.
By II(2.4) L_0 does not centralize z, whence $z \notin N(L_0)$. Let by (1) and (3)
$L \in \Sigma$ containing L_0. Then, if $[L_0, L_0^z] = 1$, we obtain $L^z \in \Lambda_L$ as $z \in D(\Sigma)$.
But then either $C_V(L_0) = C_V(L_0^z)$ or $[V, L_0] = [V, L_0^z]$, which both contradict
$L_0 \neq L_0^z$ and $L_0 = \underline{L}_0$ as $R(R) = 1$. Thus $\langle L_0, L_0^z \rangle$ is a rank one group and

$$\langle L, z \rangle = \langle L, L^z \rangle \simeq SL_2(K) \text{ by } I(1.4).$$

Let $Z \in \Sigma$ containing z. Then $Z^\# = D(\Sigma) \cap \langle L, z \rangle \cap C(A)$. Let $\overline{Y} = \langle A_0, L_0, D_0 \rangle$
and $\overline{N} = R(\overline{Y})$. Then, as $L_0^z = L_0^{ad}$, we have $\overline{Y} = \overline{N}\langle L_0, L_0^z \rangle$. ($ad \notin \overline{N}$, as
$[L_0, L_0^z] \neq 1$!). Hence by II(2.1) applied to $\overline{Y}/\overline{N}$ there exists a

$$T_0 \in \langle L_0, L_0^z \rangle \cap C(A_0) \cap \Sigma.$$

Thus by the above $T_0 \leq Z$, whence $[V, T_0] = [V, z]$ and thus $T_0 \in \ell_{A_0, C_0}$ by
(5). Since $B_0 \in C_{\Sigma_0}([V, z])$ this implies $T_0 = F_0$, which proves (6).   □

§3   Local BN-pairs

In this section we discuss applications of the theory of abstract root subgroups
for the determination of chamber transitive automorphism groups of spherical
buildings.

Let $\mathcal{B}$ be a thick, irreducible, spherical building of rank $\ell \geq 2$. Then by II§5
the group of Lie type $\mathcal{B}$ is generated by a class of abstract root subgroups
and by III§9 each group generated by a class of abstract root subgroups of
finite rank, see III(9.9), is a group of Lie type $\mathcal{B}$ for some such building $\mathcal{B}$.
So these two notions are more or less equivalent. Hence it seems sensible to
use the classification of groups generated by abstract root subgroups to obtain
other classifications of the groups of Lie type $\mathcal{B}$. One possible condition to
demand is, that G acts chamber transitively on $\mathcal{B}$. But as the example (11.14)
of [Tit74] shows, even the condition that G acts strongly transitive on $\mathcal{B}$, i.e.
transitive on the pairs $(c, \mathcal{A})$, c a chamber and $\mathcal{A}$ an apartment containing c,
is not sufficient. This motivates somehow the following:

(3.1)   Definition

Let $\mathcal{B}$ be a thick, irreducible, spherical building over $I = \{1, \ldots, \ell\}, \ell \geq 2$ and
G a chamber transitive subgroup of $\mathrm{Aut}(\mathcal{B})$. Use the notation of I§4, i.e. fix
a chamber $c \in \mathcal{B}$ and let $\Delta_i(c)$ be the residue of type i containing c, $B = G_c$
and $P_i = G_{\Delta_i(c)}$ for $i \in I$. Then we say that G has a local BN-pair (of rank
ℓ), if the following hold:

(i) Each P_i has a nilpotent normal subgroup $U_i \leq B$ acting transitively on $\Delta_j(c) - \{c\}$ for $i \neq j \in I$.

(ii) Let $U = \Pi_{i \in I} U_i$. Then $Z(U) \leq U_i$ for each $i \in I$.

I have chosen the word local BN-pair, because this condition lies somehow in between the weak BN-pair condition of Delgado-Stellmacher [DGS85] and the condition of Fong and Seitz [FS73] that G has a BN-pair with $B = U \cdot H$, where $U \trianglelefteq B$ is nilpotent. Indeed, the BN-pair condition is, since it is equivalent to strong transitivity, global. While the above condition is, apart from chamber transitivity, local. On the other hand, we have the condition that G acts on $\mathcal{B}$, which is not the case in the situation of Delgado-Stellmacher. Moreover, using (2.3) and Proposition (4.1) of a paper of K. Tent and H. Van Maldeghem [TVM], it is possible to show that condition (i) and (ii) of (3.1) hold, if G has a BN-pair of rank two, satisfying $B = U \cdot H, U \trianglelefteq B$ nilpotent.

In this section we prove the following:

(3.2)   Theorem

Suppose $\mathcal{B}$ is a thick generalized n-gon with $n \leq 4$ and $G \leq \operatorname{Aut}(\mathcal{B})$ has a local BN-pair. Then $\mathcal{B}$ is a Moufang polygon and either G contains the group of Lie type $\mathcal{B}$ (as normal subgroup) or $G \simeq A_6$ and $\mathcal{B}$ is the $\operatorname{Sp}(4, 2)$ quadrangle.

A similar theorem can be proved for $n \leq 6$. ($n = 5$ is non-existent.) The proof of this more general theorem will appear elsewhere. Indeed, from the point of view of classifying groups with a local BN-pair of rank ℓ, the problem is divided in the case $\ell \geq 3$, which should be easier (and possibly can be reduced to theorem 1 of [Tim00b]). And the more difficult case $\ell = 2$, in which $\mathcal{B}$ is not known. In the latter case even no reduction of n is known so far.

The proof of (3.2) depends on the following proposition, which may be of independent interest.

(3.3)   Proposition

Let G be a group with subgroups P_1 and P_2 and $B = P_1 \cap P_2$ satisfying:

(a) For $\{i, j\} = \{1, 2\}$ there exist nilpotent normal subgroups $U_i \trianglelefteq P_i$ with $U_i \leq B$, such that U_i acts transitively on $\{Bx \mid x \in P_j - B\}$.

(b) Let $U = U_1 U_2$. Then $Z = Z(U) \leq U_1 \cap U_2$.

(c) Set $L_i = \langle U^{P_i} \rangle$ and $V_i = \langle Z^{P_i} \rangle$ for $i = 1, 2$. Then $C_{V_i}(L_i) = 1 \neq [V_1, V_2]$.

(d) Either L_1 is finite or $C_B(\langle L_1, L_2 \rangle) = 1$.

Then there exists a division ring or a Cayley division algebra K such that L_i is the semidirect product of $SL_2(K)$ with its natural module V_i for $i = 1, 2$. (For definition of $SL_2(K)$ and its natural module see I(1.8)!)

Proof. The proof of (3.3) consists of elementary weak closure arguments together with an application of I(3.2). It will be done over a series of reductions, which we number consecutively:

By (a) and (c) L_i acts doubly transitively on $\{Bx \mid x \in P_i\}$. ((c) implies $|P_i : B| \geq 3!$)) Hence B is a maximal subgroup of P_1 and P_2 and the permutation actions of P_i and L_i on

$$\{Bx \mid x \in P_i\}, \{B^x \mid x \in P_i\} \text{ and } \{Z^x \mid x \in P_i\}$$

are equivalent. (The latter since $N_{P_i}(Z) = B$!) Thus there exist $w_i \in L_i$ with $w_i^2 \in B$ and $P_i = B \cup Bw_iB$. By (b) $V_i \leq Z(U_i)$ and by (c) V_i is a non-trivial $\mathbb{Z}L_i$-module. Moreover we have:

$$
\begin{array}{ll}
(1) & \begin{aligned}
U^{P_i} &= \{U\} \cup (U^{w_i})^U, \quad Z^{P_i} = \{Z\} \cup (Z^{w_i})^U, \\
\text{and} \quad L_i &= \langle U, x \rangle \text{ for each } x \in L_i - B.
\end{aligned}
\end{array}
$$

This follows immediately from the transitivity of U on $U^{P_i} - U$. We show next:

(2) $V_i = (V_i \cap U_j) \times Z^{w_i}$ for $\{i, j\} = \{1, 2\}$.

To simplify notation let $i = 1, j = 2$. Since $[Z^{w_1}, U] = [Z^{w_1}, U_2] \leq V_1 \cap U_2$ it follows that $(V_1 \cap U_2)Z^{w_1}$ is U invariant. Since $Z \leq V_1 \cap U_2$ (1) implies $V_1 = (V_1 \cap U_2)Z^{w_1}$. Clearly $V_1 \cap V_2 = Z$ and

$$Z \cap Z^{w_1} \leq C_{V_1}(\langle U, U^{w_1} \rangle) = C_{V_1}(L_1) = 1.$$

Hence $V_2 \not\leq B^{w_1}$, since otherwise $[V_2, Z^{w_1}] \leq V_1 \cap V_2 \cap Z^{w_1} = 1$ and thus $V_2 \leq C(V_1)$, a contradiction to (c). Thus we obtain by (1) $L_1 = \langle V_2, U^{w_1} \rangle$ and $(V_1 \cap U_2) \cap Z^{w_1} = 1$.

(3) We have $Z \cap Z^x = 1$ or Z for each $x \in P_i$, $i = 1$ or 2. Further $Z^{w_2} \cap B^{w_1} = 1 = Z^{w_1} \cap B^{w_2}$.

By (2) we have $Z \cap Z^{w_i} = 1$. Hence the first part of (3) follows from the doubly transitivity of P_i on Z^{P_i}. Now suppose $1 \neq x \in Z^{w_2} \cap B^{w_1}$. Then $[x, Z^{w_1}] \leq (V_1 \cap V_2) \cap Z^{w_1} = 1$. Hence by the first part of (3) $Z^{w_1} \leq N(Z^{w_2})$. Thus by (2) $V_1 = N(Z^{w_2})$ and $[V_1, Z^{w_2}] \leq (V_1 \cap V_2) \cap Z^{w_2} = 1$. Together with (2) this implies $V_1 \leq C(V_2)$, a contradiction to (c).

Next we show:

$$
\begin{array}{ll}
(4) & \begin{aligned}
{[x, Z^{w_1}]} &= Z \text{ for all } 1 \neq x \in Z^{w_2}, \\
\text{and} \quad [y, Z^{w_2}] &= Z \text{ for all } 1 \neq y \in Z^{w_1}.
\end{aligned}
\end{array}
$$

Clearly $[x, Z^{w_1}] \leq [V_1, V_2] \leq Z$. Suppose $Z_0 = [x, Z^{w_1}] < Z$ for some $1 \neq x \in Z^{w_2}$. Then $Z_0 \times (V_1 \cap U_2)^{w_1}$ is by (3) invariant under $L_1 = \langle x, U^{w_1} \rangle$, since by (2) $[x, V_1] = [x, Z^{w_1}] = Z_0$. This implies $Z \leq Z_0 \times (V_1 \cap U_2)^{w_1}$, a contradiction to $Z_0 < Z$ and $Z \cap (V_1 \cap U_2)^{w_1} = 1$ by (2). Now symmetry in w_1 and w_2 shows that (4) holds.

Now let $W = Z \times Z^{w_1}$ and $X = \langle Z^{w_2}, Z^{w_2 w_1} \rangle \leq L_1$. Then we show:

(5) $\overline{X} = X/C_X(W) \simeq SL_2(K)$ and W is a natural $\mathbb{Z}\overline{X}$-module.

By (3) and (4) we have $Z^X = \{Z\} \cup (Z^{w_1})^{Z^{w_2}}$ is a partition of W and $\overline{X}$ acts doubly transitively on Z^X. (Conjugating with w_1, the equations (4) also hold with $Z^{w_2 w_1}$ in place of Z^{w_2} and Z in place of Z^{w_1} (resp. Z^{w_1} in place of Z), as $Z^{w_1^2} = Z$!)

Moreover
$$(\overline{Z^{w_2}})^\# = \{\overline{x} \in \overline{X} \mid C_W(\overline{x}) = Z = [W, \overline{x}]\}.$$
(Argue as in I(1.8)(8)!) Hence $\overline{Z^{w_1}} \lhd N_{\overline{X}}(Z)$ and acts regularly on $Z^X - \{Z\}$. Thus by I(1.3) $\overline{X}$ is a rank one group. Now (5) is a consequence of I(3.2).

(6) $V_1 = W$ and $L_1 = C_{L_1}(W)X$.

By (5) and the doubly transitivity of L_1 on Z^{P_1} each element $1 \neq v \in V_1$ is contained in some $Z^\ell, \ell \in L_1$. Hence, if $V_1 \cap U_2 > Z$, then there exist a $Z^\ell \neq Z$, $\ell \in L_1$ with $Z^\ell \cap (V_1 \cap U_2) \neq 1$. But then by (1) Z^ℓ is conjugate to Z^{w_1} under U, a contradiction to (2).

Thus $V_1 \cap U_2 = Z$ and (2) implies $V_1 = W$. Now $[W, U] \leq Z \leq C_W(U)$. If now $x \in U$ with $C_W(x) > Z$, then by (3) $[Z^{w_1}, x] \leq Z \cap Z^{w_1} = 1$ and $x \in C(W)$. This immediately shows that $U = C_U(W)Z^{w_2}$. Now $L_1 = \langle U, U^{w_1} \rangle$ implies (6).

Let $Y = \langle Z^{w_1}, Z^{w_1 w_2} \rangle$. Then symmetry shows that $\overline{Y} = Y/C_Y(V_2) \simeq SL_2(\overline{K})$ and V_2 is the natural $\mathbb{Z}\overline{Y}$-module. Moreover the coodinatization procedure of I(3.2) shows that $K \simeq \overline{K}$. (If K is a division ring this is obvious, since Z is a 1-dimensional vector space for K and $\overline{K}$.) We show next:

(7) $C_{L_1}(V_1) = V_1 \times C_{L_1}(V_1 X), C_{L_2}(V_2) = V_2 \times C_{L_2}(V_2 Y)$ and $U = V_1 V_2$.

To prove (7) notice $C_{L_1}(V_1) \leq N_{L_1}(Z) = B$. Hence $[C_{L_1}(V_1), V_2] \leq C_{V_2}(V_1) = Z$. Let $1 \neq x \in Z^{w_2}$ and $a \in C_{L_1}(V_1) - V_1$. Then, as $[x, Z^{w_1}] = Z$, there exists a $b \in Z^{w_1}$ such that $[x, ab] = 1$. Now (3) implies $[Z^{w_2}, ab] \leq Z^{w_2} \cap C(V_1) = 1$ and thus $ab \in C_{L_1}(V_1 V_2)$. We have shown $C_{L_1}(V_1) = V_1 C_{L_1}(V_1 V_2)$. Conjugating with w_1, also $C_{L_1}(V_1) = V_1 C_{L_1}(V_1 V_2^{w_1})$. Hence we obtain

$$V_1 C_{L_1}(V_1 X) = (Z^{w_1} Z)(C_{L_1}(V_1 V_2) \cap C_{L_1}(V_1 V_2^{w_1}))$$
$$= Z^{w_1}(C_{L_1}(V_1 V_2) \cap V_1 C_{L_1}(V_1 V_2^{w_1})) = V_1 C_{L_1}(V_1 V_2) \cap V_1 C_{L_1}(V_1 V_2^{w_1})$$
$$= C_{L_1}(V_1).$$

This shows the first part of (7). Together with (6) we have

$$L_1 = (V_1 \times C_{L_1}(V_1 X))X.$$

Thus $U_0 = U \cap C_{L_1}(V_1 X) \lhd U$ and so $Z(U) \cap U_0 \neq 1$, if $U_0 \neq 1$, since U is nilpotent. But this is impossible since $Z(U) \leq V_1$. Hence $U_0 = 1$ and $U \cap C_{L_1}(V_1) = V_1$, which proves the second part of (7).

Since $L_i = \langle U^{L_i} \rangle$ (7) implies $C_{L_1}(V_1 X) = Z(L_1)$, $C_{L_2}(V_2 Y) = Z(Y_2)$ and $L_1 = V_1 X, L_2 = V_2 Y$. Now $Z(L_1) \leq C_B(U)$ and thus $[Z(L_1), Y] \leq V_2$ and $[Z(L_1), V_2] = 1$. Since $L_2 = \langle V_1, V_1^{w_2} \rangle$ and since $Z(L_1)$ centralizes $w_2 \bmod V_2$, this implies $Z(L_1) \leq C(L_2)$. Hence by (d) $Z(L_1) = 1 = Z(L_2)$, since L_i/V_i is a perfect central extension of $SL_2(K)$ with abelian unipotent subgroup U/V_i and if $|K| < \infty$ such a proper extension does not exist. (I.e. $L_i/V_i \simeq SL_2(K)$ and $Z(L_i) = 1$!)

It remains to show that the extension of L_i/V_i by V_i splits. If Char $K \neq 2$ this follows from the Frattini argument. (I.e. a complement is the centralizer of an involution inverting V_i.) If Char $K = 2$ this is probably also a general fact, since the extension of U/V_i by V_i splits. But we will indicate a proof in terms of chapter I.

Put $Z^{w_2} = \{a(\lambda) \mid \lambda \in K\}$ and let $b(\lambda^{-1})$ be the element of $Z^{w_2 w_1}$ satisfying:

$$a(\lambda)^{b(\lambda^{-1})} \equiv b(\lambda^{-1})^{a(\lambda)} \bmod V_1$$

(such an element exists!). Since this equation is equivalent to $o(a(\lambda)b(\lambda^{-1})) \equiv 3 \bmod V_1$ and since $C_{V_1}(a(\lambda)) \cap C_{V_1}(b(\lambda^{-1})) = 1$, we have $a(\lambda)^{b(\lambda^{-1})} = b(\lambda^{-1})^{a(\lambda)}$. Now the TI-set property of the conjugates of Z, i.e. (3), implies:

$$A^{b(\lambda^{-1})} = B^{a(\lambda)} \text{ for } A = Z^{w_2}, B = Z^{w_2 w_1}.$$

Hence X is a rank one group and thus $X \cap V_1 = 1$.   □

The condition (d) in (3.3) is necessary. Indeed, if one takes for G a proper perfect central extension of $SL_3(K)$, $|K|$ infinite, it might happen that $1 \neq Z(L_i) \leq Z(G)$.

(3.4) Lemma

Suppose $G \leq \mathrm{Aut}(\mathcal{B})$ has a local BN-pair, where $\mathcal{B}$ is the chamber system of a thick projective plane. Then the following hold:

(1) G satisfies the hypothesis of (3.3) with $P_1 = G_{\Delta_1(c)}$ (point-stabilizer), $P_2 = G_{\Delta_2(c)}$ (line-stabilizer) and $B = P_1 \cap P_2 = G_c$.
(2) $\mathcal{B}$ is a Moufang plane and $G_0 \simeq PSL_3(K)$ or E_6^K, K as in (3.3), where $G_0 = \langle L_1, L_2 \rangle$. (Clearly $G_0 \lhd G$ and $G = G_0 B$!)

Proof. By I(4.1) we have $G = \langle P_1, P_2 \rangle$. Since $P_i = L_i B$ and $L_i \trianglelefteq P_i$ this shows $G_0 \trianglelefteq G$ and $G = G_0 B$. In particular $C_B(G_0) \leq B_G = 1$. Now by (i) and (ii) of (3.1) the hypothesis (a) and (b) of (3.3) are satisfied. As $Z \leq U_1 \cap U_2$, Z consists of central elations corresponding to a point-line pair (P, ℓ). Hence $C_Z(L_i)$ either fixes all points on two different lines or all lines through two different points, whence $C_Z(L_i) = 1$ for $i = 1, 2$.

Now (up to duality) V_2 consist of central elations corresponding to the line ℓ and each $Z^x, x \in P_2$ consists of central elations corresponding to (P^x, ℓ). In particular the transitivity of P_2 on ℓ shows that we have non-identity central elations for each point on ℓ. Since the symmetric statement holds for V_1 we obtain $[V_1, V_2] \neq 1$. Thus (1) holds.

If now $g = \ell^x \neq \ell, x \in P_1$ is another line through P, then (3.3) applied to L_2^x shows that Z is transitive on $g - P$. Hence Z contains all central elations corresponding to (P, ℓ). Since $G_0 = \langle Z^G \rangle$ by (3.3), (2) now follows from the definition of $PSL_3(K)$ or E_6^K, see II(1.3) and II(5.24)(5). $\qquad\square$

(3.5) Lemma

Suppose $G \leq \mathrm{Aut}(\mathcal{B})$ has a local BN-pair, where $\mathcal{B}$ is the chamber system of a thick generalized quadrangle. Let $P_i, U_i, i = 1, 2$ and B, U be as in (3.1) and set $Z = Z(U)$, $V_i = \langle Z^{P_i} \rangle$ and $L_i = \langle U^{P_i} \rangle$. Then one of the following holds:

(1) G satisfies the hypothesis of (3.3).

(2) $1 \neq Z_i = C_Z(L_i)$ for $i = 1$ or 2. Further Z_i is (up to duality) the set of all central elations on $\mathcal{B}$ contained in Z.

Proof. As in the proof of (3.4) it is obvious that G satisfies hypothesis (a), (b) and (d) of (3.3). Assume $Z_i = C_Z(L_i) \neq 1$ for $i = 1$ or 2. Then clearly Z_i consists (up to duality) of central elations. Since $P_i = \langle Z_i^G \cap P_i \rangle B$ by the maximality of B in P_i and since central elations corresponding to collinear points commute, Z_i is the set of all central elations contained in Z. Thus, to prove (3.5) we may assume $Z_i = C_Z(L_i) = C_{V_i}(L_i) = 1$ for $i = 1, 2$ and then show $[V_1, V_2] \neq 1$. If now $V_1 \not\leq K_2 = \bigcap B^g, g \in P_2$, then $P_2 = (\langle V_1^{P_2} \rangle) B$. Hence, if $[V_1, V_2] = 1$, then $V_2 \leq Z(U)$, a contradiction to $Z_2 = 1$. This shows that $V_1 \leq K_2$ and $V_2 \leq K_1$ if $[V_1, V_2] = 1$. But then V_1 consists of central and V_2 of axial elations corresponding to some incident point-line pair, a contradiction to $1 \neq Z \leq V_1 \cap V_2$. $\qquad\square$

We now treat the two cases of (3.5) separately. We have:

(3.6) **Proposition**

Suppose G satisfies the hypothesis of (3.5) and $Z_i = 1$ for $i = 1, 2$. (I.e. G satisfies the hypothesis of (3.3)). Then $\mathcal{B}$ is the $\mathrm{Sp}(4, 2)$-quadrangle and $G \simeq A_6$.

Proof. For the whole proof we use the notation of (3.3) and the convention that P_1 is the stabilizer of a point P and P_2 the stabilizer of a line ℓ with $PI\ell$. We show first:

(1) $P_i = N(V_i)$ for $i = 1, 2$.

Suppose $P_1 < N(V_1)$ and let $x \in N(V_1) - P_1$. Then V_1 fixes all lines through the point $Q = P^x \neq P$. If P and Q are on a line, then we may by the transitivity of P_1 on the lines through P assume $QI\ell$. But then the transitivity of U on the points of ℓ different from P shows that Z consists of axial elations corresponding to ℓ. This is obviously a contradiction to $[Z^{w_1}, V_2] \neq 1$, since the groups of all axial elations corresponding to two intersecting lines must commute.

Hence P and Q are not collinear. Hence V_1 also stabilizes the intersecting points of the lines through P and Q

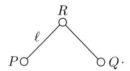

But this is impossible, since as $V_1 = U_1$, V_1 acts transitively on the points of ℓ different from P.

This proves (1), which then allows us to identify the points of $\mathcal{B}$ with the conjugates of V_1 and the lines with the conjugates of V_2. A point V_1^x lies on the line V_2^y if and only if $1 \neq [V_1^x, V_2^y] \leq V_1^x \cap V_2^y$. Next we show:

(2) Char $K = 2$.

Suppose this is false. Then we may choose $w_1 \in X$ with $w_1^2 = z_1$ inverting V_1, where X is a complement to V_1 in L_1. Similarly choose $w_2 \in Y$ with $w_2^2 = z_2$ inverting V_2, Y a complement to V_2 in L_2. Then, since all conjugates of z_1 in $V_1\langle z_1 \rangle$ are conjugate under V_1, we may by the Frattini argument assume in addition that $[z_1, z_2] = 1$. Hence $z_2 \in N(X)$ and $z_1 \in N(Y)$, since $X = C_{L_1}(z_1)$ and $Y = C_{L_2}(z_2)$. Now z_2 inverts Z, Z^{w_2} and centralizes $V_1 \cap Y = Z^{w_1}$. (We may start choosing by $Y = \langle Z^{w_1}, Z^{w_1 w_2} \rangle$. To obtain that z_2 centralizes z_1, we just have to conjugate with an element of V_1, which normalizes Z^{w_1}.) Hence z_2 normalizes $V_2^{w_1} \cap X$ and thus also inverts $V_2^{w_1} \cap X$, since it has to invert with each $a \in Z^{w_2}$ also the element $b(a) \in V_2^{w_1} \cap X$. ($X$ is a rank one group with unipotent subgroups $Z^{w_2}, V_2^{w_1} \cap X$!) Now in the notation of I(2.7) $w_1 = n(\lambda) = a(\lambda)b(a(-\lambda))a(\lambda)$ for some $\lambda \in K$ with $a(\lambda) \in Z^{w_2}$ and $b(a(-\lambda)) \in V_2^{w_1} \cap X$.

Since X is a special rank one group, i.e. $b(a(-\lambda)) = b(a(\lambda))^{-1}$ by I(2.2), this shows that $w_1^{z_2} = w_1^{-1}$ and $[w_1, z_2] = z_1$. By the same reason $[w_2, z_1] = z_2$. Hence $E = \langle z_1, z_2 \rangle \trianglelefteq \langle w_1, w_2 \rangle$ and $\langle w_1, w_2 \rangle$ induces an $SL_2(2) = \Sigma_3$ on E. This shows $\langle w_1, w_2 \rangle \simeq \Sigma_4$.

Now E stabilizes the path

$$(V_1^{w_2 w_1}, V_2^{w_1}, V_1, V_2, V_1^{w_2}, V_2^{w_1 w_2}) = \mathcal{P}$$

and $\mathcal{P}$ is contained in a unique apartment $\mathcal{A}$ of $\mathcal{B}$. Moreover, E fixes exactly two lines through V_1, namely $V_2, V_2^{w_1}$ and two points on V_2, namely V_1 and $V_1^{w_2}$. Hence the same holds for all the conjugates of V_1 and V_2 under $\langle w_1, w_2 \rangle$. But since E fixes of course all points and lines in $\mathcal{A}$, we obtain $\mathcal{A} = (V_2^{w_1 w_2 w_1}, V_1^{w_2 w_1}, V_2^{w_1}, V_1, V_2, V_1^{w_2}, V_2^{w_1 w_2}, V_1^{w_2 w_1 w_2})$. But this is impossible since $\langle w_1, w_2 \rangle / E \simeq \Sigma_3$ can not act faithfully on $\mathcal{A}$. (Indeed we have $V_1^{w_2 w_1 w_2} = V_1^{w_1 w_2 w_1} = V_1^{w_2 w_1}$ and the apartment would be a triangle!)

(3) $K = GF(2)$

Suppose (3) is false. We first treat the case that K is commutative. Then we may choose w_1, w_2 so that $w_1^2 = 1 = w_2^2$. Hence $w_i \sim Z$ in G by the structure of L_i. Let $X = \langle Z^{w_1}, Z^{w_2 w_1} \rangle$ and $Y = \langle Z^{w_1}, Z^{w_1 w} \rangle$ be complements to V_1 resp. V_2 in L_i and let $H_1 = N_X(V_2) \cap N_X(V_2^{w_1})$, $H_2 = N_Y(V_1) \cap N_Y(V_1^{w_2})$ and choose the w_i so that they invert H_i. We show:

(*) Passing to some conjugate of H_2 under Z if necessary we obtain $[H_1, H_2] = 1$.

By (1) $H_1 \leq N(L_2)$. Hence H_1 normalizes $H_2 V_2 = N_{L_2}(Z) \cap N_{L_2}(Z^{w_2})$. ($Z^{w_2} = V_2 \cap X$!). Similarly $H_2 \leq N(H_1 V_1)$. Hence

$$[H_1, H_2] \leq H_1 V_1 \cap H_2 V_2 \leq H_1 Z \cap H_2 Z.$$

Since H_1 acts by scalar multiplication on Z and normalizes Z^{w_2} we have $L_2 H_1 / V_2 \subseteq GL_2(K)$, since as shown in (3.3) $C_B(L_2) = 1$. Hence $(H_1 H_2)' \leq V_2$ and similarly $(H_1 H_2)' \leq V_1$. Thus $(H_1 H_2)' \leq Z$. Let $H = \langle H_1, H_2 \rangle$. Then $ZH = (ZC_H(Z))H_1$, since H_1 and H_2 act by scalar multiplication on Z. Now $H_1 \leq N(V_1^{w_2})$ since $V_1^{w_2}$ is the conjugate of V_1 in L_2 containing Z^{w_2}. Thus $ZH \leq P_1^{w_2}$ and $ZC_H(Z) = Z \times K$ with $K = C_{ZC_H(Z)}(L_1^{w_2}/V_1^{w_2})$. Since K is H-invariant we obtain $ZH = (Z \times K)H_1$ and $K \trianglelefteq ZH$. Now by exercise

III(6.18)(1) all complements to Z in ZH_1 are conjugate. Hence $H_2^z \leq KH_1$ for some $z \in Z$, which proves (∗).

So choose by (∗) H_2 so that $[H_1, H_2] = 1$. Then H_2 normalizes $\langle w_1 \rangle H_1 = N_{L_1}(H_1)$ and thus $[H_2, w_1] \leq H_1$. Similarly $[H_1, w_2] \leq H_2$ so that $W = \langle w_1, w_2 \rangle$ normalizes $H = H_1 H_2$. Now, as in the proof of (2) V_2 and $V_2^{w_1}$ are the only H-invariant lines through V_1 and $V_1, V_1^{w_1}$ are the only H-invariant points on V_2. Since the same applies for all conjugates of V_1 and V_2 under W, we obtain again that $\mathcal{A}$ is an H-invariant apartment. In particular W stabilizes $\mathcal{A}$ and w_1 fixes two non-collinear points $P = V_1$ and P' on $\mathcal{A}$.

On the other hand $w_1 \in Z^{w_2 x} \leq V_2^x$ for some $x \in X$. Hence w_1 fixes all points on the line $\ell^x = V_2^x$ and all lines through $P^{w_2 x}$. Now P' is collinear to some point Q on ℓ^x. Suppose $Q \neq P^{w_2 x}$.

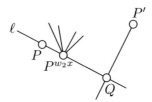

Then w_1 fixes $\overline{QP'}$. But this is impossible since by the structure of L_1 an element of $Z^\#$ fixes no line $g \neq \ell$ through a point R on ℓ different from P.

Hence $Q = P^{w_2 x}$. Now $U^{w_2 x}$ acts transitively on the points of $\overline{QP'} - \{Q\}$ and on the lines through Q different from ℓ. Thus w_1 is a central elation corresponding to the point Q. But then there is a central elation $1 \neq z \in Z$, a contradiction to $C_Z(L_1) = 1$.

In case K is not commutative choose a subfield k of K with $|k| \geq 4$ such that V_1 and V_2 are k-vector spaces. (I.e. $k \subseteq Z(K)$ if K is a Cayley division algebra.) We construct a subfiguration over k.

Coordinatize $Z, Z^{w_1}, Z^{w_2}, Z^{w_2 w_1}$ and $Z^{w_1 w_2}$ as in I(3.3) starting with $Z = \{z(\lambda) \mid \lambda \in K\}$. Let $Z_0 = \{z(\lambda) \mid \lambda \in k\}$. Then there exist $A_0 < Z^{w_2}, B_0 < Z^{w_2 w_1}$ such that $X_0 = \langle A_0, B_0 \rangle \simeq SL_2(k)$ and $Z_0[Z_0, B_0]$ is the natural X_0-module. Hence we may pick $w_1 \in X_0$ with $A_0^{w_1} = B_0$ and $w_1^2 = 1$. Then $[Z_0, B_0] = Z_0^{w_1}$. Thus there exists a $D_0 \leq Z^{w_1 w_2}$ such that $Y_0 = \langle Z_0^{w_1}, D_0 \rangle \simeq SL_2(k)$ and acts naturally on $Z_0 A_0$. As before we may choose $w_2 \in Y_0$ with $Z_0^{w_1 w_2} = D_0$ and $w_2^2 = 1$. Then we have altogether that $V_1^0 = Z_0 Z_0^{w_1}$ is a natural module for $X_0 = \langle Z_0^{w_2}, Z_0^{w_2 w_1} \rangle$ and $V_2^0 = Z_0 Z_0^{w_2}$ is a natural module for $Y_0 = \langle Z_0^{w_1}, Z_0^{w_1 w_2} \rangle$. Let $H_1 = N_{X_0}(Z_0^{w_2}) \cap N_{X_0}(Z_0^{w_2 w_1})$ and $H_2 = N_{Y_0}(Z_0^{w_1}) \cap N_{Y_0}(Z_0^{w_1 w_2})$. Then, as before, $[H_1, H_2] \leq Z$. Now Z is a k-vector space and both, H_1 and H_2, act by scalar multiplication on Z. In particular $H_1 \equiv H_2 \bmod C(Z)$. Since as before all complements to Z in ZH_1 are conjugate, we obtain arguing as before $[H_1, H_2^z] = 1$ for some $z \in Z$. Thus, replacing H_2 by H_2^z, we may assume $[H_1, H_2] = 1$.

Now H_2 normalizes $X_0 V_1 = N_{L_1}(V_1^0)$. Hence H_2 normalizes $\langle w_1 \rangle H_1 = N_{X_0 V_1}(H_1)$. Thus $[w_1, H_2] \leq H_1$ and similarly $[w_2, H_1] \leq H_2$. Hence $W = \langle w_1, w_2 \rangle \leq N(H)$, $H = H_1 H_2$ and, since H fixes exactly two lines through V_1 and two points on V_2, we get a contradiction as before. This proves (3).

Now, as $K = GF(2)$, $\mathcal{B}$ is by [VM98, (1.7.9)] the $Sp(4,2)$ quadrangle. Hence $G \simeq A_6$ by the structure of $\Sigma_6 \simeq \mathrm{Aut}(\mathcal{B})$. $\qquad\square$

In the other case in (3.5), when without loss of generality $Z_1 \neq 1$, we show that $\Sigma = Z_1^G$ is a class of abstract transvection groups of G_0. For this we need a lemma to identify rank one groups. We have:

(3.7) Lemma

Suppose X acts doubly transitively on Ω, $|\Omega| \geq 3$ with $X_\alpha = A \cdot H$, $A \trianglelefteq X_\alpha$ nilpotent and $H = X_{\alpha,\beta}$ for $\alpha \neq \beta \in \Omega$. Suppose $1 \neq A_0 \leq Z(A)$ is X_α-invariant. Let $w \in X$ with $\alpha^w = \beta, \beta^w = \alpha$ and set $B = A^w$ and $B_0 = A_0^w$. Suppose further that:

(1) $A_0 \cap X_\beta = 1$.

(2) For $a \in A_0^\#$ there exists no $b \in B^\#$ such that $ab \sim A^\#$ in X.

Then $X_0 = \langle A_0, B_0 \rangle$ is a rank one group.

Proof. Pick $a \in A_0^\#$. Then $a \notin N_A(B)$, $B = A^w$ since $N(B) \leq N(B_0)$. As A is transitive on $\Omega - \{d\}$ there exists a $b = b(a) \in B$ with $B^a = A^b$. Hence there exists a $\bar{b} \in B^\#$ with $a^b = \bar{b}^a$. We obtain $\bar{b} = a^{ba^{-1}} \in A_0^{ba^{-1}} = B_0$. (Clearly $B_0 = A_0^x$ for each $x \in X$ with $\alpha^x = \beta$!) If now $\bar{b} = b^{-1} = b(a)^{-1}$ for each $a \in A_0^\#$, then there exists for each $a \in A_0^\#$ a $b \in B_0^\#$ with $B_0^a = A_0^b$ and $a^b = b^{-a}$ which implies (3.7).

So we may without loss of generality assume $\bar{b} \neq b^{-1}$. Hence $\bar{b} = (a^b)^{a^{-1}} = ab^{-1}b^{a^{-1}}$ and

$$a^{b^{-1}} b^{a^{-1}} = bab^{-1}b^{a^{-1}} = b\bar{b} \in B^\#.$$

We obtain

$$a^{b^{a^{-1}}b} = (b\bar{b})^b = b\bar{b} \text{ since } \bar{b} \in B_0 \leq Z(B).$$

Suppose $b^{a^{-1}b} \in A$. Then $b\bar{b} \in A$ and $A^b = A^{\bar{b}^{-1}}$. As $N_{B_0}(A_0) = 1$, this implies that $\bar{b}^{-1}$ is the unique element in $B_0^\#$ with

$$B_0^a = A_0^{\bar{b}^{-1}}.$$

If this holds for each $a \in A_0^\#$ we obtain that X_0 is a rank one group, which was to be shown.

So we may assume $b^{a^{-1}b} \notin A$. Hence there exists a $y \in A$ with $(b^{a^{-1}b})^y \in B$. Thus we obtain

$$ab^{a^{-1}by} = (ab^{a^{-1}b})^y = (b\bar{b})^y \sim A^\#$$

and $b^{a^{-1}by} \in B^\#$, a contradiction to (2). □

(3.8) Proposition

Suppose the group X acts doubly transitively on a set Ω with $|\Omega| \geq 3$, such that $X_\alpha = A \cdot H$ with $A \trianglelefteq X_\alpha$ nilpotent and $H = X_{\alpha\beta}$ for $\alpha \neq \beta \in \Omega$. Suppose $1 \neq A_0 \leq Z(A)$ is X_α invariant and pick $w \in X$ with $\alpha^w = \beta, \beta^w = \alpha$ and set $B_0 = A_0^w$, $X_0 = \langle A_0, B_0 \rangle$. Then X_0 is a special rank one group if the following holds:

(1) $A_0 \cap X_\beta = 1$.

(2) There exists a $\mathbb{Z}X$-module V satisfying:

(i) $[V, X, X_0] \neq 0$.

(ii) $[V, A, A_0] = 0$.

The proof of (3.8) is, with the help of (3.7) instead of I(2.3), exactly the same as the proof of I(2.4).

(3.9) Theorem

Suppose G and $\mathcal{B}$ satisfy the hypothesis of (3.5) with $Z_1 \neq 1$. Then $\Sigma = Z_1^G$ is a class of abstract transvection groups of $G_0 = \langle \Sigma \rangle$.

Proof. Assume without loss of generality that P_1 is the stabilizer of a point P of $\mathcal{B}$ in G. Since P_1 is doubly transitive on the lines through P we have $P_1 = (\langle \Sigma \cap P_1 \rangle)B$. Since central elation groups corresponding to collinear points commute, this shows that:

(1) Z_1 is the set of all central elations corresponding to the point P contained in Z.

Let $V_2 = \langle Z_1^{P_2} \rangle$. Then, as $G = \langle P_1, P_2 \rangle$, V_2 is a non-trivial $\mathbb{Z}L_2$-module, $L_2 = \langle U_1^{P_1} \rangle$. Let $w_2 \in L_2$ interchanging the points P and $Q = P^{w_2} \neq P$ on ℓ. (Exists, since L_2 is doubly transitive on ℓ !). Then

(2) $V_2 = (V_2 \cap U_1) \times Z_1^{w_2}$.

Since $Z_1^{w_2}$ consists of central elations corresponding to Q, clearly $U_1 \cap Z_1^{w_2} = 1$. Now $Z_1^{P_2} = \{Z_1\} \cup (Z_1^{w_2})^{U_1}$ and $[Z_1^{w_2}, U_1] \le V_2 \cap U_1$. Thus $Z_1^{P_2} \subseteq (V_2 \cap U_1) Z_1^{w_2}$, which proves (2).

Now pick $w_1 \in P_1$ interchanging the lines ℓ and $\ell^{w_1} \ne \ell$ through P. Then, by the transitivity of U_2 on the lines through P different from ℓ we obtain $L_1 = \langle U_2^{P_1} \rangle = \langle U_2, U_2^{w_1} \rangle$ and we may choose $w_1 \in L_1$. Let $W_1 = \langle (V_2 \cap U_1)^{L_1} \rangle$, $T = (V_2 \cap U_1) \cap (V_2 \cap U_1)^{w_1}$ and $\widetilde{W}_1 = W_1/T$. We show:

(3) $\widetilde{W}_1$ is a non-trivial $\mathbb{Z} L_1$-module.

First, as $V_2 \le Z(U_2)$ since $Z_1 \le Z(U_2), T \le Z(L_1)$. Next $[V_2 \cap U_1, (V_2 \cap U_1)^{w_1}] \le T$. Since

$$W_1 = (V_2 \cap U_1) \langle (V_2 \cap U_1)^{w_1 U_2} \rangle,$$

this implies that $\widetilde{W}_1$ is abelian. So it remains to show that $\widetilde{W}_1$ is a non-trivial L_1-module.

Suppose $\widetilde{W}_1$ is trivial. Then $V_2 \cap U_1 = (V_2 \cap U_1)^{w_1} \le Z(L_1)$ and thus consists of central elations corresponding to P. This implies $(V_2 \cap U_1) \cap (V_2 \cap U_1)^{w_2} = 1$, since the intersection consists of central elations corresponding to two different points. But then $Z_1^{w_2} = (V_2 \cap U_1)^{w_2}$ and $Z_1 = V_2 \cap U_1$. Now V_2 contains a conjugate of Z_1 corresponding to each point on ℓ. Thus, if P' is a point opposite P and $Z_1' = Z_1^g$ is the conjugate of Z_1 corresponding to P', we have $V_2 = Z_1 C_{V_2}(Z_1')$ and $C_{V_2}(Z_1')$ is the central elation group corresponding to the point R on ℓ collinear to P'.

Let $F_1 = \langle V_2^{P_1} \rangle$. Then, since the same holds for each $V_2^x, x \in P_1$ we obtain $F_1 = Z_1 \times C_{F_1}(Z_1')$. (Clearly $Z_1 \cap C(Z_1') = 1$!). Hence $F_1 = C_{F_1}(Z_1)'$ fixes P' and thus each point opposite to P. On the other hand, since $C_{V_2}(Z_1')Z_1'$ contains central elation groups corresponding to each point on $\overline{RP'}$, symmetry shows that $F_1' = (C_{F_1}(Z_1')Z_1')'$ also fixes each point opposite to P'. Hence clearly $F_1' = 1$. Now V_2 is not in the kernel of the action of P_1 on the lines through P. Hence, by the doubly transitivity of P_1, F_1 is already transitive on the lines through P. Thus

$$F_1 = \langle V_2^{F_1} \rangle = V_2 \text{ as } F_1' = 1,$$

which is obviously impossible. This proves (3).

(4) Let $X = L_1/L_1 \cap U_1 = \widetilde{L}_1, A = \widetilde{U}_2, A_0 = \widetilde{V}_2$ and Ω the set of lines through P. Thus X satisfies, with the action on $\widetilde{W}_1$, the hypothesis of (3.9).

To prove (4) let $\alpha = \ell$ and $\beta = \ell^{w_1}$. Then, since $Z_1^{w_2}$ consists of elations corresponding to Q, we have $Z_1^{w_2} \cap (L_1)_\beta = 1$, whence $A_0 \cap X_\beta = 1$ by (2).

Since $\widetilde{U}_2$ is transitive on $\Omega - \{\alpha\}$, the Frattini argument implies $X_\alpha = AX_{\alpha\beta}$. Hence condition (2) of (3.8) remains to be shown. Now $[W_1, V_2] \leq V_2 \cap W_1$ and $[W_1, V_2, U_2] = 1$. Since $[V_2, U_2] = 1$ this proves (2)(ii) of (3.9).

Let $B_0 = A_0^{w_1}$. Then, as $[W_1, B_0] \leq (V_2 \cap W_1)^{w_1}$ and $X = \langle A, B_0 \rangle$ by the transitivity of A on $\Omega - \alpha$, we have:

$$(*) \qquad\qquad \widetilde{W}_1 = [\widetilde{W}_1, A] + (\widetilde{V_2 \cap W_1})^{w_1}.$$

Since $[W_1, U_2] \leq C_{W_1}(V_2)$ and $(V_2 \cap W_1)^{w_1} \leq C_{W_1}(U_2^{w_1})$ we also have $[W_1, U_2] \cap (V_2 \cap W_1)^{w_1} \leq T$. Hence the sum in $(*)$ is direct and both summands are by (3) non-trivial. Moreover $[\widetilde{W}_1, A] = C_{\widetilde{W}_1}(A_0)$ since $\widetilde{T} = 1$. Hence $(\widetilde{V_2 \cap W_1}) \leq [\widetilde{W}_1, A] \leq [\widetilde{W}_1, X]$ and thus $\widetilde{W}_1 = [\widetilde{W}_1, X]$. This shows that also $(3.8)(2)(i)$ holds.

Now (3.8) shows that $X_0 = \langle A_0, B_0 \rangle$ is a rank one group. Hence $X_1 = \langle Z_1^{w_2}, Z_1^{w_2 w_1} \rangle$ is $\mathrm{mod}\, U_1$ a rank one group. But on the other hand, if

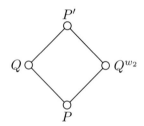

is a quadrangle, then X_1 fixes P' and thus $X_1 \cap U_1$ a point on each line through P. In particular, if for $g \in P_1$, $Z_1^{w_2 g}$ is the elation group corresponding to the point on ℓ^g collinear to P', then $X_1 \cap U_1$ centralizes $R_1 = \langle Z_1^{w_2 g} \mid g \in P_2 \rangle$, since $[Z_1^{w_2}, X_1 \cap U_1] \leq Z_1^{w_2} \cap U_1 = 1$. We claim that $Z_1^{w_2}$ is strongly closed in $Z_1^{w_2}(X_1 \cap U_1)$. Indeed, if $1 \neq z \sim Z_1^{w_2}$ in X_1 with $z \in Z_1^{w_2}(X_1 \cap U_1)$, then z is a central elation corresponding to some point on ℓ. (Since z fixes P and Q!) Hence $z \in V_2$ and thus $S = V_2 \cap X_1 \cap U_1 \neq 1$ if $z \notin Z_1^{w_2}$. But S fixes all points on ℓ and thus all points on all lines through P, since $S \leq Z(R_1)$ and R_1 is already transitive on the lines through P since $\langle V_2^{P_1} \rangle \leq U_1 R_1$. Thus $S = 1$, $z \in Z_1^{w_2}$ and $Z_1^{w_2}$ is strongly closed in $Z_1^{w_2}(X_1 \cap U_1)$.

This implies that X_1 is a rank one group, since $X_1/X_1 \cap U_1$ is a rank one group. We have shown that, if Z_1^g, Z_1^h are conjugates of Z_1 corresponding to non-collinear points, then $\langle Z_1^g, Z_1^h \rangle$ is a rank one group. Hence (1) implies (3.9).

$\square$

(3.10)   Proof of (3.2)

Assume that the hypothesis of (3.2) holds. Then by (3.3)–(3.6) either the conclusion holds or $\mathcal{B}$ is a generalized quadrangle and, without loss of generality, $1 \neq Z_1 = C_Z(L_1)$. In the latter case by (3.9) $\Sigma = Z_1^G$ is a class of abstract transvection groups of $G_0 = \langle \Sigma \rangle \trianglelefteq G$. Moreover, there is a 1-1-correspondence between the points of $\mathcal{B}$ and the elements of Σ. Hence $R(G_0) = 1$ by definition of $R(G_0)$ in II§4 and $\mathcal{P}(\Sigma) \simeq \mathcal{B}$, considering $\mathcal{B}$ as point-line geometry. Hence by the proof of III(1.6) $M_A = R(\langle C_\Sigma(A)\rangle)$ is transitive on Ω_A for $A \in \Sigma$. (The proof of III(1.6) depends only on the fact that $\mathcal{P}(\Sigma)$ is a thick polar space and on II(2.9), but not on the condition (H) of II§3.) Now the proof of III(1.7) shows that G_0 contains both types of root groups on $\mathcal{B}$. Hence by definition G_0 is the group of Lie type $\mathcal{B}$. (The proof of III(1.7) depends only on III(1.6) and II(3.1), but also not on condition (H). (H) is used only to obtain that $\mathcal{P}(\Sigma)$ is a thick polar space, which we know already!) $\qquad \square$

Bibliography

We split the references in two parts, I. Books and II. Journal Articles, since, as mentioned in the introduction, Chapters I–III do depend only on references available in books.

I. Books

[Asc86] M. Aschbacher, *Finite group theory.*, Cambridge Studies in Advanced Mathematics, vol. 10, Cambridge University Press, Cambridge, 1986.

[Asc97] M. Aschbacher, *3-transposition groups.*, Cambridge Tracts in Mathematics, vol. 124, Cambridge University Press, Cambridge, 1997.

[Bou81] N. Bourbaki, *Groupes et algebres de Lie. Chapitres 4, 5 et 6.*, Elements de Mathematique. Paris, 1981 (French).

[Car72] R. W. Carter, *Simple groups of Lie type.*, Pure and Applied Mathematics., vol. XXVIII, John Wiley & Sons, a Wiley Interscience Publication. London, 1972.

[Coh95] A. M. Cohen, *Point-line spaces related to buildings.*, Handbook of incidence geometry: buildings and foundations. (F. Buekenhout, ed.), North-Holland, Amsterdam, (1995), 647–737.

[DGS85] A. Delgado, D. Goldschmidt, and B. Stellmacher, *Groups and graphs: new results and methods.*, DMV Seminar, vol. 6, Birkhäuser Verlag, Basel-Boston-Stuttgart, 1985.

[Gor68] D. Gorenstein, *Finite groups*, Harper's Series in Modern Mathematics., Harper and Row, New York-Evanston-London, 1968.

[HO89] A. Hahn and O. O'Meara, *The classical groups and K-theory. Foreword by J. Dieudonne.*, Grundlehren der Mathematischen Wissenschaften, vol. 291, Springer-Verlag, Berlin, 1989.

[HP73] D. Hughes and F. Piper, *Projective planes.*, Graduate Texts in Mathematics, vol. 6, Springer-Verlag, New York-Heidelberg-Berlin, 1973.

[Hum90] J. E. Humphreys, *Reflection groups and Coxeter groups.*, Cambridge Studies in Advanced Mathematics, vol. 29, Cambridge University Press, Cambridge, 1990.

[Hup67] B. Huppert, *Endliche Gruppen. I.*, Die Grundlehren der mathematischen Wissenschaften in Einzeldarstellungen, vol. 134, Springer-Verlag, Berlin-Heidelberg-New York, 1967 (German).

[Pas94] A. Pasini, *Diagram geometries.*, Clarendon Press, Oxford, 1994.

[Pic55] G. Pickert, *Projektive Ebenen.*, Die Grundlehren der Math. Wissenschaften, vol. 80, Springer-Verlag, Berlin-Goettingen-Heidelberg, 1955 (German).

[Ron89] M. A. Ronan, *Lectures on buildings.*, Perspectives in Mathematics, vol. 7, Academic Press, Inc., Boston, MA, 1989.

[Sch95] R. Scharlau, *Buildings.*, Handbook of incidence geometry: buildings and foundations. (F. Buekenhout, ed.), North-Holland, Amsterdam, (1995), 477–645.

[Spr98] T. Springer, *Linear algebraic groups. 2nd ed.*, Progress in Mathematics (Boston, Mass.), vol. 9, Birkhäuser, Boston, MA, 1998.

[Stg67] R. Steinberg, *Lectures on Chevalley groups.*, Lecture Notes, Yale University, 1967.

[Tit74] J. Tits, *Buildings of spherical type and finite BN-pairs.*, Lecture Notes in Mathematics., vol. 386, Springer-Verlag, Berlin-Heidelberg-New York, 1974

[TW]   J. Tits and R. Weiss, *Moufang polygons*, to appear.

[VM98] H. Van Maldeghem, *Generalized polygons.*, Monographs in Mathematics, vol. 93, Birkhäuser, Basel, 1998.

II. Journal Articles

[AG66] J. Alperin and D. Gorenstein, *The multiplicators of certain simple groups*, Proc. Am. Math. Soc. **17** (1966), 515–519.

[Asc72] M. Aschbacher, *On finite groups generated by odd transpositions. I-IV.*, Math. Z. **127** (1972), 45–56, J. Algebra **26** (1973), 451–491.

[Asc73]   M. Aschbacher, *A condition for the existence of a strongly embedded subgroup.*, Proc. Am. Math. Soc. **38** (1973), 509–511.

[Asc75]   M. Aschbacher, *On finite groups of component type.*, Ill. J. Math. **19** (1975), 87–115.

[Asc76]   M. Aschbacher, *Tightly embedded subgroups of finite groups.*, J. Algebra **42** (1976), 85–101.

[Asc77a] M. Aschbacher, *A characterization of Chevalley groups over fields of odd order. I., II.*, Ann. Math., II. Ser. **106** (1977), 353–398, 399–468.

[Asc77b] M. Aschbacher, *On finite groups in which the generalized fitting group of the centralizer of some involution is extraspecial.*, Ill. J. Math. **21** (1977), 347–364.

[Asc80]   M. Aschbacher, *Groups of characteristic 2-type.*, Finite groups, Santa Cruz Conf. 1979, Proc. Symp. Pure Math., **37**, (1980), 29–36.

[Asc90]   M. Aschbacher, *The 27-dimensional module for E_6. IV.*, J. Algebra **131** (1990), no. 1, 23–39.

[Ben71]   H. Bender, *Transitive Gruppen gerader Ordnung, in denen jede Involution genau einen Punkt fest lässt.*, J. Algebra **17** (1971), 527–554 (German).

[BS74]   F. Buekenhout and E. Shult, *On the foundations of polar geometry.*, Geom. Dedicata **3** (1974), 155–170.

[CH91]   P. J. Cameron and J. Hall, *Some groups generated by transvection subgroups.*, J. Algebra **140** (1991), no. 1, 184–209.

[CC83]   A. M. Cohen and B. N. Cooperstein, *A characterization of some geometries of Lie type.*, Geom. Dedicata **15** (1983), 73–105.

[Che55]   C. Chevalley, *Sur certains groupes simples.*, Tohoku Math. J., II. Ser. **7** (1955), 14–66 (French).

[Coo79]   B. N. Cooperstein, *The geometry of root subgroups in exceptional groups. I., II.*, Geom. Dedicata **8** (1979), 317–381, **15** (1983), 1–45.

[Coo81]   B. N. Cooperstein, *Subgroups of exceptional groups of Lie type generated by long root elements. I., II.*, J. Algebra **70** (1981), 270–298.

[CH92] H. Cuypers and J. Hall, *The classification of 3-transposition groups with trivial center.*, Groups, combinatorics and geometry. Proceedings of the L.M.S. Durham symposium, held July 5-15, 1990 in Durham, UK. (Martin Liebeck et al., ed.), Lond. Math. Soc. Lect. Note Ser., **165**, Cambridge University Press, Cambridge, (1992), 121–138.

[CJP92] H. Cuypers, P. Johnson, and A. Pasini, *On the embeddability of polar spaces.*, Geom. Dedicata **44** (1992), no. 3, 349–358.

[CS99] H. Cuypers and A. Steinbach, *Linear transvection groups and embedded polar spaces.*, Invent. Math. **137** (1999), no. 1, 169–198.

[Fis64] B. Fischer, *Distributive Quasigruppen endlicher Ordnung*, Math. Z. **83** (1964), 267–303 (German).

[Fis66] B. Fischer, *A characterization of the symmetric groups on 4 and 5 letters*, J. Algebra **3** (1966), 88–98.

[Fis69] B. Fischer, *Finite groups generated by 3-transpositions.*, Mim. Notes, University of Warwick, 1969.

[Fis71] B. Fischer, *Finite groups generated by 3-transpositions. I.*, Invent. Math. **13** (1971), 232–246.

[Fon67] P. Fong, *Some Sylow subgroups of order 32 and a characterization of $U(3,3)$*, J. Algebra **6** (1967), 65–76.

[FS73] P. Fong and G. M. Seitz, *Groups with a (B,N)-pair of rank 2. I., II.*, Invent. Math. **21** (1973), 1–57, **24** (1974), 191–239.

[Gla72] G. Glauberman, *Quadratic elements in unipotent linear groups.*, J. Algebra **20** (1972), 637–654.

[HKS72] C. Hering, W. M. Kantor, and G. M. Seitz, *Finite groups with a split BN-pair of rank 1. I.*, J. Algebra **20** (1972), 435–475.

[Hig68] G. Higman, *Odd characterizations of finite simple groups*, Lecture Notes, University of Michigan, 1968.

[Ho76] C.-Y. Ho, *On the quadratic pairs.*, J. Algebra **43** (1976), 338–358.

[Hua49] L.-K. Hua, *On the automorphisms of a field.*, Proc. Natl. Acad. Sci. USA **35** (1949), 386–389.

[Jan69] Z. Janko, *Some new simple groups of finite order. I.*, Sympos. Math., Roma 1, Teoria Gruppi, Dic. 1967 e Teoria Continui Polari, Aprile 1968, (1969), 25–64.

[Kan79] W. M. Kantor, *Subgroups of classical groups generated by long root elements.*, Trans. Am. Math. Soc. **248** (1979), 347–379.

[LS94]   M. W. Liebeck and G. M. Seitz, *Subgroups generated by root elements in groups of Lie type.*, Ann. Math., II. Ser. **139** (1994), no. 2, 293–361.

[McL67] J. McLaughlin, *Some groups generated by transvections.*, Arch. Math. **18** (1967), 364–368.

[McL69] J. McLaughlin, *Some subgroups of $SL_n(F_2)$*, Ill. J. Math. **13** (1969), 108–115.

[MT83]   T. Meixner and F. G. Timmesfeld, *Chamber systems with string diagrams.*, Geom. Dedicata **15** (1983), 115–123.

[PS83]   A. Premet and I. Suprunenko, *Quadratic modules for Chevalley groups over fields of odd characteristics.*, Math. Nachr. **110** (1983), 65–96.

[Ron80] M. A. Ronan, *A geometric characterization of Moufang hexagons.*, Invent. Math. **57** (1980), 227–262.

[Smi79] S. D. Smith, *Large extraspecial subgroups of widths 4 and 6.*, J. Algebra **58** (1979), 251–281.

[Smi80a] S. D. Smith, *A characterization of finite Chevalley and twisted groups of type E over $GF(2)$.*, J. Algebra **62** (1980), 101–117.

[Smi80b] S. D. Smith, *The classification of finite groups with large extraspecial 2-subgroups.*, Finite groups, Santa Cruz Conf. 1979, Proc. Symp. Pure Math., **37**, (1980), 111–120.

[Stb92] A. Steinbach, *Gruppen, die von k-Transvektionen erzeugt werden. (Groups which are generated by k-transvections).*, Mitt. Math. Semin. Gießen **211** (1992), 27–48 (German).

[Stb97] A. Steinbach, *Subgroups of classical groups generated by transvections or Siegel transvections. I., II.*, Geom. Dedicata **68** (1997), no. 3, 281–357.

[Stb98a] A. Steinbach, *Generalized quadrangles arising from groups generated by abstract transvection groups.*, Groups and geometries. Proceedings of the conference, Siena, Italy, September 1-7, 1996. (Di Martino, Lino et al., ed.), Trends in Mathematics, Birkhäuser, Basel, (1998), 189–199.

[Stb98b] A. Steinbach, *Subgroups isomorphic to $G_2(L)$ in orthogonal groups.*,
          J. Algebra **205** (1998), no. 1, 77–90.

[Stb00]   A. Steinbach, *Groups of Lie type generated by long root elements in
          $F_4(K)$*, Habilitationsschrift, Mathematisches Institut, Justus-Liebig-
          Universität Gießen, Germany, 2000.

[SVM99]   A. Steinbach and H. Van Maldeghem, *Generalized quadrangles
          weakly embedded of degree > 2 in projective space.*, Forum Math.
          11 (1999), no. 2, 139–176.

[SVM00]   A. Steinbach and H. Van Maldeghem, *Generalized quadrangles
          weakly embedded of degree 2 in projective space.*, Pac. J. Math. **193**
          (2000), no. 1, 227–248.

[Ste78]   E. Stensholt, *Certain embeddings among finite groups of Lie type.*,
          J. Algebra **53** (1978), 136–187.

[Suz62]   M. Suzuki, *On a class of doubly transitive groups*, Ann. Math., II.
          Ser. **75** (1962), 105–145.

[Suz65]   M. Suzuki, *Finite groups in which the centralizer of any element of
          order 2 is 2- closed*, Ann. Math., II. Ser. **82** (1965), 191–212.

[Suz69]   M. Suzuki, *A simple group of order 448,345,497,600.*, Theory finite
          Groups, Symp. Harvard Univ. **1968** (1969), 113–119.

[TVM]   K. Tent and H. Van Maldeghem, *On irreducible BN-pairs of rank
          two*, Preprint.

[Thm70]   G. Thomas, *A characterization of the Steinberg groups $D_4^2(q^3), q =
          2^n$*, J. Algebra **14** (1970), 373–385.

[Tho]   J. Thompson, *Quadratic Pairs*, unpublished.

[Tho70]   J. Thompson, *Quadratic Pairs*, Proc. of the International Congress
          of Mathematicians, Nice, (1970).

[Tim70]   F. G. Timmesfeld, *Eine Kennzeichnung der linearen Gruppen über
          $GF(2)$*, Math. Ann. **189** (1970), 134–160 (German).

[Tim73]   F. G. Timmesfeld, *A characterization of the Chevalley- and Stein-
          berg-groups over F_2.*, Geom. Dedicata **1** (1973), 269–321.

[Tim75a] F. G. Timmesfeld, *Groups generated by root-involutions. I., II.*, J.
          Algebra **33** (1975), 75–134, **35** (1975), 367–441.

[Tim75b] F. G. Timmesfeld, *Groups with weakly closed TI-subgroups.*, Math. Z. **143** (1975), 243–278.

[Tim78a] F. G. Timmesfeld, *Finite simple groups in which the generalized Fitting group of the centralizer of some involution is extraspecial.*, Ann. Math., II. Ser. **107** (1978), 297–369.

[Tim78b] F. G. Timmesfeld, *On the structure of 2-local subgroups of finite groups.*, Math. Z. **161** (1978), 119–136.

[Tim83] F. G. Timmesfeld, *Tits geometries and parabolic systems in finitely generated groups. I., II.*, Math. Z. **184** (1983), 377–396, 449–487.

[Tim90a] F. G. Timmesfeld, *Groups generated by k-transvections.*, Invent. Math. **100** (1990), no. 1, 167–206.

[Tim90b] F. G. Timmesfeld, *On the identification of natural modules for symplectic and linear groups defined over arbitrary fields.*, Geom. Dedicata **35** (1990), no. 1–3, 127–142.

[Tim91] F. G. Timmesfeld, *Groups generated by k-root subgroups.*, Invent. Math. **106** (1991), no. 3, 575–666.

[Tim94a] F. G. Timmesfeld, *Moufang planes and the groups E_6^K and $SL_2(K)$, K a Cayley division algebra.*, Forum Math. **6** (1994), no. 2, 209–231.

[Tim94b] F. G. Timmesfeld, *Subgroups generated by root-elements of groups generated by k-root subgroups.*, Geom. Dedicata **49** (1994), no. 3, 293–321.

[Tim98] F. G. Timmesfeld, *Presentations for certain Chevalley groups.*, Geom. Dedicata **73** (1998), no. 1, 85–117.

[Tim99] F. G. Timmesfeld, *Abstract root subgroups and quadratic action. With an Appendix by A.E. Zalesskii.*, Adv. Math. **142** (1999), no. 1, 1–150.

[Tim00a] F. G. Timmesfeld, *On the structure of special rank one groups*, to appear in Adv. Stud. Pure Math., 2001.

[Tim00b] F. G. Timmesfeld, *Structure and Presentations of Lie-type groups*, Proc. Lond. Math. Soc. **81** (2000), 428–484.

[Tit66] J. Tits, *Classification of algebraic semisimple groups.*, Proc. Sympos. Pure Math., **9**, (1966), 33–62.

[Tit77] J. Tits, *Endliche Spiegelungsgruppen, die als Weylgruppen auftreten.*, Invent. Math. **43** (1977), 283–295 (German).

[Tit82]   J. Tits, *A local approach to buildings.*, The geometric vein, The Cox-
           eter Festschr., (1982), 519–547.

[Tit92]   J. Tits, *Twin buildings and groups of Kac-Moody type.*, Groups, com-
           binatorics and geometry. Proceedings of the L.M.S. Durham sympo-
           sium, held July 5-15, 1990 in Durham, UK. (Martin Liebeck et al.,
           ed.), Lond. Math. Soc. Lect. Note Ser., **165**, Cambridge University
           Press, Cambridge, (1992), 249–286.

[Tit94]   J. Tits, *Moufang polygons. I: Root data.*, Bull. Belg. Math. Soc. **1**
           (1994), no. 3, 455–468.

Symbol Index

In this index we first define standard symbols used without definition in the book. Then we list for special symbols the page of definition. Sometimes the same symbol may occur twice in different, but related, situations. (For example E_d when D is a class of root involutions and $d \in D$ and E_σ when $\mathcal{Q}$ is a set of quadratically acting elements and $\sigma \in \mathcal{Q}$.

$\langle \ldots \rangle$	group generated by a given set	—
$\| \ldots \|$	cardinality of a finite set	—
$[a, b]$	$= a^{-1}b^{-1}ab$, commutator of a and b	—
$[A, B]$	$= \langle [a,b] \mid a \in A, b \in B \rangle$, commutator of A and B	—
$[A, B, C]$	$= [[A, B], C]$	—
$A^{\#}$	set of all elements of A different from 1	—
a^b	$= b^{-1}ab$, conjugate of a under b	—
A^b	$= b^{-1}Ab$, conjugate of A under b	—
A^B	$= \{A^b \mid b \in B\}$	—
$\mathrm{Aut}(G)$	set of automorphisms of G	—
$\mathbb{C}$	complex numbers	—
$C_G(A)$	centralizer of A in G	—
$cl(U)$	nilpotence class of a nilpotent group U	—
$\mathrm{End}(G)$	$= \mathrm{Hom}(G, G)$	—
$G * H$	central product of G and H	—
$G \cdot H$	product of two subgroups of a group	—
$G \times H$	direct product of G and H	—
G'	$= [G, G]$ commutator subgroup	—
$\mathrm{Gal}(K : L)$	Galois group of K over L	—
G_α	stabilizer of α in G	—
$\Gamma L(V)$	group of all semilinear maps of V	—

Index